RESEARCH METHODS IN PHYSICAL ACTIVITY

Fifth Edition

RESEARCH METHODS IN PHYSICAL ACTIVITY

Fifth Edition

Jerry R. Thomas, EdD

Jack K. Nelson, EdD

Stephen J. Silverman, EdD

Human Kinetics

Library of Congress Cataloging-in-Publication Data

Thomas, Jerry R.
   Research methods in physical activity / Jerry R. Thomas, Jack K. Nelson, Stephen
J. Silverman.-- 5th ed.
      p. cm.
   Includes bibliographical references and index.
   ISBN 0-7360-5620-3 (hardcover)
  1. Physical education and training--Research. 2. Health--Research. 3.
Recreation--Research. 4. Dance--Research. I. Nelson, Jack K. II. Silverman,
Stephen J. III. Title.
   GV361.T47 2005
   613.7'1'072--dc22

                                   2005005754

ISBN: 0-7360-5620-3

The Web addresses cited in this text were current as of February 22, 2005, unless otherwise noted.

Acquisitions Editor: Loarn D. Robertson, PhD; **Developmental Editor:** Holly Gilly; **Assistant Editors:** Amanda M. Eastin and Bethany J. Bentley; **Copyeditor:** Patsy Fortney; **Proofreader:** Kathy Bennett; **Indexer:** Betty Frizzéll; **Permission Manager:** Dalene Reeder; **Graphic Designer:** Fred Starbird; **Graphic Artist:** Denise Lowry; **Photo Manager:** Sarah Ritz; **Cover Designer:** Jack W. Davis; **Photographers (interior):** Dan Wendt. Photo on page 282 by Sarah Ritz; **Art Manager:** Kelly Hendren; **Illustrators:** Accurate Art, Gretchen Walters, and Kelly Hendren; **Printer:** Edwards Brothers

We thank the University of Illinois, HDC Engineering, and Follett's Bookstore in Champaign, Illinois, for assistance in providing the location for the photo shoot for this book.

Printed in the United States of America   10   9   8   7   6   5   4   3   2   1

Human Kinetics
Web site: www.HumanKinetics.com

United States: Human Kinetics, P.O. Box 5076, Champaign, IL 61825-5076
800-747-4457
e-mail: humank@hkusa.com

Canada: Human Kinetics, 475 Devonshire Road Unit 100, Windsor, ON N8Y 2L5
800-465-7301 (in Canada only)
e-mail: orders@hkcanada.com

Europe: Human Kinetics, 107 Bradford Road, Stanningley, Leeds LS28 6AT, United Kingdom
+44 (0) 113 255 5665
e-mail: hk@hkeurope.com

Australia: Human Kinetics, 57A Price Avenue, Lower Mitcham, South Australia 5062
08 8277 1555
e-mail: liaw@hkaustralia.com

New Zealand: Human Kinetics, Division of Sports Distributors NZ Ltd., P.O. Box 300 226 Albany, North Shore City, Auckland
0064 9 448 1207
e-mail: info@humankinetics.co.nz

Contents

Chapter 19   Qualitative Research   345

- *Contrasting Characteristics of Qualitative and Quantitative Research   346* ▪ *Procedures in Qualitative Research   347* ▪ *Analysis of the Data   352* ▪ *Concluding Remarks   362* ▪ *Summary   363*

▪ Part IV   Writing the Research Report   365

Chapter 20   Completing the Research Process   367

- *Research Proposal   367* ▪ *Developing a Good Introduction   367* ▪ *Describing the Method   368* ▪ *The Proposal Process   369* ▪ *Preparing and Presenting Qualitative Research Proposals   372* ▪ *Writing Proposals for Granting Agencies   373* ▪ *Submitting Internal Proposals   374* ▪ *Completing Your Thesis or Dissertation   374* ▪ *Results and Discussion   374* ▪ *How to Handle Multiple Experiments in a Single Report   378* ▪ *How to Use Tables and Figures   379* ▪ *Summary   386*

Chapter 21   Ways of Reporting Research   389

- *Basic Writing Guidelines   389* ▪ *A Brief Word About Acknowledgments   390* ▪ *Thesis and Dissertation Format: Traditional Versus Journal   390* ▪ *Helpful Hints for Successful Journal Writing   397* ▪ *Writing Abstracts   398* ▪ *Making Oral and Poster Presentations   401* ▪ *Summary   404*

▪ Appendixes   405

A   Statistical Tables   405

B   A Brief Historical Overview of Research   418
in Physical Activity in the United States

C   Sample Consent Forms   422

Preface

With the publication of this fifth edition, 20 years have passed since the first edition was published in 1985. We want to take this opportunity to thank all the people who have used this book over the years. We hope you have learned a lot about research methods in the study of physical activity. Maybe you've even enjoyed a few of the humorous stories, jokes, and pictures we've included to enliven the reading. We were particularly pleased to note the paper by Silverman and Keating (2002) in *Research Quarterly for Exercise and Sport* that indicated that 71% of the people responding to their survey about the research methods course used our book. Their survey results clearly show that the content of this book is well aligned with the topics teachers of research methods believe are important. We are also delighted that many of the other English-speaking countries also use this book—*Good on ya' mates*!! In addition we appreciate that earlier editions have been translated into Chinese, Japanese, and Portuguese.

Dr. Stephen Silverman has joined us as a coauthor on this edition. Because Dr. Jack Nelson retired, but continued to provide input and assistance in this revision, we needed to provide for a transition of authorship. Dr. Silverman is a well-known scholar and methodologist in physical education pedagogy, and at the writing of this edition, he is the editor-in-chief of *Research Quarterly for Exercise and Sport*. He brings considerable expertise and research experience to this fifth edition.

The main use of this text still appears to be in the first graduate-level research methods courses, although it is also being used in undergraduate research methods and as a resource for research planning and analysis. Our use of the term *physical activity* in the book title is meant to convey the broadly conceived field of study often labeled kinesiology, exercise science, exercise and sport science, human movement, sport studies, or physical education as well as related fields such as physical therapy, rehabilitation, and occupational therapy. We hope everyone who reads, understands, plans, carries out, writes, or presents research will find the book a useful tool to enhance their efforts.

The fifth edition remains organized as the fourth edition was, as follows:

- *Part I* is an overview of the research process including developing the problem, using the literature, preparing a research plan, and understanding ethical issues in research and writing.

- *Part II* provides an introduction to statistical and measurement issues in research, including statistical descriptions, power, relationships among variables, differences among groups, nonparametric procedures, and measurement issues in research.

- *Part III* presents the various types of, or approaches to, research including historical, philosophical, research synthesis, survey, descriptive, epidemiological, experimental, and qualitative.

- *Part IV* is designed to help you complete the research process, including writing the results and discussion, organizing the research paper, developing good figures and tables, and presenting research in written and verbal forms.

- Appendixes include statistical tables, a brief historical overview of the field, and sample forms for securing permission to use humans and animals in research.

Although the format of the book remains similar to that of the fourth edition, we have made a number of changes that we hope improve and update the text. Various people have read and reviewed the previous edition and provided helpful comments, including the many students we have taught and faculty who use the book to teach research methods. We truly do pay attention to the things you tell us. However, sometimes when we read reviews, we feel like Day (1983, p. xi), who said that a reviewer described his book as both good and original, but then went on to say that "the part that is good is not original and the part that is original is not good." Following is a short review of the changes for this fifth edition:

• *Part 1: Overview of the Research Process.* Each chapter has minor revisions with updates of information and more recent reports. We strive in each new edition to keep the chapter about using library techniques up-to-date and always ask our university librarian to review and evaluate what we have presented.

• *Part II: Statistical and Measurement Concepts.* We strive in each edition to increase the relevance of the examples and provide students with easy-to-understand calculations for basic statistics. Using the learning activities in the instructor guide and the approaches to statistical analyses in the appendix, students should grasp basic understandings about statistical techniques. In the fourth edition we changed to a more unified approach to parametric and nonparametric techniques, and we maintain that approach here. We have developed a new chapter in this section called "Using Power to Plan and Interpret Research." In the fourth edition this was part of chapter 6, but because of its importance, we have separated it out and expanded the information.

• *Part III: Types of Research.* We have continued our use of expert authors to present coherent views of historical research (Nancy Struna), philosophical research (Scott Kretchmar), and epidemiological research in physical activity (Barbara Ainsworth and Chuck Matthews). These three types of research are outside our expertise, and we wanted them presented effectively by expert scholars. Another significant revision in this section is the qualitative research chapter. The many changes in the field have prompted us to expand and update our discussion on qualitative research. We also have expanded the sections on quasi-experimental and single-subject research. In addition, we have made minor revisions and updates to all of the other chapters in this part.

• *Part IV: Writing the Research Report.* The two chapters in this section remain essentially the same, with minor changes and updates of material.

We are also grateful to the literary executor of the late Sir Ronald A. Fisher, F.R.S., to Dr. Frank Yates, F.R.S., and to Longman Group Ltd., London, for permission to reprint tables A.3 and A.5 from their book *Statistical Tables for Biological, Agricultural and Medical Research* (6th ed., 1974).

As we have said in each edition, we are grateful for the help of our friends, both for help that we acknowledge in various places in the book, and in other places where we have inadvertently taken an idea without giving credit. "After the passage of time, one can no longer remember who originated what idea. After the passage of even more time, it seems to me that all of the really good ideas originated with me, a proposition which I know is indefensible" (Day, 1983, p. xv).

We believe this book provides the necessary information for both the consumer and the producer of research. Although no amount of knowledge about the tools of research can replace expertise in the content area, good scholars of physical activity cannot function apart from the effective use of research tools. Researchers, teachers, clinicians, technicians, health workers, exercise leaders, sport managers, athletic counselors, and coaches need to understand the research process. If they do not, they are forced to accept information at face value or on the recommendation of others. Neither is necessarily bad, but the ability to carefully evaluate and reach a valid conclusion based on data, method, and logic is the mark of a professional.

Inserted into various chapters are humorous stories, anecdotes, sketches, laws, and corollaries. These are intended to make a point and enliven the reading without distracting from the content. Research processes are not mysterious events that graduate students should fear. To the contrary, they are useful tools that every academic and professional should have access to; they are, in fact, the very basis on which we make competent decisions.

Jerry R. Thomas
Jack K. Nelson
Stephen J. Silverman

Dear Student of Research Methods:

Before you begin this course, your potential as a research methods student should be evaluated. *Score one point for each of the following statements that applies to you.*

You might be a research methods student if . . .

your library carrel is better decorated than your apartment.

you have ever brought a scholarly article to a bar or coffee shop.

you rate coffee shops on the availability of outlets for your laptop.

you have ever discussed academic matters at a sporting event.

you actually have a preference between microfilm and microfiche.

you always read the reference lists in research articles.

you think the sorority sweatshirt Greek letters are a statistical formula.

you need to explain to children why you are in the 20th grade.

you refer to stories as "Snow White et al."

you wonder how to cite talking to yourself in APA style.

Scoring Scale

5 or 6—Definitely ready to be a student in research methods

7 or 8—Probably a master's student

9 or 10—Probably a doctoral student

Humorously yours,
Professors of Research Methods

Acknowledgments

As in any work, there are numerous people who contributed to this book and whom we should recognize. Many of these individuals are former students and colleagues who have said or done things that better developed our ideas as expressed in these pages. Also, a number of faculty members who have used previous editions have either written reviews or made suggestions to us that have improved the book. While we cannot list or even recall all of these contributions, we do know you made them, and we thank all of you.

In particular, we thank Karen French at the University of South Carolina, Dick Magill at Louisiana State University, Brad Cardinal at Oregon State University, and Kathi Thomas at Iowa State University for allowing us to use materials that were published jointly with them. Scott Kretchmar, Nancy Struna, Barb Ainsworth, and Chuck Matthews made an invaluable contribution with their chapters on research methods in philosophy, history, and exercise epidemiology, topics that we simply could not write about effectively.

If you adopt this book for your class, we hope you will make use of the class teaching materials available on the Human Kinetics Web site. Included are over 300 Microsoft PowerPoint® slides, learning activities, test questions, and other course materials. We thank Phil Martin from Pennsylvania State University for his contributions to these materials.

Finally, we thank the staff at Human Kinetics, in particular Holly Gilly, our developmental editor for this edition, for their support and contributions. They have sharpened our thinking and improved our writing.

Jerry R. Thomas
Jack K. Nelson
Stephen J. Silverman

PART I

Overview of the Research Process

The researches of many have thrown much darkness
on the subject and if they continue,
soon we shall know nothing at all about it.
—attributed to Mark Twain

This part provides you with an overall perspective of the research process. The introductory chapter defines and reviews the various types of research done in physical activity and gives you some examples. We define science as systematic inquiry, and we discuss the steps in the scientific method. This logical method answers the following four questions (Day, 1983, 4), which constitute the parts of a typical thesis, dissertation, or research report:

1. What was the problem? Your answer is the introduction.
2. How did you study the problem? Your answer is the materials and methods.
3. What did you find? Your answer is the results.
4. What do these findings mean? Your answer is the discussion.

We also present alternative approaches for doing research relative to a more philosophic discussion of science and ways of knowing. In particular, we address qualitative research, the use of field studies, and methods of introspection as strategies for answering research questions instead of relying on the traditional scientific paradigm as the only approach to research problems.

Chapter 2 suggests ways of developing a problem and using the literature to clarify the research problem, specify hypotheses, and develop the methodology. In particular, we propose a system for searching, reading, analyzing, synthesizing, organizing, and writing the review of literature.

The next two chapters in part I present the format of the research proposal with examples. This information is typically required of the master's or doctoral student before collecting data for the thesis or dissertation. Chapter 3 defines and delimits the research problem, including the introduction, statement of the problem, research hypotheses, operational definitions, assumptions and limitations, and significance. Chapter 4 covers methodology, or how to do the research. Included are the topics of participant selection, instrumentation or apparatus, procedures, and design and analysis. We emphasize the value of pilot work conducted before the research and how cause and effect may be established.

Chapter 5 discusses ethical issues in research and scholarship. We include information on misconduct in science; ethical considerations in research writing, working with advisors, and copyright; and the use of humans and animals in research.

Once you have completed part I, you should better understand the research process. Then comes the tricky part: learning all the details. We consider these details in part II ("Statistical and Measurement Concepts in Research"), part III ("Types of Research"), and part IV ("Writing the Research Report").

Introduction to Research in Physical Activity

Research is to see what everybody else has seen, and to think what nobody else has thought.

—Szent-Gyorgyi

Mention the word *research,* and depending on his or her background, each person will conjure up a different picture. One might think of searching the Internet or going to the library; another might visualize a lab filled with test tubes, vials, and perhaps little white rats. It is important, then, as we begin a text on the subject, to establish a common understanding of research. In this chapter we introduce you to the nature of research. We do this by discussing methods of problem solving and types of research. We explain the research process and relate it to the parts of a thesis. By the time you reach the end of chapter 1, you should understand what research really involves.

The Nature of Research

The object of research is to determine how things are as compared to how they might be. To achieve this, research implies a careful and systematic means of solving problems and involves five characteristics (Tuckman, 1978):

• **Systematic.** Problem solving is accomplished through the identification and labeling of variables and is followed by the design of research that tests the relationships among these variables. Data are then collected that, when related to the variables, allow the evaluation of the problem and hypotheses.

• **Logical.** Examination of the procedures used in the research process allows researchers to evaluate the conclusions drawn.

• **Empirical.** The researcher collects data on which to base decisions.

• **Reductive.** The researcher takes individual events (data) and uses them to establish general relationships.

• **Replicable.** The research process is recorded, enabling others to test the findings by repeating the research or to build future research on previous results.

Problems to be solved come from many sources and can entail resolving controversial issues, testing theories, and trying to improve present practice. For example, a popular topic of concern is obesity and methods of losing weight. Suppose we want to investigate this by comparing the effectiveness of two exercise programs in reducing fat. Of course, we know that caloric expenditure results in the loss of fat, so we will try to find out which program does this better under specified conditions. (Note: Our approach here is to give a simple, concise overview of a research study. We do not intend it to be a model of originality or sophistication.)

applied research

Type of research that has direct value to practitioners but in which the researcher has limited control over the research setting.

This study is definitely an example of **applied research** (more on this in the next section). Rather than try to measure the calories expended and so on, we approach it strictly from a programmatic standpoint. Say that we are operating a health club and that we offer aerobic dance and jogging classes for people who want to lose weight. Our research question is: Which program is more effective in reducing fat?

Suppose that we have a pool of participants to draw from and that we can randomly assign two thirds of them to the two exercise programs and one third to a control group. We have their scout's honor that no one is on a drastic diet or engaging in any other strenuous activities while the study is in progress. Both the aerobic dance and the jogging classes are one hour long and are held five times a week for 10 weeks. The same enthusiastic and immensely qualified instructor teaches both classes.

hydrostatic weighing

Technique that measures body composition in which body density is computed by the ratio of an individual's weight in air and the loss of weight underwater.

Our measure of fatness is the sum of skinfold measurements taken at eight body sites. Of course, we could use other measures, such as percentage of fat estimated from **hydrostatic weighing** (or total body water or some other estimate of fatness). However, we can defend our measures as valid and reliable indicators of fatness, and skinfolds are functional field measures. We measure all the participants, including those in the control group, at the beginning and the end of the 10-week period. During the study we try to ensure that the two programs are similar in procedural aspects, such as motivational techniques and the aesthetics of the surroundings. In other words, we do not favor one group by cheering them on but not encouraging the other; nor do we have one group exercise in an air-conditioned, cheerful, and healthful facility while the other has to sweat it out in some dingy room or parking lot. It is very important that we try to make the programs as similar as possible in every respect except the experimental treatments. The control group does not engage in any regular exercise.

After we have measured all the participants on our criterion of fatness at the end of the 10-week program, we are ready to analyze our data. We want to see how much change in skinfold thickness has occurred and whether there are differences between the two types of exercise. Because we are dealing with samples of people (from a whole universe of similar people), we need to use some type of statistics to establish how confident we can be in our results. In other words, we need to determine the significance of our results. Suppose the mean (average) scores for the groups are as follows:

- Aerobic dance = −21 mm
- Jogging = −25 mm
- Control = +8 mm

These values (which we made up) represent the average change in the combined skinfold thicknesses of the eight body sites. The two experimental groups lost fat, but the control group actually showed increased skinfold thicknesses over the 10-week period.

We decide to use the statistical technique of analysis of variance with repeated measures. We find a significant F ratio, indicating that significant differences exist among the three groups. Using a follow-up test procedure, we discover that both exercise groups are significantly different from the control group. But we find no significant difference between the aerobic dance and the jogging groups. (Many of you may not have the foggiest idea what we are talking about with the statistical terms F ratio and significance, but don't worry about it. All this is explained later. This book is about these kinds of things.)

Our conclusion from this study is that both aerobic dance and jogging are effective (apparently equally so) in bringing about a loss in fatness of overweight people (like the ones in our study) over 10 weeks. Although these results are reasonable, please remember that this is only an example. We can also pretend that this study was published in a prestigious journal and that we won the Nobel Prize.

Research Continuum

Research in our field can be placed on a continuum with applied research at one extreme and **basic research** at the opposite extreme. The research extremes have certain characteristics generally associated with them. Applied research tends to address immediate problems, to use so-called real-world settings, to use human participants, and to have limited control over the research setting, but to give results that are of direct value to practitioners. At the other extreme, basic research usually deals with theoretical problems. It uses the laboratory as the setting, frequently uses animals as subjects, carefully controls conditions, and produces results that have limited direct application. Christina (1989) suggested that basic and applied forms of research are useful in informing each other as to future research directions. Table 1.1 demonstrates how research problems in motor learning might vary along a basic to applied continuum depending on their goal and approach.

basic research

Type of research that may have limited direct application but in which the researcher has careful control of the conditions.

Table 1.1   Levels of Relevance of Motor Learning Research for Finding Solutions to Practical Problems in Sport

Level 1 Least direct relevance Basic research	Level 2 Moderate direct relevance Applied research	Level 3 Most direct relevance Applied research
Ultimate goal	**Ultimate goal**	**Ultimate goal**
Develop theory-based knowledge appropriate for understanding motor learning in general with no requirement to demonstrate its relevance for solving practical problems	Develop theory-based knowledge appropriate for understanding the learning of sport skills in sport settings with no requirement to find immediate solutions to learning problems in sport	Find immediate solutions to learning problems in sport with no requirement to demonstrate or develop theory-based knowledge at either level 1 or level 2
Main approach	**Main approach**	**Main approach**
Test hypotheses in a laboratory setting using experimenter-designed motor tasks	Test hypotheses in a sport setting or in a laboratory setting similar to it using sport skills or motor tasks that have properties of those skills	Test solutions to specific learning problems in sport in the settings described under the applied research at level 2

From "Whatever Happened to Applied Research in Motor Learning?" by R.W. Christina. In *Future Directions in Exercise and Sport Science Research* (p. 418) by J.S. Skinner et al. (Eds.), 1989, Champaign, IL: Human Kinetics. Copyright 1989 by James S. Skinner. Reprinted with permission.

To some extent, the strengths of applied research are the weaknesses of basic research, and vice versa. Considerable controversy exists in the literature on psychology, education, and physical activity (for examples, see Christina, 1989; Martens, 1979, 1987; Siedentop, 1980; Thomas, 1980) about whether research should be more basic or more applied. This issue, labeled **ecological validity,** deals with two concerns: Is the research setting perceived by the research participant in the way intended by the experimenter? Does the setting have enough of the real-world characteristics to allow generalizing to reality?

Of course, most research is neither purely applied nor purely basic but incorporates some degree of both. We believe that systematic efforts are needed in the study of physical activity to produce research that moves back and forth across Christina's (1989) levels of research (table 1.1). Excellent summaries of this type of research and the accumulated knowledge are provided in three edited volumes representing exercise physiology, sport psychology, and motor behavior: *Physical Activity, Fitness, and Health* (Bouchard, Shepard, & Stephens, 1994), *Handbook of Sport Psychology* (Singer, Hausenblas, & Janelle, 2001), and *Cognitive Issues in Motor Expertise* (Starkes & Allard, 1993). An expert prepared each chapter in these books to summarize theories as well as basic and applied research about areas related to

ecological validity

The extent to which research emulates the real world.

exercise physiology, motor expertise, and sport psychology. The novice researcher would do well to read several of these chapters as examples of how knowledge is developed and accumulated in the study of physical activity. We need more efforts to produce a related body of knowledge in the study of physical activity. Although the research base has grown tremendously in our field over the past 25 years, much remains to be done.

There is a great need to prepare proficient consumers and producers of research. To be proficient requires a thorough understanding of the appropriate knowledge base (e.g., exercise physiology, motor behavior, pedagogy, and the social and biological sciences) and of research methods. In this book we attempt to explain the tools necessary to consume and produce research. Many of the same methods are used in the various areas of physical education, exercise science, and sport science (as well as in psychology, sociology, education, and physiology). Quality research efforts always involve some or all of the following components:

- Identification and delimitation of a problem
- Searching, reviewing, and effectively writing about relevant literature
- Specifying and defining testable hypotheses
- Designing the research to test the hypotheses
- Selecting, describing, testing, and treating the participants
- Analyzing and reporting the results
- Discussing the meaning and implications of the findings

Practicality and Accessibility

We recognize that not everyone will be a researcher. Many people in our profession have little interest in research per se. In fact, some have a decided aversion to it. Researchers are sometimes viewed as strange people who deal with insignificant problems and who are out of touch with the real world (although we know that none of you feel that way). In a very informative yet entertaining book on writing scientific papers, Day (1983) related the story about two men who, while riding in a hot-air balloon, encountered some cloud coverage and lost their way. When they finally descended, they did not recognize the terrain and had not the faintest idea where they were. It so happened that they were drifting over the grounds of one of our more famous scientific research institutes. When the balloonists saw a man walking alongside a road, one of them called out, "Hey, mister, where are we?" The man looked up, took in the situation, and after a few moments of reflection said, "You're in a hot-air balloon." One balloonist turned to the other and said, "I'll bet that man is a researcher." The other balloonist said, "What makes you think so?" The first replied, "His answer is perfectly accurate and totally useless" (p. 152).

All kidding aside, the need for research in any profession just cannot be denied. After all, one of the primary distinctions between a discipline or profession and a trade is that the trade deals only with how to do something, whereas the discipline or profession concerns itself not only with how but also with why something should be done in a certain manner (and why it should even be done). However, even though most people in a discipline or profession recognize the need for research, most of those people do not read research results. This situation is not unique to our field. It has been reported that only 1% of chemists read research publications, fewer than 7% of psychologists read psychological research journals, and so on. The big question is why. We guess that most professionals who do not read research believe that doing so is not necessary. The research is not practical enough or does not directly pertain to their work. Another reason given by practitioners for not reading research publications is that they cannot understand them. The language is too technical, and the terminology unfamiliar and confusing. This is a valid complaint; we could argue, however, that if the professional preparation programs were more scientifically oriented, this would not be such a problem. Nevertheless, the research literature is extremely difficult for the nonresearcher to understand.

Reading Research

Someone once said (facetiously) that a scientific paper was not meant to be read but was meant to be published. Unfortunately, we find considerable truth in that observation. We as writers are often guilty of trying to use language to dazzle the reader and perhaps to give the impression that our subject matter is more esoteric than it really is. We tend to write for the benefit of a rather small number of readers, that is, other researchers in our particular field.

We have the problem of jargon, of course. In any field, whether it is physics, football, or cake baking, jargon confounds the outsider. The use of jargon serves as a kind of shorthand. It provides meaning to the people within the field because everyone uses those words in the same context. Research literature is famous for using a three-dollar word when a nickel word would do. As Day (1983, 147) asked, what self-respecting writer would use a three-letter word like "now" when one can use the elegant expression "at this point in time"? Researchers never "do" anything, they "perform" it; they never "start," they "initiate"; and they "terminate" instead of "end." Day further remarked that an occasional author slips and uses the word "drug," but most salivate like Pavlov's dogs in anticipation of using "chemotherapeutic agent."

For years there has been a recognized need to try to bridge the gap between the researcher and the practitioner. The American Alliance for Health, Physical Education, Recreation and Dance (AAHPERD) launched a series of publications titled *What Research Tells the Coach* [of a particular sport]. *The Journal of Physical Education, Recreation and Dance* has a feature called "Research Works," which is designed to disseminate applied research information to teachers, coaches, and fitness and recreation leaders. Yet despite these and other attempts to bridge the gap between researchers and practitioners, the gap is still imposing.

It goes without saying that if you are not knowledgeable about the subject matter, you cannot read the research literature. Conversely, if you know the subject matter, you can probably wade through the researcher's jargon more effectively. For example, if you know baseball and the researcher is recommending that by shortening the radius, one can increase the angular velocity, you can figure out that the researcher means to choke up on the bat.

One of the big stumbling blocks is the statistical analysis part of research reports. Even the most ardent seeker of knowledge can be turned off by such descriptions as this: "The tetrachoric correlations among the test variables were subjected to a centroid factor analysis, and orthogonal rotations of the primary axes were accomplished by Zimmerman's graphical method until simple structure and positive manifold were closely approximated." Please note that we are not criticizing the authors for such descriptions, as reviewers and editors usually require them. We are just acknowledging that it is frightening to someone who is trying to read a research article and who does not know a factor analysis from a volleyball. The widespread use of computers and "computerese" probably compounds the mystery associated with statistics. Many people believe anything that comes out of a computer. Others are more old-fashioned and check the computer's accuracy with their calculators. A classic case of a computer mistake occurred in a high school where the computer printed the students' locker numbers in the column where their IQs were supposed to go. It is classic because no one noticed the error at the time, but at the end of the year the students with the highest locker numbers got the best grades.

How to Read Research

Despite all the hurdles that loom in the practitioner's path when reading research, we contend that you can read and profit (not materially, unfortunately) from the research literature even if you are not well grounded in research techniques and statistical analysis. We would like to contend that after you read this book you can read any journal in any field, but the publisher would not let us. We offer the following suggestions to the practitioner on reading the research literature.

- Become familiar with a few publications that contain pertinent research in your field. You might get some help on this from a professor or librarian.

- Read only studies that are of interest to you. That may sound too trite to mention, but some people feel obligated to wade through every article.

- Read it as a practitioner would. Don't look for eternal truths. Look for ideas and indications. No study is proof of anything. Only when it has been verified time and again does it constitute knowledge.

- Read the abstract first. This saves time by helping you determine whether you wish to read the whole thing. If you are still interested, then you can read the study to better understand the methodology and the interpretations, but do not get bogged down with details.

- Do not be too concerned about statistical significance. It certainly helps to understand the concept of significance, but a little common sense serves you about as well as knowing the difference between the .02 and the .01 levels, or a one-tailed test versus a two-tailed test. Think in terms of meaningfulness. For example, if two methods of teaching bowling result in an average difference of 0.5 pins, what difference does it make whether it is significant? On the other hand, if there is a big difference that is not significant, further investigation is warranted, especially if the study involved a small number of participants. It is certainly helpful to know the concepts of the different types of statistical analysis, but it is not crucial to being able to read a study. Just skip that part.

- Be critical but objective. You can usually assume that a national research journal selects studies for publication by the jury method. Two or three qualified individuals read and judge the relevance of the problem, the validity and reliability of the procedures, the efficacy of the experimental design, and the appropriateness of the statistical analysis. It is true that some studies are published that should not be. Yet if you are not an expert in research, you do not need to be suspicious about the scientific worth of a study that appears in a recognized journal. If it is too far removed from any practical application to your situation, do not read it.

You will find that the more you read, the easier it becomes to understand, simply because you become more familiar with the language and the methodology, like the man who was thrilled to learn he had been speaking prose all his life.

An Example of Practical Research

To illustrate our research consumer suggestions, consider the lighthearted account of a young physical education teacher and coach named Sonjia Roundball (J.K. Nelson, 1988) In a moment of weakness, Sonjia glanced through the table of contents of the *Research Quarterly for Exercise and Sport,* which had been left in her car by a graduate student friend of hers (they had used it to keep their tacos from dripping on the upholstery). She experienced a spark of interest when she noticed an article titled "The Effects of a Season of Basketball on the Cardiorespiratory Responses of High School Girls." With some curiosity, she turned to the article and began to read. In its introductory passages, the article stated that relatively little specific information was available on the physiological changes in girls due to sport participation. A short review of literature cited a few studies on swimmers and other sport participants, and the upshot was that female athletes possess higher levels of cardiorespiratory fitness than do nonathletes. The author emphasized that no studies had tried to detect changes in girls' fitness during a season of basketball.

The next section of the study dealt with methods. It noted the length of the season, the number of games, the number of practices and their length, and the breakdown of time devoted to drills, scrimmaging, and individual practice. The participants were 12 girls on the high school basketball team and 14 girls in the nonparticipant group who were from physical education classes and who had similar academic and activity schedules as the participants. All participants were tested at the beginning and end of the season on maximal oxygen consumption and various other physiological measurements dealing with ventilation, heart rate, and blood pressure. Sonjia remembered those things from her exercise physiology course a number of years ago and was willing to accept these as good indicators of cardiorespiratory fitness.

The results were then presented in tables. She did not understand these things but was willing to trust the author as to their appropriateness. The author discovered no significant increases in any of the cardiorespiratory measures from the preseason test to the postseason test for either group. This jolted Sonjia to the quick! Surely, a strenuous sport such as basketball should produce improvements in fitness. Something must be wrong here, she thought. She further read (with small consolation) that the basketball players had higher values of maximal oxygen consumption than the nonparticipants at both the beginning and the end of the season. Sonjia then read the **discussion,** which mentioned things such as that the values were higher than similar values in other studies. ("So what?" Sonjia thought.) She read with more interest the observations by the author that boys' basketball programs were more strenuous in terms of length and number of workouts. Sonjia began to think about this. The author admitted that the number of participants was small and that some changes might not have been detected, and the author offered other speculations. The author concluded, however, that the training program used in this study was not strenuous enough to induce significant improvement in cardiorespiratory fitness.

Sonjia was sophisticated enough to realize the limitations of one study. Nevertheless, it was very similar to her schedule and general practice routines. She noticed in the references for the article three studies from a journal called *Medicine and Science in Sports and Exercise.* She had never read this journal, but she decided to drive over to the university the next weekend to look up this publication in the library. When she located the journal, the latest issue happened to have an article on conditioning effects of swimming on college women. Although this was a different sport and a different age group, she reasoned that the review of literature might prove fruitful. She was right. It cited a recent study on aerobic capacity, heart rate, and energy cost during a season of girls' basketball. Sonjia quickly located this study and now read with the excitement that comes from the personal discovery of ideas. She also was pleasantly surprised to find that it was easier reading than the first study because she was now more familiar with the terminology and the general organization of the article.

This study also reported no improvement in aerobic capacity during the season. This study involved monitoring heart rates during games by telemetry, and researchers infrequently observed heart rates of over 170 beats per minute (bpm). They concluded that the practice sessions were apparently too moderate in intensity and that the training should be structured to meet both the skill and the fitness demands of the sport.

Sonjia returned to her school determined to take a more scientific approach to her basketball program. To start with, she had one of her managers chart the number of minutes that players were actually engaged in movement in the practice sessions. Sonjia also had the players take their pulses at various intervals during the sessions. She was surprised to find that the heart rates rarely surpassed 130 bpm. As an outgrowth of her recent literature search, she remembered that there is an intensity threshold necessary to bring about improvement in cardiorespiratory fitness. She knew that for this age group a heart rate of about 160 bpm was needed to provide a significant training effect. Consequently, she initiated some changes in her practice sessions (including more conditioning drills) and made the scrimmages more intensive and gamelike. To end this saga of Sonjia, you will be happy to know that Coach Roundball's team went on to win all its games, the district and the state championships, and the world games.

Summarizing the Nature of Research

Thomas Huxley wrote that **science** is simply common sense at its best. Moments of discovery can be very rewarding, whether that discovery is finding research that applies to and can improve your situation or discovering new knowledge through the research of your thesis or dissertation. Research should be viewed more as a method of problem solving than as some dark and mysterious realm inhabited by impractical people who speak and write in baffling terms. We think practitioners can read research literature, and we are dedicated in this text to trying to facilitate the process of becoming a research consumer.

discussion

Chapter or section of a research report that explains what the results mean.

science

A process of careful and systematic inquiry.

Unscientific Versus Scientific Methods of Problem Solving

Although there are many definitions of research, nearly all characterize research activity as some sort of structured problem solving. The word *structured* refers to the fact that a number of research techniques can be used as long as the techniques are considered acceptable by scholars in the field. Thus, research is concerned with problem solving, which then may lead to new knowledge.

The problem-solving process involves several steps whereby the problem is developed, defined, and delimited; hypotheses are formulated; data are gathered and analyzed; and the results are interpreted with regard to the acceptance or rejection of the hypotheses. These steps are often referred to as the **scientific method of problem solving.** The steps also constitute the chapters, or sections, of the research paper, thesis, and dissertation. Consequently, we devote much of this text to the specific ways these steps are accomplished.

Some Unscientific Methods of Problem Solving

Before we go into more detail concerning the scientific method of problem solving, it is important to recognize some other ways by which humankind has acquired knowledge. All of us have used these methods, so they are recognizable. Helmstadter (1970) labeled the methods as tenacity, intuition, authority, the rationalistic method, and the empirical method.

Tenacity

People sometimes cling to certain beliefs despite the lack of supporting evidence. Our superstitions are good examples of this method called **tenacity.** Coaches and athletes are notoriously superstitious. A coach may wear a particular sport coat, hat, tie, or pair of shoes because the team won the last time he wore it. Athletes frequently have a set pattern that they consider lucky for dressing, warming up, or entering the stadium. Even though they acknowledge no logical relationship between the game's outcome and their particular routine, they are afraid to break the pattern.

For example, take the man who believed that black cats brought bad luck. One night when he was returning to his ranch, a black cat started to cross the road. The man swerved off onto the prairie to keep the cat from crossing in front of him and hit a hard bump that caused the headlights to go off. Unable to see the black cat in the dark night, he sped frantically over rocks, mounds, and holes until he came to a sudden stop in a ravine, wrecking his car and sustaining moderate injuries. Of course, this just confirmed his staunch belief that black cats do indeed bring bad luck. Obviously, tenacity has no place in science. It is the least reliable source of knowledge.

Intuition

Intuitive knowledge is sometimes considered to be common sense or self-evident. However, many self-evident truths are subsequently found to be false. That the earth is flat is a classic example of the intuitively obvious; that the sun revolved around the earth was once self-evident; that no one could run a mile in less than 4 min once was self-evident. Furthermore, for anyone to shot-put more than 70 ft (21 m) or pole-vault more than 18 ft (5.5 m) or for a woman to run distances over a half mile (0.8 km) was impossible. One fundamental tenet of science is that we must be ever cognizant of the importance of substantiating our convictions with factual evidence.

Authority

Reference to some authority has long been used as a source of knowledge. Although this is not necessarily invalid, it does depend on the authority and on the rigidity of adherence. However, the appeal to authority has been carried to absurd lengths. Even personal observation and experience have been deemed unacceptable when they dispute authority. Supposedly, people refused to look through Galileo's telescope when he disputed Ptolemy's

scientific method of problem solving

Method of solving problems that uses the following steps: defining and delimiting the problem, forming a hypothesis, gathering data, analyzing data, and interpreting the results.

tenacity

An unscientific method of problem solving in which people cling to certain beliefs regardless of the lack of supporting evidence.

explanation of the world and of the heavens. Galileo was later jailed and forced to recant his beliefs. Bruno also rejected Ptolemy's theory and was burned at the stake. (Scholars read and believed Ptolemy's book on astrology and astronomy for 1,200 years after his death!) In 1543, Vesalius wrote a book on anatomy, much of which is still considered correct today. However, because his work clashed with Galen's theories, he met with such ridicule that he gave up his study of anatomy.

Perhaps the most crucial aspect of the appeal to authority as a means of obtaining knowledge is the right to question and to accept or reject the information. Furthermore, the authority's qualifications and the methods by which the authority acquired the knowledge also determine the validity of this source of information.

The Rationalistic Method

In the rationalistic method, we derive knowledge through reasoning. A good example is the following classic syllogism:

All men are mortal. (major premise)

The emperor is a man. (minor premise)

Therefore, the emperor is mortal. (conclusion)

Although you probably would not argue with this reasoning, the key to this method is the truth of the premises and their relationship to each other. For example,

Basketball players are tall.

Tom Thumb is a basketball player.

Therefore, Tom Thumb is tall.

In this case, however, Tom is very short. The conclusion is trustworthy only if derived from premises (assumptions) that are true. Also, the premises may not in fact be premises but rather descriptions of events or statements of fact. The statements are not connected in a cause-and-effect manner. For example,

There is a positive correlation between shoe size and mathematics performance among elementary school children (i.e., children with large shoe sizes do well in math).

Herman is in elementary school and wears large shoes.

Therefore, Herman is good in mathematics.

Of course, in the first statement the factor common to both mathematics achievement and shoe size is age. Older children tend to be bigger and thus have bigger feet than younger

Her study carrel is nicer than her apartment.

children. Older children also have higher achievement scores in mathematics, but there is no cause-and-effect relationship. You must always be cognizant of this when dealing with correlation. Reasoning is fundamental in the scientific method of problem solving but cannot be used by itself to arrive at knowledge.

The Empirical Method

empirical

Describes data or a study that is based on objective observations.

The word **empirical** denotes experience and the gathering of data. Certainly, data gathering is part of the scientific method of solving problems. However, there can be pitfalls in relying too much on your own experience (or data). First, your own experience is very limited. Furthermore, your retention depends substantially on how the events agree with your past experience and beliefs, on whether things "make sense," and on your state of motivation to remember. Nevertheless, the use of data (and the empirical method) is high on the continuum of methods of obtaining knowledge as long as you are aware of the limitations of relying too heavily on this method.

The Scientific Method of Problem Solving

The methods of acquiring knowledge previously discussed lack the objectivity and control that characterize the scientific approach to problem solving. Several basic steps are involved in the scientific method. Some authors list seven or eight steps, and others condense these steps into three or four. Nevertheless, all the authors are in general agreement as to the sequence and processes that are involved. The steps are briefly described next. Greater detail concerning the basic processes are covered in other chapters.

Step 1: Developing the Problem (Defining and Delimiting It)

This step may sound a bit contradictory, for how could the development of the problem be a part of solving it? Actually, the discussion here is not about finding a problem to study (ways of locating a problem are discussed in chapter 2); the assumption is that the researcher has already selected a topic. However, to design and execute a sound investigation, the researcher must be very specific about what is to be studied and to what extent it will be studied.

independent variable

The part of the experiment that the researcher is manipulating; also called the *experimental* or *treatment variable*.

Many ramifications constitute this step, an important one being the identification of the independent and the dependent variables. The **independent variable** is what the researcher is manipulating. If, for example, two methods of teaching a motor skill are being compared, then the teaching method is the independent variable; this is sometimes called the experimental, or treatment, variable.

dependent variable

The effect of the independent variable; also called the *yield*.

The **dependent variable** is the effect of the independent variable. In the comparison of teaching methods, the measure of skill is the dependent variable. If you think of an experiment as a cause-and-effect proposition, the cause is the independent variable and the effect is the dependent variable. The latter is sometimes referred to as the yield. Thus, the researcher must define exactly what will be studied and what will be the measured effect. When this is resolved, the experimental design can be determined.

Step 2: Formulating the Hypothesis

hypothesis

The anticipated outcome of a study or experiment.

The **hypothesis** is the expected result. When a person sets out to conduct a study, he or she generally has an idea as to what the outcome will be. This anticipated solution to the problem may be based on some theoretical construct, on the results of previous studies, or perhaps on the experimenter's past experience and observations. The last source is probably least likely or defensible because of the weaknesses of the unscientific methods of acquiring knowledge discussed previously. Regardless, the research should have some experimental hypothesis about each subproblem in the study.

We enjoyed "Calvin and Hobbes," the cartoon strip by Bill Watterson. In a clever strip, Calvin is talking to his friend Susie in the lunch room:

> Calvin: Curiosity is the essence of the scientific mind. For example, you know how milk comes out your nose if you laugh while drinking? Well, I'm going to see what happens when I inhale milk into my nose and laugh!

Susie (as she leaves): Idiocy is the essence of the male mind.

Calvin: I'm guessing it will shoot out my ears. Don't you want to see?

Calvin has developed a testable hypothesis: "If I inhale milk into my nose and laugh, it will shoot out of my ears." Susie has an untestable hypothesis (at least in our view): "Idiocy is the essence of the male mind."

One of the essential features about the hypothesis is that it be testable. The study must be designed in such a way that the hypothesis can be either supported or refuted. It should be obvious to you, then, that the hypothesis cannot be a type of value judgment or an abstract phenomenon that cannot be observed.

For example, you might hypothesize that success in athletics depends solely on fate. In other words, if a team wins, it is because it was meant to be; similarly, if a team loses, a victory was just not meant to be. There is no way to refute this hypothesis because there is no evidence that could be obtained to test it.

Step 3: Gathering the Data

Of course, before step 2 can be accomplished, the researcher must decide on the proper methods of acquiring the necessary data to be used in testing the research hypothesis. The reliability of the measuring instruments, the controls that are employed, and the overall objectivity and precision of the data-gathering process are crucial to the problem's solution.

In terms of difficulty, gathering data may be the easiest step because in many cases it is routine. However, planning the method is one of the most difficult steps. Good methods attempt to maximize both the **internal validity** and the **external validity** of the study.

Internal validity and external validity relate to the research design and controls that are used. Internal validity refers to the extent to which the results can be attributed to the treatments used in this study. In other words, the researcher must try to control all other variables that could influence the results. For example, Jim Nasium wants to scientifically assess the effectiveness of his exercise program in developing physical fitness in young boys. He tests his participants at the beginning and then at the end of a nine-month training program and concludes that the program brought about significant improvement in fitness. What is wrong with Jim's conclusion? His study contains several flaws. The first is that Jim gave no consideration to maturity. Nine months of maturation produced significant changes in size and in accompanying strength and endurance. Also, what else were the participants doing during this time? How do we know that their other activities were not responsible, or partly so, for the changes in their fitness levels? Chapter 18 deals with these threats to internal validity.

External validity pertains to the generalizability of the results. To what extent can the results apply to the real world? This often produces a paradox for research in the behavioral sciences because of the controls required for internal validity. In motor learning studies, for example, the task is often something novel so as to control for past experience. Furthermore, it is desirable to be able to measure the performance objectively and reliably. Consequently, the learning task is frequently a maze, a rotary pursuit meter, or a linear position task, all of which may meet the demands for control with regard to internal validity. But then you are faced with the question of external validity: How does performance in a laboratory setting with a novel, irrelevant task apply to learning gymnastics or basketball? These questions are important and sometimes vexing, but they are not insurmountable. (They are discussed later.)

Step 4: Analyzing and Interpreting Results

The novice researcher finds this step to be the most formidable for several reasons. First, this step usually involves some statistical analysis, and the novice researcher (particularly the master's student) often has a limited background in and a fear of statistics. Second, analysis and interpretation require considerable knowledge, experience, and insight, which the novice may lack.

That analysis and interpretation of results are the most challenging step goes without question. It is here that the researcher must provide evidence for the support or the rejection of the research hypothesis. In doing this, the researcher also compares the results with those of others (the related literature) and perhaps attempts to relate and integrate the

internal validity

The extent to which the results of a study can be attributed to the treatments used in the study.

external validity

The generalizability of the results of a study.

results into some theoretical model. Inductive reasoning is employed in this step (whereas deductive reasoning is primarily used in the statement of the problem). The researcher attempts to synthesize the data from his or her study along with the results of other studies to contribute to the development or substantiation of a theory.

Alternative Models of Research

In the preceding section we summarized the basic steps in the scientific method of problem solving. Science is a way of knowing, often defined as structured inquiry. One basic goal of science is to explain things or to be able to generalize and build a theory. When a scientist develops a useful model to explain behavior, scholars often test predictions from this model using the steps of the scientific method. The model and the approaches used to test the model are called a paradigm.

Normal Science

normal science

An objective manner of study grounded in the natural sciences that is systematic, logical, empirical, reductive, and replicable.

For centuries, the scientific approaches used in studying problems in both the natural and the social sciences have been what Thomas Kuhn (1970), a noted science historian, has termed **normal science.** It is characterized by the elements we listed at the beginning of this chapter (i.e., systematic, logical, empirical, reductive, and replicable). Its basic doctrine is objectivity. Normal science is grounded in the natural sciences, which have long adhered to the idea of the orderliness and reality of matter, that nature's laws are absolute and discoverable by objective, systematic observations and investigations that are not influenced by (in other words, independent of) humans. The experiments are theory driven and have testable hypotheses.

Normal science received a terrific jolt with Einstein's theory of relativity and the quantum theory, which indicated that nature's laws could be influenced by humans (that is, that reality depends to a great extent on how one perceives it). Moreover, some things, such as the decay of a radioactive nucleus, happen for no reason at all. The fundamental laws that had been believed to be absolute were now considered to be statistical rather than deterministic. Phenomena could be predicted statistically but not explained deterministically (Jones, 1988).

Challenges to Normal Science

paradigm crisis phenomenon

Development of discrepancies in a paradigm leading to proposals of a new paradigm that better explains the data.

Relatively recently (since about 1960), there have been serious challenges regarding normal science's concept of objectivity (i.e., that the researcher can be detached from the instruments and conduct of the experiment). Two of the most powerful challengers to the idea of objective knowledge have been Thomas Kuhn (1970) and Michael Polanyi (1958). They contend that objectivity is a myth. From the first inception of the idea for the hypothesis through the selection of apparatus to the analysis of the results, the observer is involved. The conduct of the experiment and the results can be considered expressions of the researcher's point of view. Polanyi has been especially opposed to the adoption of normal science for the study of human behavior.

Kuhn (1970) maintains that normal science does not really evolve in systematic steps the way that scientific writers describe it. Kuhn discusses the **paradigm crisis phenomenon,** in which researchers who have been following a particular paradigm begin to find discrepancies in it. The findings no longer agree with the predictions, and a new paradigm is advanced. Interestingly, the old paradigm does not die completely but only develops varicose veins and fades away. Many researchers with a great deal of time and effort invested in the old paradigm are reluctant to change, so it is usually a new group of researchers who propose the new paradigm. Thus, normal science progresses by revolution, with a new group of scientists breaking away and replacing the old. Kuhn and Polanyi concur that the doctrine of objectivity is simply not a reality. Nevertheless, normal science has been and will continue to be successful in the natural sciences and in certain aspects of the study of humans. However, Martens (1987) contends that it has failed miserably in the study of human behavior, especially in the more complex functions.

As a sport psychologist, Martens has asserted that laboratory experiments have limited use in answering questions about complex human behavior in sport. He considers his role as a practicing sport psychologist to have been far more productive in gaining knowledge about athletes and coaches and the solutions to their problems. Other workers in the so-called helping professions have made similar observations about both the limitations of normal science and the importance of alternative sources of knowledge in forming and shaping professional beliefs. Schein (1987), a noted scholar of social psychology, related an interesting (some might call it shocking) revelation concerning the relative impact of published research results versus practical experience. At a conference, he and a number of his colleagues were discussing what they relied on most for their classroom teaching. There seemed to be widespread agreement among these professors that the data they really believed in and used in the classroom came from personal experience and information learned in the field. Schein was making the point that different categories of knowledge can be obtained by different methods. In effect, some people are more influenced by sociological and anthropological research models than by the normal science approach.

For some time many scholars in education, psychology, sociology, anthropology, sport psychology, physical education, and other disciplines have proposed methods of studying human behavior other than those of conventional normal science. Anthropologists, sociologists, and clinical psychologists have used in-depth observation, description, and analysis of human behavior for nearly three quarters of a century. For over 40 years, researchers in education have used participant and nonparticipant observation to obtain comprehensive, firsthand accounts of teacher and student behaviors as they occur in real-world settings. More recently, physical educators, sport psychologists, and exercise specialists have been engaged in this type of field research. A number of names given to this general form of research are ethnographic, qualitative, grounded, naturalistic, and participant observational research. Regardless of the names, the commitments, and the beliefs of the researchers, this type of research was at first not well received by the adherents of normal science and the scientific method. In fact, this form of research (we include all its forms under the name **qualitative research**) has often been labeled by normal scientists as superficial, lacking in rigor, and just plain unscientific. As qualitative research methods have evolved, so has the thinking of many of these people. As you will see in chapter 19, many of the research tenets listed by Kuhn (1970) are found in contemporary qualitative research.

Martens (1987) referred to such adherents of normal science as the gatekeepers of knowledge because they are the research journal editors and reviewers who decide who gets published, who serves on the editorial boards, and whose papers are presented at conferences. Studies without internal validity do not get published, yet studies without external validity lack practical significance. Martens (1987) charged that normal science (in psychology) prefers publishability to practical significance.

The debates between qualitative and normal (often classified as quantitative) research have been heated and prolonged. The qualitative proponents have gained confidence and momentum in recent years, and there is no question that this point of view is now recognized as a viable method of addressing problems in the behavioral sciences. Credibility is established by systematically categorizing and analyzing causal and consequential factors. The naturalistic setting of qualitative research both facilitates analysis and precludes precise control of so-called extraneous factors as does much other research occurring in field settings. The holistic interrelationship among observations and the complexity and dynamic processes of human interaction make it impossible to limit the study of human behavior to the sterile, reductionistic approach of normal science. **Reductionism,** a characteristic of normal science, assumes that complex behavior can be reduced, analyzed, and explained as parts that can then be put back together as a whole and understood. Critics of the conventional approach to research believe the central issue is the unjustified belief that normal science is the only source of true knowledge.

Implications of These Challenges

There are many implications of the challenges to normal science. For example, when we study simple movements, such as linear positioning in a laboratory to reflect cognitive processing of information, do we learn anything about movements in real-world settings such as the performance of sport skills? When we evaluate EMG activity in specific muscle

qualitative research

Research method that often involves intensive, long-time observation in a natural setting; precise and detailed recording of what happens in the setting; interpretation and analysis of the data using description, narratives, quotes, charts, and tables. Also called *ethnographic, naturalistic, interpretive, grounded, phenomenological, subjective,* and *participant observational.*

reductionism

A characteristic of normal science that assumes that complex behavior can be reduced, analyzed, and explained as parts that can then be put back together to understand the whole.

groups during a simple movement, does it really tell us anything about the way the nervous system controls movements in natural settings such as athletics? Can we study the association of psychological processes related to movement in laboratory settings and expect the results to apply in sport and exercise situations? When we conduct these types of experiments, are we studying nature's phenomena or laboratory phenomena?

Do not misinterpret the intent of these questions. They do not mean that nothing important can be discovered about physical activity from laboratory research. What they suggest is that these findings do not necessarily model accurately the way humans plan, control, and execute movements in natural settings associated with exercise and sport.

Kuhn's (1970) descriptions about how science advances and the limitations of applying normal science to natural settings demonstrate that scientists need to consider the various ways of knowing and that the strict application of the normal scientific method of problem solving may sometimes hinder rather than advance science. If the reductionistic approach of the scientific method has not well served the natural scientists who developed it, then certainly researchers in human behavior need to carefully assess the relative strengths and weaknesses of conventional and alternative research paradigms for their particular research questions.

Alternative Forms of Scientific Inquiry

Martens (1987, 52) has suggested that we view knowledge not as being either scientific or unscientific or either reliable or unreliable but rather as existing on a continuum, as illustrated in figure 1.1. This continuum, labeled "DK" for degrees of knowledge, ranges from "don't know" to "damn konfident." Considered in this way, varying approaches to disciplined inquiry are useful in accumulating knowledge. As examples, Martens (1979, 1987) has urged sport psychologists to consider the idiographic approach, introspective methods, and field studies instead of relying on the paradigm of normal science as the only answer to research questions in sport psychology. Thomas, French, and Humphries (1986) detailed how to study children's sport knowledge and skills in games and sports. Costill (1985) discussed the study of physiological responses in practical exercise and sport settings. Locke (1989) presented a tutorial on the use of qualitative research in physical education and sport. In later chapters we give greater detail about some of these alternative strategies for research.

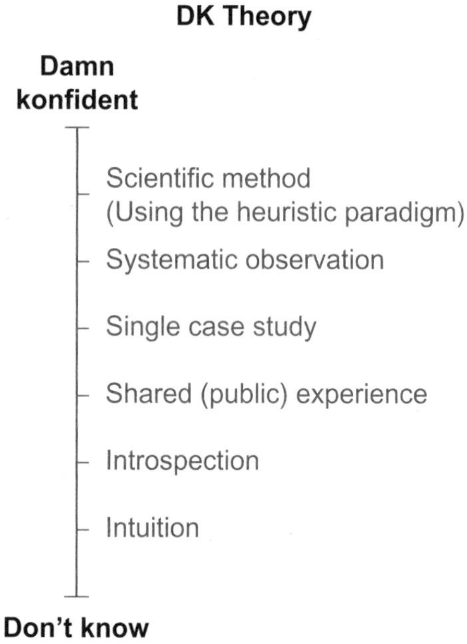

What we hope you gain from this section is that science is disciplined inquiry, not a set of specific procedures. Although advocates of alternative methods of research are often very persuasive, we certainly do not want you to conclude that the study of physical activity should abandon the traditional methods of normal science. We have learned much from these techniques and will continue to do so. Furthermore, we certainly do not want you to toss away this book as being pointless. We have not even begun to tell you all the fascinating things we have learned over the years (it is hard to tell whether some of these things should be classified as normal or abnormal science). In addition, we have many funny stories yet to tell (abnormal humor). Aside from these compelling reasons for continuing with the book, we want you to realize and appreciate that even though so-called normal science may not be the solution to all questions raised in our field, it is the recognized model for research, and it is often taught as the only approach in graduate study. Furthermore, none of the alternative methods of research denounce the scientific method of problem solving. The main bones of contention are with the methods, the setting, the controls, the types of data, and the analysis.

The bottom line is that different problems require different solutions. As we said before, science is disciplined inquiry, not a set of specific procedures. We need to em-

DK Theory

Damn konfident

Scientific method
(Using the heuristic paradigm)

Systematic observation

Single case study

Shared (public) experience

Introspection

Intuition

Don't know

Figure 1.1   The degrees of knowledge theory with examples of different types of methods varying in degree of reliability.

Reprinted, by permission, from R. Martens, 1987, "Science, knowledge, and sport psychology," *The Sport Psychologist* 1(1): 46.

brace all systematic forms of inquiry. Rather than argue about the differences, we should capitalize on the strengths of both methods to provide useful knowledge about human movement. The nature of the research questions and setting should drive the selection of approaches to acquiring knowledge. In fact, just as Christina (1989) suggested that researchers might move among levels of research (basic to applied), so researchers might move among paradigms (quantitative to qualitative) to acquire knowledge.

Types of Research

Research is a structured way of solving problems. There are different kinds of problems in the study of physical activity; thus, different types of research are used to solve these problems. This text concentrates on these four types of research: analytical, descriptive, experimental, and qualitative. A brief description of each follows.

Analytical Research

As the name implies, **analytical research** involves in-depth study and evaluation of available information in an attempt to explain complex phenomena. The different types of analytical research are historical, philosophic, reviews, and research synthesis.

analytical research

Type of research that involves in-depth study and evaluation of available information in an attempt to explain complex phenomena; can be categorized in the following way: historical, philosophic, review, and meta-analysis.

Historical Research

Obviously, historical research deals with events that have already occurred. Historical research focuses on events, organizations, institutions, and people. In some studies, the researcher is interested mostly in preserving the record of events and past accomplishments. In other investigations, the writer attempts to discover facts that will provide more meaning and understanding of past events to explain the present state of affairs. Some historians have even attempted to use information from the past to predict the future. The research procedures associated with historical studies are addressed in considerable detail in chapter 12.

Philosophic Research

Critical inquiry characterizes philosophic research. The researcher establishes hypotheses, examines and analyzes existing facts, and synthesizes the evidence into a workable theoretical model. Many of the most important problem areas must be dealt with by the philosophic method. Problems dealing with objectives, curricula, course content, requirements, and methodology are but a few of the important issues that can be resolved only through the philosophic method of problem solving.

Although some authors emphasize the differences between science and philosophy, the philosophic method of research follows essentially the same steps as other methods of scientific problem solving. The philosophic approach uses scientific facts as the bases for formulating and testing research hypotheses.

An example of such philosophic research was Morland's 1958 study, in which he analyzed the educational views held by leaders in American physical education and categorized them into educational philosophies of reconstructionism, progressivism, essentialism, and perennialism.

Having an opinion is not the same as having a philosophy. In philosophic research, beliefs must be subjected to rigorous criticism in light of the fundamental assumptions. Academic preparation in philosophy and a solid background in the fields from which the facts are derived are necessary. Other examples and a more detailed explanation of philosophic research are given in chapter 13.

Reviews

A **review** is a critical evaluation of recent research on a particular topic. The author must be very knowledgeable about the available literature as well as the research topic and procedures. A review involves analysis, evaluation, and integration of the published literature,

review

A research paper that is a critical evaluation of research on a particular topic.

often leading to important conclusions concerning the research findings up to that time (for good examples of reviews, see Blair's 1993 review, "Physical Activity, Physical Fitness, and Health," or Silverman and Subrumanian's 1999 review, "Student Attitude Toward Physical Education and Physical Activity: A Review of Measurement Issues and Outcomes").

Certain publications consist entirely of reviews, such as the *Psychological Review,* the *Annual Review of Physiology,* and the *Review of Educational Research.* A number of journals publish reviews periodically, and some occasionally devote entire issues to reviews. For example, the 75th anniversary issue of the *Research Quarterly for Exercise and Sport* (Silverman, 2005) contains some excellent reviews on various topics.

Research Synthesis

Reviews of literature are difficult to write because they require that a large number of studies be synthesized to determine common underlying findings, agreements, or disagreements. To some extent this is like trying to make sense of data collected on a large number of participants by simply looking at the data. Glass (1977) and Glass, McGaw, and Smith (1981) proposed a quantitative means of analyzing the findings from numerous studies; this is called meta-analysis. Findings between studies are compared by changing results within studies to a common metric called effect size. Over the years many meta-analyses have been reported in the physical activity literature (Feltz & Landers, 1983; Payne & Morrow, 1993; Sparling, 1980; Thomas & French, 1985). This technique is discussed in more detail in chapter 14.

Descriptive Research

Descriptive research is concerned with status. The most prevalent descriptive research technique is the survey, most notably the questionnaire. Other forms of surveys include the interview (personal and via telephone) and the normative survey. Chapter 15 provides detailed coverage of these techniques.

The following paragraphs briefly identify three different types of survey research techniques.

The Questionnaire

The main justification for using a questionnaire is the need to obtain responses from people, often from a wide geographical area. The questionnaire usually strives to secure information about present practices, conditions, and demographic data. Occasionally, a questionnaire asks for opinions or knowledge.

The Interview

The interview and the questionnaire are essentially the same technique insofar as planning and procedures are concerned. Obviously, the interview has certain advantages over the questionnaire. The researcher can rephrase questions and ask additional ones to clarify responses and secure more valid results. Becoming a skilled interviewer requires training and experience. Telephone interviewing has become increasingly common in recent years. It costs half as much as face-to-face interviews and can cover a wide geographical area, which is generally a limitation in personal interviews. We discuss some other advantages of the telephone interview technique in chapter 15.

The Normative Survey

There have been a number of notable normative surveys in the fields of physical activity and health. The normative survey generally seeks to gather performance or knowledge data on a large sample from a population and to present the results in the form of comparative standards, or norms. The development of the norms for the *AAHPER Youth Fitness Test Manual* (American Association for Health, Physical Education and Recreation, 1958) is an outstanding example of a normative survey. Thousands of boys and girls ages 10 to 18 throughout the United States were tested on a battery of motor fitness items. Percentiles were then established for comparative performances to provide information for students, teachers, administrators, and parents. Actually, the Youth Fitness Test was developed in response to another survey, the Kraus-Weber test (Kraus & Hirschland, 1954), which revealed that American children scored dramatically lower on a test battery of minimum muscular fitness when compared with European children.

Other Descriptive Research Techniques

There are several other descriptive research techniques. Other forms of descriptive research include the case study, the job analysis, observational research, developmental studies, and correlational studies. Chapter 16 provides detailed coverage of these descriptive research procedures.

The Case Study

The case study is used to provide detailed information about an individual (or institution, community, etc.). It aims to determine unique characteristics about the subject or condition. This descriptive research technique is used widely in such fields as medicine, psychology, counseling, and sociology. The case study is also a technique used in qualitative research.

The Job Analysis

This type of research is a special form of case study. It is done to describe the nature of a particular job, including the duties, responsibilities, and preparation required for success in the job.

Observational Research

Observational research is a descriptive technique in which behaviors are observed in the participants' natural setting, such as the classroom or play environment. The observations are frequently coded, and then their frequency and duration are analyzed.

Developmental Studies

In developmental research, the investigator is usually concerned with the interaction of learning or performance with maturation. For example, a researcher may wish to assess the extent to which the ability to process information about movement can be attributed to maturation as opposed to strategy, or the researcher may desire to determine the effects of growth on a physical parameter such as aerobic capacity.

Developmental research can be undertaken by what is called the longitudinal method, whereby the same participants are studied over a period of years. Obvious logistical problems are associated with longitudinal studies, so an alternative is to select samples of participants from different age groups to assess the effects of maturation. This is called the cross-sectional approach.

Correlational Studies

The purpose of correlational research is to examine the relationship between certain performance variables, such as heart rate and ratings of perceived exertion; the relationship between traits, such as anxiety and pain tolerance; or the correlation between attitudes and behavior, as in the attitude toward fitness and the amount of participation in fitness activities. Sometimes correlation is employed to predict performance. For example, a researcher may wish to predict percentage of body fat from skinfold measurements. Correlational research is descriptive in that you cannot presume a cause-and-effect relationship. All that can be established is that there is (or is not) an association between two or more traits or performances.

Epidemiologic Research

Another form of descriptive research that has become a viable approach to studying problems dealing with health, fitness, and safety concerns is the epidemiologic research method. This type of research pertains to the frequencies and distributions of health and disease conditions among various populations. Rate of occurrence is the basic concept in epidemiologic studies. The size of the population being studied is an important consideration in examining the prevalence of such things as injuries, illnesses, or health conditions in a specified at-risk population.

Although cause and effect cannot be established by incidence and prevalence data, a strong inference of causation can often be made through association. Chapter 17 is devoted to epidemiologic research.

Experimental Research

experimental research

Type of research that involves the manipulation of treatments in an attempt to establish cause-and-effect relationships.

Experimental research has a major advantage over other types of research in that the researcher can manipulate treatments to cause things to happen (i.e., a cause-and-effect situation can be established). This contrasts with other types of research in which already existing phenomena or data from the past are observed and analyzed. As an example of an experimental study, assume that Virginia Reel, a dance teacher, hypothesizes that students would learn more effectively through the use of a videotape. First, she randomly assigns students to two sections. One section is taught by the so-called traditional method (explanation, demonstration, practice, and critique). The other section is taught in a similar manner, except the students are filmed while practicing and can thus observe themselves as the teacher critiques their performances. After nine weeks, a panel of dance teachers evaluates both sections. In this study, method of teaching is the independent variable and dance performance (skill) is the dependent variable. After the groups' scores are compared statistically, Virginia can conclude whether her hypothesis can be supported or not.

In experimental research, the researcher attempts to control all factors except the experimental (or treatment) variable. If the extraneous factors can be successfully controlled, then the researcher can presume that the changes in the dependent variable are due to the independent variable. Chapter 18 is devoted to experimental and quasi-experimental research.

Qualitative Research

In the study of physical activity, qualitative research is the new kid on the block. Actually, qualitative research has been used for many years in other fields, such as anthropology and sociology. Researchers in education have been engaged in qualitative methods longer than researchers in our field. As previously mentioned, several names are given to this type of research (ethnographic, naturalistic, interpretive, grounded, phenomenological, subjective, and participant observational). Some are simply name differences, whereas some have different approaches and points of focus. We have arbitrarily lumped them all under the heading of qualitative research, as that seems to be the most common term used in our field.

Qualitative research is different from other research methods. It is a systematic method of inquiry, and it follows the scientific method of problem solving to a considerable degree; however, it deviates in certain dimensions. Qualitative research rarely establishes hypotheses at the beginning of the study, but instead uses more general questions to guide the study. It proceeds in an inductive process in developing hypotheses and theory as the data unfold. The researcher is the primary instrument in the data collection and analysis. Qualitative research is characterized by intensive firsthand presence. The tools of data collection are observation, interviews, and researcher-designed instruments (Goetz & LeCompte, 1984). Qualitative research is described in chapter 19.

Overview of the Research Process

A nice overview of the research methods course, as well as an introduction to this book, is provided in figure 1.2. This flowchart provides a linear way to think about planning a research study. Once the problem area is identified, reading and thinking about relevant theories and concepts, as well as a careful search of the literature for relevant findings, lead to the specification of hypotheses or questions. Operational definitions are needed in a research study so that the reader knows exactly what the researcher means by certain terms. Operational definitions describe observable phenomena that enable the researcher to examine empirically whether the predictions can be supported. The study is designed, and the methods are made operational. The data are then collected and analyzed, and the findings identified. Finally, the results are related back to the original hypotheses or questions and discussed in relation to theories, concepts, and previous research findings.

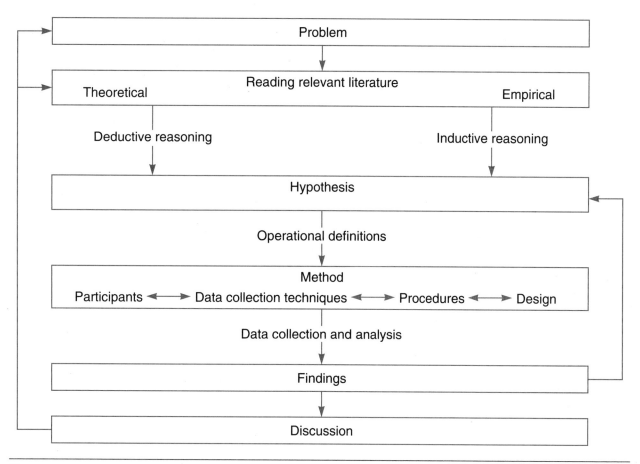

Figure 1.2 The total research setting.

The Parts of a Thesis: A Reflection on the Steps in the Research Process

In this chapter, you have been introduced to the research process. The theme has been the scientific method of problem solving. Generally speaking, a thesis or research article has a standard format. This is for expediency, so that the reader knows where to find the different information, such as purpose, methods, and results. The format also reflects the steps in the scientific method of problem solving. We now look at a typical thesis format and see how the parts correspond to the steps in the scientific method.

Sometimes theses and dissertations are done in a chapter format where each chapter represents a separate part of the research report (e.g., introduction). That has been a common model over the years. We believe it is more appropriate for graduate students to prepare their theses or dissertations in a form suitable for journal publication because that is an important part of the research process. In chapter 21 we provide considerable detail about how to use a journal format for a thesis or dissertation and the value of doing so. Throughout this book, we refer to the typical parts of a research report. These can be considered either as parts of a journal paper or as chapters, depending on the format selected for the thesis or dissertation.

Introduction

In the introduction, the problem is defined and delimited. The researcher specifically identifies the problem and often states the research hypotheses. Certain terms critical to the study are operationally defined for the reader, and limitations and perhaps some basic assumptions are acknowledged.

The literature review may be in the first part, or it may warrant a separate section. When it is in the first part, it more closely adheres to the steps in the scientific method of problem solving; that is, the literature review is instrumental in the formulation of hypotheses and the deductive reasoning leading to the problem statement.

Method

The purpose in the method section is to make the thesis format parallel to the data-gathering steps of the scientific method. First, the researcher explains how the data were gathered. The participants are identified, the measuring instruments are described, the measurement and treatment procedures are presented, the experimental design is explained, and the methods of analyzing the data are summarized. The major purpose of the method section is to describe the study in such detail and with such clarity that a reader could duplicate it.

research proposal

A formal preparation that includes the introduction, review of literature, and proposed method for conducting a study.

The first two parts often comprise the **research proposal** and are presented to the student's thesis committee prior to the research being undertaken. For the proposal, these two parts are often written in future tense, then changed to past tense when the final version of the thesis is completed.

Results

The results present the pertinent findings from the data analysis and represent the contribution to new knowledge. Results correspond to the step in the scientific method in which the meaningfulness and reliability of the results are scrutinized.

Discussion and Conclusions

In this last step in the scientific method, the researcher employs inductive reasoning in an effort to analyze the findings, to compare these findings with previous studies, and to integrate them into a theoretical model. In this part of a research paper, the acceptability of the research hypotheses are judged. Then, on the basis of the analysis and discussion, conclusions are usually made. The conclusions should address the purpose and subpurposes that were specified in the first part.

Summary

Research is simply a way of solving problems. Questions are raised, and methods are devised to try to answer them. There are different ways of approaching problems (research methods). Sometimes the nature of the problem dictates the method of research. For example, if one wants to discover the origins of a sport, the historical method of research is used. Sometimes one wishes to look at a problem from a particular angle and selects a research method that can best answer the question from that angle.

Research on the topic of teaching effectiveness, for example, can be approached in several ways. An experimental study could be conducted in which the effectiveness of teaching methods in bringing about measurable achievement is compared. Or a study could be designed in which teachers' behaviors are coded and evaluated using some observational instrument. Or another form of descriptive research could be used that employs the questionnaire or the interview technique to examine teachers' responses to questions concerning their beliefs or practices. Or perhaps a qualitative study could be undertaken to systematically observe and interview one teacher in one school over an extended period to portray the teacher's experiences and perceptions in the natural setting.

The point is that there is not just one way to do research. It is true that some people do only one type of research, and, being human, some are critical of the methods used by others. However, anyone who believes his or her type of research is the only "scientific" way to solve problems is narrow-minded and downright foolish. Science is disciplined inquiry, not a set of specific procedures.

Basic research deals primarily with theoretical problems, and the results are not intended to have immediate application. Applied research, on the other hand, strives to answer questions that have direct value to the practitioner. There is a need to prepare proficient consumers of research as well as researchers. Thus, one purpose of a book on research methods is to help the reader understand the tools necessary both to consume and to produce research.

We presented here an overview of the nature of research. The scientific method of problem solving was contrasted with "unscientific" methods by which people acquire information. Multiple research models were discussed to emphasize that there isn't just one way to approach problems in our discipline and profession. We identified the four major types of research used in the study of physical activity: analytical, descriptive, experimental, and qualitative. These categories and the different techniques that they encompass are covered in detail in later chapters.

☑ Check Your Understanding

1. Look through some recent issues of *Research Quarterly for Exercise and Sport*. Find and read a research article of interest that is quantitative in nature and another that you think is qualitative. Which of these did you find easier to understand? Why?

2. Find a research article that you would classify as an applied research study. Also find one you think is more of a basic research study. Defend your choices.

3. Think of two problems needing research in your field. From the descriptions of types of research in this chapter, suggest how each problem might be researched.

Developing the Problem and Using the Literature

I find that a great part of the information I have was acquired by looking up something and finding something else on the way.—Franklin P. Adams

■ ■ ■ ■ ■ ■ ■

Getting started is the hardest part of almost any new venture, and research is no exception. You can't do any significant research until you have identified the area you want to investigate, learned what has been published in that area, and figured out how you are going to conduct the investigation. In this chapter, we discuss ways to identify researchable problems, search for literature, and write the literature review.

Identifying the Research Problem

Of the many major issues facing the graduate student, a primary one is the identification of a research problem. Problems may arise from real-world settings or be generated from theoretical frameworks. Regardless, a basic requirement for proposing a good research problem is in-depth knowledge about the area of interest. However, sometimes as one becomes more knowledgeable about a content area, everything seems to be already known. Thus, although you want to become expert, do not focus too narrowly. Relating your knowledge base to other areas often provides insight into significant areas for research.

Sometimes it seems ironic that we ask students to start thinking about possible research topics in their research methods course, because they typically take the course in the first semester (or quarter) of graduate school—before they have had the opportunity to acquire the necessary in-depth knowledge. As a result, many of their research problems are trivial, lack a theoretical base, and replicate earlier research. Although this is a considerable shortcoming, the advantages of taking the research methods course early in the program are substantial in terms of success in other graduate courses. This is because the student learns

- to approach and solve problems in a scientific way,
- to search the literature,

- to write in a clear, scientific fashion,
- to understand basic measurement and statistical issues,
- to use an appropriate writing style,
- to be an intelligent consumer of research, and
- to appreciate the wide variety of research strategies and techniques used in an area of study.

How, then, does a student without much background select a problem? It seems that the harder you try to think of a topic, the more you are inclined to think that all the problems in the field have already been solved. Adding to this frustration is the pressure of time. To assure you that important questions still wait to be addressed, we have provided a list of 10 such provocative questions below.

Top 10 Problems That Have Not Been Resolved By Humankind

10. Why is it that inside every older person is a younger person wondering "What the heck happened"?

9. In winter why do we try to keep the house as warm as it was in summer when we complained about the heat?

8. Why is *phonics* not spelled the way it sounds?

7. Why do we sing "Take Me Out to the Ball Game" when we are already there?

6. Why do *slow down* and *slow up* mean the same thing?

5. If love is blind, why is lingerie so popular?

4. If one synchronized swimmer drowns, do the rest drown too?

3. Why do we say something is "out of whack"? What is a *whack?*

2. How come *abbreviated* is such a long word?

Drum roll ....

1. Why doesn't glue stick to the inside of the bottle?

Guidelines for Finding a Topic

To help alleviate the topic-finding problem, we offer the following suggestions. First, be aware of the research being done at your institution, for research spawns other research ideas. Often, a researcher has a series of studies planned. Second, be alert for any controversial issues in some area of interest. Lively controversy prompts research in efforts to resolve the issue. In any case, be sure to talk to professors and advanced graduate students in your area of interest and to use their suggestions to focus on a topic (note in box on page 27 why students, when seeking a topic, should look for causes, not effects). Third, read a review paper (possibly in a review journal, research journal, or recent textbook). From there, read several research studies in the reference lists and locate other current research papers on the topic. Using all this information, make a list of either research questions that appear unanswered or logical extensions of the material you have read. Try to pick problems that are neither too hard nor too easy. The hard ones will take you forever, and you'll never get your thesis done. No one cares about the easy ones.

Of course, no single problem necessarily meets all the criteria perfectly. For example, some theoretical problems may have limited direct application; however, theoretical problems should be directed toward issues that may ultimately prove useful to practitioners. By honestly answering the questions set forth by McCloy (1930), a practical evaluation of the selected problem is possible (see box on page 28).

Look for Causes, Not Effects

In a very interesting paper, Salzinger (2001) points out that scientists should not just look at the effects they observe, but try to find the causes underlying them. Finding surprising results is always interesting, but the causes behind them are what we should seek. We often hear that the purpose of the scientific method is to "disprove" theories, that we can never prove a theory. However, a single experiment seldom disproves a theory, "Old theories, like old soldiers just fade away—when better theories come to take their place" (Salzinger, 2001, B14). On the other hand, confirming a theory leads us toward underlying mechanisms; results explain why something happened.

We often read research results like these:

- ESP is shown to work
- Listening to Mozart improves spatial reasoning
- People who go to church live longer

When we see headlines like these, we should ask ourselves why these things might happen. What are the explanations? Does religion actually have anything to do with living longer, or is it the fact that people who go to church receive emotional reinforcement, network to learn about better physicians, get out of the house, are likely to walk to church, or any other number of explanations?

Salzinger (2001) suggests we should spend more time thinking about what causes the results and the possible mechanisms—identifying the critical ingredient that allows results to be applied in other contexts.

Reference

Salzinger, K. (February 16, 2001). Scientists should look for basic causes, not just effects. *Chronicle of Higher Education, 157*(23), B14.

An intriguing way to develop a research problem is to see how experts develop problems. Snyder and Abernethy (1992) organized *The Creative Side of Experimentation,* in which they had well-established scholars in motor control, motor development, and sport psychology give first-person accounts of factors that influenced their career research programs. The editors also have deduced themes that seem to run through these scholars' research programs. Their analysis focuses on such questions as these:

- What are common personal and professional characteristics of expert researchers?
- What types of experimentation do expert researchers perform?
- What strategies do expert experimenters use to enhance their ability to ask important questions?

In addition, Locke, Spirduso, and Silverman (2000) provide a 20-step approach to help graduate students identify topics and develop a proposal.

Using Inductive and Deductive Reasoning

closed-loop theory

Theory of motor skill learning advanced by Adams (1971) in which information received as feedback from a movement is compared with some internal reference of correctness.

The means for identifying specific research problems come from two methods of reasoning: inductive and deductive. Figure 2.1 provides a schema of the inductive reasoning process. Individual observations are tied together into specific hypotheses, which are grouped into more general explanations that are united into theory. To move from the level of observations to that of theory requires many individual studies that test specific hypotheses. But even beyond the individual studies, someone must see how all the findings relate and then offer a theoretical explanation that encompasses all the individual findings.

An example of this process can be found in the motor learning and control area. Adams (1971) proposed a **closed-loop theory** of motor skill learning. Basically, the

Criteria for Selecting a Research Problem

In 1930, the first volume (issue #2) of *Research Quarterly (now Research Quarterly for Exercise and Sport)* had a paper by one of our historically famous scholars, Dr. C. H. McCloy, in which he provided a list of questions by which a researcher can judge the quality and feasibility of a research problem. Those questions (listed below) are as valid and useful today as they were in 1930.

- Is the problem in the realm of research?
- Does it interest you?
- Does it possess unity?
- Is it worthwhile?
- Is it feasible?
- Is it timely?
- Can you attack the problem without prejudice?
- Are you prepared in the techniques to address the problem?

Our thanks to Jim Morrow, University of North Texas, for pointing out this information.

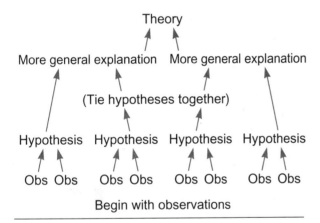

Figure 2.1   Inductive reasoning.
Reprinted, by permission, from R.L. Hoenes and B. Chissom, 1975, *A student guide for educational research* (Statesboro, GA: Vog Press), 22.

schema theory

Theory of motor skill learning advanced by R.A. Schmidt (1975) as an extension of Adams' (1971) closed-loop theory. The theory proposed to unify two more general explanations under one theoretical explanation.

closed-loop theory proposes that information received as feedback from a movement is compared to some internal reference of correctness (assumed to be stored in memory). Then the discrepancies between the movement and the intended movement are noted. Finally, the next attempt at the movement is adjusted to more nearly approximate the movement goal. Adams' theory ties together many previous observations about movement response. The theory was tightly reasoned but limited to slow-positioning responses. This limitation really makes it a more general explanation, according to figure 2.1.

Schmidt (1975) proposed a **schema theory,** which extended Adams' reasoning to include more rapid types of movements, frequently called ballistic tasks. (Schema theory also deals with several other limitations of Adams' theory that are not important to this discussion.) The point is that schema theory proposed to unify two general explanations, one about slow movements and the other about ballistic (rapid) movements, under one theoretical explanation—clearly an example of inductive reasoning.

Reasoning must be careful, logical, and causal; otherwise, one of our examples of inappropriate induction may result:

A researcher spent several weeks training a cockroach to jump. The bug became well trained and would leap high in the air on the command "Jump." The researcher then began to manipulate his independent variable, which was to remove the bug's legs one at a time. Upon removing the first leg, the researcher said, "Jump," and the bug did. He then removed the second, third, fourth, and fifth leg and said, "Jump" after each leg was removed, and the bug jumped every time. Upon removing the sixth leg and giving the "Jump" command, the bug just lay there. The researcher's conclusion from this research was this: "When all the legs are removed from a cockroach, the bug becomes deaf." (Thomas, 1980, 267)

Figure 2.2 presents a model of deductive reasoning. Deductive reasoning moves from a theoretical explanation of events to specific hypotheses that are tested against (or compared with) reality to evaluate whether the hypotheses are correct. Using the previously presented notions from his schema theory (to avoid explaining another theory), Schmidt advanced a hypothesis frequently called **variability of practice.** Essentially, this hypothesis (reasoned or deduced from the theory) says that practice of a variety of movement experiences (within a movement class), when compared with practicing a single movement, facilitates transfer to a new movement (still within the same class). Several researchers have tested this hypothesis, identified by deductive reasoning, and found it viable. In fact, within any given study, both inductive and deductive reasoning are useful. The total research setting was presented in the last chapter in figure 1.2. Review how the deductive and inductive processes operate; that is, at the beginning of a study, the researcher deduces hypotheses from relevant theories and concepts and induces hypotheses from relevant findings in other research.

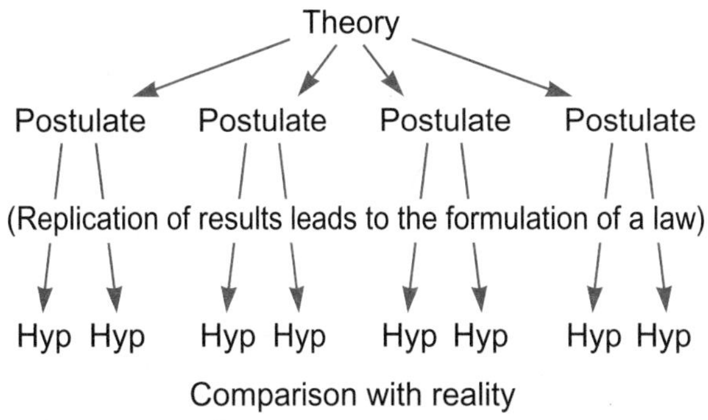

Figure 2.2 Deductive reasoning.
Reprinted, by permission, from R.L. Hoenes and B. Chissom, 1975, *A student guide for educational research* (Statesboro, GA: Vog Press), 23.

variability of practice

Tenet of motor skill learning advanced by R.A. Schmidt in which the practice of a variety of movement experiences facilitates transfer to a new movement when compared with practicing a single movement.

Purpose of the Literature Review

A major part of developing the research problem is reading what has already been published about the problem. There may already have been much research done on the problem in which you are interested. In other words, the problem has been pretty much "fished out." We hope this won't be the case when you start this phase of the research process. (In many instances, your major professor can steer you away from a saturated topic.) Whatever the topic, past research is invaluable in planning new research.

Browsing in a library confirms that information—and lots of it—exists. The dilemma lies in knowing how to locate and evaluate the information you want and, ultimately, how to use the information once you've found it.

Reviews of literature serve several purposes. Frequently, they stimulate inductive reasoning. A scholar may seek to locate and synthesize all the relevant literature on a particular topic to develop a more general explanation or a theory to explain certain phenomena. An alternative way of analyzing the literature, mentioned in chapter 1, is meta-analysis (Glass, McGaw, & Smith, 1981), which is discussed in detail in chapter 14.

The major problem of literature reviews is how all those studies can be related to one another in an effective way. Most frequently, authors attempt to relate studies by similarities and differences in theoretical frameworks, problem statements, methodologies (participants, instruments, treatments, designs, and statistical analyses), and findings. Results are then determined by counting votes. For example, you would write, "From the eight studies with similar characteristics, five found no significant difference between the treatments; thus, this treatment has no consistent effect."

This procedure is most easily accomplished through use of a summary sheet, such as the one in table 2.1, which relates the frequency and intensity of exercise to the percentage of change in body fat. The conclusion from looking at these studies might be, "Exercising 20 min per day for 3 days per week for 10–14 weeks at 70% of the maximum heart rate produces moderate losses of body fat (4% to 5%). However, more frequent exercise bouts produce minimal increases, but less frequent or less intense exercise is substantially less effective in eliminating body fat."

Techniques of this type lend themselves to the development of the literature review around central themes or topics. Not only does this approach allow synthesis of the relevant findings, but it also makes the literature review interesting to read.

Table 2.1 Sample Form for Synthesizing Studies (Hypothetical Example)

	Characteristics of studies				
Study	Problem statement	Participant description	Instrument	Procedure and design	Finding
Smith (1985)	Effects of exercise on body fat	30 college-age males	Underwater weighing	Exercise 3 d/wk at 70% of (220 − age) for 12 wk	4% reduction in body fat
Johnson (1978)	Effects of exercise on body fat	45 college-age males	Underwater weighing	Jog 3 d/wk at 70% or 50% of (220 − age) for 10 wk	5% for 70% gp 2% for 50% gp
Andrews (1989)	Effects of frequent and intense exercise on body fat	36 college-age males	Skinfold calipers	Jog 2, 4, 6 d/wk at 75% of (220 − age) for 12 wk	1% for 2 d 4% for 4 d 5% for 6 d
Mitchell (1980)	Effects of workload on body fat	24 high school males	Skinfold calipers	Pedaled at 30, 45, 60 rpm with 2 KP resistance for 20 min, 3 d/wk for 14 wk	1% at 30 rpm 3% at 45 rpm 4% at 60 rpm

Identifying the Problem

As we have already discussed, the literature review is useful in identifying the specific problem. Of course, after locating a series of studies, the first task is to decide which studies are related to the topic area. This can frequently be accomplished by reading the abstract and, if necessary, some specific parts of the paper. Once a few key studies are identified, a careful reading usually produces several ideas and unresolved questions. You will find it useful to discuss these questions with a professor or advanced graduate student from your area of specialization. Doing so can eliminate unproductive approaches or dead ends. After the problem is specified, an intensive library search begins.

Developing Hypotheses or Questions

Hypotheses are deduced from theory or induced from other empirical studies and real-world observations. These hypotheses are based on logical reasoning and, when predictive of the study's outcome, are labeled research hypotheses. For example, after spending a good deal of time at registration as undergraduates, graduate students, and faculty members, we are able to put forth the following hypothesis for you to test: The shortest line at registration is always the slowest. If you change lines, the one you left will speed up, and the one you enter will suddenly stop. In qualitative studies, more general questions often serve the same purpose as hypotheses do for quantitative studies.

Developing the Method

Although considerable work is involved in identifying the problem and specifying hypotheses and questions, one of the more creative parts of research is developing the method to answer the hypotheses. If the method is planned and pilot-tested appropriately, the study's outcome allows the hypotheses and questions to be evaluated. We believe that the researcher fails when the results of a study are blamed on methodological problems. Post hoc methodological blame results from lack of (or poor) planning and pilot work before undertaking the research.

The review of literature can be extremely helpful in identifying methods that have been successfully used to solve particular types of problems. Valuable elements from other studies may include the characteristics of the participants, data collection instruments and testing

She's doing her literature review.

procedures, treatments, designs, and statistical analyses. All or parts or combinations of the previously used methods are quite helpful as the researcher plans the study, but these should not limit the researcher in designing the study. Creative methodology is a key to good hypothesis testing. But neither other scholars' research nor creativity ever replaces the need to conduct thorough pilot work.

Basic Literature Search Strategies

The prospect of beginning a literature search can sometimes be frightening or depressing. How and where do you begin? What kind of sequence or strategy should you use in finding relevant literature? What services does the library offer in your search?

Authors of research texts have advanced various strategies for finding pertinent information on a topic. We know of no single "right" way of doing it. The search process depends considerably on your initial familiarity with the topic. In other words, if you have virtually no knowledge about a particular topic, your starting point and sequence will be different from that of someone already quite familiar with the literature.

Some people are inclined to jump into the fray via a computer search. This can prove very fruitful, to be sure. However, this strategy is not foolproof in that only the more recent references are available for computer searching, depending on the database. We hasten to point out, however, that the latest is usually the best, and using electronic information sources enables the researcher to obtain the most up-to-date information on a topic (Edge & Claxton, 2000). A drawback of just considering the most recent literature is that it may be difficult to obtain the necessary background and broad overview of the problem from individual studies. For students who are not well grounded in the topic, certain preliminary (general) sources may be helpful in locating **secondary sources** with which to become familiar with the topic at hand, and they better prepare the student to acquire and understand **primary sources.**

An index, such as the *Education Index,* is a preliminary (not primary) source. An index can provide the researcher with books and articles that relate to the problem. Textbooks are valuable *secondary* sources that can give the reader an overview of the topic and what has been done in the way of research. In fact, secondary sources such as encyclopedias and scholarly books may be the starting point for the search, by which the student becomes aware of the problem in the first place. *Primary* sources are ultimately the most valuable for the researcher in that the information is firsthand. Most primary sources in a literature review are journal articles. Theses and dissertations are also primary sources, but they are usually more difficult to locate if they are written at another university.

secondary sources

Source of data in research in which an author has evaluated and summarized previous research.

primary sources

Firsthand source of data in research; the original study.

Steps in the Literature Search

There are six steps you should follow when reviewing the literature. These steps ensure thoroughness and make the search more productive.

Step 1: Write the Problem Statement

We discuss the formal writing of the problem statement in the next chapter. At this point, you are merely trying to specify what research questions you are asking. For example, a researcher wants to find out whether the student-teaching experience influences attitudes toward teaching. More specifically, the researcher wishes to examine the attitudes toward teaching before and after the student-teaching experience of students in a teacher preparation program in physical education. By carefully defining the research problem, the researcher can keep the literature search within reasonable limits. Write the statement as completely (but concisely) as you can at this time.

Step 2: Consult Secondary Sources

This step helps you gain an overview of the topic, but it can be omitted if you are knowledgeable about the topic. Secondary sources such as textbooks and encyclopedias are helpful when students have very limited knowledge about a topic and can profit from background information and a summary of previous research. A review paper on the topic of interest is especially valuable.

Encyclopedias

Encyclopedias provide an overview of information on research topics and summarize knowledge about subject areas. General encyclopedias provide broad information about an entire field. Specialized encyclopedias offer much narrower topics. Examples are the *Encyclopedia of Sport Sciences and Medicine; Encyclopedia of Physical Fitness; Encyclopedia of Physical Education, Fitness and Sports; Encyclopedia of Educational Research;* and *Handbook of Research on Teaching.*

Because a rather lengthy period (years) often elapses from the time the authors submit their contributions until the publication date, you should be aware that the information in an encyclopedia is dated. Still, you can get important background information about a subject, become familiar with basic terms, and note references to some pertinent research journals.

Research Reviews

Reviews of research are an excellent source of information for three reasons:

- Some knowledgeable person has spent a great deal of time and effort in compiling the latest literature on the topic.
- The author has not only found the relevant literature but also critically reviewed and synthesized it into an integrated summary of what is known about the area.
- The reviewer often suggests areas of needed research, for which the graduate student may be profoundly grateful.

Some actual review publications are the *Annual Reviews of Medicine, Annual Review of Psychology, Review of Educational Research, Physiological Reviews, Psychological Review,* and *Exercise and Sport Science Reviews.* Some of these sources, such as *Annual Review of Psychology* and *Physiological Reviews,* are available electronically as well as in print.

A number of reviews have been published by the American Alliance for Health, Physical Education, Recreation and Dance (AAHPERD). One is a series called *What Research Tells the Coach,* which is about various sports including baseball, football, sprinting, distance running, swimming, tennis, and wrestling. Another AAHPERD series includes *Kinesiology Reviews I, II,* and *III.* Many other research journals regularly publish reviews, to which some occasionally devote entire issues; for example, see *International Journal of Sport Psychology, 30*(2), 1999, "The development of expertise in sport: Nature and nurture."

Step 3: Determine Descriptors

Descriptors are terms that help to locate sources pertaining to a topic. For the topic of the effect of student teaching on attitudes toward teaching in physical education, obvious descriptors are attitude (toward teaching), changes in attitudes, student teaching, and physical education. Descriptors can be classified as major and minor. It is the combination of descriptors that helps the researcher pinpoint pertinent related literature. Obviously, terms such as "attitude," "student teaching," and "physical education" by themselves are too broad. Various databases have their own descriptors for topics. We discuss this further when we describe computer searching later in this chapter.

Step 4: Search Preliminary Sources

Use preliminary (general) sources to find primary sources via computer-aided searches. Preliminary sources primarily consist of abstracts and indexes. Preliminary sources that are helpful to researchers in physical education, exercise science, and sport science are described in the following paragraphs.

Abstracts

Concise summaries of research studies are valuable sources of information. Abstracts of papers presented at research meetings are available at national, district, and most state conventions. Abstracts of symposia sponsored by the AAHPERD Research Consortium, free communications, and **poster sessions** presented at the national AAHPERD convention are published each year in the *Research Quarterly for Exercise and Sport Supplement.* *Medicine and Science in Sports and Exercise* publishes a special supplement of abstracts each year for papers to be presented at the annual meeting of the American College of Sports Medicine.

Other abstract sources are *Dissertation Abstracts International,* which contains abstracts of dissertations from most colleges and universities in the United States and Canada, and the *Index and Abstracts of Foreign Physical Education Literature,* which provides abstracts from journals outside the United States. Sources of abstracts in related fields include *Biological Abstracts, Psychological Abstracts, Sociological Abstracts, Resources in Education,* and *Current Index to Journals in Education.*

poster sessions

Method of presenting research at a conference in which the author places summaries of his or her research on the wall or on a poster stand and answers questions from passersby.

Indexes

Several indexes provide references to magazine and journal articles concerning specific topics. Some general indexes commonly used in physical education, exercise science, and sport science include the *Education Index,* the *Reader's Guide to Periodical Literature,* the *New York Times Index,* the *Web of Science,* and the *Physical Education Index.* Despite the title, this last source provides a comprehensive subject index to domestic and foreign periodicals in the fields of dance, health education, recreation, sports, physical therapy, and sports medicine. Researchers in the different areas within physical education, exercise science, and sport science tend to use specific indexes that pertain to their topics of interest, such as *Index Medicus, PsycINFO, ERIC, Current Contents,* and *SPORTDiscus.* All these indexes are available electronically, and this is the format that is preferred in most research libraries.

- *Index Medicus.* Widely used in exercise science, the *Index Medicus* provides access to more than 2,500 biomedical journals around the world. It is published monthly, and each issue has subject and author sections and a bibliography of medical reviews. *Index Medicus* can also be searched via computer through *MEDLINE.*

- *PsycINFO.* A computer index in the field of behavioral science, *PsycINFO* selects keywords and lists appropriate titles, identifying authors and journals.

- *ERIC.* The acronym *ERIC* stands for Educational Resources Information Center. It is the world's largest source of education information. The *ERIC* system collects, sorts, classifies, and stores thousands of documents on various topics concerning education and related fields. Its basic indexes are *Resources in Education (RIE)* and the *Current Index to Journals in Education (CIJE).* Besides containing the abstracts, *RIE* provides information on how you can obtain a document: whether you can purchase it on microfilm or as a .pdf file, order an *ERIC* copy, or request the original copy from the publisher. Both *RIE* and *CIJE* provide valuable assistance in the search for information on specific topics. In addition, *ERIC* produces a thesaurus containing thousands of index terms that can be used in locating references and in conducting a computer search for information.

- *Current Contents.* This small weekly magazine, published by the Institute for Scientific Information, contains the table of contents of journals published recently within a general content area (e.g., social and behavioral sciences) and divides journals by subarea (e.g., psychology, education, rehabilitation, and special education). *Current Contents* indexes journal titles by topic and author and provides authors' addresses, so that you can obtain reprints. It also publishes a section on current books in each issue as well as weekly editions for life sciences; physical, chemical, and earth sciences; social and behavioral sciences; agriculture, biology, and environmental sciences; clinical medicine; engineering, technology, and applied sciences; and arts and humanities. *Current Contents* is now available electronically as the *Web of Science* in most university libraries.

- *SPORTDiscus.* This index is a valuable resource for both practical and research literature on sport, physical fitness, and physical education topics. Topics include sports medicine, exercise psychology, biomechanics, psychology, training, coaching, physical education and fitness, and other sport- and fitness-related topics. *SPORTDiscus* also provides comprehensive indexing of the *Microform Publications* collection of dissertations and theses of the International Institute for Sport and Human Performance. It also includes all citations from the previous *Sport and Leisure* bibliographic database of *SIRLS,* which covers the research literature for the social and psychological aspects of sport and leisure, including play, games, and dance. *SPORTDiscus* is available only electronically; there is no print equivalent. Coverage begins with 1975.

Bibliographies

annotated bibliography

List of resources that provides a brief description of the nature and scope of each article or book.

Bibliographies list books and articles about specific topics. They come in many forms, depending on how the information is listed. All contain the authors, titles of books or articles, journal names, and publishing information. Some bibliographies are **annotated,** meaning that a brief description of the nature and scope of the article or book is included with each reference.

The bibliography of a recent study on the topic in question is an invaluable aid to the researcher. Some authors have stated that one of the most valuable contributions of a dissertation is the review of literature and bibliography. However, you cannot simply lift the literature review from a previous study. Just because someone else has reviewed pertinent sources does not relieve you of the responsibility to read and evaluate each source yourself. Keep in mind that (a) the previous author may have been careless and cited the source or sources incorrectly and (b) the previous author may have taken the results of a study out of context or from a point of view different from the original author's or your own.

We have found incorrect bibliographic entries on numerous occasions. A study by Stull, Christina, and Quinn (1991) revealed that of 973 citations in the 1988 and 1989 volumes of the *Research Quarterly for Exercise and Sport,* 457 contained one or more errors. This is an error rate of 47%. Such findings are by no means unique to this journal. A 50% error rate was reported for *JAMA: The Journal of the American Medical Association* (Goodrich & Roland, 1977). Stull, Christina, and Quinn (1991) emphasized that every component in a bibliographic citation is important. Moreover, the ultimate responsibility for accuracy rests with the author. In the publishing process, mistakes can often creep in, starting with copying the source, typing the article, revising the manuscript, and checking page proofs.

A good search strategy is to look for the most recent sources of information and then work backward. You will save much time by consulting the most recent studies, and you will profit from the searches of others. Some examples of bibliographies are the *Annotated Bibliography on Movement Education; Bibliography of Research Involving Female Subjects; Bibliography on Perceptual Motor Development; Bibliography of Medical Reviews in Index Medicus; Annotated Bibliography in Physical Education, Recreation and Psychomotor Function of Mentally Retarded Persons;* and *Social Sciences of Sports.*

The Library Information System

The traditional card catalog with little trays of cards containing bibliographic information by author and subject has all but disappeared. Nearly all university libraries have gone to a computerized catalog system, and it seems inevitable that all libraries will adopt this method as finances permit.

Computerized catalogs abound. Usually, the searcher first selects the type of search from a menu, such as author, title, keyword, or call number. When the source is found, the full display for the reference includes author, title, publication information, all the index terms, and the call number. Most libraries in colleges and universities can also be accessed via the Internet by faculty and students using personal computers.

Some people who have used the older system for years are nostalgically reluctant to see it disappear. They maintain that while browsing through the cards they often experienced serendipity, finding something of value by chance when searching for something else. Nevertheless, as you become familiar with the computer operations, the search process becomes much faster and more productive. Remember, if you have questions about any library operations, ask a librarian. They tend to be remarkably helpful and courteous.

Computer Searches

Computer service facilities can greatly expedite the literature search. Automated searching provides more effective and efficient access to indexes and information than does manual searching. Some databases available for computer searches are listed on page 36. The computer search covers many abstracting and indexing services in the sciences, humanities, and social sciences.

Increasingly, libraries have electronic databases that enable individuals to do the literature search from a desktop computer. Often the indexes allow the user to link directly to the full text of an article.

Adjusting the Scope of the Search

As we have mentioned before, in many university libraries, an individual can do the computer search from any computer connected to the Internet. The information displayed on the screen can then be printed. Remember, the key to a successful literature search is careful planning. Therefore, write down your problem statement, and then formulate cogent descriptors and keywords. If the database you are using has a thesaurus, by all means use it. It is also beneficial to find one or more journal articles pertaining to your topic to see how they are indexed to find productive keywords or subject terms to assist you in the search strategy.

The scope of your search can be narrowed or broadened by using special words called Boolean operators. The two most common operators, or connectors, are the words *and* and *or*. To narrow the search, you add another term with the word *and*. For example, in the proposed study on investigating changes in attitude toward teaching following the student-teaching experience, 2,237 items were listed under the descriptor "attitude change." For "attitude + teaching" there were 186 items, and for "student teacher attitudes" there were 100 items. When all were connected with the word *and,* there were 44 references, which is a manageable number to examine. Figure 2.3 illustrates the Boolean logic with this example.

Some Databases Available for Computer Searches

ERIC (Educational Resources Information Center) (1966—)

Descriptor Guidelines: *Thesaurus of ERIC Descriptors*

Type of Information: All topics concerning education from preschool to adult education are covered in areas of administration, curriculum, teaching, and learning. Published articles are accessed from the *Current Index to Journals in Education,* and unpublished materials can be obtained from the *Resources in Education* part of *ERIC.* These two databases can be searched simultaneously or separately. *ERIC* computer search is also available on CD-ROM.

MEDLINE (1966—)

Descriptor Guidelines: *Medical Subject Headings*

Type of Information: The vast coverage of journal articles in biomedicine, health care, gerontology, and other areas in *Index Medicus* can be accessed through *MEDLINE,* which is available on CD-ROM.

PsycINFO (1967—)

Descriptor Guidelines: *Thesaurus of Psychological Index Terms*

Type of Information: Journal articles, dissertations, and technical reports (book and book-chapter citations have been available since 1992) in psychology and related behavioral sciences. This database comes from *Psychological Abstracts. PsycLit* (1974—) is the CD-ROM version of *PsycINFO.*

Sociological Abstracts (1963—)

Descriptor Guidelines: *Thesaurus of Sociological Indexing Terms.*

Type of Information: This database contains comprehensive coverage of journal articles, conference publications, and books in social, developmental, and clinical psychology. Several database files are combined in this coverage. *Sociofile* (1974—) is the CD-ROM version.

Dissertation Abstracts Online (1861—)

Descriptor Guidelines: None. Keywords are listed by searcher.

Type of Information: Dissertation abstracts and master's theses (since 1962) from *Dissertation Abstracts International* and from *American Doctoral Dissertations.* It includes dissertations from nearly all American doctorate-granting institutions and numerous dissertations from Canada and other countries. This service on CD-ROM.

Ingenta (1988—)

Descriptor Guidelines: None. Keywords or names are listed by searcher.

Type of Information: Journal articles and journal tables of contents from over 15,000 journal titles with thousands of citations added daily. This is a fee-based service that offers the opportunity to acquire a fax or .pdf file of most articles, usually within 24 hours.

SPORTDiscus (1975—)

Descriptor Guidelines: None. Keywords are listed by searcher.

Type of Information: Wide coverage of literature on sport, physical fitness, and physical education topics. Also a comprehensive index of dissertations and theses plus citations on social and pyschological aspects of sport and leisure research. This service is available on CD-ROM.

Web of Science (1965–)

Descriptor Guidelines: Key words, author's name, journal name.

Type of Information: The Web of Science is a searchable electronic database that is most often used through a library. Access to three large indexes of journals and other information is provided: *Science Citation Index* (1965–), *Social Sciences Citation Index* (1965–), and *Arts and Humanities Citation Index* (1975–). The Web of Science provides a list of all the journals that are covered (by general area, field of work, and alphabetically). Users can locate specific published papers, authors of papers, and journals. In addition, data about impact factors for various journals and citation rates for various papers and authors is searchable.

The word *or* broadens the search. Additional related terms can be connected with *or* so that the computer searches for more than one descriptor (see figure 2.4). For example, a researcher who seeks information about "practice teaching" (133 items) could broaden the search by using "practice teaching" (133 items) *or* "internship" (495 items) to get a total of 628 related items.

It may be that at your library you cannot conduct the computer search yourself, or you may prefer to have someone do it for you (which may be the most efficient and thorough course of action). In this case, you have to consult with a librarian, and the search must be scheduled and carried out by library personnel. You will most likely be asked to complete a form on which the research statement, keywords, and database are specified. The library personnel can be extremely helpful in selecting descriptors and in the entire literature search process. Search results often can be downloaded onto your diskette so that you can print them at your convenience. Although there may be a fee for such searches, in most cases, the resulting bibliography is well worth the expense, especially in light of its comprehensiveness and the reduced time and energy you expend.

Obtaining the Primary Sources

After you have a list of related references, you must obtain the actual studies and read them. Many references have abstracts in addition to the bibliographic information. The abstract is extremely helpful in making the decision as to whether the article is worth retrieving.

Keep in mind that a computer search does not replace the hand search. The computer is remarkably fast and effective in identifying references that may be pertinent to your topic. However, perhaps the most valuable step in the search process is finding a recent, closely related study and reading that study's review of literature. Then, you can find the cited studies and read them, which leads you to other sources. Remember, as we stressed in the section on bibliographies, you are responsible for reading those studies. Don't rely on someone else's critique of the literature. Moreover, although the computer search is helpful, it is not infallible or magical. It is not unusual to have a pertinent article in your possession that is not listed on the computer search printout. This happens because somewhere in the process mistakes were made by people, not computers. It may be that the author of an article did not suggest the proper keywords, or maybe the abstract did not adequately describe the study. Or perhaps someone responsible for indexing the article had a different perspective on its subject matter. In any case, the computer search can never replace the hand search.

Your library may not have all the journals that are on the list of references (see table 2.2). You must consult your library's information system to see whether your library carries a particular journal. If it doesn't, you should use interlibrary loan or possibly obtain the fax of the article through the service *UnCover.*

Using the Internet: The Good, the Bad, and the Ugly

A lot of information may be found online when searching the Internet. However, the correctness and accuracy of information is often difficult to verify. In fact, much of it is wrong and misleading. You can search the Internet using key words with one of the standard search engines. You will likely find many sites when you conduct the search. Separating

Entries dealing with changes in attitude and attitude toward teaching and student teaching (44 items)

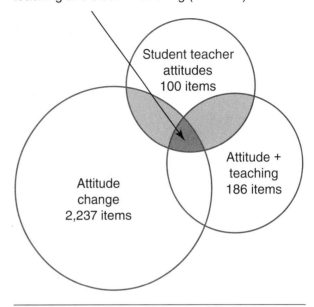

Figure 2.3   An illustration of the *and* connector to narrow a search.

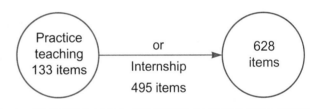

Figure 2.4   An illustration of the *or* connector to broaden the search.

Table 2.2 Journals With Impact Factors*

Kinesiology journals

Rank	Journal title	Impact factor
1	*Movement Disorders*[1]	2.895
2	*Journal of Applied Physiology*	2.720
3	*Medicine and Science in Sports and Exercise*	2.600
4	*Sports Medicine*	2.281
5	*American Journal of Sports Medicine*	2.270
6	*Journal of Biomechanics*[1]	1.889
7	*Gait and Posture*	1.753
8	*Clinical Journal of Sport Medicine*	1.686
9	*Journal of Sport and Exercise Psychology*	1.676
10	*Research Quarterly for Exercise and Sport*	1.619
11	*Journal of Motor Behavior*	1.549
12	*Clinics in Sports Medicine*	1.546
13	*Journal of Electromyography and Kinesiology*	1.425
14	*European Journal of Applied Physiology*	1.417
15	*International Journal of Sports Medicine*	1.348
16	*Archives of Physical Medicine and Rehabilitation*	1.327
17	*Exercise Immunology Review*	1.273
18	*British Journal of Sports Medicine*	1.202
19	*Adapted Physical Activity Quarterly*	1.200
20	*Canadian Journal of Applied Physiology-Revue Canadienne De Physiologie Appliquee*	1.163
21	*Motor Control*	1.119
22	*Scandinavian Journal of Medicine and Science in Sports*	1.117
23	*Scandinavian Journal of Rehabilitation Medicine*	1.113
24	*Journal of Sports Sciences*	1.082
25	*Knee Surgery Sports Traumatology Arthroscopy*	1.051
26	*Journal of Applied Sport Psychology*	1.029
27	*Journal of Athletic Training*	1.028
28	*Journal of Rehabilitation Medicine*	1.000
29	*Clinical Biomechanics*	0.996
30	*Human Movement Science*	0.987
31	*Pediatric Exercise Science*	0.982
32	*Journal of Shoulder and Elbow Surgery*	0.952
33	*Journal of Orthopaedic Trauma*	0.929
34	*American Journal of Physical Medicine & Rehabilitation*	0.877
35	*International Journal of Sport Nutrition and Exercise Metabolism*	0.857
36	*Quest*	0.806

Rank	Journal title	Impact factor
37	*International Journal of Sport Psychology*	0.778
38	*Journal of Strength and Conditioning Research*	0.762
39	*Journal of Aging and Physical Activity*	0.742
40	*Sport Psychologist*	0.731
41	*Aviation, Space, and Environmental Medicine*	0.721
42	*Journal of Orthopaedic Sport Physical Therapy*	0.690
43	*Journal of Sports Medicine and Physical Fitness*	0.634
44	*Sport Education and Society*	0.591
45	*Journal of Applied Biomechanics*	0.545
46	*Sociology of Sport Journal*	0.511
47	*Journal of Sport Rehabilitation*	0.510
48	*Physician and Sports Medicine*	0.492
49	*Journal of Teaching in Physical Education*	0.453
50	*Physical and Medical Rehabilitation Kuror*	0.441
51	*Wilderness and Environmental Medicine*	0.417
52	*Journal of Sport Management*	0.414
53	*Operative Techniques in Sports Medicine*	0.390
54	*Australian Journal of Physiotherapy*	0.375
55	*Sports Medicine and Arthroscopy Review*	0.355
56	*Isokinetics and Exercise Science*	0.354
57	*ACSMS Health and Fitness Journal*	0.348
58	*Knee*	0.333
59	*Sportverletzung-Sportschaden*	0.295
60	*Strength and Conditioning Journal*	0.269
61	*Journal of Human Movement Studies*	0.263
62	*Perceptual and Motor Skills[1]*	0.258
63	*Science and Sports*	0.158
64	*Biology of Sport*	0.157
65	*Athletic Therapy Today*	0.129
66	*Journal of the Philosophy of Sport*	0.091
67	*Journal of Back and Musculoskeletal Rehabilitation*	0.089
68	*Japanese Journal of Physical Fitness and Sports Medicine*	0.079
69	*Journal of Sport History[1]*	0.062
70	*Journal of Sports Traumatology and Related Research*	0.056
71	*Medicina Dello Sport*	0.055
72	*Sports Medicine Standards and Malpractice Reporter*	0.028
73	*Exercise Standards and Malpractice Reporter*	0.000

(continued)

Table 2.2   *(continued)*

Journals from related fields

Rank	Journal title	Impact factor
1	*Epidemiology*	3.962
2	*American Journal of Physiology–Heart and Circulation*	3.369
3	*Child Development*	3.272
4	*Journal of Experimental Psychology–Human Perception and Performance*	2.335
5	*Experimental Brain Research*	2.300
6	*American Journal of Mental Retardation*	1.840
7	*Physical Therapy*	1.658
8	*American Educational Research Journal*	1.438
9	*Human Biology*	1.242
10	*Ergonomics*	0.832
11	*Human Factors*	0.786
12	*Journal of Curriculum Studies*	0.390

* Impact factors represent the average number of times all journals are cited during one year. They are often considered as a way to judge the quality of journals.

**Not in ISI's list of journals in sport sciences

Reprinted with permission from *75th Anniversary of Research Quarterly for Exercise and Sport,* Vol. 70, pp. 11-23, Copyright 2005 by the American Alliance for Health, Physical Education, Recreation and Dance, 1900 Association Drive, Reston, VA 20191.

the good and valid ones from the useless and biased ones is a difficult task. However, there are sites that will consistently contain useful information on topics students want to find.

- Scout Report (http://scout.cs.wisc.edu/index.html) is a joint project from the National Science Foundation and the University of Wisconsin at Madison. It includes a considerable amount of useful scientific information on the biological sciences, physical sciences, and math, engineering, and technology.

- SOSIG stands for the Social Science Information Gateway (http://sosig.ac.uk/) and connects to more than 50,000 social science Web pages in such areas as anthropology, business, economics, education, environmental sciences, European studies, geography, government, law, philosophy, politics, psychology, sociology, and other areas.

- 1400/The Virtual Library (http://vlib.org/) is organized using the Library of Congress subject headings, thus making it similar to your local college or university library. Headings include agriculture, business and economics, computing, communications and media, education, engineering, humanities, international affairs, law, recreation, regional studies, science, and society.

Personal Computers

We have already mentioned that faculty and students often can access library sources and conduct a literature search from personal computers in their own living quarters or office. Your library has information about the recommended software and the log-on and log-off procedures.

You should also be aware of the capabilities of microcomputers for storing bibliographic entries, abstracts, and even reprints of studies. A number of commercial programs are available for this purpose. The information can be retrieved by the use of pertinent keywords, author names, or journal names. Additions, changes, and deletions are easily accomplished, and hundreds of items can be stored on a single disk. Each new entry is automatically stored in alphabetical order, and the complete bibliography is always instantly accessible. Often laboratories (e.g., for exercise science and physical education) keep files of reprints and use a microcomputer software library program to catalog these materials.

Other Library Services

The library services available depend mostly on the size of the institution; larger schools generally provide more financial support for the library. However, some relatively small institutions have excellent library resources that provide outstanding support for the institutional and research aspects of the school.

Besides the usual services of the library information system (the computerized card catalog, reference and circulation departments, bibliographic collections, and stack areas), libraries also offer resources and services such as copy services, newspaper rooms, government document sections, telephone directories, college catalogs, and many other special features and services. One such valuable service is the interlibrary loan, which enables you to get books, theses and dissertations, and photocopies of articles in journals that your library does not carry. An interlibrary loan usually takes two to four weeks from the date it is requested.

Universities typically have extensive collections of resources available on **microform,** which is a general term embracing microfilm, microfiche, microcard, and microprint. Microforms are simply miniaturized photographic reproductions of the contents of a printed page. You must use special machines called readers to enlarge the information so it can be read.

An obvious advantage of microforms is economy of space. In addition, they are useful for acquiring material that would otherwise not be available. Rare books and manuscripts and deteriorating materials can be preserved on microform. Some journals are on microform because it represents an economical means for the library to obtain them. Collections of government documents are often contained on microform, as are *ERIC* publications. The library usually has a reader/printer that can make paper copies of microform material. Microform holdings are normally listed in the library's electronic card catalog system and periodical listings. Many microform collections (e.g., *ERIC*) are being converted to .pdf files for easier access and storage.

Many libraries offer guided tours, short courses for orientation to the library, self-guided tour information, and tutorials in the use of electronic resources. Get acquainted with your library. It will be the wisest investment of the time you spend as a graduate student.

microform

A general term that encompasses microfilm, microfiche, and any form of data storage where the pages of a book, journal, or newspaper are photographed and reduced in size.

Step 5: Read and Record the Literature

Collecting related literature is a major undertaking, but the next step is even more time-consuming. You must read, understand, and record the relevant information from the literature, keeping in mind one of the many (anonymous) Murphy's Laws: "No matter how many years you save an item, you will never need it until after you have thrown it away."

It is likely that the person originating this quote was a researcher working on a literature review. You can count on the fact that if you throw away one note from your literature search, that paper will be cited incorrectly in your text or reference list, or if you throw away your notes on an article because you do not believe it is relevant to your research, your major professor, a committee member, or the journal to which you submit the paper will request inclusion of the article. Then, when you go back to locate the article in the library, either it will have been ripped out of the journal, the whole journal will be missing, or a professor will have checked it out and never returned it (and the librarian will not reveal the professor's name). Therefore, when you find a particular paper, take careful and complete notes, including exact citation information.

A Key to Understanding Scientific Research Literature

What was said	What was meant
It has long been known that . . .	I haven't bothered to look up the original reference but . . .
Of great theoretical and practical importance.	Interesting to me.
While it has not been possible to provide definite answers to those questions . . .	The experiment didn't work out, but I figured I could at least get a publication out of it.
The W-PO system was chosen as especially suitable to show the predicted behavior.	The researcher in the next lab had some already made up.
Three of the samples were chosen for detailed study.	The results on the others didn't make sense.
Accidentally strained during mounting.	Dropped on the floor.
Handled with extreme care throughout the experiment.	Not dropped on the floor.
Typical results are shown.	The best results are shown.

Agreement with the predicted curve is:
   excellent.
   good.
   satisfactory.
   fair.
	
	fair.
	poor.
	doubtful.
	imaginary.

It is suggested that . . .	I think.
It is believed that . . .	I think.
It may be that . . .	I think.
It is clear that much additional work will be required before a complete understanding . . .	I don't understand it.
Unfortunately, a quantitative theory to account for these results has not been formulated.	Neither does anybody else.
Correct with an order of magnitude . . .	Wrong.
Thanks are due to Joe Glotz for assistance with the experiments and to John Doe for valuable discussion.	Glotz did the work and Doe explained what it meant.

To help you understand the literature you read, try to decipher the scientific phrases in the box on this page. Once you understand what each phrase really means, you should have little difficulty understanding authors of research articles. Seriously, you, as a researcher, should note the following information from research studies that you read:

- Statement of the problem (and maybe hypotheses)
- Characteristics of the participants
- Instruments and tests used (including reliability and validity information if provided)
- Testing procedures
- Independent and dependent variables
- Treatments applied to participants (if an experimental study)
- Design and statistical analyses

Criteria for Critiquing a Research Paper

I. Overall impression (most important): Is the paper a significant contribution to knowledge about the area?

II. Introduction and review of literature

    A. Is the research plan developed within a reasonable theoretical framework?

    B. Is current and relevant research cited and properly interpreted?

    C. Is the statement of the problem clear, concise, testable, and derived from the theory and research reviewed?

III. Method

    A. Are relevant participant characteristics described, and are the participants appropriate for the research?

    B. Is the instrumentation appropriate?

    C. Are testing or treatment procedures described in sufficient detail?

    D. Are the statistical analyses and research design sufficient?

IV. Results

    A. Do the results evaluate the stated problems?

    B. Is the presentation of results complete?

    C. Are the tables and figures appropriate?

V. Discussion

    A. Are the results discussed?

    B. Are the results related back to the problem, theory, and previous findings?

    C. Is there excessive speculation?

VI. References

    A. Are all references in the correct format, and are they complete?

    B. Are all references cited in the text?

    C. Are all dates in the references correct, and do they match the text citation?

VII. Abstract

    A. Does it include a statement of the purpose; description of participants, instrumentation, and procedures; and a report of meaningful findings?

    B. Is the abstract the proper length?

VIII. General

    A. Are keywords provided?

    B. Are running heads provided?

    C. Does the paper use nonsexist language and provide for protection and appropriate labelling of human participants?

- Findings
- Questions raised for further study
- Citations to other relevant studies not located

When studies are particularly relevant to the proposed research, make a photocopy. Write the complete citation on the title page if the journal does not provide this. Many students (and faculty members) make the mistake of simply photocopying an article without looking to see whether the pages copied contain all the necessary information for a citation. Frequently, they do not, which means you have to go back to the library and try to find the journal or volume from which the article was taken. Also, indicate on a note card the studies that are photocopied.

Another useful way to learn to read and understand related literature is to critique a few studies. Use the series of questions on this page when critiquing a study. Page 44 provides

Form for a Critique

I. Basic information
    A. Name of journal.
    B. Publisher of journal.
    C. How many articles in this issue?
    D. Are there publication guidelines for authors? Which issue?
    E. Who is editor?
    F. Is there a yearly index? Which issue?
    G. What writing style is used (APA, *Index Medicus,* etc.)?
    H. Complete reference in APA style for article you will review.

II. Summary of the article

III. Critique of the article
    A. Introduction and review.
    B. Method.
    C. Results.
    D. Discussion.
    E. References.
    F. Overall.

IV. Photocopy of the article

V. Information about critique
    A. It must be typed (double-spaced) in APA style.
    B. Do not put it in any type of folder; just staple the pages in upper left-hand corner.
    C. Use a cover page identifying course, purpose, and yourself.
    D. Critique may not exceed five typewritten pages exclusive of cover page.

Critique of Schubert's Productivity

A company CEO was given a ticket for a performance of Schubert's Unfinished Symphony. Unable to go, he passed the invitation to the company TQM (total quality management) coordinator. The next morning the CEO asked him how he enjoyed it. Instead of voicing a few plausible observations, he handed the CEO a memorandum which read as follows:

- For considerable periods, the oboe players had nothing to do. Their number should be reduced, and their work spread over the whole orchestra, thus avoiding peaks of inactivity.

- All 12 violins were playing identical notes. This seems unnecessary duplication, and the staff of this section should be drastically cut. If a large volume of sound is really required, this could be obtained through the use of an amplifier.

- Much effort was involved in playing the demi-semiquavers. This seems an excessive refinement, and it is recommended that all notes should be rounded up to the nearest semiquaver. If this were done, it should be possible to use trainees instead of craftsmen.

- No useful purpose is served by repeating with horns the passage that has already been handled by the strings. If all such redundant passages were eliminated, the concert could be reduced from two hours to 20 minutes.

In light of the above, one can only conclude that had Schubert given attention to these matters, he would probably have had the time to finish his symphony.

a sample form that we use for reporting critiques. A few critiquing attempts should aid you in focusing on the important information contained in research studies.

Sometimes reviewers can go overboard in their critiques, much like the critique of Schubert's Unfinished Symphony we present, also on page 44.

To summarize, the best system for recording relevant literature is probably a combination of note taking and photocopying. By using index cards or an equivalent computerized system, the important information about most studies can be recorded and indexed by topic. Always be sure to record the complete and correct citation on the card with the appropriate citation style used by your institution (e.g., American Psychological Association [APA], *Index Medicus*).

Step 6: Write the Literature Review

After you have located and read the necessary information and have recorded the appropriate bibliographic data, you are ready to begin to write the literature review. The literature review has three basic parts.

1. Introduction
2. Body
3. Summary and conclusions

The introduction should explain the purpose of the review and the how and why of its organization. The body of the review should be organized around important topics. Finally, the review should summarize important implications and suggest directions for future research. The review's purpose is to demonstrate that your problem needs investigation and that you have considered the value of relevant past research in developing your hypotheses and methods; that is, that you know and understand what other people have done and how that relates to and supports what you plan to do.

The introduction to the review (or to topical areas within the review) is very important. If these paragraphs are not well done and interesting, the reader may skip the entire section. Attempt to attract the reader's attention by identifying in a provocative way the important points to be covered.

The body of the literature review requires considerable attention. Relevant research must be organized, synthesized, and written in a clear, concise, and interesting way. There is no unwritten law dictating that literature reviews must be boring and poorly written, although we suspect that some graduate students work from that assumption. Part of the problem stems from graduate students' perceptions that they must find a way to make their scientific writing complex and circuitous as opposed to simple and straightforward. Apparently, the rule is to never use a short and simple word when a longer, more complex one can be substituted. In table 2.3, Day (1983) has provided a useful aid to potential research writers. In fact, we strongly recommend Day's book, which can easily be read in four to six hours. It provides many excellent and humorous examples that are valuable in writing for publication and for theses and dissertations.

In addition to removing as much jargon as possible (using Day's suggestions), you should be clear and to the point. We advocate the KISS principle (Keep It Simple, Stupid) as a basic tenet for writing. Many grammatical errors can be avoided by using simple declarative sentences. The list on page 47 (Day 1983) rephrases the commandments of good writing from a 1968 Council of Biology Editors newsletter. The first 10 items are Day's, but we have included a few of our own comments as well.

Proper syntax (the way words and phrases are put together) is the secret of successful writing. Some examples of improper syntax and other unclear writing are highlighted on page 48.

As mentioned previously, the literature review should be organized around important topics. These topics serve as subheadings in the paper to direct the reader's attention. The best way to organize the topics and the information within topics is to develop an outline. The more carefully the outline is planned, the easier the writing will be. A good task is to select a review paper from a journal or from a thesis or dissertation review of literature and reconstruct the outline the author must have used. In looking at older theses and

Table 2.3   Words and Expressions to Avoid

Jargon	Preferred usage	Jargon	Preferred usage
a considerable amount of	much	in order to	to
a majority of	most	in relation to	toward, to
a number of	many	in respect to	about
absolutely essential	essential	in some cases	sometimes
accounted for by the fact	because	in terms of	about
along the lines of	like	in the event that	if
an order of magnitude faster	10 times faster	in the possession of	has, have
are of the same opinion	agree	in view of the fact that	because, since
as a consequence of	because	inasmuch as	for, as
as a matter of fact	in fact (or leave out)	initiate	begin, start
as is the case	as happens	is defined as	is
as of this date	today	it has been reported by Smith	Smith reported
as to	about (or leave out)	it has long been known that	I haven't bothered to look up the reference
at an earlier date	previously		
at the present time	now	it is apparent that	apparently
at this point in time	now	it is believed that	I think
based on the fact that	because	it is clear that	clearly
by means of	by, with	it is clear that much additional work will be required before a complete understanding	I don't understand it
completely full	full		
consensus of opinion	consensus		
definitely proved	proved	it is doubtful that	possibly
despite the fact that	although	it is evident that *a produced b*	*a produced b*
due to the fact that	because	it is of interest to note that	(leave out)
during the course of	during, while	it is often the case that	often
elucidate	explain	it is suggested that	I think
end result	result	it is worth pointing out in this context that	note that
entirely eliminate	eliminate		
fabricate	make	it may be that	I think
fewer in number	fewer	it may, however, be noted that	but
finalize	end	it should be noted that	note that (or leave out)
first of all	first	it was observed in the course of the experiments that	we observed
for the purpose of	for		
for the reason that	since, because	lacked the ability to	couldn't
from the point of view of	for	large in size	large
give rise to	cause	let me make one thing perfectly clear	(a snow job is coming)
has the capability of	can		
having regard to	about	militate against	prohibit
in a number of cases	some	needless to say	(leave out, and consider leaving out whatever follows it)
in a position to	can, may		
in a satisfactory manner	satisfactorily		
in a very real sense	in a sense (or leave out)	of great theoretical and practical importance	useful
in case	if		
in close proximity	close, near	on a daily basis	daily
in connection with	about, concerning	on account of	because
in many cases	often	on behalf of	for
in my opinion it is not an unjustifiable assumption that	I think	on the basis of	by
		on the grounds that	since, because

Jargon	Preferred usage	Jargon	Preferred usage
on the part of	by, among, for	the opinion is advanced that	I think
our attention has been called to the fact that	we belatedly discovered	the question as to whether	whether
owing to the fact that	since, because	the reason is because	because
perform	do	there is reason to believe	I think
pooled together	pooled	this result would seem to indicate	this result indicates
prior to	before	through the use of	by, with
protein determinations were performed	proteins were determined	ultimate	last
quite unique	unique	utilize	use
rather interesting	interesting	was of the opinion that	believed
red in color	red	ways and means	ways, means (not both)
referred to as	called	we have insufficient knowledge	we don't know
relative to	about	we wish to thank	we thank
resultant effect	result	with a view to	to
smaller in size	smaller	with reference to	about (or leave out)
subsequent to	after	with regard to	concerning, about (or leave out)
sufficient	enough		
take into consideration	consider	with respect to	about
terminate	end	with the possible exception of	except
the great majority of	most	with the result that	so that

The Ten Commandments of Good Writing—Plus a Few Others

1. Each pronoun should agree with their antecedent.
2. Just between you and I, case is important.
3. A preposition is a poor word to end a sentence with. (Incidentally, did you hear about the streetwalker who violated a grammatical rule? She unwittingly approached a plainclothes officer, and her proposition ended with a sentence.)
4. Verbs has to agree with their subjects.
5. Don't use no double negatives.
6. A writer mustn't shift your point of view.
7. When dangling, don't use participles.
8. Join clauses good, like a conjunction should.
9. Don't write a run-on sentence it is difficult when you got to punctuate it so it makes sense when the reader reads what you wrote.
10. About sentence fragments.
11. Don't use commas, which aren't necessary.
12. Its' important to use apostrophe's right.
13. Check to see if you any words out.
14. As far as incomplete constructions, they are wrong.
15. Last but not least, lay off cliches.

Examples of Unclear Writing

Sentences with improper syntax:

- The patient was referred to the hospital for repair of a hernia by a social worker.
- As a baboon who grew up wild in the jungle, I realized that Wiki had special nutritional needs.
- No one was injured in the blast, which was attributed to a buildup of gas by one town official.
- Table 1 contains a summary of responses pertaining to suicide and death by means of a questionnaire. (We are aware that some questionnaires can be ambiguous, irrelevant, and trivial, but we had no idea that they were fatal.)

Sentences taken from letters received by government agencies:

- I am forwarding my marriage certificate and six children. I had seven but one died which was baptized on a half sheet of paper.
- I am writing to say that my baby was born two years old. When do I get my money?
- I am glad to report that my husband who is missing is dead.
- This is my eighth child. What are you going to do about it?
- In accordance with your instructions, I have given birth to twins in the enclosed envelope.
- I want my money as quick as I can get it. I have been in bed with a doctor for two weeks and he doesn't do me any good. If things don't improve, I will have to send for another doctor.

Sentences from doctors' notes on hospital charts:

- Discharge status: Alive but without my permission.
- The patient refused autopsy.
- She is numb from her toes down.
- Occasional, constant infrequent headaches.
- The patient was to have a bowel resection. However, he took a job as a stock broker instead.

Sentences taken from job recommendations that are not quite clear:

- In my opinion you will be very fortunate to get this person to work for you.
- All in all, I cannot say enough good things about this candidate or recommend him too highly.
- I am pleased to say this candidate is a former colleague of mine.
- I can assure you that no person would be better for this job.

Unclear excuses for missing school:

- Please excuse Mary for being absent. She was sick and I had her shot.
- Please excuse Fred from being absent yesterday. He had diarrhea and his boots leak.
- Please excuse Mary from Jim [Gym?] yesterday. She is administrating.

A couple of others:

- This horse is an eight-year-old gelding trained by the owner who races him with his wife.
- She rode 106 miles on a bicycle with 1,400 other people.

dissertations, we find that the literature review tends to be a historical account, often presented in chronological order. We suggest that you not select one of these older studies, as this style is cumbersome and usually poorly synthesized.

To write a literature review effectively, you should write as you like to read. No one wants to read abstracts of study after study presented in chronological order. A more interesting and readable approach is to present a concept and then discuss the various findings about that concept, documenting findings by references to the various research reports related to it. In this way, consensus and controversy can be identified and discussed in the literature review. More relevant and important studies can be presented in greater detail, and several studies with the same outcome can be covered in one sentence.

In a thesis or dissertation, the two important aspects of the literature review are criticism and completeness. The various studies should not simply be presented relative to a topic. The theoretical, methodological, and interpretative aspects of the research should be criticized not necessarily study by study, but rather across studies. This criticism demonstrates your grasp of the issues and identifies problems that should be overcome in the study you are planning. Frequently, the problems identified by criticism of the literature provide justification for your research.

Completeness (not in the sense of the length of the review but rather of reference completeness) is the other important aspect of the literature review. You should demonstrate to your committee that you have located, read, and understood all the related literature. Many studies may be redundant and need only appropriate citing, but they must be cited. The thesis or dissertation is your passport to graduation because it demonstrates your competence; therefore, never fail to be thorough. This, however, applies only to the thesis or dissertation. Writing for publication or using alternate thesis or dissertation formats does not require emphasizing the completeness of the literature cited (note "cited," not "read"). Journals do not have unlimited space and usually want the introduction and literature review to be integrated and relatively short.

Summary

Identifying and formulating a researchable problem are often difficult tasks, especially for the novice researcher. Some suggestions were given to help the graduate student find suitable topics. Inductive and deductive reasoning for formulating research hypotheses were discussed. Inductive reasoning moves from observations to specific hypotheses to a more general theoretical model. Deductive reasoning moves from a theoretical explanation to specific hypotheses to be tested.

The steps in a literature search include writing the problem statement, consulting secondary sources, determining descriptors, searching preliminary sources to locate primary sources, reading and recording the literature, and writing the review.

There are no shortcuts to locating, reading, and indexing the literature and then writing the literature review. If you follow our suggestions, you can do it more effectively, but much hard work is still required. A good scholar is careful and thorough. Do not depend on what others report, as they are often incorrect. Look it up yourself.

No one can just sit down and write a good literature review. A careful plan is necessary. First, outline what you propose to write, write it, and then write it again. When you are convinced that the review represents your best effort, have a knowledgeable graduate student or faculty member read it, then welcome their suggestions. Next, have a friend who is not as knowledgeable read it. If your friend can understand it, then your review is probably in good shape. Of course, your research methods professor will find something wrong or at least something that he or she thinks should be different. Just remember that professors feel obligated to find errors in graduate students' work.

☑ Check Your Understanding

Our suggestions for exercises that can help you locate, synthesize, organize, critique, and write the literature review are summarized below. You will need to refer to various text and tables in the chapter to complete these exercises.

1. Critique a research study in your area of interest. Use the questions in the sidebar titled "Criteria for Critiquing a Research Paper" (page 43) to help you. Report the critique using the "Form for a Critique" on page 44 as your model.

2. Select a review paper from a journal or the literature review from a thesis or dissertation. Construct the outline that the author probably used to write this paper.

Presenting the Problem

Truth in science can be defined as the working hypothesis best suited to open the way to the next better one.—Konrad Lorenz

■　■　■　■　■　■　■

In a thesis or dissertation, the first section or chapter serves to introduce the problem. Indeed, it is often titled "Introduction." Its purpose is to do just that: to introduce the reader to the problem being studied. Several sections in the introduction serve to convey the significance of the problem and set forth the dimensions of the particular study. This chapter discusses each of the following sections, which are frequently required in the first part of a thesis or dissertation:

- Title
- Introduction
- Problem statement
- Hypothesis
- Definitions
- Assumptions and limitations
- Significance

Not all thesis advisors subscribe to the same thesis format, for there is no universally accepted one. Moreover, because of the nature of the research problem, there are differences in format. For example, a historical study would not adhere to the same format as used in an experimental study and these section titles might differ in descriptive or qualitative studies. We merely present sections, each with a purpose and with specific characteristics, typically found in the introduction.

Choosing the Title

Although discussing the title first may seem logical, it might surprise you to learn that titles are often not determined until after the study has been written. However, at the proposal meeting, you must have a title (even though it may be provisional), so we discuss it first.

Some writers claim that there is a trend toward shortening titles (e.g., Day, 1983). However, an analysis of more than 10,000 dissertations in seven areas of education failed to demonstrate such a trend (Coorough & Nelson, 1997). Many titles are, in essence, the statement of the problem (in fact, some even include the methods section). Here is an example of a too lengthy title (Note: The examples we use as representing poor practices are fictional. Frequently, they have been suggested by actual studies, but any similarity to a real study is purely coincidental.):

> An Investigation of a Survey and Analysis of the Influence of PL 94-142 on the Attitudes, Teaching Methodology, and Evaluative Techniques of Randomly Selected Male and Female Physical Education Teachers in Public High Schools in Cornfield County, State of Confusion

Simply too much information is in such a title. Day (1983) humorously addressed this problem by reporting a conversation between two students. When one asked whether the other had read a certain paper, the reply was, "Yes, I read the paper, but I haven't finished the title yet" (p. 10). A better title for the study mentioned previously would be "PL 94-142's Influence on Physical Education Teachers' Attitudes, Methodology, and Evaluations."

The purpose of the title is to convey the content, but this should be done as succinctly as possible. For example, "The 12-Minute Swim as a Test for Aerobic Endurance in Swimming" (Jackson, 1978) is a good title because it tells the reader exactly what the study is about. It defines the specific purpose, which is the validation of the 12-minute swim, and it delimits the study to the assessment of aerobic endurance for swimmers.

However, do not go to the other extreme in striving for a short title. A title such as "Professional Preparation" is not very helpful. It does not include the field or the aspects of professional preparation studied. The key to the effectiveness of a short title is whether it reflects the study's contents. A title that is specific is more easily indexed and retrieved through electronic databases, and is more meaningful for a potential reader who is searching for literature on a certain topic.

Avoid "waste" words and phrases such as "An Investigation of," "An Analysis of," and "A Study of." They simply increase the length of the title and contribute nothing to the description of the content. Consider this title: "A Study of Three Teaching Methods." Half the title consists of waste words: "A Study of." The rest of the title is not specific enough to be indexed effectively: Teaching what? What methods?

Furthermore, always be aware of your audience. You can assume that your audience is reasonably familiar with the field, the accepted terminology, and relevant problem areas. An outsider can question the significance of studies in any field. Some titles of supposedly scholarly works are downright humorous. For a rousing good time, peruse the titles of theses and dissertations completed at a university in any given year; for example, "The Phospholipid Distribution in the Testes of the House Cricket." How weird can you get? We are joking, of course. The point is that there is a tendency to criticize studies done in other disciplines simply because the critic is ignorant about the discipline. An example of this in our field could be a scholarly article by Grabe and Widule (1988) titled "Comparative Biomechanics of the Jerk in Olympic Weight Lifting." To someone unfamiliar with the sport, this might seem to be a case study about some unpopular weight lifter in the Olympic Games.

Writing the Introduction

The introductory portion of a thesis or research article is designed to create interest in the problem. You use the introduction to persuade readers of the significance of the problem, provide background information, bring out areas of needed research, and then skillfully and logically lead to the specific purpose of the study.

How to Write a Good Introduction

A good introduction requires literary skill because it should flow smoothly yet be reasonably brief. Be careful not to overwhelm the reader with technical jargon, for the reader must be able to understand the problem to gain an interest in the solution. Therefore, an important rule is this: Do not be too technical. A forceful, simple, and direct vocabulary is more effective for purposes of communication than scientific jargon and worship of polysyllables. Day (1983, 147–148) related a classic story of the pitfalls of scientific jargon:

> This reminds me of the plumber who wrote the Bureau of Standards saying he had found hydrochloric acid good for cleaning out clogged drains. The Bureau wrote back, "The efficacy of hydrochloric acid is indisputable, but the corrosive residue is incompatible with metallic permanence." The plumber replied that he was glad the Bureau agreed. The Bureau tried again, writing, "We cannot assume responsibility for the production of toxic and noxious residues with hydrochloric acid and suggest that you use an alternative procedure." The plumber again said that he was glad the Bureau agreed with him. Finally, the Bureau wrote to the plumber, "Don't use hydrochloric acid. It eats the hell out of pipes."

Audience awareness is important. Again, you can assume that readers are reasonably informed about the topic (or they probably would not be reading it in the first place). However, even an informed reader needs some refresher background information to understand the nature of the problem, to be sufficiently interested, and to appreciate your rationale for studying the problem. You must remember that your audience has not been as completely and recently immersed in this particular area of research as you have been.

The introductory paragraphs must create interest in the study; thus, your writing skill and knowledge of the topic are especially valuable in the introduction. The narrative should introduce the necessary background information quickly and explain the rationale behind the study. A smooth, unified, well-written introduction should lead to the problem statement with such clarity that the reader could state the study's purpose before specifically reading it.

The following introductions were selected from research journals for their brevity of presentation and for their effectiveness. This is not to say that brevity in itself is a criterion, for some topics require more comprehensive introductions than others. For example, studies developing or validating a theoretical model usually need longer introductions than do applied research topics. Furthermore, theses and dissertations (in the traditional format) almost always have longer introductions than journal articles simply because of the page-cost considerations in the latter.

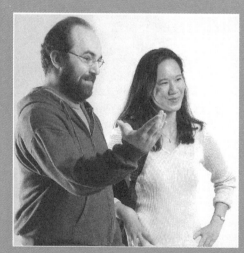

They told her she needed a mouse for her research. They should have operationally defined their terms.

Examples of Good Introductions

The following examples specify some desirable features in an introduction, including a general introduction, background information, a mention of gaps in the literature and areas of needed research, and a logical progression leading to the problem statement. After you've read them, see if you can write the purpose for each study.

Example 3.1 (From Sundgot-Borgen, 1994)

[General Introduction.] In recent years, there has been growing interest in eating disorders in athletes (7, 30). Studies have shown that athletes are more prone to developing eating disorders than nonathletes (4, 5, 21, 28). In addition, the highest prevalence of eating disorders is in female athletes competing in sports where leanness and/or a specific weight are considered important for either performance or appearance (4, 27, 28).

[Background Information.] There has been considerable speculation about why athletes are at increased risk for eating disorders. Predisposing personality or family interaction variables might be primary (3, 26, 32), so participation in sports favoring leanness could be a consequence of preexisting eating problems or could be coincidental. Alternatively, participation in certain sports could be related causally to the onset of eating disorders. In all likelihood these factors interact.

[Lead-In.] One important area of inquiry, therefore, is to identify risk factors to help determine which athletes are most vulnerable, or which conditions or trigger factors elicit the pathological behavior.

Example 3.2 (From Dolgener et al., 1994)

Reprinted with permission from *Research Quarterly for Exercise and Sport*, Vol. 65, pgs. 152-158, Copyright 1994 by the American Alliance for Health, Physical Education, Recreation and Dance, 1900 Association Drive, Reston, VA 20191.

[General Introduction.] Cardiorespiratory fitness is generally recognized as a major component of physical fitness. Indeed, cardiorespiratory fitness is the most significant component of physical fitness in the relationship to health. Direct measurement of maximal oxygen uptake ($\dot{V}O_2$max) is the single best measure of cardiorespiratory fitness or aerobic capacity (Åstrand & Rodahl, 1986; Mitchell, Sproule, & Chapman, 1958; Taylor, Buskirk, & Henschel, 1955). However, direct measurement is time-consuming, requires extensive laboratory equipment, and does not lend itself to testing large numbers of subjects in field settings. Because of the limitations of direct measures, numerous field tests have been developed to estimate $\dot{V}O_2$max.

[Background Information.] Prediction of $\dot{V}O_2$max from field tests requires performing at either a maximal or submaximal effort, commonly running, stepping, or bicycling. Recently, Kline, Porcari, Hintermeister et al. (1987) developed a submaximal field test for predicting $\dot{V}O_2$max using a 1-mile walk protocol. This test, which has become known as the Rockport Fitness Walking Test (RFWT), was developed on a broad age range (30–69 years) of males and females who were heterogeneous in terms of aerobic capacity. [Note: We are omitting some information concerning specific correlations and errors of estimate to save space.] These data indicate that the regression equations developed by Kline, Porcari, Hintermeister et al. are valid for adults between the ages of 30 and 69 years.

The RFWT has been cross-validated in samples of subjects 65 to 79 years old (Fenstermaker, Plowman, & Looney, 1992; O'Hanley et al., 1987) and 30 to 39 years old (Zwiren, Freedson, Ward, Wilke, & Rippe, 1991).

[Lead-In.] The Kline, Porcari, Hintermeister et al. (1987) equations, however, have not been validated for use with groups younger than those in the original sample.

Stating the Research Problem

The problem statement follows the introduction. We should point out that the literature review is often included in the introductory section and thus precedes the formal statement of the problem. If this is the case, then a brief problem statement should appear fairly soon in the introductory section before the literature review.

The problem statement in example 3.1 from Sundgot-Borgen (1994) was to examine risk factors for eating disorders along with trigger factors that may be responsible for precipitating the onset or exacerbation of eating disorders in elite female athletes. The problem statement in the Dolgener et al. (1994) study was also obvious from the introduction. The purposes stated were (a) to validate the Kline et al. 1-mile walk test in a sample of male and female college students and (b) to develop prediction equations for $\dot{V}O_2$max on this college sample if their equations proved invalid.

Identifying the Variables

The problem statement should be succinct. However, when the study has several subpurposes, this is not always easily accomplished. The statement should identify the different variables in the study, including the independent variable, the dependent variable, and the categorical variable (if any) for an experimental or quasi-experimental study. Usually, some **control variables** (which could possibly influence the results and are kept out of the study) can also be identified here.

The independent and dependent variables were mentioned in chapter 1. The independent variable is the experimental, or treatment, variable; it is the cause. The dependent variable is what is measured to assess the effects of the independent variable; it is the effect. A **categorical variable,** sometimes called a **moderator variable** (Tuckman, 1978), is a kind of independent variable except that it cannot be manipulated, for example, age, race, or sex. It is studied to determine whether the cause-and-effect relationship of the independent and dependent variables is different in the presence of the categorical variable or variables.

Following is an actual study in which the independent, dependent, and categorical variables can be identified. Anshel and Marisi (1978) studied the effect of synchronous and asynchronous movement to music on endurance performance. One group performed an exercise in synchronization to background music, one group exercised with background music that was not synchronized to the pace of the exercise, and a third group exercised with no background music.

The independent variable was the background music condition. There were three levels of this variable: synchronous music, asynchronous music, and no music. The dependent variable was endurance performance, which was reflected by the amount of time the participant could exercise on a bicycle ergometer until exhaustion. In this study, the endurance performances of men and women under the synchronous, asynchronous, and no-music conditions were compared. The authors thus sought to determine whether men responded differently than women to the exercise conditions. Sex, then, represented a categorical variable. Not all studies have categorical variables.

The researcher decides which variables to manipulate and which variables to control. One can control the possible influence of some variable by keeping it out of the study. Thus, the researcher chooses not to assess a variable's possible effect on the relationship between the independent and dependent variables, so this variable is controlled. For example, suppose a researcher is comparing stress-reduction methods on the competitive state anxiety of gymnasts before dual meets. The gymnasts' years of competitive experience might have a bearing on their anxiety scores. The researchers have a choice. They can include it as a categorical variable by requiring that half the participants have had a certain number of years of experience and that the other half have had less, or they can control the variable of experience by requiring that all participants have similar experience.

The decision to include or exclude some variable depends on several considerations, such as whether the variable is closely related to the theoretical model and how likely there is to be an interaction. Practical considerations include how difficult it is to make a variable a categorical variable or to control it (such as availability of participants having a particular trait) and how much control the researcher has over the experimental situation.

In the study of the effects of synchronized music on endurance (Anshel & Marisi, 1978), the factor of fitness level was controlled by giving all the participants a physical working capacity test. Then, on the basis of this test, each person exercised at a workload that would

control variable

A factor that could possibly influence the results and that is kept out of the study.

categorical variable

A kind of independent variable that cannot be manipulated, such as age, race, sex, and so on; also called *moderator variable.*

moderator variable

See *categorical variable.*

cause a heart rate of 170 bpm. Thus, even though the ergometer resistance settings would be different from person to person, all participants would be exercising at approximately the same relative workload; the differences in fitness were controlled in this manner. Another way of controlling fitness as a variable would be to test participants on a fitness test and select just those persons with a certain level of fitness.

Extraneous variables are factors that could affect the relationship between the independent and dependent variables but are not included or controlled. The possible influence of an extraneous variable is usually brought out in the discussion section. Anshel and Marisi (1978) speculated that some differences in the performances of men and women might be due to the women's reluctance to exhibit maximum effort in the presence of a male experimenter. Consequently, this would be an extraneous variable. (All the kinds of variables are discussed in more detail in chapter 18.)

Rarely are the variables labeled as such in the actual problem statement. Occasionally, the researcher identifies the independent and dependent variables, but mostly these variables are just implied.

In summary, an effectively constructed introduction leads smoothly to the study's purpose. This is expressed as the statement of the problem and should be as clear and concise as the subpurposes, or variables, allow it to be.

Structuring the Problem Statement

To achieve clarity in the statement of the problem, a final but important aspect to consider is sentence structure, or syntax. For example, suppose a researcher conducted a study "to compare sprinters and distance runners on anaerobic power, as measured by velocity in running up a flight of stairs." Observe the difference in meaning if the researcher had worded the purpose as "to compare the anaerobic power of sprinters and distance runners while running up a flight of stairs." The researcher would have to be in good shape to make those comparisons while running up stairs. Another example of faulty syntax was the case in which the purpose of the study was "to assess gains in quadriceps strength in albino mice using electrical stimulation." Those mice had to be awfully clever to use electrical instruments.

Presenting the Research Hypothesis

After you have stated the research problem, you must present the hypothesis (or present a question for a qualitative study). The formulation of hypotheses was discussed in chapters 1 and 2. The discussion here is on the statement of the hypotheses and the distinction between research hypotheses and the null hypothesis. Remember that **research hypotheses** are the expected results. In the study by Anshel and Marisi (1978), a research hypothesis might be that endurance performance would be enhanced by exercising to synchronized music. The introduction produces a rationale for that hypothesis. Another hypothesis might be that exercise to asynchronous music would be more effective than exercising with no background music (because of the pleasurable sensory stimuli's blocking the unpleasant stimuli associated with the fatiguing exercise). As a further example, a researcher in cardiac rehabilitation might hypothesize that distance from the exercise center is more influential as a factor in patients' exercise adherence than the type of activities offered in the cardiac rehabilitation program. In the example given in chapter 1, a dance teacher hypothesized that the use of videotape in the instructional program would enhance the learning of dance skills.

In contrast, the **null hypothesis** is used primarily in the statistical test for the reliability of the results; the null hypothesis says that there are no differences between treatments (or no relationships between variables). For example, any observed difference or relationship is due simply to chance (see chapter 7). The null hypothesis is usually not the research hypothesis and the research hypothesis is what usually is presented. Generally, the researcher expects one method to be better than others or anticipates a relationship between two variables. One does not embark on a study if nothing is expected to happen. On the other hand, a

extraneous variable

A factor that could affect the relationship between the independent and dependent variables, but that is not included or controlled.

research hypothesis

Hypothesis deduced from theory or induced from empirical studies that is based upon logical reasoning and predicts the outcome of the study.

null hypothesis

Hypothesis used primarily in the statistical test for the reliability of the results that says that there are no differences among treatments (or no relationships among variables).

researcher sometimes hypothesizes that one method is just as good as another. For example, in the multitude of studies done in the 1950s and 1960s on isometric versus isotonic exercises, it was often hypothesized that the "upstart" isometric exercise was just as effective as the traditional isotonic exercise, provided that there was regular specific knowledge of results. In a study on the choice of recreational activities of mentally retarded children, Matthews (1979) showed that most research in this area, which reported differences between children with and without mental retardation, failed to consider socioeconomic status. Consequently, he hypothesized that there were no differences in frequency of participation in recreational activities between children who were mildly mentally retarded and who were not mentally retarded when socioeconomic status was held constant.

Furthermore, sometimes the researcher does not expect differences in some aspects of the study but does expect a difference in others. A researcher might hypothesize that children of high aptitude in learning would do better with one style of teaching, whereas children of low aptitude would fare better with another style. In a study of age differences in the strategy for recall of movement (Thomas, Thomas, Lee, Testerman, & Ashy, 1983), the authors hypothesized that because location is automatically encoded in memory, there would be no real difference between younger and older children in remembering a location (where an event happened during a run). However, they hypothesized that there would be a difference in remembering distance because the older child spontaneously uses a strategy for remembering and the younger child does not. The formulation of hypotheses is an important aspect of defining and delimiting the research problem.

Operationally Defining Your Terms

Another task in the preparation of the first section of a thesis or dissertation is operationally defining certain terms so that the researcher and the reader can adequately evaluate the results. It is imperative that the dependent variable be operationally defined.

An **operational definition** describes an observable phenomenon, as opposed to a synonym definition or dictionary definition. To illustrate, a study such as Anshel and Marisi's (1978), which investigated the effects of music on forestalling fatigue, must operationally define fatigue. The author cannot use a synonym, such as *exhaustion,* because that is not concrete enough. We all might have our own ideas of what fatigue is, but if we are going to say that some independent variable affects fatigue, we must supply some observable evidence of changes in fatigue. Therefore, fatigue must be operationally defined. Anshel and Marisi did not use the term *fatigue,* but from their description of procedures we can infer its operational definition as being when the participant was unable to maintain the pedaling rate of 50 revolutions per minute for 10 consecutive seconds.

Another researcher might define fatigue as the point when a maximal heart rate is achieved; still another might define fatigue as the point of maximal oxygen consumption. In all cases, though, it must be an observable criterion.

A study dealing with dehydration must provide an operational definition such as a loss of 5% of body weight. The term *obesity* in males could be defined as having 25% body fat. A study of different teaching methods on learning must operationally define the term *learning.* To use the old definition, "a change in behavior" is meaningless in providing evidence of learning. Learning might be demonstrated by five successful maze traversals or some other observable performance criterion.

You may not always agree with the investigator's definitions, but at least you know how a particular term is being used. A common mistake of novice researchers is thinking that every term needs to be defined. (We have seen master's students define terms not even used in their studies!) Consider an example of an unnecessary definition in a study of the effects of strength training on changes in self-concept. *Self-concept* would need to be defined (probably as represented by some scale), but *strength* would not. The strength-training program used would be described in the methods section. Basically, operational definitions are directly related to the research hypotheses because, if you predict that some treatment will produce some effect, you must define how that effectiveness will be manifested.

operational definition

Observable phenomenon that enables the researcher to empirically test whether or not the predicted outcomes can be supported.

Basic Assumptions, Delimitations, and Limitations

Besides writing the introduction, stating the research problem and hypothesis, and operationally defining your terms, you may be required to outline the basic assumptions and limitations under which you performed your research.

Assumptions

Every study has certain fundamental premises that it could not proceed without. In other words, you must assume that certain conditions exist and that the particular behaviors in question can be observed and measured (along with various other basic suppositions). A study in pedagogy that compares teaching methods must assume that the teachers involved are capable of promoting learning; if this assumption is not made, the whole study is worthless. Furthermore, in a learning study, the researcher must assume that the sample selection (e.g., random selection) results in a normal distribution with regard to learning capacity.

A study designed to assess an attitude toward exercise is based on the assumption that this attitude can be reliably demonstrated and measured. Furthermore, you can assume that the participants will respond truthfully, at least for the most part. If you cannot assume those things, you should not waste your time conducting the study.

Of course, the experimenter does everything possible to increase the credibility of the premises. The researcher takes great care in selecting measuring instruments, in sampling, and in gathering data concerning such things as standardized instructions and motivating techniques. Nevertheless, the researcher still must rely on certain basic assumptions.

Consider the following studies. Johnson (1979) investigated the effects of different levels of fatigue on visual recognition of previously learned material. Among his basic assumptions were that (a) the mental capacities of the participants were within the normal range for university students, (b) the participants understood the directions, (c) the mental task typified the types of mental tasks encountered in athletics, and (d) the physical demands of the task typified the levels of exertion commonly experienced by athletes.

Lane (1983) compared skinfold profiles of black and white girls and boys and tried to determine which skinfold sites best indicated total body fatness with regard to race, sex, and age. She assumed that (a) the skinfold caliper is a valid and reliable instrument for measuring subcutaneous fat, (b) skinfold measurements taken at the body sites indicate the subcutaneous fat stores in the limbs and trunk, and (c) the sum of all skinfolds represents a valid indication of body fatness.

In some physiological studies, the participants are instructed (and agree) to fast or to refrain from smoking or drinking liquids for a specified period before testing. Obviously, unless the study is conducted in some type of prison environment, the experimenter cannot physically monitor the participants' activities. Consequently, a basic assumption is that the participants will follow instructions.

limitation

A possible shortcoming or influence that either cannot be controlled or is the result of the delimitations imposed by the investigator.

delimitation

A limitation imposed by the researcher in the scope of the study; a choice the researcher makes to define a workable research problem.

Delimitations and Limitations

Every study also has **limitations.** Limitations are possible shortcomings or influences that either cannot be controlled or are the results of the restrictions imposed by the investigator. Some limitations refer to the scope of the study, which is usually set by the researcher. These are often called **delimitations.** Kroll (1971) described delimitations as choices the experimenter makes to define a workable research problem, such as the use of one particular personality test in the assessment of personality characteristics. Moreover, in a study dealing with individual-sport athletes, the researcher may choose to restrict the selection of participants to athletes in just two or three sports, simply because all individual sports could not be included in one study. Thus, the researcher delimits the study. You probably notice that these delimitations resemble operational definitions. Although they are similar, they are not alike. For example, the size of the sample is a delimitation but is not included under operational definitions.

You can also see that basic assumptions are entwined with delimitations as well as with operational definitions. The researcher must proceed on the assumption that the restrictions imposed on the study will not be so confining as to destroy the external validity (generalizability) of results.

Remember, theses or dissertations do not have one "correct" format. An examination of studies shows numerous variations in organization. Some studies have delimitations and limitations described in separate sections. Some use a combination heading, some list only one heading but include both in the description, and still others embed these in other sections of the thesis or dissertation. As with all aspects of format, much depends on how the advisor was taught. Graduate colleges and schools often allow great latitude in format as long as the study is internally consistent. You will find considerable differences in format even within the same department.

In Kroll's (1971) example of delimiting the scope of the study to just two sports to represent individual-sport athletes, there is an automatic limitation of how well these represent all individual sports. Moreover, if the researcher is studying personality traits of these athletes and delimits the measurement of personality to just one test, this results in a limitation. Furthermore, there is at least one inherent limitation in all self-report instruments in which the participant responds to questions about his or her behavior, likes, or interests as to the truthfulness of the responses.

Thus, you can see that limitations accompany the basic assumptions to the extent that the assumptions fail to be justified. As with assumptions, the investigator tries as much as possible to reduce limitations that might stem from faulty procedures. In R.L. Johnson's 1979 study of fatigue effects on the visual recognition of previously learned material, he had to have the participants first learn the material. He established criteria for learning (operational definition) and tried to control for overlearning (i.e., differences in the level of learning). Despite his efforts, however, he recognized that a limitation was that there may have been differences in the degree of learning, which could certainly influence recognition.

In the study of skinfold profiles by Lane (1983), she had to delimit the study to a certain number of participants in one part of Baton Rouge, Louisiana. Consequently, a limitation was that the children were from only one geographic location. She also recognized that there are changes in body fatness associated with the onset of puberty, but she was unable to obtain data on puberty or other indices of maturation, and this therefore posed a limitation. Still another limitation was the inability to control possible influences on skinfold measurement, such as dehydration and other diurnal variations. Finally, because there are no internationally recognized standard body sites for skinfold measurement, generalizability may have been limited to the body sites used in this study.

You should not be overzealous in searching for limitations, or you can apologize away the worth of the study. For example, one of our advisees who was planning to meet with his proposal committee was overly apologetic with these anticipated limitations:

- The sample size may be too small.
- The tests may not represent the parameter in question.
- The training sessions may be too short.
- The investigator lacks adequate measurement experience.

As a result, the proposal was revised and the method was reassessed.

Remember, there is no perfect study. You must carefully analyze the delimitations to determine whether the resulting limitations outweigh the delimitations. In addition, careful planning and painstaking methodology increase the validity of the results, thus greatly reducing possible deficiencies in a study.

Justifying the Significance of the Study

The inevitable question you face at both the proposal meeting and the final oral exams deals with the worth, or significance, of the study, which may be asked in different ways, such as, So what? or What good is it? or How is this of any importance to your profession? Regardless of the manner in which the question is asked, it must be answered. Perhaps because of the inevitability of the question, most students are required to include a section titled "Significance of the Study" or sometimes "The Need for the Study," "Importance of the Study," "Rationale for the Study."

Basic and Applied Research Revisited

To a large extent, the worth of the research study is judged by whether it is basic or applied research. In chapter 1, we explained that basic research does not have immediate social significance; it usually deals with theoretical problems and is conducted in a very controlled laboratory setting. Applied research addresses immediate problems for improving practice. There is less control but ideally more real-world application. Consequently, basic and applied research cannot be evaluated by the same criteria.

The significance of a basic research study obviously depends on the specific purpose of the study, but usually the criteria focus on the extent to which the study contributes to the formulation or validation of some theory. The worth of applied research must be evaluated on the basis of its contribution to the solution of some immediate problem.

Writing the Section on the Significance of the Study

The significance section is often difficult to write, probably because the student thinks only in terms of the practicality of the study—how the results can be immediately used to improve some aspect of the profession. Kroll (1971) emphasized the importance of maintaining continuity of the significance section with the introduction. Too often the sections are written with different frames of reference instead of a continuous flow of thought. The significance section should focus on such things as contradictory findings of previous research and gaps in knowledge in particular areas and how the study might contribute to practice. Difficulties in measuring aspects of the phenomenon in question are sometimes emphasized. The rationale for verifying existing theories may be the focus of the section in some studies, whereas in others the practical application is the main concern. Generally, both theoretical and practical reasons are expressed, but the emphasis will vary according to the study.

Just as the length of the introduction varies, the length of the significance section varies considerably from study to study. A sample of a significance section from Lane (1983) may serve to illustrate an approach that focuses on some conflicts between previous research findings and present practice:

> **S**ince the measurement of body composition has become an important aspect of physical fitness testing, the validity, reliability, and administrative feasibility of the measurements are of paramount concern. Adult formulas for estimating percent fat are not considered valid for children, thus skinfolds are in themselves used as measures of body composition.
>
> The *AAHPERD Health Related Physical Fitness Test Manual* (1980) describes two skinfold measurements: the triceps and subscapular. Norms for the total of these two measures are provided for boys and girls ages 6 to 17 years. Abbreviated norms are also given for the triceps skinfolds only. Norms were taken from HES data (Johnston et al., 1974). There is minimal evidence for why these two body sites were selected, especially when one, the subscapular, poses some problem with regard to modesty. If the triceps and subscapular were selected to represent one site from the limbs and one from the trunk, are they the most predictive of total fatness as indicated by the sum of several skinfolds from the limbs and trunk? Furthermore, if some sites are equally predictive, the ease of administration needs to be considered.
>
> Of major significance in this study is whether the skinfolds that best represent body fatness in white children are equally suitable for black children. Authors (Cronk & Roches, 1982; Harsha et al., 1974; Johnston et al., 1974) have reported that there are differences in skinfold thicknesses between black and

white children, yet the AAHPERD norms make no distinction. If norms are to be of value, they must represent the population for which they are intended. Moreover, there may be greater differences in fatness between blacks and whites at different ages.

Similarly, it may be that a different combination of skinfolds would be more valid for girls than boys. It does not seem to be of any great administrative advantage to use the same sites for both sexes if other sites are equally valid indicators of total fatness.

A final word of warning: In being asked the inevitable question in the final oral exam about the significance of the study, do not reply, "It was necessary to get my degree." The stony silence will be unnerving.

The Differences Between the Thesis and the Research Article

A mere glance at research articles in journals reveals that a number of the sections found in the traditional thesis or dissertation described in this chapter are missing. At least two reasons account for this. The first is financial: Periodicals are concerned about publishing costs, so brevity is emphasized. Second, a kind of novice–expert ritual seems to be in operation. The novice is required to explicitly state the hypotheses, define terms, state assumptions, recognize limitations, and justify the worth of the study in writing. Certainly these steps are all part of defining and delimiting the research problem, and it is undoubtedly a worthwhile experience to address each step formally.

Research journal authors, on the other hand, need not explain the step-by-step procedure they used in developing the problem. Typically, a research journal has an introduction that includes a short review of literature. The length varies considerably, and some journals insist on brief introductions.

The purpose of the study is nearly always given but is usually not designated by a heading; rather, it is often the last sentence or so in the introduction. For example, in 30 articles in one volume of the *Research Quarterly for Exercise and Sport,* only one had a section titled "Purpose of the Study." Twenty-four ended their introductions with sentences that began with the words, "The purpose of this study was . . . " Four indicated the purpose with sentences beginning "This study was designed to . . . " or "The intent of the study was . . . " One study did not state the purpose at all. In 29 cases, the authors and editors felt that the purpose was evident from the title and introduction.

Research hypotheses and questions are sometimes given but with little uniformity. Operational definitions, assumptions, limitations, and significance of the study are rarely if ever stated in research articles. Apparently, for the "expert" researcher these steps were accomplished in the development of the problem and are understood or obvious, or both, and need not be stated. If the article is well written, you should be able to discern the operational definitions, the assumptions and limitations, and the independent, dependent, and categorical variables even though they are not specifically stated. Moreover, the significance of the study should be implicitly obvious if the author has written a good introduction.

Summary

This chapter discussed the information that is typically presented in the first section or chapter of a thesis or dissertation (excluding the review of literature). First, we considered the length and substance of the study title. The importance of a good, short, descriptive title for indexing and searching the literature is sometimes overlooked.

The introduction of a research study often proves difficult to write. It requires a great deal of thought, effort, and skill to convey to the reader the study's significance. If it is poorly done, the reader may not bother to read the rest of the study.

The problem statement and the research hypotheses or questions commonly appear in most research studies, whether they are theses or dissertations, journal articles, or research grants. Operational definitions, assumptions, limitations and delimitations, and the significance of the study usually appear explicitly only in theses and dissertations. Their purpose is to help (or force) the researcher to succinctly define and delimit the research problem. Operational definitions specifically describe how certain terms (especially the dependent variables) are being used in a particular study. Assumptions identify the basic conditions that must be assumed to exist for the results to have credibility. Delimitations relate to the scope of the study imposed by the researcher, such as the number and characteristics of the participants, the treatment conditions, and the specific dependent variables that are used and how they are measured. Limitations are possible influences on the results that are consequences of the delimitations or that cannot be completely controlled.

The section on the significance of the study forces the researcher to address the inevitable question of the study's worth. It should be a continuation of the introduction in terms of the contextual flow. It usually calls attention to the relationship (and differences) between this study and previous ones, controversies and gaps in the literature, and the contribution that this study might make to the practitioner, to existing theoretical models, or to both.

✔ Check Your Understanding

1. For each of the following brief descriptions of studies, write a title, the purpose or purposes, and three research hypotheses.

The researchers assessed the following:

   a. Skill acquisition of three groups of fourth-grade boys and girls who had been taught by different teaching styles (A, B, and C).

   b. Self-concept of two groups of boys (a low-strength group and a high-strength group) before and after a strength-training program.

   c. Body composition (estimated percentage of fat) using the electrical impedance analysis method on participants at normal hydration and again after they had dehydrated.

   d. Grade-point averages of male and female athletes of major and minor (club) sports from large universities and small private colleges.

2. Locate five articles from research journals, and for *each* try to determine (a) the hypotheses, if not stated, (b) the independent variable(s), (c) the dependent variable(s), (d) an operational definition for the dependent variable, (e) at least two delimitations, (f) at least two limitations, and (g) a basic assumption.

Formulating the Method

*The difference between failure and success is doing
a thing nearly right and doing a thing exactly right.*
—Edward Simmons

■　■　■　■　■　■　■

The previous chapter provided an overview of the introduction in the proposal for a thesis or dissertation. As already indicated, the journal format typically includes the literature review (see chapter 2 in this book) as part of the introduction. (If the chapter format is used, the literature review may appear as a separate chapter or as part of the introduction.) Regardless, once the introduction has been completed, the researcher must describe the methodology for the research. Typically, this section is labeled "Method." We give an overview of the four parts of the method section here:

1. Participants
2. Instruments or apparatuses
3. Procedures
4. Design and analysis

For our purposes, let's assume that the journal format is used and that the literature review is included in the introduction of your thesis or dissertation, followed by the method section. Much of the remainder of this book is focused on the method:

- Important aspects of the study: who the participants are, instruments, procedures, design and analysis (described in this chapter)
- How to measure and analyze the results (part II)
- How to design the study (part III)

The purpose of the method section is to explain how the study was conducted. The standard rule is that the description should be thorough enough for a competent researcher to reproduce the study.

Science as we know it today grew out of the murky lore of the Middle Ages (e.g., sorcery and religious ritual).

> **B**ut while witches, priests, and chiefs were developing taller and taller hats, scientists worked out a method for determining the validity of their experimental results: they learned to ask, Are they reproducible?—that is, would anyone using the same materials and methods arrive at the same results? (Scherr, 1983, p. ix)

For example, if a 10-kg and a 5-kg iron ball were dropped from the Empire State Building by King Kong 40 years ago or by Arnold Schwarzenegger today, they should strike the ground simultaneously on both occasions.

How to Present Methodological Details

Dissertations and theses differ considerably from published articles in the methodological details provided. However, when using the journal format, the additional materials should be placed in an appendix. Journals try to conserve space, but space is no issue in a dissertation or thesis. Thus, while standard techniques in a journal article are referenced only to another published study (in an easily obtainable journal), a thesis or dissertation should provide considerably greater detail in the appendix. Note that we indicated that a technique could be referenced to an easily obtainable journal. When writing for publication, use common sense in this regard. Consider, for example, this citation:

> Farke, F.R., Frankenstein, C., & Frickenfrack, F. (1921). Flexion of the feet by foot fetish feet feelers. *Research Abnormal: Perception of Feet, 22,* 1–26.

By most standards, this citation would not be considered easily obtainable. Therefore, if you are in doubt, give the details of the study or technique.

Furthermore, because theses and dissertations have appendixes, much of the detail that would clutter and extend the method section should be placed there. Examples include exact instructions to **participants,** samples of tests and answer sheets, diagrams and pictures of equipment, sample data-recording sheets, and informed-consent agreements.

participants

Individuals who are used as subjects in the study. In APA style, the term *participants* is used rather than *subjects.*

Why Planning the Method Is Important

The purpose of planning the method is to eliminate any alternative or rival hypotheses. This really means that when you design the study correctly and the results are as predicted, the only explanation is what you did in the research. Using a previous example to illustrate, our hypothesis is, "Shoe size and mathematics performance are positively related during elementary school." To test this hypothesis, we go to an elementary school, measure shoe sizes, and obtain standardized mathematics performance scores of the children in grades 1–5. When we plot these scores, they appear as in figure 4.1, each dot representing a single child. Moving from the dot to the x axis shows shoe size, while the y axis shows math performance. "Look!" we say. "We are correct. As shoe sizes get larger, the children's mathematics performance increases. Eureka! All we need to do is buy the children bigger shoes and their mathematics performance will improve." But wait a minute! We have overlooked two things. First, there is a rival explanation: Both shoe size and mathematics performance are related to age. That really explains the relationship. As the kids get older, their feet get bigger and they perform better on mathematics tests. Second, just because two things are related does not mean that one causes the other. Correlation does not prove causation. Obviously, we cannot improve children's math performance by buying them bigger shoes.

In research we want to use the **MAXICON principle:** Maximize true variance, or, increase the odds that the real relationship or explanation will be discovered; minimize error variance, or, reduce all the mistakes that could creep into the study to disguise the true relationship; and control extraneous variance, or, make sure that rival hypotheses are not the real explanations of the relationship.

MAXICON principle

A method of controlling any explanation for the results except the hypothesis the researcher intends to evaluate. This is done by maximizing true variance, minimizing error variance, and controlling extraneous variance.

Two Principles for Planning Experiments

In an interesting paper Cohen (1990) puts forward two principles that make good sense when planning experiments. The first is *less is more.* Of course, this seldom applies to the number of participants for a study, but it does apply to other aspects. For example, graduate students want to conduct meaningful studies that address and solve important problems. To do this, they often plan complex studies with many independent and dependent variables. From one perspective, this is good: The world of physical activity is truly complex. Students frequently start with useful ideas to study, but the ideas become so cumbersome that the study often fails because of sheer complexity. Carefully evaluate the number of independent and dependent variables that are practically and theoretically important to your study. Don't let anyone convince you (except your major professor, of course, who can convince you of anything) to add additional variables just to see what happens. This complicates your study and causes all types of measurement and statistical problems.

This idea leads to the second principle from Cohen (1990): *Simple is better.* This statement is true from the design to the treatments to the analysis to displaying data to interpreting results. Keep your study straightforward so that when you find something, you can understand and interpret what it means. Understanding your data is really an important concept. Although fancy statistical programs are nice and informative, there is no substitute for plotting data graphically and evaluating it carefully. Summary statistics (e.g., mean, standard deviation) are very helpful and informative, but they are no substitute for looking at the original data and how it is distributed. Summary statistics may not show us what we really need to know: things that become evident when we look at a graphic display of the data.

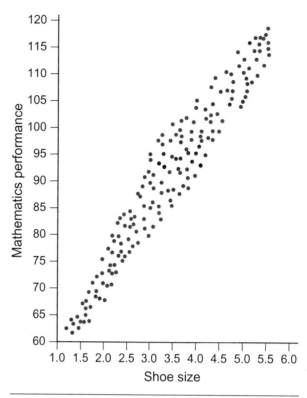

Figure 4.1 Relationship of academic performance and shoe size of children in grades 1–5.

Describing Participants

This section of the method of a thesis or dissertation describes how and why the participants were selected and which of their characteristics are pertinent to the study. These are questions to consider when selecting participants:

- Are participants with special characteristics necessary for your research?
  - Age (children, elderly)
  - Sex (females, males, or both)
  - Level of training (trained or untrained)
  - Level of performance (experts or novices)
  - Size (weight, fatness)
  - Special types (athletes, cyclists, runners)
- Can you obtain the necessary permission and cooperation from the participants?
- Can you find enough participants?

Of course, you want to select participants who will potentially respond to the treatments and measures used in the study. For example, if you want to see the results of training a group of children in overhand throwing, selecting expert 12-year-old baseball pitchers as participants will not likely produce a change in measures of throwing outcome. It would take an intense, long-term training program to have any influence on these participants. However, selecting children who are soccer players and have never played organized baseball would offer better odds for a training program to produce changes.

In experimental research the interaction among participants, measures, and the nature of the treatment program is essential in allowing the treatment program to have a chance to work (Thomas, Lochbaum, et al., 1997). If you select participants who have high levels of physical fitness, subjecting them to a moderate training program will not produce changes in fitness. Also, participants high in physical fitness will have a small range of scores on a measure of cardiovascular endurance (e.g., $\dot{V}O_2$max). For example, you will not find a significant correlation between $\dot{V}O_2$max and marathon performance in world-class marathon runners. This is because their range of scores in $\dot{V}O_2$max is small, as is their range of scores in marathon performance times. Because the range of scores is small in both measures, no significant correlation will be found. This does not mean that running performance and cardiovascular endurance are not related. It means that you have restricted the range of participants' performance so much that the correlation cannot be exhibited. If you had selected moderately trained runners (e.g., women who jog three times per week for 40 min each time), there would be a significant correlation between running performance on a 3-mile (5-km) run and $\dot{V}O_2$max. (We discuss procedures for selecting sample participants in chapter 6.)

What to Tell About the Participants

The exact number of participants should be given, as should any loss of participants during the time of study. In the proposal, some of this information may not be exact. For example, the following might describe the potential participants:

> ***P****articipants:* For this study 48 males, ranging in age from 21 to 34 years, will be randomly selected from a group ($N = 147$) of well-trained distance runners ($\dot{V}O_2$max = 60 ml · kg · min^{-1} or higher) who have been competitive runners for at least 2 years. Participants will be randomly assigned to one of four groups ($n = 12$).

Once the study is completed, details are available on the participants, so now this section might read as follows:

> ***P****articipants:* In this study 48 males, ranging in age from 21 to 34 years, were randomly selected from a group ($N = 147$) of well-trained runners ($\dot{V}O_2$max = 60 ml · kg · min^{-1} or higher) who had been competitive runners for at least 2 years. The participants had the following characteristics (standard deviations in parentheses): age, $M = 26$ years (3.3); height, $M = 172.5$ cm (7.5); weight, $M = 66.9$ kg (8.7); and $\dot{V}O_2$max, $M = 65$ ml · kg · min^{-1} (4.2). Participants were randomly assigned to one of four groups ($n = 12$).

The participant characteristics listed are extremely pertinent in an exercise physiology study but not at all, for example, in a study of equipment used by children on the playground. The nature of the research dictates the participant characteristics of interest to the researcher. Carefully think through the important characteristics you will report in your research. Look at related studies for ideas about important characteristics to report.

The characteristics of participants that you identify and report must be clearly specified. Note in the example that "well-trained runners" were exactly defined; that is, their $\dot{V}O_2$max must be equal to 60 ml · kg · min^{-1} or higher. Where participants of different ages are to be used is another good example. It is not sufficient just to say that 7-, 9-, and 11-year-olds will be the participants. How wide an age range is 7 years old: ±1 month, ±6 months, or what? In the proposal, you may say that 7-, 9-, and 11-year-olds will be included in the study. At the time of testing, each age will be limited to a range of ±6 months. Then, when the thesis or dissertation is actually written, it may read as follows:

> ***A****t* each age level 15 children were selected for this study. The mean ages are as follows (standard deviations in parentheses): the youngest group, 7.1 years (4.4 months); 9-year-olds, 9.2 years (3.9 months); and the oldest group, 11.2 years (4.1 months).

Protecting Participants

Most research in the study of physical activity deals with humans, often children, but it also includes animals. Therefore, the researcher must be concerned about any circumstances in the research setting or activity that could harm the humans or animals. In chapter 5, "Ethical Issues in Research and Scholarship," we provide considerable details on what the researcher must do to protect both humans and animals used in research. It is particularly important to obtain informed consent from humans and ensure the protection and care of animals. Sample forms for the use of humans and animals in research are provided in appendix C. (Such forms may be slightly different at your college or university.)

Describing Instruments

Information about the instruments, apparatuses, or tests used to collect data is used to generate the dependent variables in the study. Consider the following points when selecting tests and instruments:

- What is the validity and reliability of the measures?
- How difficult is it to obtain the measures?
- Do you have access to the instruments, tests, or apparatuses needed?
- Do you know (or can you learn) how to administer the tests or use the equipment?
- Do you know how to evaluate participants' test performance?
- Will the tests, instruments, or apparatuses yield a reasonable range of scores for the participants you have selected?
- Will the participants be willing to spend whatever time is required for you to administer the tests or instruments?

For example, in a sport psychology study, you are interested in how a group of university football players will be affected by a lecture on steroid use. In addition, you suspect that the players' attitudes might be modified by certain personality traits. So you select three tests—a steroid knowledge test, an attitude inventory about responsible drug use, and a trait personality measure—and administer them to all participants. The knowledge and attitude tests will probably be given before and after the lecture, and the trait personality test only before the lecture (traits should not change, and this test is being used to stratify players in some way). In the instrument section, you should describe the three tests and probably put complete copies of each in the appendix (see chapter 5 about the ethical use of standardized tests). You also should describe the reliability (consistency) and validity (what the test measures) information that is available on each test with appropriate citations. Then you should explain the scoring sheets (place a sample in the appendix) and the scoring methods (however, don't use inappropriate conversions for measurements like the ones in the box at right).

Another example is a motor behavior study in which participants' reaction and movement times are measured under various conditions. In the instruments section, you should describe the testing apparatus and provide a diagram or picture. If the apparatus interfaces with a computer to control the testing situation and data collection, you should describe the computer (brand name and model) and how the interface was made. At least a description of how the computer program operates should be included in the appendix (if not a complete copy

Incorrect Metric Conversions

1 trillion microphones = 1 megaphone

1 million bicycles = 2 megacycles

2,000 mockingbirds = 2 kilomockingbirds

1/2 lavatory = 1 demijohn

1 millionth of a fish = 1 microfiche

453.6 graham crackers = 1 pound cake

10 rations = 1 decoration

10 millipedes = 1 centipede

10 monologs = 5 dialogs

of the program). You should explain the dependent variables generated for reaction and movement time and give reliability estimates for these characteristics. All the necessary information could be presented by the appropriate use of both the instrument (or apparatus) part of the method section and the appendix, thus allowing the method section to flow smoothly.

Describing Procedures

In the procedures section, you should describe how the data were obtained, including all testing procedures for obtaining scores on the variables of interest. How tests were given and who gave them are important features. You should detail the setup of the testing situation and instructions given to the participants (although you may place some of this information in the appendix). If the study is experimental, then you should describe the treatments applied to the different groups of participants. Consider these points when planning procedures:

- Collecting the data
  - When? Where? How much time is required?
  - Do you have pilot data to demonstrate your skill and knowledge in using the tests and the equipment and how participants will respond?
  - Have you developed a scheme for data acquisition, recording, and scoring? (These are often computer controlled; to check whether your computer is female or male, see below.)
- Planning the treatments (in quasi-experimental and experimental studies)
  - How long? How intense? How often?
  - How will participants' adherence to treatments be determined?
  - Do you have pilot data to show how participants will respond to the treatments and that you can administer these treatments?
  - Have you selected appropriate treatments for the type of participants to be used?

One of our favorite summaries of the problems encountered and solutions proposed is presented on page 69. These statements are extracted from an article by Martens (1973).

Is Your Computer Female or Male?

Reasons to believe your computer is female:

1. No one but the Creator understands its internal logic.
2. "Bad command or file name" is about as informative as "If you don't know why I'm mad at you, I'm certainly not going to tell you!"
3. The language used to communicate with other computers is incomprehensible.
4. When you commit to one, all your additional money is spent on accessories.

Reasons to believe your computer is male:

1. It has a lot of data but is still clueless.
2. Rather than solving problems, it often *is* the problem.
3. When you commit to one, you find that if you had waited a little longer you could have obtained a better model.
4. In order to get its attention, you have to turn it on.

Errors in Experiments

Martens' method derives from the basic premise that

> In people experiments, people errors increase disproportionately to the contact people have with people.

It is obvious that the most logical deduction from this premise is

> To reduce people errors in people experiments, reduce the number of people.

Although this solution might be preferred for its elegant simplicity, its feasibility can be questioned. Therefore, the following alternative formulation warrants consideration:

> The contact between people testers and people subjects in people experiments should be minimized, standardized, and randomized

Reprinted, by permission, from R. Martens, 1973, "People errors in people experiments," *Quest* 20: 22.

The procedures section contains most of the details that allow another researcher to replicate the study. (However, these details must be useful, unlike the examples below.) Tuckman (1978) outlined these details, which generally include

- the specific order in which steps were undertaken,
- the timing of the study (e.g., time taken for different procedures and time between different procedures),
- instructions given to participants, and
- briefings, debriefings, and safeguards.

Details About Materials That Are Not Useful

The following were found on product labels.

On a bar of soap: Directions—use like regular soap.

On frozen dinners: Serving suggestions—defrost.

On a dessert box (printed on the bottom): Do not turn upside down.

On bread pudding: Product will be hot after heating.

On an electric iron: Do not iron clothes on body.

On a sleeping aid: Warning, may cause drowsiness.

On a can of peanuts: Warning, contains nuts.

On an airline packet of nuts: Open packet, eat nuts.

On a child's Superman costume: Wearing this garment does not enable you to fly.

Avoiding Methodological Faults

Unless you carefully pilot all your procedures, "quirk theory" (p. 70) will apply to your research. No single item in this book is more important than our advice to pilot all your procedures. Physical education, kinesiology, exercise science, and sport science have produced thousands of studies in which the discussions centered on methodological faults that caused the research to lack validity. It is worth repeating that placing post hoc blame on the methodology for inadequate results is unacceptable. Every thesis or dissertation proposal should present **pilot work** that verifies that all instruments and procedures will function as specified on the type of participants for which the research is intended. In addition, you must demonstrate that you can use these procedures and apparatuses accurately and reliably.

pilot work

Verifying that you can correctly administer the tests and treatments for your study using appropriate participants.

Law of Experiment: Quirk Theory

First Law: In any field of scientific endeavor, anything that can go wrong will go wrong.

Corollary 1: Everything goes wrong at one time.

Corollary 2: If there is a possibility of several things going wrong, the one that will go wrong is the one that will do the most damage.

Corollary 3: Left to themselves, things always go from bad to worse.

Corollary 4: Experiments must be reproducible; they should fail in the same way.

Corollary 5: Nature always sides with the hidden flaw.

Corollary 6: If everything seems to be going well, you have overlooked something.

Second Law: It is usually impractical to worry beforehand about interference; if you have none, someone will supply some for you.

Corollary 1: Information necessitating a change in design will be conveyed to the designer after, and only after, the plans are complete.

Corollary 2: In simple cases presenting one obvious right way versus one obvious wrong way, it is often wiser to choose the wrong way so as to expedite subsequent revisions.

Corollary 3: The more innocuous a modification appears to be, the further its influence will extend, and the more plans will have to be redrawn.

Third Law: In any collection of data, the figures that are obviously correct, beyond all need of checking, contain the errors.

Corollary 1: No one whom you ask for help will see the errors.

Corollary 2: Any nagging intruder who stops by with unsought advice will spot them immediately.

Fourth Law: If in any problem you find yourself doing a transfinite amount of work, the answer can be obtained by inspection.

To assist in the research suggested, the following rules have been formulated for the use of those new to this field.

Rules of Experimental Procedure

1. Build no mechanism simply if a way can be found to make it complex and wonderful.

2. A record of data is useful; it indicates that you have been busy.

3. To study a subject, first understand it thoroughly.

4. Draw your curves; then plot your data.

5. Do not believe in luck; rely on it.

6. Always leave room when writing a report to add an explanation if it doesn't work. (Rule of the way out.)

7. Use the most recent developments in the field of interpretation of experimental data.

   a. Items such as Finagle's constant and the more subtle Bougeurre factor (pronounced "Bugger") are loosely grouped, in mathematics, under constant variables or, if you prefer, variable constants.

   b. Finagle's constant, a multiplier of the zero-order term, may be characterized as changing the universe to fit the equation.

   c. The Bougeurre factor is characterized as changing the equation to fit the universe. It is also known as the "Soothing" factor; mathematically, somewhat similar to the damping factor, it has the characteristic of dropping the subject under discussion to zero importance.

   d. A combination of the two, the Diddle coefficient, is characterized as changing things so that the universe and the equation appear to fit without requiring any change in either.

Illinois Technograph, 1968.

Don't let your procedures become so complex that you end up like Santa Claus on Christmas Eve, with more to do than it seems possible to accomplish.

According to our calculations, Santa has about 31 hours to work on Christmas (this is because of different time zones and the earth's rotation). If he travels from east to west and there are 91.8 million households to visit, Santa must make 822.6 visits per second, allowing him 1/1,000 of a second to park, hop out of the sleigh, jump down the chimney, fill the stockings, distribute the remaining presents under the tree, eat whatever snacks have been left, get back up the chimney, get back into the sleigh, and move on to the next house.

Importance of Pilot Work

During our years as professors, editors, and researchers, we have seen abstracts of thousands of master's theses and doctoral dissertations. More than 75% of these research efforts are not publishable and make no contribution to theory or practice because of major methodological flaws that could have been easily corrected with pilot work. Sadly, this reflects negatively not only on the discipline and profession but also on the graduate students who conducted the research and the faculty who directed it. Yet nearly all the problems could have been corrected by increased knowledge of the topic, better research design, and pilot work on the procedures.

Graduate students frequently seek information about appropriate procedures from related literature (and they should). Procedures for intensity, frequency, and duration of experimental treatments are often readily available, as is information about testing instruments and procedures. However, it is important to remember that procedures in one area do not necessarily work well in another, as the example below illustrates.

Research Procedures May Not Generalize

Dr. I.M. Funded was a good life scientist who studied the biochemistry of exercise in a private research laboratory. He had also done several studies with a colleague in sport psychology to determine whether some biochemical responses he had found were factors in psychological responses to exercise. Thus, he had a firm grasp of some of the social science techniques as well as those of life science.

Unfortunately, Dr. Funded's funding ran out, and he lost his job. A friend of his was the superintendent of a large school district. Dr. Funded went to his friend, Dr. Elected, and said, "I am a good scientist well trained in problem-solving techniques. Surely, you must need someone like me in your administrative structure. In addition I have an undergraduate degree in physical education, so I am certified to teach, although I never have." Dr. Elected agreed to hire him as his teaching effectiveness supervisor because the school system was having difficulty identifying good teaching. Dr. Elected thought that perhaps a scientist with good problem-solving skills and the ability to make careful measurements could find a solution.

Dr. Funded decided that his first task was to identify some good teachers so he could determine the characteristics they possessed. He would use some of the techniques he had acquired from his colleague in sport psychology to identify good teachers. He had learned that questionnaires were effective in surveying large groups but that interviews were more valid. Dr. Funded drew a random sample of 6 schools from the 40 in the district. Then, he randomly selected six teachers in each school and interviewed them. He used a direct interview question: "Are you a good teacher?" All 36 indicated they were. So he went back to Dr. Elected, explained what he had done, and said, "You don't have a teaching problem. All of your teachers are good." (Of course, he noted there could be

(continued)

Research Procedures May Not Generalize   *(continued)*

some sampling error, but he was certain of his results.) Dr. Elected was not very happy with Dr. Funded's procedures and results and suggested that perhaps he needed more sophisticated techniques and strategies to identify good teachers.

Dr. Funded was slightly distraught but thought to himself, "I have always questioned the techniques of those psychologists anyway—I will return to my life science techniques to determine the answer." He went back to the previously selected 36 teachers with a plan to draw blood, sample urine, and do muscle biopsies (at four sites) once per week for four weeks. Immediately, 34 teachers said no, but 2 who were triathletes agreed to participate. Dr. Funded noted that the participant mortality rate was about normal for biopsy studies, so the data should be generalizable. He collected all the data, did the correct chemical analyses, and reported back to Dr. Elected. He indicated that effective teachers had 84% slow-twitch fiber, higher-than-average amounts of hemoglobin per deciliter of blood, and a specific profile of catecholamines (epinephrine and norepinephrine) in the urine. In addition, good teachers trained for at least 100 miles per week on the bicycle, 50 miles running, and 7,500 meters swimming. Dr. Funded sat back smugly and said, "Techniques for the life sciences can be applied to solve many problems." Dr. Elected said, "You are fired."

Describing Design and Analysis

Design is the key to controlling the outcomes from experimental and quasi-experimental research. The independent variables are manipulated in an attempt to judge their effects on the dependent variable. A well-designed study is one in which the only explanation for change in the dependent variable is how the participants were treated (independent variable). The design and theory have enabled the researcher to eliminate all rival or alternative hypotheses. The design requires a section heading in the method for experimental and quasi-experimental research. The plans for data analysis must also be reported. In most studies some type of statistical analysis is used, but there are exceptions (e.g., historical or qualitative research where other types of analysis occur).

Typically, the researcher explains the proposed application of the statistics. In nearly all cases, descriptive statistics are provided, such as means and standard deviations for each variable. If correlational techniques (relationships among variables) are used, then the variables to be correlated and the techniques are named, for example, "The degree of relationship between two estimates of percentage fat will be established by using Pearson *r* to correlate the sum of three skinfolds with underwater weighing." In experimental and quasi-experimental studies, descriptive statistics are provided for the dependent measures, and the statistics for establishing differences among groups are reported, for example, "A *t* test was used to determine whether youth league hockey players watching professional games produced more violent actions during their games than youth league players who did not watch professional hockey games."

The major problem that graduate students encounter in the description of statistical techniques is the tendency to inform everyone of their knowledge of statistics. Of course, that is not much of a problem for the new graduate student. But if your program of studies is a research-oriented one in which you take several statistics courses, your attitude may change rapidly.

> **H**iawatha, who at college majored in applied statistics, consequently felt entitled to instruct his fellow men on any subject whatsoever. (Kendall, 1959, 23)

The point is for you to describe your statistical analyses but not to instruct in their theoretical underpinnings and proper use.

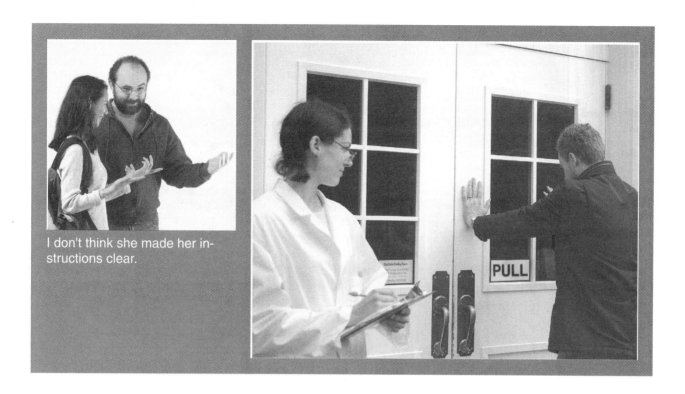

I don't think she made her instructions clear.

In terms of scientific progress, any statistical analysis whose purpose is not determined by theory, whose hypothesis and methods are not theoretically specified, or whose results are not related back to theory must be consid-ered, like atheoretical fishing and model building, to be hobbies. (Serlin, 1987, 371)

Establishing Cause and Effect

The establishment of cause and effect in an experimental study is much more than a statistical and design issue. The issue is one of logic. If the null hypothesis is not true, then what hypothesis is true? Of course, the scholar hopes the research hypothesis is true, but this is difficult to establish. In science the researcher seeks to explain that certain types of effects normally happen given specific circumstances or causes. For example, an effect may occur in the presence of something but not in its absence. Water boils in the presence of a high enough temperature but not in the absence of this specific temperature (given a specific air pressure).

People can (and do) differ in their opinions about what can be a cause–effect relationship or even whether such a relationship can exist. For example, whether you believe in universal laws, in destiny, in free will, or in an omnipotent deity is likely to influence your view of cause and effect.

Must causes be observable? If yes, then two criteria are needed to establish cause and effect. First is the method of agreement. If an effect occurs when both A and B are present, and A and B have only C in common, then C is the likely cause (or at least part of it). Second is the method of disagreement. If the effect does not occur in E and F when C is the only common element absent, then C is the cause (or part of it). Thus, it is clear that the researcher's reasoning influences the establishment of cause and effect because the researcher's beliefs set the stage for what may be considered a cause or even whether one can exist. This suggests that the researcher's theoretical beliefs as well as the study's design and analysis are essential factors in establishing cause and effect (for a more detailed discussion of causation, see White, 1990).

Manipulation Effects

When using a treatment in experimental or quasi-experimental research, the participant is involved in the treatment on a regular basis. For example, if participants in a home exercise program are to exercise daily for at least 40 minutes, how can the researcher be sure they are doing so? Some sort of quantitative manipulation effect needs to be carried out to be certain. Maybe the researcher will ask each participant to wear a movement recording device every day during the training. Amount of movement is recorded on this device so the researcher can determine that a vigorous program of exercise was undertaken for the correct amount of time. Even a manipulation check such as described has flaws: the participant could fail to exercise but ask someone else who was exercising to wear the device.

In any experimental study, researchers must have a plan as part of the methodology to check the manipulation of the independent variable in relation to the dependent variable. In a laboratory exercise study where there is one participant and one researcher, this is relatively easy. However, the more the research is moved into real-world settings, the more important manipulation checks become. In pedagogical, psychological, and sociological types of research related to physical activity, manipulation checks are nearly always essential to verify that participants perceived and responded to the treatments as the researcher intended.

Sometimes manipulation checks can be done in a qualitative way by questioning participants either during or following the individual treatment sessions or after the treatment period is over. While a quantitative manipulation check is preferred, a qualitative one is sometimes all that is possible or reasonable.

Fatal Flaws in Research

Every study proposal, particularly its methodology, should be carefully evaluated for "fatal flaws." That is, does this study lack a characteristic that will cause it to be rejected for publication regardless of the way the study is conducted or the outcomes? These are examples of "fatal flaws" questions to ask:

- Are all hypotheses logical when considering theory and study characteristics?
- Are all the assumptions you are making about the study good ones?
- Are the "right" participants in sufficient numbers selected for the study, especially in relation to the treatments and measurements used?
- Are the treatments intense enough and of sufficient length to produce the desired changes (especially considering the participants selected)?
- Are all measurements valid and reliable as well as appropriate for the planned study, especially considering the characteristics of participants and treatment levels in experimental studies?
- Are data collection and storage procedures well planned and carefully done?
- Are all extraneous variables being controlled?

Interaction of Participants, Measurements, and Treatments

Correlational studies use selected participants to evaluate the relationship among two or more measures, while experimental studies use selected participants to evaluate whether a specific treatment causes a specific outcome. In both instances, it is extremely important to select the "correct" participants, measurements, and treatments. In correlational studies, the measurements must assess the characteristics of interest and produce an appropriate range of scores on the measures; if they do not, then the researcher has reduced her or his chances of discovering important relationships. For example, if the question is to discover the relationship between anxiety and motor performance, then participants must exhibit a range of both anxiety and motor performance. If all participants have low levels of anxiety and have low motor skill, little opportunity exists to find a relationship.

The same idea is true in an experimental study to discover the influence of weight training on jumping performance. If all the participants are regular weight lifters who work on leg strength and all the participants are excellent jumpers, a weight-training program will not increase jumping performance. Said another way, the participants must show a range in performance on the measures of interest in correlational studies, and the measures must capture the characteristics that the researcher hopes to assess. In experimental studies, the participants must be at the "right" level to respond to the treatments, the treatments must be intense enough and frequent enough to produce the changes that the researcher hopes to see, and the measuring instruments must be responsive to the treatment changes in the participants.

Summary

This chapter provided an overview of the method for the research study. We identified the major parts as participants, instruments or apparatuses, procedures, and design and analysis. The four parts of the method section and their major purposes are to eliminate alternative or rival hypotheses and to eliminate any explanation for the results except the hypothesis that the researcher intends to evaluate. The MAXICON principle shows the way to accomplish this: (a) Maximize the true or planned sources of variation, (b) minimize any error or unplanned sources of variation, and (c) control any extraneous sources of variation. In later chapters, we detail how to do this from the viewpoints of statistics and measurement (part II) and of design (part III). We explain the final sections of the thesis or dissertation in part IV.

☑ Check Your Understanding

1. Find an experimental study in *Research Quarterly for Exercise and Sport* and critique the method section. Comment on the degree to which the author provided sufficient information concerning the participants (and informed consent), instruments, procedures, and design and analysis.

2. Locate a survey study and compare and contrast its method description with that in the study in problem 1.

Ethical Issues in Research and Scholarship

When choosing between two evils, I always like to try the one I've never tried before.—Mae West

■   ■   ■   ■   ■   ■   ■

As a graduate student, you will encounter a number of ethical issues in research and scholarship. In this chapter we draw your attention to many of these issues and provide a framework for discussion and decision making. However, the choices are not always clear-cut. The most important aspect of making good decisions is to have good information and to obtain advice from trusted faculty members. The major topics to be presented include misconduct in science, working with faculty and other graduate students, and using humans and animals as participants in research.

Seven Areas of Scientific Dishonesty

The White House Office of Science and Technology Policy has defined *scientific misconduct* for all U.S. agencies:

> Scientific misconduct is fabrication, falsification, or plagiarism in proposing, performing, or reviewing research, or in reporting research results. (*Federal Register,* October 14, 1999)

In this section we discuss issues in scientific misconduct that are generally pertinent to all scholarly areas of physical activity. We have also supplied a list of references on scientific misconduct as well as some situations and questions about honesty in science.

Shore (1991) identified seven areas in which scientific dishonesty might occur; each is discussed in the following subsections. A 1993 issue of *Quest* (Thomas & Gill [Eds.]) includes several thought-provoking articles on ethics in the study of physical activity, and Drowatzky (1996) is a valuable resource on the topic. The list on pages 78–79 provides additional readings.

Plagiarism

Plagiarism means using the ideas, writings, or drawings of others as your own. Of course, this is completely unacceptable in the research process (including writing). Plagiarism

References on Misconduct in Science: Readings for Graduate Students

American Association of University Professors. (1981). Statement on professional ethics. *New Directions for Higher Education, 33,* 83–85.

Anderson, A. (1988). First scientific fraud conviction. *Nature, 335,* 389.

Association of American Medical Colleges. (1992). *Beyond the "framework": Institutional considerations in managing allegations of misconduct in research.* Washington, DC: Association of American Medical Colleges.

Bell, R. (1992). *Impure science: Fraud, compromise, and political influences in scientific research.* New York: Wiley.

Callahan, J.C. (Ed.). (1988). *Ethical issues in professional life.* New York: Oxford University Press.

Chalmers, I. (1990). Underreporting research is scientific misconduct. *Journal of the American Medical Association, 263,* 1405–1408.

Dickersin, K. (1990). The existence of publication bias and risk factors for its occurrences. *Journal of the American Medical Association, 263,* 1385–1389.

Engler, R.L., Covell, J.W., Friedman, P.J., Kitcher, P.S., & Peters, R.M. (1987). Misrepresentation and responsibility in medical research. *New England Journal of Medicine, 317,* 1383–1389.

Ethics in the study of physical activity. (1993). Special issue. *Quest 45*(1).

Evans, J.T., Nadjari, H.I., & Burchell, S.A. (1990). Quotational and reference accuracy in scientific journals: A continuing peer review problem. *Journal of the American Medical Association, 263,* 1353–1357.

Federal Register. (1991). Federal policy for the protection of human subjects: Notices and rules. *56,* 28001–28032.

Friedman, P.J. (1988). Research ethics, due process, and common sense. *Journal of the American Medical Association, 260,* 1937–1938.

Friedman, P.J. (1990). Correcting the literature following fraudulent publication. *Journal of the American Medical Association, 263,* 1416–1419.

Garfield, E., & Welljams-Dorof, A. (1990). The impact of fraudulent research on the scientific literature: The Stephen E. Breuning case. *Journal of the American Medical Association, 263,* 1424–1426.

Goodstein, D. (1992). What do we mean when we use the term "science fraud"? *Scientist,* March 2, 11–12. Reprinted from *Windows,* fall 1991, p. 7.

Jayarama, K.S. (1990). Scientific ethics: Accusations of "paper recycling." *Nature, 334,* 187.

Kimmel, A.J. (1988). *Ethics and values in applied social research.* Newbury Park, CA: Sage.

Klotz, I.M. (1986). *Diamond dealers and feather merchants: Tales from the sciences.* Boston: Birkhauser.

Kohn, A. (1988). *False prophets: Fraud and error in science and medicine.* New York: Basil Blackwell.

LaPidus, J.B., & Mishkin, B. (1991). Values and ethics in the graduate education of scientists. In W.W. May (Ed.), *Ethics and higher education,* pp. 283–298. New York: Macmillan.

Loeb, J.M., Hendee, W.R., Smith, S.J., & Schwartz, M.R. (1989). Human vs. animal rights: In defense of animal research. *Journal of the American Medical Association, 262,* 2716–2720.

Mallon, T. (1989). *Stolen words: Forays into the origins and ravages of plagiarism.* New York: Ticknor and Fields.

Mishkin, B. (1988). Responding to scientific misconduct: Due process and prevention. *Journal of the American Medical Association, 260,* 1932–1936.

Mooney, C.J. (1992). Critics question higher education's commitment and effectiveness in dealing with plagiarism. *Chronicle of Higher Education, 38*(23), A16, A18.

Mooney, C.J. (1992). Plagiarism charges against a scholar can divide experts, perplex scholarly societies, and raise intractable questions. *Chronicle of Higher Education, 38*(23), A1, A14, A16.

National Academy of Sciences, National Academy of Engineering, and Institute of Medicine. (1993). *Responsible science: Ensuring the integrity of the research process.* Vol. 2. Washington, DC: National Academy Press.

Pfeifer, M.P., & Snodgrass, G.L. (1990). The continued use of retracted, invalid scientific literature. *Journal of the American Medical Association, 263,* 1420–1423.

Pope, K.S., & Vetter, V.A. (1992). Ethical dilemmas encountered by members of the American Psychological Association. *American Psychologist, 47,* 397–411.

Responsibilities of awardee and applicant institutions for dealing with and reporting possible misconduct in science. (1989). *Federal Register, 54*(151), 32446–32451.

Rudolph, J., & Brackstone, D. (1990). Too many scholars ignore the basic rules of documentation. *Chronicle of Higher Education, 36,* 11 April, A56.

Schurr, G.M. (1982). Toward a code of ethics for academics. *Journal of Higher Education, 53,* 318–334.

Taubes, G. (1993). *Bad science: The short life and weird times of cold fusion.* New York: Random House.

Whistleblowing and scientific misconduct. (1993). Special issue. *Ethics and Behavior, 3*(1).

Wrather, J. (1987). Scientists and lawyers look at fraud in science. *Science, 238,* 813–814.

carries severe penalties at all institutions. A researcher who plagiarizes work carries a stigma for life in his or her profession. No reward is worth the risk involved.

On occasion a graduate student or faculty member can inadvertently be involved in plagiarism. This generally occurs on work that is coauthored. If one author plagiarizes material, the other could be equally punished even though he or she is unaware of the plagiarism. Although there is no surefire means of protection (except not working with anyone else), never allow a paper with your name on it to be submitted (or revised) unless you have seen the complete paper in its final form.

In scientific writing, originality is also important. Common practice is to circulate manuscripts and drafts of papers among scholars (which are often shared with graduate students) who are known to be working in a specific area. If ideas, methods, findings, and so on are borrowed from these, proper credit should always be given.

plagiarism

Using ideas, concepts, writings, or drawings of others as your own; cheating.

Fabrication and Falsification

There are approximately 40,000 journals that together publish more than one million papers annually (Henderson, 1990). Thus, it is not surprising to learn that scientists have occasionally been caught making up or altering research data. Of course, this is completely unethical, and severe penalties are imposed on individuals who are caught. Pressures have been particularly intense in medical and health-related research because such research is often expensive, requires external funding, and involves risk. It seems easy to make a little change here or there or to make up data because "I only need a few more participants, but I am running out of time." The odds of being detected in these types of actions are high, but even if you should get away with it, you will always know you did it, and you will probably put other people at risk because of your actions.

Although graduate students and faculty may knowingly produce fraudulent research, established scholars are sometimes indirectly involved in scientific misconduct. This may occur when they work with other scientists who produce fraudulent data that follow the predicted outcomes (e.g., as in a funded grant where the proposal suggests what outcomes are probable). In these instances, the established scholar sees exactly what he or she expects to see in the data. Because this verifies the hypotheses, the data are assumed to be acceptable. One of the most famous cases was when Nobel laureate David Baltimore coauthored a paper with principal authors Thereza Imanishi-Kari and David Weaver that was published in *Cell* in April 1986. Baltimore checked the findings, but he saw in the data the expected outcomes and agreed to submit the paper. The fact that the data were not accurate subsequently led to Baltimore's resignation as president of Rockefeller University. Thus, even though Baltimore was not the paper's principal author, his career was seriously damaged by this incident.

Falsification can also occur with cited literature. Graduate students should be careful how they interpret what an author says. Work of other authors should not be "bent" to fit projected hypotheses. This is also a reason that graduate students should read original sources instead of relying on the interpretations of others, as those interpretations may not follow the original source closely.

Nonpublication of Data

Some data are not included because they do not support the desired outcome. This has sometimes been called "cooking" data. There is a very thin line between eliminating "bad" data and "cooking" data. Bad data should be caught, if possible, at the time of data acquisition. For example, if a test value seems too large or small and the researcher checks the instrument and finds it out of calibration, eliminating this bad data is good research practice. However, deciding that a value is inappropriate when data are being analyzed and changing it is "cooking" data.

outlier

Unrepresentative score; a score that lies outside of the normal scores.

Another term applied to unusual data is **outlier.** Some people have called such data "outright liars," suggesting that the data are bad. However, extreme values now are sometimes "trimmed." Just because a score is extreme does not mean it is based on bad data. Although extreme scores can create problems in data analysis, trimming them *automatically* is a poor practice (see chapter 10 on nonparametric analysis for one solution to this problem).

The most drastic instance in this category is the failure to publish results that do not support projected hypotheses. Journals are often accused of a publication bias, meaning that only significant results are published, but authors should publish the outcomes of solid research regardless of whether findings support projected hypotheses. The results from well-planned studies based on theory and previous empirical data have important meaning regardless of whether the predicted outcomes are found or not.

Faulty Data-Gathering Procedures

A number of unethical activities can occur at the data-gathering stage of a research project. In particular, graduate students should be aware of issues such as these:

- Continuing with data collection from participants who are not meeting the requirement of the research (e.g., poor efforts, failure to adhere to agreements about diet, exercise, rest)
- Malfunctioning equipment
- Inappropriate treatment of participants (e.g., failure to follow the guidelines from the Human Subjects Committee)
- Recording data incorrectly

For example, a doctoral student we know was collecting data on running economy in a field setting. Participants returned several times to be videotaped while repeating a run at varying stride lengths and rates. On the third day of testing, a male runner performed erratically during the run. When the experimenter questioned him, she learned he had been out drinking with his buddies very late on the previous evening and he was really hungover. Of course, she wisely discarded the data and scheduled him for another run several days later. Had she not noted the unusual nature of his performance and questioned him carefully, she would have included data that would probably have skewed her results because the participant was not adhering to previously agreed-upon conditions of the study.

Poor Data Storage and Retention

Data must be stored and maintained as originally recorded and not altered. All original records should be maintained so that the original data are available for examination. Federal agencies and most journals suggest that original data should be maintained for at least three years following publication of the results.

Misleading Authorship

A major ethical issue among researchers involves joint research projects or, more specifically, the publication and presentation of joint research efforts. Generally, the order of authorship for presentations and publications should be based on the researchers' contributions to the project. The first, or senior, author is usually the researcher who developed the idea and the plan for the research. Second and third authors are normally listed in the order of their contributions (see Fine & Kurdek, 1993, for a detailed discussion and case studies).

Although this sounds easy enough, authorship decisions are difficult at times. Sometimes researchers make equal contributions and decide to flip a coin to determine who is listed first. In fact, the (Thomas and Nelson) order of authorship for the first several editions of this book was decided in that manner. We team-taught a research methods course for several years while we were on the faculty at Louisiana State University. We contributed equally to this book. But before we began, we tossed a coin to decide who would be listed as first author (at the time, Nelson was unaware of Thomas' extreme skill in games of chance; he subsequently learned that lesson well). Note the phrase "before we began." This is a good procedure to follow. Decide the order of authors at the beginning of a collaborative effort. This saves hard feelings later, when everyone may not agree on whose contribution was most important. If the contributions of various authors change over the course of the research project, discuss a change at that time. In fact, we have continued to decide the authorship order for this latest edition as we added a new author (Silverman), and we have agreed on changes well into the future.

A second issue is who should be an author (see Crase & Rosato, 1992, for a discussion of the changing nature of authorship). Studies occasionally have more authors than participants. In fact, sometimes the authors are also the participants. When you look at what participants must go through in some research studies, you can see why only a major professor's graduate students would allow such things to be done to them. Even then, they insist on being listed as authors as a reward. More seriously, these two rules should help define authorship:

- **Technicians do not necessarily become joint authors.** Graduate students sometimes feel that, because they collect the data, they should be coauthors. Only when graduate students contribute to the planning, analysis, and writing of the research report are they entitled to be listed as coauthors. Even this rule does not apply to grants that pay graduate students for their work. A good major professor involves his or her graduate students in all aspects of his or her research program; thus, these students frequently may serve as coauthors and technicians.

- **Authorship should involve only those who contribute directly to the specific research project.** This does not necessarily include the laboratory director or a graduate student's major professor. The only thing we advocate by the chain letter on page 82 is the humor.

Chain Letter to Increase Publications

Dear Colleague:

We are sure you are aware of the importance of publications in establishing yourself and procuring grants, awards, and well-paying academic positions or chairpersonships. We have devised a way in which your curriculum vitae can be greatly enhanced with very little effort.

This letter contains a list of names and addresses. Include the top two names as coauthors on your next scholarly paper. Then remove the top name and place your own name at the bottom of the list. Send the revised letter to five colleagues.

If these instructions are followed, by the time your name reaches the top of the list, you will have claim to coauthorship of 15,625 refereed publications. If you break this chain, your next 10 papers will be rejected as lacking in relevance to "real-world" behavior. Thus, you will be labeled as ecologically invalid by your peers.

Sincerely,

List as coauthors: Jerry R. Thomas, Professor
Jerry R. Thomas   Jack K. Nelson, Professor
Jack K. Nelson
I.M. Published
U.R. Tenured
C.D. Raise

Unacceptable Publication Practices

The final concern deals with joint publications, specifically those between the major professor and the graduate student. Major professors do (and should) immediately begin to involve graduate students in the major professor's research program (see Zelaznik, 1993, for a discussion). When this happens, the general guidelines we suggested earlier apply. However, two conflicting forces are at work. A professor's job is to foster and develop students' scholarly ability. However, pressure is increasing on faculty to publish so that they can obtain the benefits of promotions, tenure, outside funding, and merit pay. Being the first (senior) author is a benefit in these endeavors. As a result, faculty members want to be selfless and to assist students, but they feel the pressure to publish. This may not be a major issue for senior faculty, but it certainly is for untenured assistant professors. As mentioned previously, there are no hard-and-fast rules other than everyone agreeing before the research is undertaken.

The thesis or dissertation is a special case. By definition, this is how graduate students demonstrate their competence to receive a degree. Frequently, for the master's thesis, the major professor supplies the idea, design, and much of the writing and editing. In spite of this, we believe it should be regarded as the student's work. The dissertation should always be regarded as the student's work. However, second authorship for the major professor on either the thesis or the dissertation is acceptable under certain circumstances. The American Psychological Association (2001) has defined these circumstances adequately, and we recommend the use of their guidelines:

- Only second authorship is acceptable for the dissertation supervisor.

- Second authorship may be considered obligatory if the supervisor designates the primary variables, makes major interpretive contributions, or provides the database.

- Second authorship is a courtesy if the supervisor designates the general area of concern, is substantially involved in the development of the design and measurement procedures, or substantially contributes to the writing of the published report.

- Second authorship is not acceptable if the supervisor provides only encouragement, physical facilities, financial support, critiques, or editorial contributions.

- In all instances, agreement should be reviewed before the writing for publication is undertaken and at the time of the submission. If disagreements arise, they should be resolved by a third party using these guidelines.

Authors must also be careful about **dual publication.** Sometimes this is legitimate; for example, a scientific paper published by one journal may be reprinted by another journal or in a book of readings (this should always be noted). Authors may not, however, publish the same paper in more than one copyrighted original research journal. But what constitutes the "same paper"? Can more than one paper be written from the same set of data? The line is rather hazy. For example, Thomas (1986, iv–v) indicated that

Frequently, new insights may be gained by evaluating previously reported data from a different perspective. However, reports of this type are always classed as research notes whether the reanalysis is undertaken by the original author or someone else. This does not mean that reports which use data from a number of studies (e.g., meta-analyses, power analyses) are classed as research notes.

Generally, good scientific practice is to publish all the appropriate data in a single primary publication. For example, if both psychological and physiological data were collected as a result of a specific experiment on training, publishing these separately may not be appropriate. Frequently, the main finding of interest is the interaction between psychological and physiological responses. But in other cases, the volume of data may be so large as to prohibit an inclusive paper. Sometimes the papers can be published as a series; at other times they may be completely separate. Another example is a large-scale study in which a tremendous amount of information is collected (e.g., exercise epidemiologic or pedagogical studies). Usually, data are selected from the computer records (or videotapes) to answer a specific set of related questions for a research report. Researchers may then use a different part (or even an overlapping part) of the data set to address another set of questions. This results in more than one legitimate publication from the same data set. However, authors should identify that more than one paper has been produced from these data. Researchers should follow these general types of rules, or they may be viewed as lacking scientific objectivity in their work and certainly as lacking in modesty (Day, 1988).

Most research journals require that the author include a statement that the paper has not been previously published or submitted elsewhere while the journal is considering it. Papers published in one language may not be published as an original paper in a second language.

> **dual publication**
>
> Having the same scientific paper published in more than one journal or other publication is generally unethical.

Ethical Issues Regarding Copyright

Graduate students should be aware of copyright regulations and the concept of fair use as it applies to educational materials. Copyrighted material is often used in theses and dissertations, and this is acceptable if the use is fair and reasonable. Often graduate students want to use a figure or table in their thesis or dissertation. If you use a table or figure from another source, you must seek permission (see page 84 for a sample letter) from the copyright holder (for published papers, this is usually the author but sometimes the research journal) and cite it appropriately (e.g., "used by permission of . . . ").

The concept of fair use has four basic rules:

1. *Purpose.* Is the use to be commercial or educational? More leeway is given for educational use such as theses, dissertations, and published research papers.

2. *Nature.* Is copying expected or not? Copying a journal article for your personal use is expected and reasonable. However, copying a complete book or standardized test is not expected and probably is a violation of the fair-use concept.

3. *Amount.* How much is to be copied? The significant issue is how important the copied part is.

4. *Effect.* How does copying affect the market for the document? Making a single copy of a journal article has little effect on the market for the journal, but copying a book (or maybe a book chapter) or a standardized test reduces royalties to the author and income for the publisher. That is not fair use.

Sample Copyright Permission Letter

Following is a sample copyright permission letter (altered for use with a thesis or dissertation) recommended by Human Kinetics, the publisher of this textbook.

Date

Permissions Editor (or name of an individual author)
Publisher (not needed if an individual)
Address

Dear _____:

I am preparing my thesis/dissertation, tentatively titled _____.

I would like permission to use the following material:

Title of article in journal, book, book chapter: _____

Author of article, book, or book chapter: _____

Title of journal or book (include volume and issue number of journal): _____

_____

Editor if edited book: _____

Year of publication: _____

Place of publication and name of publisher of book/journal: _____

_____

Copyright year and holder: _____

Page number(s) on which material appears: _____ ☐ to be reprinted ☐ to be adapted

A COPY OF THE MATERIAL IS ATTACHED.

Table, figure, or page number in my thesis/dissertation: _____

I request permission for nonexclusive rights in all languages throughout the world. I will, of course, cite a standard source line, including complete bibliographic data. If you have specific credit line requirements, please make them known in the space provided below.

_____

A duplicate copy of this form is for your files. Your prompt cooperation will be appreciated.

If permission is granted, please sign the release below and return to:

YOUR SIGNATURE: _____

YOUR NAME AND ADDRESS: _____

_____

- -

Permission granted, Signature: _____

Address: _____

_____ Date: _____

There are few standard answers about fair use. Fair use is a flexible idea (or alternatively, a statement that can't be interpreted rigidly). Better to be safe than sorry when using material in your thesis or dissertation. If you have any doubt, seek permission. University Microfilms (which produces *Dissertation Abstracts International,* the national warehouse for dissertations) reports that about 15% of dissertations require follow-up on copyright issues.

Model for Considering Scientific Misconduct

Intention is often used as the basis for differentiating between scientific misconduct and mistakes. Drowatzky (1993, 1996) noted that the following model is often suggested:

Scientific misconduct ➤ Sanctions

Scientific mistakes ➤ Remedial activities

Sanctions for Scientific Misconduct

Sanctions are often imposed on individuals who are fraudulent in their scientific work. Internal sanctions imposed by the researcher's university have included the following:

- Restriction of academic duties
- Termination of work on the project
- Reduction in professorial rank
- Fines to cover costs
- Separation from the university (either with or without loss of benefits)
- Salary freeze
- Promotion freeze
- Supervision of future grant submissions
- Verbal reprimands
- Letter of reprimand (either included or not included in the permanent record)
- Monitoring research with prior review of publications

In addition to internal university sanctions for scientific fraud, sanctions may be imposed by the agency that has funded the research, scholarly journals that published the work, and related scholarly or professional groups. In recent years, external sanctions have included the following:

- Revocation of prior publications
- Letters to offended parties
- Prohibition of obtaining outside grants
- Discontinuance of service to outside agencies
- Release of information to agencies and professional organizations
- Referral to legal system for further actions
- Fines to cover costs

Responsibilities of Graduate Students

As a graduate student, you must become concerned with ethical issues. Of course, the issues are important for much more than just the biological science areas (such as exercise physiology, biomechanics, motor behavior, health promotion and exercise, and exercise psychology) or behavioral science areas (such as sport sociology, sport philosophy, sport history, sport psychology, and physical education pedagogy). Fraud, misrepresentation, and inaccurate interpretation of data, plagiarism, unfair authorship issues, and unethical publications practices are problems that extend to any area of scholarship. Although these practices sometimes occur simply because the person is unethical, often pressures that exist in our system of higher education tempt researchers to behave unethically:

- The need to obtain external funding for research
- Pressure to publish scholarly findings
- The need to complete graduate degree work
- The desire to obtain rewards in higher education (e.g., promotion, merit raises)

Academic units should be encouraged to develop mechanisms for discussing ethical issues in scholarship with graduate students. This might include seminars for graduate students focused on these issues. But some systematic means is needed to bring the issues to the forefront for discussion.

Philosophical Positions Underlying Ethical Issues

One's basic philosophic position on ethical issues drives the decision-making process in research. Drowatzky (1993, 1996) has summarized the different ethical views that underpin the decision-making process:

- The individual is precious, and the individual's benefit takes precedence over the society.
- Equality is of utmost importance, and everyone must be treated equally.
- Fairness is the overriding guide to ethics, and all decisions must be based on fairness.
- The welfare of society takes precedence over that of the individual, and all must be done for the benefit of society.
- Truth, defined as being true, genuine, and conforming to reality, is the basis for decision making.

Of course, several statements in this list are in direct conflict with each other and will lead to substantially different decisions depending on one's view. Discussing and evaluating these statements and what they imply about decisions in scholarship should enhance graduate students' understanding of important issues.

Finally, reading some of the literature on fraud and misconduct in research is a sobering experience for anyone. A notable example is the special issue "Whistleblowing and Scientific Misconduct" of *Ethics and Behavior* (vol. 3, no. 1, 1993). This issue gives a fascinating account of the David Baltimore and Herbert Needleman cases, including overviews and responses by the whistleblowers and those accused of scientific misconduct. Drowatzky (1996) also offers numerous examples of misconduct and the problems associated with it.

Working With Faculty

Ethical considerations among researchers and ethical factors in the graduate student–major professor relationship are two important topics (see Roberts, 1993, for a detailed discussion). Major professors should treat graduate students as colleagues. If we want our students to be scholars when they complete their graduate work, then we should treat them like scholars from the start, for graduate students do not become scholars on receipt of a degree. By the same token, graduate students must act like responsible scholars. This means producing careful, thorough, and high-quality work.

Selecting a Major Professor

Students should try to select major professors who share their views in their area of interest. Master's students frequently choose an institution based on location or the promise of financial aid. Doctoral students, on the other hand, should select the institution they attend based on the program's quality and the faculty in their area of specialization (see Baxter, 1993–94, for a discussion). A listing and electronic link to doctoral programs in physical activity is available at www.aakpe.org. Do not choose your major professor hastily (or he or she may turn out like someone in the sidebar below). If you are already at the institution, carefully evaluate the specializations available in your interest areas. Ask questions about faculty and whether they publish in these areas. Read some of these publications and determine your interest. What financial support, such as laboratories and equipment, is given to these areas? Also, talk to fellow graduate students. Finally, talk with the faculty members to determine how effectively you will be able to work with them.

Ethics of the Workplace

accountant—Someone who knows the cost of everything and the value of nothing.

auditor—Someone who arrives after the battle and bayonets all the wounded.

banker—Someone who lends you his or her umbrella when the sun is shining and wants it back when it begins to rain.

consultant—Someone who takes the watch off your wrist and tells you the time.

economist—An expert who will know tomorrow why the things he or she predicted yesterday didn't happen today.

professor—A person who talks in someone else's sleep.

psychologist—A man who watches everyone else when a beautiful woman enters the room.

statistician—Someone who is good with numbers but lacks the personality to be an accountant.

We advocate a mentor model for preparing graduate students (particularly doctoral students) in kinesiology, physical education, exercise science, and sport science. For students to become good researchers (or good clinicians) requires a one-on-one student–faculty relationship. This means several things about graduate students and graduate faculty.

First, graduate students need to be full-time students to develop the research and clinical skills needed for success in research and teaching. They need to work with a mentor in their ongoing research program. This lends continuity to research efforts and pulls graduate students together into effective research teams. Theses and dissertation topics arise naturally from these types of settings. Additionally, more senior students become models and can offer assistance to novice graduate students. Expertise is acquired by watching experts, working with them, and then practicing the techniques acquired.

Conversely, for faculty members to be good mentors, they must have active research programs. This means that appropriate facilities and equipment must be available as well as time for faculty members to devote to research and graduate student mentoring. Potential graduate students should carefully investigate the situations into which they will place themselves, especially if they have a major interest in research (for a good description of mentors, see Newell, 1987).

If you are not yet enrolled at an institution, find out which institutions offer the specialties you are interested in. Request information from their graduate schools and departments and look at their Web site on the Internet. Read the appropriate journals (over the past 5–10 years) and see which faculty are publishing. After you narrow your list, explore financial support and plan a visit. Speak to the graduate coordinator for the department and the faculty in your area of interest. Sometimes you can meet faculty at conventions, such as those of AAHPERD (national or district meetings), the American College of Sports Medicine, the International Society for Biomechanics, and the North American Society for Psychology of Sport and Physical Activity.

After you select a major professor, you must select a committee. Normally the master's or doctoral committee is selected in consultation with your major professor. The committee selected should be one that can contribute to the planning and evaluation of your work, not one that might be the easiest. It is preferable to wait a semester or quarter or two (if you can) before selecting a final committee. This gives you the opportunity to have several potential committee members as teachers and to better evaluate your common interest.

Changing Your Major Professor

What happens, however, if you have a major professor (or committee member) who is not ideal for you? First, evaluate the reason. You need not be best friends, but it is important that you and your major professor are striving for the same goals. Sometimes students' interests change. Sometimes people just cannot get along. If handled professionally, however, this situation should not be a problem. Go to your major professor and explain the situation as you perceive it and offer him or her an opportunity to respond. Of course, the conflict may be more personal. If so, use an objective and professional approach. If you cannot, or if this does not produce satisfactory results, the best recourse may be to seek the advice of the graduate coordinator or department chairperson.

Protecting Human Participants

Most research in the study of physical activity deals with humans, often children. Therefore, the researcher must be concerned about any circumstances in the research setting or activity that could harm the participants. Harm should be interpreted to mean to frighten, embarrass, or negatively affect the participants (Tuckman, 1978). Of course, researchers always run the risk of creating a problem. What must be balanced is the degree of risk, the participants' rights, and the potential value of the research in contributing to knowledge, to the development of technology, and to the improvement of people's lives.

What Should Research Participants Expect?

Tuckman (1978) has summarized the participants' rights that experimenters must consider:

• **The right to privacy or nonparticipation.** This includes researchers' not asking for unnecessary information and their obtaining direct consent from adults and consent from parents for children (as well as the consent of the children themselves where appropriate).

• **The right to remain anonymous.** The researcher should explain that the study focuses on group data and that an ID number (rather than the participant's name) will be used to record data.

• **The right to confidentiality.** Participants should be told who will actually have access to original data by which the participants might be identified (keep this to as few people as possible).

• **The right to expect experimenter responsibility.** The experimenter should be well-meaning and sensitive to human dignity. If the participant is not told the purpose of the study (or is misled), the participant must be informed immediately after the completion of testing.

Qualitative research (discussed in detail in chapter 19) lends itself to some potential ethical problems because of the close, personal interaction with participants. The researcher often spends a great deal of time with participants, getting to know them and asking them to share their thoughts and perceptions. Griffin and Templin (1989) raised ethical concerns about whether to share field notes, how to protect a participant's self-esteem without compromising accuracy in the research report when the two are in conflict, and what to do if you are told about (or observe) something illegal or immoral while collecting data. Locke, Spirduso, and Silverman (2000) discuss these issues further.

There are no easy answers to situational ethics in **fieldwork.** Qualitative research sometimes deals with so-called deviants, such as drug addicts and unlawful motorcycle gangs. Informed consent is impossible in some circumstances. Punch (1986, 36) made this point when he described his research with police. The patrol car he was riding in was directed to a fight. As the policemen jumped out and started wrestling with the combatants, Punch wondered whether he was supposed to yell "freeze," thrust his head between the entangled limbs, and, Miranda-like, chant out the rights of the participants. Similarly, when Powermaker (cited by Punch) came face to face with a lynch mob, should she have flashed her academic identity card and explained to the crowd the nature of her presence? By these two examples we are not implying that qualitative researchers are exempt from considerations such as informed consent and deception. We are simply pointing out that certain types of qualitative research situations present special ethical problems. We invite you to read the discussion by Punch (1986) and consult some of the sources he cites concerning this issue.

People with disabilities present a special issue as research participants. A participant with a disability is protected under the Right to Privacy Act. Thus, institutions are prohibited from releasing the names of people with disabilities as potential research participants. The researcher must contact the institution about possible participants. The institution must then request permission from the participant or his or her parents to release the participant's name and the nature of the disability to the researcher. If the participant or parents approve, the institution allows the researcher to contact the participant or parents to seek approval for the particular research to be undertaken. Although this procedure is rather cumbersome and varies from state to state, individuals clearly have the right not to be cited in studies and labeled as "disabled" unless they so choose.

> **fieldwork**
>
> Methodology common in qualitative research in which data are gathered in natural settings.

Informed Consent

Consideration must be given to the protection of human participants. The researcher is required to protect the rights and well-being of participants in his or her study. The regulations detailing the procedures are published by the U.S. Department of Health and Human Services (45 CFR 46.101). Most institutions regulate this protection in two ways. First, researchers are required to complete some type of form describing their research. Sample form C.1 in appendix C is used for conducting research with human participants at Arizona State University.

It may be useful to computerize the research application form so that graduate students or faculty can answer the questions and print out the completed form. The researcher completes this form, attaches an abstract, and has it approved before beginning any work, including pilot work. The source of approval may vary. For example, some institutions may require that all forms be approved by a central committee. Other institutions may delegate approval for so-called standard types of research to a lower level (e.g., a college committee).

Typically, institutions require that you include your informed-consent form with your application to conduct human-participant research. Sample form C.2 is the model for the form used for adults at Arizona State University. If the participants are minors, then you must obtain their parents' permission (sample form C.3) and the children's assent (sample form C.4) if they are old enough to understand. Every university has slightly different guidelines for research with human subjects and for obtaining consent. In many instances these have been updated recently. You should get these guidelines early and make sure your informed-consent forms conform to local requirements. While we present these examples here, your university guidelines will be the ones you will have to meet before permission is granted to do the study.

Following are basic elements of informed consent as specified by the advisory committee of the *Research Quarterly for Exercise and Sport* (Thomas, 1983, 221):

- A fair explanation of the procedures to be followed, including an identification of those that are experimental.

She's pondering the premise that 100 monkeys in a room full of typewriters in enough time would come up with the complete works of Shakespeare, and she's wondering if the monkeys would be charged with plagiarism.

- A description of the attendant discomforts and risks.
- A description of the benefits to be expected.
- A disclosure of appropriate alternative procedures that would be advantageous for the participant.
- An offer to answer any inquiries concerning the procedures.
- An instruction that the participant is free to withdraw consent and to discontinue participation in the project or activity at any time. In addition, the agreement should contain no exculpatory language through which the participant is made to waive, or appear to waive, any legal right or to release the institution or its agents from liability or negligence.

The researcher is required to comply with any institutional guidelines for the protection of human participants and for informed consent. A description of this compliance should be included under the participant heading in the method section of the thesis or dissertation. Most journals also require a statement with regard to this issue. The form used for informed consent is normally placed in an appendix of the thesis or dissertation.

Protecting Animal Subjects

Matt (1993) in her article "Ethical Issues in Animal Research" points out that this is not a new issue, having been discussed in Europe for over 400 years and in the United States for over 100 years. The ground rules were established long ago when Descartes indicated it was justifiable to use animals in research because they cannot reason and are therefore "lower" in the order of things than humans (Matt, 1993). However, Bentham (1970) said the issue is not whether animals can think and reason but whether they perceive pain and suffer.

As Matt (1993) argues, far fewer animals are used in research than are slaughtered for food, held in zoos, and killed as unwanted pets in animal shelters. In fact, animal studies may have more stringent criteria for approval than studies with humans. Institutional review boards typically require that investigators demonstrate that animal studies add significant knowledge to the literature and are not replications, a requirement not placed on studies using humans as participants.

If animals are well treated, is their use in research justified? Matt (1993, 46-47) says yes if the purposes of the research fall into one of five categories:

> (1) drug testing, such as the development and testing of AIDS drugs; (2) animal models of disease, such as the development of animal models of arthritis, diabetes, iron deficiency, auto-immune dysfunction, and aging; (3) basic research, focused on examining and elucidating mechanisms at a level of definition not possible in human models; (4) education of undergraduate and graduate students in laboratories and lectures, with experience and information gained from the use of animal models; and (5) development of surgical techniques, used extensively in the training of medical students and the testing of new surgical devices and procedures.

A careful consideration of these categories suggests few suitable alternatives (but see Zelaznik, 1993, for a discussion).

If animals are used as subjects for research studies in exercise science, institutions require adherence to the *Guide for the Care and Use of Laboratory Animals,* published by the U.S. Department of Health and Human Services, as detailed in the Animal Welfare Act (PL 89-544, PL 91-979, and PL 04-279). Most institutions also support the rules and procedures for recommended care of laboratory animals as outlined by the American Association for Accreditation of Laboratory Animal Care.

All these documents recognize that for advancements to be made in human and animal research, animals must be used. These animals must be well tended, and if their use results in the animals being incapacitated or sacrificed, this must be done humanely. Sample form C.5 is the form used at Arizona State University to ensure compliance with all regulations involving the use of animals in research.

Summary

We discussed ethical issues that affect graduate students in their research and scholarly activities. We identified ethical issues and set the stage for you to think about and discuss these values as they influence your graduate and scholarly life.

Points that typically arise in scientific misconduct include plagiarism, fabrication and falsification of data, nonpublication of data, data-gathering problems, data storage and retention issues, authorship controversies, and publication practices. We discussed copyright issues in research and publication. We also presented the model most frequently used to deal with scientific misconduct and some of the internal and external sanctions that have been imposed on individuals found guilty of scientific misconduct.

We discussed ethical and procedural issues in working relationships. How should graduate students select a major professor and committee members? Should students seek new major professors or committee members if they are unable to work effectively or amicably with their original choices?

Finally, we discussed ethics and procedures in the use of human and animal participants. This included remembering participants' rights, protecting human and animal participants, and obtaining informed consent from human participants.

✔ Check Your Understanding

Our thanks to Dr. Philip Martin, professor of kinesiology at Pennsylvania State University, for developing these examples and allowing us to use them. Names and institutions used should not be taken seriously; we're just trying to bring humor to our friends.

Plagiarism

What is plagiarism? Can you recognize it when you see it?

> *Case Study.* In preparing his thesis introduction, Graduate Assistant Christina periodically takes multiple sentences verbatim from some of his sources (his attitude is, "I couldn't have written it better myself!").

Is he wrong to do this?

If he provides a reference to his sources at the end of his paragraph, is he still wrong?

Fabrication or Falsification of Data

> *Case Study.* Professor Wade has strength training data on 20 elderly participants. As he was madly processing data in an attempt to meet the ACSM abstract deadline, Professor Wade realized that his sample did not show a significant increase in strength. Being quite disappointed about this, he examined his data more closely and noticed that 15 participants appeared to increase strength substantially whereas 5 participants actually showed strength declines. Professor Wade concluded that these 5 participants must not have adhered to the training program, and he decides to drop them from the study. Using only data on the remaining 15 participants, he now can demonstrate statistically significant improvements in strength, and he writes his abstract based on those 15 participants.

Has Professor Wade acted ethically?

What is an outlier? How do you define an outlier?

How long should you keep raw data available for others to review following publication?

Are you obligated to provide your raw data for examination upon request?

Authorship of Presentations and Publications

> *Case Study.* Professor Singer is well-known for her research on the effects of exercise on bone density. In 1995 she was awarded a five-year research grant from NIH to study bone density. Graduate Assistant (GA) Martin begins work under Professor Singer's mentorship in 1997 and is immediately assigned the task of running one of the experiments on bone density that was outlined in the grant proposal. Professor Singer can never be found in the laboratory for data collection, but she regularly holds lab meetings during which she discusses progress on data collection with GA Martin. Upon completing data collection, GA Martin organizes and presents the data to Professor Singer, who is quite pleased with the data. Professor Singer then assigns GA Martin the task of drafting a manuscript based on the data. After several iterations, both are pleased with the final product and conclude that the manuscript is ready for submission. Unfortunately, they have had no conversations about authorship.

Should the authorship be Singer and Martin or Martin and Singer?

> *P*rofessor Singer indicates to GA Martin during the manuscript preparation phase that she would like to include GAs Powers, Cauraugh, Stelmach, and Thomas (the other four GAs in Singer's lab) as coauthors because they have been involved in other aspects of the funded research (after all, they need publications on their curricula vitae in order to secure prestigious postdoc positions).

Is this a reasonable request by Professor Singer? [As a side note, it is sometimes an interesting exercise to compute some oddball descriptors for manuscripts with long lists of authors. Consider (1) the ratio of words to author, (2) the ratio of lines to author, and (3) the ratio of participants to author, and then consider just how significant each author's contributions might be.]

If the project for which GA Martin collected the data was for his dissertation topic (an offshoot of Professor Singer's funded research line), should the authorship be Singer and Martin, Martin and Singer, or just Martin?

If Professor Singer hired Technician Magill to assist GA Martin with data processing (for the original scenario), should the authorship be Singer and Martin; Singer, Martin, and Magill; Martin and Singer; or Martin, Singer, and Magill?

How important is having some understanding about authorship at the beginning of the writing process?

Other Publication and Presentation Issues

Case Study. Professor Sharp, an expert in forensic biomechanics, presented a research paper on the relationship between footprint spacing and body size at the 1999 annual meeting of ACSM. Shortly after ACSM, during a moment of free time, he saw an announcement in one of his research magazines for a meeting titled "The Science of Forensic Biomechanics" sponsored by the Society of Police Detectives. He submits the same abstract used at ACSM and subsequently presents the same research paper at the SFB meeting. Right or wrong?

Case Study. Professor Sharp submits a manuscript titled "The Relationship Between Footprint Spacing and Body Size" to the *Journal of Biomechanics*. It is accepted and scheduled to be published in the March 2005 issue of *JoB*. He also has had an abstract on the same subject accepted for presentation at the May 2005 ACSM meeting. Is there anything wrong with presenting published data?

Case Study. Professor Sharp wants to change jobs and needs to strengthen his curriculum vitae. He submits his manuscript titled "The Relationship Between Footprint Spacing and Body Size" to the *Journal of Biomechanics*. In an effort to ensure publication of his research, Professor Sharp decides to submit the same manuscript to the *Journal of Forensic Science* while his *JoB* manuscript is still under review (after all, two chances are better than one). Right or wrong?

Case Study. Professor Sharp submits his manuscript titled "The Relationship Between Footprint Spacing and Body Size" to the *Journal of Biomechanics*. He then uses the same data but interprets them from a somewhat different perspective. He prepares another manuscript titled "The Relationship Between Footprint Spacing and Walking Speed" and submits it to the *Journal of Forensic Science* while his *JoB* manuscript is still under review. Is there anything wrong with this?

Case Study. Assistant Professor Roberts, a new faculty member at the University of Reallycold (Norway), has just completed a broad, multidisciplinary dissertation on the benefits of endurance training on psychological and physiological markers of health and well-being and biomechanical aspects of the running pattern. In each of three major areas (psychology, physiology, and biomechanics), she has four major dependent variables. Knowing that she needs a healthy publication record when she reaches the promotion and tenure review in five years, she decides to pursue 12 publications from her dissertation research ("The Benefits of Endurance Training on . . . "). How do you react to her strategy for dealing with publish or perish?

Case Study. Assistant Professor French, a new faculty member at the University of South Columbia, completed his doctorate at Big Time University under the direction of Dr. Samoht. His dissertation topic fit within the general scope of research activities being pursued by Dr. Samoht but reflected a unique focus, one that Dr. Samoht had not considered prior to French's work as a doctoral student. French submits a manuscript to *Medicine and Science in Sports and Exercise,* listing himself as the sole author of the paper and the University of South Columbia as his affiliation. Has French behaved ethically?

Has he given appropriate credit to Professor Samoht or Big Time University?

> *Case Study.* Professor Conan Barbarian is an icon in the field of gerontology. He is the director of the Institute of Gerontological Research at Jellystone University, a highly funded research lab in which several faculty, postdocs, and graduate students work. Professor Barbarian requires that he be listed as an author on all manuscripts based on research completed in the IGR.

Is Professor Barbarian justified in his demand, or is this an example of "ego gone wild"?

Changing Your Major Professor

> *Case Study.* Graduate Assistant (GA) Lee has strong interests in the mechanical behavior of muscle and was accepted into the exercise science doctoral program to study with Professor Silverman, an expert on muscle mechanics. After one year in the program, the chemistry between GA Lee and Professor Silverman is not so great. In addition, GA Lee notices that one of his GA colleagues is getting some travel support from her mentor, Professor Moran, an expert on muscle energetics. GA Lee wants to continue to study muscle mechanics but thinks he would like to do so under Professor Moran's direction rather than Professor Silverman's. Should GA Lee pursue a mentor change? If so, how should he go about doing so? What are GA Lee's obligations to Professor Silverman?

Juggling Multiple Job Offers

> *Case Study.* Postdoc King is pursuing several tenure-track job opportunities in exercise science, following years of training as a graduate student and post-doctoral fellow. Her first choice is a faculty position at the University of Minnesota. UM has an excellent reputation in exercise science, the job description seems to match King's interests well, great lab facilities are already in place, and her family lives within a few hours' drive in Wisconsin. While King waits to hear from UM, she gets an invitation for an interview at Gator University in central Florida. She accepts the invitation, and the interview goes well. She likes the faculty at Gator and thinks there is good potential to build a solid research program at Gator despite the fact that the exercise science program there and the university are not considered "top tier." Gator calls and offers her a faculty position with a modest start-up package. She is not excited about living in Florida (she is not fond of gators, mosquitoes, and humidity), but no other offers are pending, and it's getting late in the job search process. Thus, she accepts in writing the faculty offer from Gator. Two days after mailing her response to Gator, King receives an invitation to interview at the University of Minnesota. What should she do?

Assume for the moment that King accepted the invitation from UM to interview. While visiting UM, she is confronted by Professor Smith, who has a very good friend and colleague at Gator. Professor Smith had heard from his Gator colleague during a friendly telephone conversation that King had interviewed at Gator and had accepted their job offer. How would you expect the faculties at UM and Gator to react to this situation?

If King had not yet received a written offer from Gator University, but rather had only verbally accepted a verbal offer from Gator (i.e., she was waiting for the official offer to reach her by mail), would the situation be any different?

PART
II

Statistical and Measurement Concepts in Research

Of all statistics, 42.7% are made up on the spot.

In the following six chapters, we present some basic statistical and measurement techniques that are frequently used in physical activity research. We give more attention here to the basic statistical techniques than to the more complex methods. The underlying concepts of statistical techniques have been emphasized rather than any derivation of formulas or extensive computations. Because an understanding of how the basic statistical techniques work facilitates a grasp of the more advanced procedures, we have provided the computational procedures for most of the basic statistics and examples of their use.

Chapter 6 discusses the need for statistics. We describe different types of sampling procedures and summarize the basic statistics used in describing data, such as distributions, measures of central tendency, and measures of variation. Statistics can reveal two things about data: reliability and meaningfulness.

Chapter 7 introduces the concepts of probability, effect size, and—most important—power. Using power analysis to evaluate studies you read and studies you plan is essential.

Chapter 8 pertains to relationships among variables. We review different correlational techniques, such as the Pearson r for the relationship between only two variables. We explain partial correlation as a technique by which one can determine the correlation between two variables while holding the influence of additional variables constant. The use of correlation for prediction is discussed when using more than one variable to predict a criterion (multiple regression). Finally, we briefly present multivariate techniques of correlation: canonical correlation, factor analysis, and structural modeling.

Chapter 9 focuses on statistical techniques for comparing treatment effects on groups, such as different training methods or different samples. The simplest comparison of differences is between two groups: the t test. Next, we describe analysis of variance as a means of testing the significance of the differences among two or more groups. We also discuss the use of factorial analysis of variance, by which two or more independent variables can be compared. The use of repeated measures is common in our field, and we describe how they are analyzed. In addition, we provide an overview of multivariate techniques in this area, including discriminant analysis and multivariate analysis of variance.

Chapter 10 provides information on nonparametric techniques for data analysis. These are procedures in which the data fail to meet one or more of the basic assumptions of parametric techniques described in chapters 8 and 9. We present a new approach to nonparametrics that parallels the procedures described in chapters 8 and 9. Thus, you do not need to learn a new set of statistics to use the techniques.

Finally, chapter 11 reviews many of the measurement issues that apply when conducting research. We focus on the validity and reliability of dependent variables. We present a short overview of measurement issues regarding data collected about physical performance, although complete coverage of this topic is not possible. In much research on physical activity, dependent variables may be affective and knowledge measures. Thus, we discuss these types of dependent variables.

After reading the six chapters in part II, you will not be a statistician or a measurement expert (unless you were one before you started). However, if you read and study these chapters carefully and perhaps explore some of the references further, you should be able to comprehend the statistical analysis and measurement techniques of most research studies.

Becoming Acquainted With Statistical Concepts

All of life is 6 to 5 against.

—Damon Runyon

■ ■ ■ ■ ■ ■ ■

The idea of learning statistics frightens many people. If you are intimidated, you needn't be. Statistics is one of the few ways data can be reported uniformly to allow relevant, accurate conclusions and comparisons to be made. Statistics are methodical, logical, and necessary, not random, inconsistent, or terrifying. Our approach to statistics in this book is to acquaint you with the basic concepts and give you a working knowledge; it is not our purpose to make you a statistician, especially given this well-known quotation: "There are liars, damn liars, and statisticians."

Why We Need Statistics

Remember that throughout the process in which you conceive, plan, execute, and write up research, it is your informed judgment as a scientist on which you must rely, and this holds as much for the statistical aspects of the work as it does for all the others (Cohen, 1990, 1310).

Statistics is simply an objective means of interpreting a collection of observations. Various statistical techniques are necessary to describe the characteristics of data, test relationships between sets of data, and test the differences among sets of data. For example, if height and a standing-long-jump score were measured for each student in a seventh-grade class, you could sum all the heights and then divide the sum by the number of students. The result (a statistic to represent the average height) is the **mean,** $M = \Sigma X/N$, where Σ means sum, X = each student's height, and N = number of students; read this as "the sum of all Xs divided by N." The mean (M) describes the average height in the class; it is a single characteristic that represents the data.

An example of testing relationships between sets of data would be to measure the degree of association between height and the scores on the standing long jump. You might hypothesize that taller people can jump farther. By plotting the scores (figure 6.1), you can see that taller people generally do jump farther. But note that the relationship is not perfect. If it were, the scores would begin in the lower left-hand corner of the figure and proceed diagonally in a straight line toward the upper right-hand corner. One measure of

mean

A statistical measure of central tendency that is the average score of a group of scores.

Pearson product moment coefficient of correlation

The most commonly used method of computing correlation between two variables; also called *interclass correlation, simple correlation,* or *Pearson* r.

the degree of association between two variables is called the **Pearson product moment coefficient of correlation** (also known as Pearson *r*, interclass correlation, or simple correlation). When two variables are unrelated, their correlation is approximately zero. In figure 6.1, the two variables (height and standing-long-jump scores) have a moderately positive correlation (*r* is probably between .40 and .60). Relationships and correlation are discussed in greater detail in chapter 8, but for now you should see that researchers frequently want to investigate the relationship between variables.

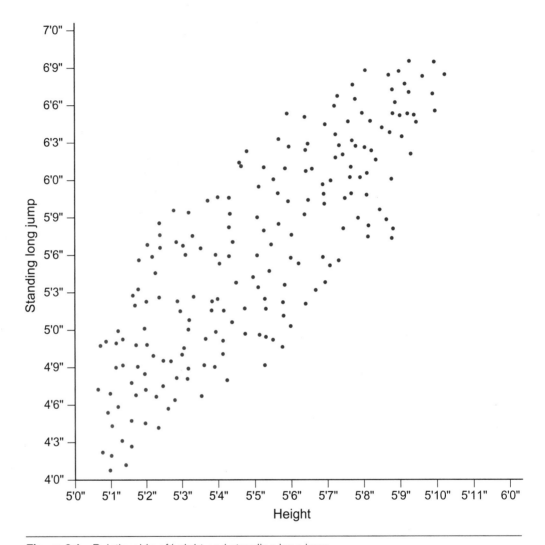

Figure 6.1 Relationship of height and standing long jump.

Besides descriptive and correlational techniques, a third category of statistical techniques is used to measure differences among groups. Suppose you believe that weight training of the legs will increase the distance a person can jump. You divide a seventh-grade class into two groups and have one group participate for eight weeks in a weight-training program designed to develop leg strength. The other students continue their regular activities. You want to know whether the independent variable (weight training versus regular activity) produces a change in the dependent variable (standing-long-jump score). Therefore, you measure the two groups' scores in the standing long jump at the end of the eight weeks (treatment period) and compare their average performances. Here, a statistical technique to assess differences between two independent groups, a *t* **test,** would be used. By calculating *t* and comparing it with a value from a *t* table, you can judge whether the two groups were significantly different on their average long-jump scores. Ways for assessing differences among groups are discussed in chapter 9.

t test

A statistical technique to assess differences between two groups.

Use of Computers in Statistical Analysis

Computers are very helpful in the calculation of statistics. Computers do not make the mistakes that can occur in hand calculations, and they are many times faster. Microcomputers (desktop or personal computers) are most frequently used for statistical analysis in laboratories, offices, and homes. The computer and its attachments (disk drives, monitors, printers, hard disks, and modems) are called **hardware,** and the computer programming is called **software.** Numerous software programs have been written to calculate statistics.

Two popular software packages for use on computers include the Statistical Package for the Social Sciences (SPSS) and the Statistical Analysis System (SAS). Most colleges and universities have one or more of these packages, which can be used on a personal computer (with proper hardware and software). Your computer center has information about available equipment and services. Many computer centers provide user services or user consultant centers where you can get advice about the hardware and software available and instructions in how to use it.

Most institutions teach statistics courses in which these software packages are used. Statistics departments or other departments may also offer consulting services on the appropriate use of statistics for research projects. You should investigate the services your institution offers. Senior graduate students and faculty can also advise you on available services.

hardware

The physical units of a computer, such as the printed circuit boards, monitor, keyboard, disk drive, and printer.

software

The programs or instructions used to make computers function in the desired manner.

Description and Inference Are Not Statistical Techniques

At the beginning of this chapter we stated that statistical techniques allow the description of data characteristics, the testing of relationships, and the testing of differences. When we discuss description and **inference,** though, we are not discussing statistical techniques, although those two words are sometimes confused with statistical techniques. This confusion is the result of saying that correlations describe relationships and that cause and effect are inferred by techniques for testing differences between groups. These statements are not necessarily true. Any statistic describes the **sample** of participants for which it was calculated. If the sample of participants represents some larger group, then the findings can be inferred (or generalized) to the larger group. However, the statistic used has nothing to do with inference. The method of selecting the sample, procedures, and context is what does or does not allow inference.

inference

Generalization of results to some larger group.

sample

A group of participants, treatments, or situations selected from a larger population.

Ways to Select a Sample

The sample is the group of participants, treatments, and situations on which the study is conducted. A key issue is how these samples are selected. In the following sections we discuss the types of sampling typically used in designing studies and using statistical analysis.

Random Selection

The sample of participants might be randomly selected from some larger group, or a **population.** For example, if your college or university has 10,000 students, you could randomly select 200 for a study. You would assign each of the 10,000 students a numeric ID. The first numeric ID would be 0000, the second 0001, the third 0002, up through the last (the 10,000th), who would be 9999. Then a **random numbers table** (see table A.1 in appendix A) would be used to select the sample. The numbers in this table are arranged in two-digit sets so that any combination of rows or columns is unrelated. In this instance (IDs 0000–9999), you need to select 200 four-digit numbers. Because the rows

population

The larger group from which a sample is taken.

random numbers table

A table in which numbers are arranged in two-digit (or greater) sets so that any combination of rows or columns is unrelated.

and columns are unrelated, you can choose any type of systematic strategy to go through the table. Enter the table at random (close your eyes and put your finger on the page). Suppose the place of entering the table is the sixth column of a two-digit number on row 8. The numeric ID here is 9953 (including columns 6 and 7 to get a four-digit number). You select the student with that numeric ID (9953). From here you can go down, across, or diagonally. Going down, to the next row, the four-digit combination of columns 6 and 7 is 9386, then 1846, and so on until you have selected 200 students. After completing this column, go to the next four-digit column. Column 7 is not used because it was included with column 6 to yield a four-digit number.

The system used in the random numbers table is not the only one. You can use any systematic way of going through the table. Of course, the purpose of all this is to select a sample randomly so that the sample represents the larger population; that is, the findings in the sample can be inferred back to the larger population. From a statistical viewpoint, inference means that a characteristic, relationship, or difference found in the sample is likely also to be present in the population from which the sample was selected.

Computer programs can also be used to generate a set of random numbers. You tell the program the population size and how many random numbers to select.

Stratified Random Sampling

stratified random sampling

Method of stratifying a population on some characteristic prior to random selection of the sample.

In **stratified random sampling,** the population is divided (stratified) on some characteristic before random selection of the sample. Returning to the previous example, the selection was of 200 students from a population of 10,000; suppose your college is 30% freshmen, 30% sophomores, 20% juniors, and 20% seniors. You could stratify on class before random selection to make sure the sample was exact in terms of class representation. Here, you would randomly select 60 students from the 3,000 freshmen, 60 from the 3,000 sophomores, 40 from the 2,000 juniors, and 40 from the 2,000 seniors. This still yields a total sample of 200.

Stratified random sampling might be particularly appropriate for survey or interview research. Suppose you suspect that attitudes toward exercise participation change over the college years. You might use a stratified random sampling technique for interviewing 200 college students to test this hypothesis. Another example would be to develop normative data on a physical fitness test for grades 4–8 in a school district. Because performance would be related to age, you should stratify the population by age before randomly selecting the sample on which to collect normative data.

Systematic Sampling

If the population from which the sample is to be selected is very large, assigning a numeric ID to each potential participant is time-consuming. Suppose you want to sample a town with a population of 50,000 concerning the need for new sport facilities. One approach would be to use systematic sampling from the telephone book. You might decide to call a sample of 500 people. To do so, you would select every 100th name in the phone book (50,000/500 = 100). Of course, you are assuming that the telephone book represents the population or, said another way, that everyone you need to sample has a listed telephone number. This turned out to be a very bad assumption in the 1948 presidential election (Dewey vs. Truman). The pollsters had predicted Dewey to win by a substantial margin. However, the pollsters had sampled from telephone books in key areas. Unfortunately for Dewey, many people without telephones voted for Truman. His victory was called an upset, but it was an upset only because of poor sampling procedures. Systematic sampling can yield a good sample and should be equivalent to random sampling if the sample is fairly large. However, researchers generally prefer random sampling.

Random Assignment

In experimental research, groups are formed within the sample. The issue here is not how the sample is selected but how the groups are formed within the sample. Chapter 18 discusses experimental research and true experimental designs. All true designs require that the groups within the sample be randomly assigned or randomized. Although this requirement has nothing to do with selecting the sample, the procedures used for random assignment are the same. Each person in the sample is given a numeric ID. If the sample has 30 participants, the numeric IDs range from 00 to 29. In this case, suppose three equal groups ($n = 10$) are to be formed. Enter a table of random numbers; the first numeric ID encountered between 00 and 29 goes in group 1, the second in group 2, and so on until each group has 10 participants. This process allows the researcher to assume that the groups are equivalent at the beginning of the experiment, which is one of several important features of good experimental design that is intended to establish cause and effect.

Computer programs are also available to randomize groups. You supply the sample size and the number of groups and decide whether the groups are to have an equal number of participants. The computer program then randomly assigns participants.

Justifying Post Hoc Explanations

Frequently, the sample for research is not randomly selected; rather, the researcher attempts a post hoc justification that the sample represents some larger group. A typical example is showing that the sample does not differ in average age, racial balance, or socioeconomic status from some larger group. Of course, the purpose is to allow the findings on the sample to be generalized to the larger group. A post hoc attempt at generalization may be better than nothing, but it is not the equivalent of random selection, which allows the assumption that the sample does not differ from the population on the characteristics measured (as well as any other characteristics). In a post hoc justification, only the characteristics measured can be compared. Whether those are the ones that really matter is open to speculation.

Post hoc justification is also used to compare intact groups, or groups within the sample that are not randomly formed, except in this case, the justification is that because the groups did not differ on certain measured characteristics before the study began, they can be judged equivalent. Of course, the same point applies: Are the groups different on some unmeasured characteristic that affects the results? This question cannot be answered satisfactorily. But, as before, a good post hoc justification of equivalence does add strength to comparisons of intact groups.

Difficulty of Random Sampling and Assignment: How Good Does It Have to Be?

In many studies of physical activity, random sampling procedures are not possible. For example, when comparing experts with novice performers, typically neither group is randomly sampled, nor are group memberships (expert and novice) randomly formed. The same is true when studying trained versus untrained runners or cyclists. Often we are interested in comparing the responses of different age groups, ethnic groups, and sexes to training in skill or exercise. Obviously, these groups are seldom randomly selected and cannot be randomly formed. In many studies, sampling is not done at all; researchers are happy to have any participants who volunteer.

Sampling also applies to the treatments used for different groups of participants. How are the treatments selected? Do they represent some population of potential treatments? What about the situational context under which participants are tested or receive the experimental treatment? Are these sampled in some way? Do they represent the potential situations?

The real answers to these questions about sampling are that we seldom randomly sample a population at any level: participants, treatments, or situations. Yet we hope to be able to use statistical techniques based on sampling assumptions and to infer that what we find applies to some larger group than the participants used in the study. What is really needed

She's random sampling.

is a sample that is "good enough for our purpose" (Kruskal & Mosteller, 1979, 259). This concept is extremely important for research in physical activity. If we cannot disregard the strict requirements of random sampling that allow study outcomes to be generalized to some similar group, we will never be able to generalize beyond the characteristics of any particular study—location, age, race, sex, time, and so on. Strictly speaking, there is no basis for the findings from a sample to be similar to a population that differs in any way from the sample (even one hour after the sampling). "A good-enough principle of sampling, however, would allow generalizations to any population for which the sample is representative enough" (Serlin, 1987, 366).

For the sample of participants, treatments, and situations to be "good enough" to generalize, it must be selected on some theoretical basis. For example, if the theory proposes that the cardiovascular system responds to training in a specific way in all untrained adults, using volunteer untrained undergraduates in a training study may be acceptable. "It is only on the basis of theory that one decides whether the experimental results can be generalized to the responder population, to the stimulus or ecological population, or both" (Serlin, 1987, 367). The best possible generalization statement is to say that the findings may be "plausible" in other participants, treatments, and situations, depending on their similarity to the study characteristics.

Unit of Analysis

unit of analysis

The concept, related to sampling and statistical analysis, that refers to what is considered the most basic unit from which data can be produced.

A question related to sampling and the statistical analyses discussed in subsequent chapters is labeled the **unit of analysis.** This concept refers to what can be considered the most basic unit from which data can be produced. We typically think of this as the individual participant; that is, we measure the change in level of physical activity of a specific elementary school child.

However, the unit of analysis concept is also connected to the level of the treatment given. For example, consider a large study about how a specific program of intervention in physical activity influences students in 16 elementary schools within a school district. We would like to randomly assign students within elementary schools to the two levels of intervention. The experimental treatment is a physical education class that focuses on all children being physically active and understanding why total physical activity (outside and inside school programs) is very important to health. The control condition is children continuing their regular physical education program (assume one that stresses competitive games and sports more than total physical activity). As you might guess, assigning students at random to these two conditions is very difficult. How can different programs be offered at the same time to children within a single class?

The researcher decides that children cannot be randomly assigned to the two treatments, but that classes can. Fortunately for our researcher, each elementary school has two classes at each grade level (it is easy to plan clean experiments when you make up the plan like the authors can; it is much harder in the real world where situations are not so accommodating). So within an elementary school, one of the two classes at each grade level is randomly assigned to either the experimental or control treatment. In this instance, the individual child is no longer the unit of analysis, but the class is because it is what has been randomly assigned. Thus, a mean physical activity score is calculated for each class rather than for each child, and the class score is used in the statistical analysis. You can see how drastically that influences the number of data points for the statistical analysis. If we consider each *child* as the unit of analysis, and each grade of K-5 has two classes of 25 children each, then the total sample size for a school (n) would be 300 and the total sample size (N) for the 16 schools is 4,800. However, if we consider *class* the unit of analysis, then the sample size is 12 for each of the 16 elementary schools, giving a total sample size (N) of 192.

But consider that giving the two different treatments within an elementary school might confound the data; that is, children receiving one treatment might well talk with students receiving the other even though they are in different classes. Maybe the children receiving the control treatment decide to exercise at home because a friend in another class (receiving the experimental treatment) is doing so. This contaminates the data and increases variability. To avoid this, the researcher might decide to randomly assign the 16 schools to each of the two treatments. Now the unit of analysis becomes the school, with the experimental treatment having an n of 8 and the control treatment having an n of 8.

You can see that deciding the appropriate unit of analysis drastically influences the sample size for statistical analysis (in the previous example it ranges from a low of 16 to a high of 4,800). As you will see in the next chapter, appropriate statistical analysis demands statistical *power*. Power is influenced by sample size, with larger samples generally having more power to detect relationships or differences. In planning studies where intact units (classes or schools in the example above) are assigned to a treatment, the effect of the appropriate unit of analysis on power should be considered early. Otherwise, during analysis the researcher will be faced with two bad choices—using the wrong unit of analysis and inflating N or having to collect further data to have sufficient power to determine whether the treatment was effective.

Measures of Central Tendency and Variability

Some of the more easily understood statistical and mathematical calculations are those that find **central tendency** and **variability** of scores. When you have a group of scores, one number may be used to represent the group. That number is generally the mean, median, or mode. These terms are ways of expressing central tendency. Within the group of scores, each individual score differs to some degree from the central tendency score. The degree of difference is the score's variability. Two terms that describe the variability of the scores are **standard deviation** and **variance.**

Central Tendency Scores

The statistic for the central tendency score with which you are probably familiar is the mean (M), or average:

$$M = \Sigma X/N \tag{6.1}$$

Thus, if you have the numbers 6, 5, 10, 2, 5, 8, 5, 1, and 3, then

$$M = (6 + 5 + 10 + 2 + 5 + 8 + 5 + 1 + 3)/9$$

$$= 45/9 = 5$$

The number 5 is the mean and represents this series of numbers.

central tendency (measure of)

A single score that best represents all of the scores.

variability

The degree of difference between each individual score and the central tendency score.

standard deviation

An estimate of the variability of the scores of a group around the mean.

variance

The square of the standard deviation.

Sometimes the mean may not be the most representative or characteristic score. Suppose we replace the number 10 in the previous set of scores with the number 46. The mean is now 9, a number larger than all but one of the scores. It is not very representative because one score (46) made the average high. In this case, another measure of central tendency is more useful. The **median** is defined as the value in the middle; the middle value is the value that occurs in the place $(N + 1)/2$ when the values are put in order. In our example, arrange the numbers from lowest to highest—1, 2, 3, 5, 5, 5, 6, 8, 46—then count $(9 + 1)/2 = 5$ places from the first score; the median is the value 5, which is a much more representative score. If N is even, the median value may be a decimal. For example, for scores 1, 2, 3, 4, the median is 2.5.

Most often you are interested in the mean of a group of scores. You may occasionally be interested in the median or perhaps in another measure of central tendency, the **mode,** which is defined as the most frequently occurring score. In the previous example, the mode is also 5, as it occurs three times. Some groups of scores may have more than one mode.

median

A statistical measure of central tendency that is the middle score in a group.

mode

A statistical measure of central tendency that is the most frequently occurring score of the group.

Variability Scores

Another characteristic of a group of scores is the variability. An estimate of the variability, or spread, of the scores can be calculated as the standard deviation (s):

$$s = \sqrt{\Sigma(X - M)^2 / (N - 1)} \qquad (6.2)$$

This formula translates as follows. Calculate the mean by equation 6.1, subtract the mean from each person's score ($X - M$), square the answer, sum the squared scores, divide by the number of scores minus one ($N - 1$), and take the square root of the answer. Table 6.1 provides an example.

The mean and standard deviation together are good descriptions of a set of scores. If the standard deviation is large, the mean may not be a good representation. Roughly 68% of a set of scores fall within ±1s, about 95% of the scores fall within ±2s, and about 99% of the scores fall within ±3s. This is called a normal distribution (discussed later in the chapter).

Equation 6.2 was used to help you understand the meaning of the standard deviation. For use with a hand calculator, equation 6.3 is simpler:

$$s = \sqrt{\left[N\Sigma X^2 - (\Sigma X)^2\right] / \left[N(N - 1)\right]} \qquad (6.3)$$

One final point for later consideration is that the square of the standard deviation is called the variance, or s^2.

Range of Scores

Sometimes the range of scores (highest and lowest) may also be reported, particularly when the median rather than the mean is used. The median and the mean may be used in connection with each other. For example, 15 people might be given 10 blocks of 10 trials (100 total trials) on a reaction-time task. The experimenter may decide to use the median reaction time of a person's 10 trials as the most representative score in each block. Thus, each person would have 10 median scores, one for each of the 10 trial blocks. Here, the range of scores from which the median was selected should be reported. Both the mean and the standard deviation should be reported for the 15 participants' median scores at trial block 1, trial block 2, and so on. Thus, the range is reported for the selection of the median of each trial block, whereas the standard deviation is reported for the mean of each trial block.

Table 6.1   Calculations of Mean and Standard Deviation

X	X – M	(X – M)²
6	1	1
5	0	0
10	5	25
2	–3	9
5	0	0
8	3	9
5	0	0
1	–4	16
3	–2	4
Σ = 45	0	64

$$M = \Sigma X/N = 45/9 = 5$$

$$s = \sqrt{\Sigma(X - M)^2 / (N - 1)} = \sqrt{64 / 8} = \sqrt{8} = 2.83$$

Confidence Intervals

Confidence intervals represent an effective technique used by researchers to help interpret a variety of statistics such as means, medians, and correlations. They are also used in hypothesis testing. A confidence interval provides an expected upper and lower limit for a statistic at a specified probability level, usually either 95% or 99%. The size, or length, of a confidence interval is affected by the size of the sample, the homogeneity of values within the sample, and the level of confidence selected by the researcher. Confidence intervals are based on the fact that any statistic possesses sampling error. This error relates to how well the statistic represents the target population. When we compute a mean for a sample, we are making an estimate of the target population's mean. A confidence interval provides a band within which the estimate of the population mean is likely to fall instead of a single point.

A confidence interval (CI) of a statistic such as the mean employs the following information:

$$\text{CI} = \text{observed statistic} \pm (\text{standard error} \times \text{specified confidence level value}) \quad (6.4)$$

We will construct a confidence interval for a mean of a sample with the following characteristics: $n = 30$, $M = 40$, $s = 8$.

The observed statistic is the mean ($M = 40$). The **standard error** represents the variability of the sampling distribution of a statistic. Imagine that instead of one sample of $n = 30$, you had selected 100 samples of this size. The means for all the samples would not be the same; in fact, they would start to approximate a normal distribution. If you calculated the standard deviation of all of these means, you would have an estimate of the standard error. Fortunately, we do not have to draw hundreds of samples to calculate standard error; we simply divide our sample standard deviation by the square root of the sample size. Don't ask why; just accept it. (Have we ever lied to you before?) So, in our example, the standard error of the mean is $s_M = 8 / \sqrt{30} = 1.46$.

The last information we need to construct a confidence interval is the value for the specified confidence level. This is easy: We just look it up in a table. Recall from the discussion of the normal curve that 95% of the scores are contained within approximately $\pm 2s$. Actually, it is $1.96s$ instead of 2.00, and 99% of the scores are encompassed by $\pm 2.576s$ instead of 3.00 (see table A.2 in appendix A). Hence, if we knew that our sample had a normal distribution, we would use 1.96 for the 95% confidence level value or 2.58 for 99%. However, since we can't assume that our sample is normally distributed, we should use a distribution table that takes into account the size of the sample. So we use the t distribution in table A.5, appendix A. The first column is labeled df (degrees of freedom). For our purposes here, this just means $n - 1$, or $30 - 1 = 29$. We are going to use the values under the two-tailed test (to reflect the spread above and below the mean). The value for the 95% confidence level is under the .05 column. Thus, reading across from 29 df, we see that the value to use is 2.045. If we were using the 99% level of confidence, we would use the .01 column, and the value would be 2.756.

We can now construct the confidence interval (95% confidence) using equation 6.4:

$$\text{CI} = \text{sample mean} \pm (\text{standard error} \times \text{confidence level table value})$$

$$= 40 \pm (1.46 \times 2.045)$$

$$= 40 \pm 2.99$$

We subtract and add 2.99 to 40 to obtain a confidence interval of 37.01 to 42.99. Thus, we are 95% confident that this interval includes the target population mean.

Confidence intervals are also used in hypothesis testing, such as applying the null hypothesis to the difference between two or more means. If we were comparing the means of two groups, the difference between the two means would be the observed statistic. The standard error would be the standard error of the difference (which combines the standard errors of the two means). The table value would be taken from the t table (table A.5) using df of $(n_1 - 1) + (n_2 - 1)$ or $(N - 2)$ for the selected level of confidence (e.g., 95%). The calculated confidence interval would give an estimate of the range within which the true difference between these two population means would fall 95% of the time.

standard error

The variability of the sampling distribution.

Frequency Distribution and the Stem-and-Leaf Display

frequency distribution

A distribution of scores including the frequency with which they occur.

frequency intervals

Small ranges of scores within a frequency distribution into which scores are grouped.

stem-and-leaf display

A method of organizing raw scores by which score intervals are shown on the left side of a vertical line and individual scores falling into each interval are shown on the right side.

A common technique for summarizing data is to produce a "picture" of the distribution of scores by means of a **frequency distribution.** A simple frequency distribution just lists all the scores and the frequency of each score. If there is a wide range of values, a grouped frequency distribution is used in which scores are grouped into small ranges called **frequency intervals.** For example, all individuals having scores of 91, 92, 93, 94, or 95 might be grouped into an interval of (91–95). Individuals with scores 86 through 90 would be in the next interval (86–90), and so on. The frequency of scores in each interval are displayed in an adjacent column as follows:

Score	f (frequency)
91–95	4
86–90	6
81– 85	7

A major drawback to a grouped frequency distribution is that there is a loss of information, that is, a reader doesn't know how many individuals achieved each score within a given interval.

Another effective way to provide information about the distribution of a set of data is called a **stem-and-leaf display.** This is similar to a grouped frequency distribution, but there is no loss of information.

To illustrate, a researcher decides to set up intervals of 10 scores (20–29, 30–39, etc.). The digit in the tens place, which forms the stem, appears on the left side of the vertical line in figure 6.2. Note that the last digit of the scores falling into each interval is not shown in the stem. Then, to the right of the vertical line are the last digits, the leaves.

The stem-and-leaf display in figure 6.2 shows the distribution of 40 scores between 25 and 86 (the range). You can see that in the 20–29 interval (the first line of the stem-and-leaf display) there are scores of 25 and 27. In the next interval there are four scores: 30, 32, 34, and 38. When the intervals are ordered from lowest to highest, you can turn the page on its side and see a type of graph. This helps you to visualize the normality (or skewness) of the distribution (discussed in the next section). If desired, one can display more narrow intervals, such as 5 digits. For example, instead of a stem of 2 representing scores from 20 to 29, as in figure 6.2, it would include only scores 20 through 24, and the stem of 2* would contain scores 25 through 29. The stem 3 would contain scores 30 through

Stem	Leaves	f
2	5 7	2
3	0 2 4 8	4
4	1 1 3 3 5 7 7 9	8
5	0 0 1 2 4 4 5 5 7 8 8	11
6	0 1 2 2 6 6 7	7
7	1 3 5 5 8	5
8	0 4 6	3
		n = 40

Figure 6.2   Stem-and-leaf display.

34, 3* would include 35 through 39, and so on. Another advantage of the stem-and-leaf display is that it is easy to prepare a rank-order set of data since the individual scores are shown from low to high.

Basic Concepts of Statistical Techniques

Besides measures of central tendency and variability, there are other, slightly more complicated statistical techniques. Before we explain each in detail, however, you must understand some general information about statistical techniques.

Two Categories of Statistical Tests

There are two general categories of statistical tests: **parametric** and **nonparametric.** The use of the various tests in each category depends on meeting the assumptions for those tests. The first category, parametric statistical tests, has three assumptions about the distribution of the data:

- The population from which the sample is drawn is normally distributed on the variable of interest.
- The samples drawn from a population have the same variances on the variable of interest.
- The observations are independent.

Certain parametric techniques have additional assumptions. The second category, nonparametric statistics, is called **distribution-free** because the previous assumptions need not be met.

Whenever the assumptions are met, parametric statistics are often said to have more **power,** although there is some debate on this issue. To have power means to increase the chances of rejecting a false null hypothesis. You frequently assume that the three criteria for use of parametric statistics are met. The assumptions can be tested by using estimates of **skewness** and **kurtosis.** (Only the meaning of these tests is explained here. Any basic statistics textbook provides considerably more detail. For a helpful discussion on skewness and kurtosis, see Newell & Hancock, 1984.)

To understand skewness and kurtosis, first consider the normal distribution in figure 6.3. This is a **normal curve,** which is characterized by the mean, median, and mode being at

parametric statistical test

Test based on data assumptions of normal distribution, equal variance, and independence of observation.

nonparametric statistical test

Any of a number of statistical techniques used when the data do not meet the assumptions required to perform parametric tests.

distribution-free statistics

See *nonparametric statistical test.*

power (statistical)

The probability of rejecting a false null hypothesis.

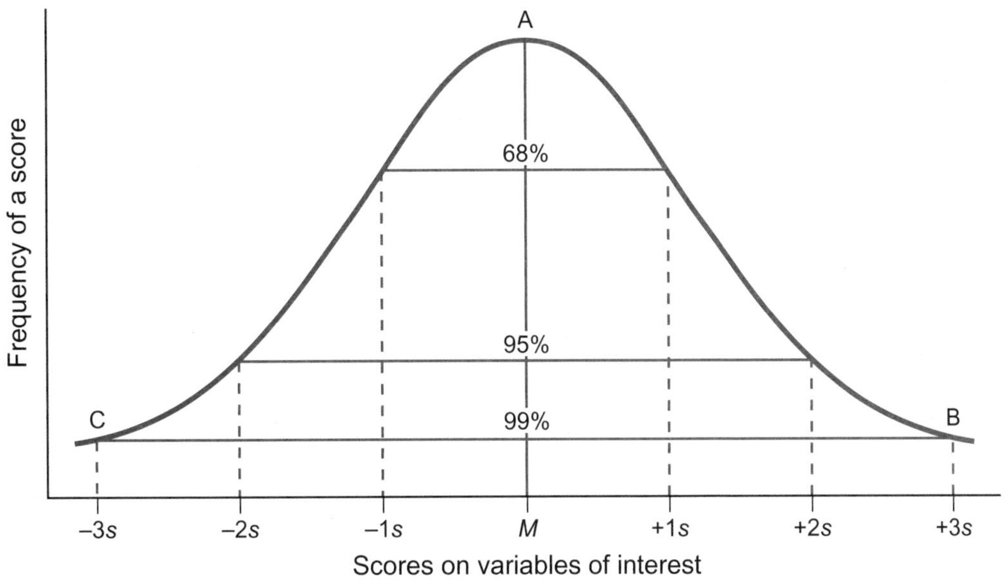

Figure 6.3   The normal curve.

skewness

Description of the direction of the hump of the curve of distribution of data and the nature of the tails of the curve.

kurtosis

Description of the vertical characteristic of the curve showing the data distribution, for example, whether the curve is more peaked or flatter than the normal curve.

normal curve

Distribution of data in which the mean, median, and mode are at the same point (center of the distribution) and in which ±1s from the mean includes 68% of the scores, ±2s from the mean includes 95% of the scores, and ±3s includes 99% of the scores.

the same point (center of the distribution). In addition, ±1s from the mean includes 68% of the scores, ±2s from the mean includes 95% of the scores, and ±3s includes 99% of the scores. Thus, data distributed as in figure 6.3 would meet the three assumptions for use of parametric techniques. Skewness of the distribution describes the direction of the hump of the curve (labeled A in figure 6.4) and the nature of the tails of the curve (labeled B and C). If the hump (A) is shifted to the left and the long tail (B) to the right (figure 6.4a), the skewness is positive. If the shift of the hump (A) is to the right and the long tail (C) to the left (figure 6.4b), the skewness is negative. Kurtosis describes the vertical aspect of the curve, for example, whether the curve is more or less peaked than the normal curve. Figure 6.5a shows a more peaked curve, and figure 6.5b a flatter curve.

Table A.2 in appendix A shows a unit normal distribution (z) for a normal curve. The column z shows the location of the mean. When the mean is in the center of the distribution, its z is equal to .00; thus, .50 (50%) of the distribution is beyond (to the right of) the mean, leaving .50 (50%) of the distribution as a remainder (to the left of the mean). As the mean of the distribution moves to the right in a normal curve (say to a z of +1s), .8413 (84%) of the distribution is to the left of the mean (remainder) and .1587 is to the right of (beyond) the mean. This table allows you to determine the percentage of the normal distribution included by the mean plus any fraction of a standard deviation. Suppose you want to know what percentage of the distribution would be included by the mean plus one half (.50) of a standard deviation. Using table A.2 you can see that it would be .6915 (remainder), or 69%.

For chapters 8 and 9, consider that the three basic assumptions for parametric statistical tests have been met. This is done for two reasons. First, the assumptions are very robust to violations, meaning that the outcome of the statistical test is relatively accurate even with severe violations of the assumptions. Second, most of the research in physical activity uses parametric tests.

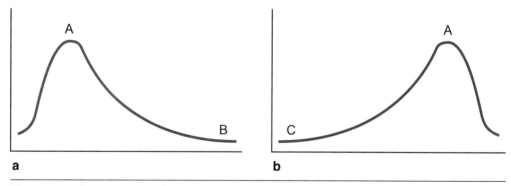

a   **b**

Figure 6.4   Skewed curves: (a) positive skewness; (b) negative skewness.

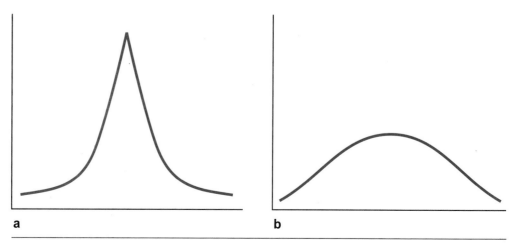

a   **b**

Figure 6.5   Curves with abnormal kurtosis: (a) more peaked; (b) more flat.

What Statistical Techniques Tell About Data

The statistical techniques presented in the next four chapters answer the following two questions about the data to which they are applied:

1. Is the effect or relationship of interest reliable? In other words, if the research is repeated, will the effect or relationship be there again (is it significant)?

2. How strong (or meaningful) is the effect or relationship of interest? This refers to the magnitude, or size, of the effect or relationship.

Two facts are important about these two statements. First, question 1 is usually answered before question 2, for the strength of the relationship or effect may be of minimal interest until it is known to be reliable (significant). Second, question 2 is always of interest if the effect or relationship is significant and occasionally of interest without a significant effect. Sometimes, in elating over the significance of effects and relationships, researchers lose sight of the need to look at the strength or meaningfulness of these relationships. This is particularly true in research in which differences among groups are compared. The experimenter frequently forgets that relatively small differences can be significant. That means only that the differences are reliable or that the same answer is likely to be obtained if the research is repeated. The experimenter then needs to look at the size of the differences to interpret whether the findings are meaningful. For each technique presented in the next three chapters, we first look at whether the relationship or effect is significant (reliable). Then we suggest ways to evaluate the strength (meaningfulness) of the relationship or effect.

The Correlation Versus Differences Distinction in Statistical Techniques

It is practical to divide statistical techniques into two categories: (a) statistical techniques used to test relationships between or among several variables in one group of participants (regression or correlation) and (b) techniques used to test differences between or among groups of participants (*t* tests and analysis of variance). This division is, strictly speaking, inaccurate because both sets of techniques are based on the general linear model and involve only different ways of entering data and manipulating variance components. However, an introduction to research methods is neither the place to reform the world of statistics nor the place to confuse you. Thus, the techniques as two distinct groups are considered. Chapter 8 discusses relationships among variables, and chapter 9 discusses differences among groups. Learning the simple calculations underlying the easier techniques and then building on them helps you understand the more complex ones. Do not panic because statistics involves manipulating numbers. You can escape from this section with a reasonable grasp of how, why, and when the various statistical techniques are used in the study of physical activity.

Remember that correlation between two variables does not indicate causation (recall the discussion in chapter 4). Causation is not determined from any statistic or correlation. Cause and effect are established by theory, logic, and the total experimental situation, of which statistics is a part. As summarized by Pedhazur (1982, 579),

> **"C**orrelation is no proof of causation." Nor does any other index prove causation, regardless of whether the index was derived from data collected in experimental or in non-experimental research. Covariations or correlations among variables may be suggestive of causal linkages. Nevertheless, an explanatory scheme is not arrived at on the basis of the data, but rather on the basis of knowledge, theoretical formulations and assumptions, and logical analyses. It is the explanatory scheme of the researcher that determines the type of analysis to be applied to data, and not the other way around.

For example, Descartes has been credited with the logical statement, "I think, therefore I am." However, the famous philosopher Edsall Murphy "recognized it as a syllogism with an unstated major premise" (Morgenstern, 1983, 112):

A nonexistent object cannot think. (major premise)

I think. (minor premise)

Therefore, I am. (conclusion)

Murphy was not satisfied with this and so tried to find deeper meaning and a better logical analysis.

Q: How can you be sure that you exist?

A: I think.

Q: How can you be sure you are thinking?

A: I can't, but I do think that I think.

Q: Does that make you sure that you exist?

A: I think so.

This should make it clear that Descartes went too far. He ought to have said, "I think I think, therefore I am," or possibly, "I think I think, therefore I think I am, I think. . . . " You do not really exist unless others are aware of your existence. Murphy proclaimed, "I stink, therefore I am" (Morgenstern, 1983, 112).

Summary

Statistics are used to describe data, to determine relationships among variables, and to test for differences among groups. In this chapter we have tried to make the point that the type of statistics used does not determine whether findings can be generalized; rather, it is sampling that permits (or limits) inference. Whenever possible, random sampling is the method of choice, but in behavioral research the more important question may be whether the sample is "good enough." In some types of research, such as surveys, stratified random sampling is desirable for the study to represent certain segments of a population. In experimental research, random assignment of participants to groups is definitely desirable so that the researcher can assume equivalence at the beginning of the experiment.

We began our coverage of statistical techniques with basic concepts such as measures of central tendency, variability, and normal distribution. It is important to remember that statistics can do two things: establish significance and assess meaningfulness. Significance means that a relationship or difference is reliable—that you could expect it to happen again if the study were repeated. Meaningfulness refers to the importance of the results.

✔ Check Your Understanding

The following scores represent the number of pull-ups attained by 15 students:

8	4	13
11	6	2
6	7	5
5	0	4
8	6	5

1. Use equation 6.1 to calculate the mean.

2. Use equation 6.2 to calculate the standard deviation.

3. Use equation 6.3 to calculate the standard deviation, and see whether the answer is the same as that derived from equation 6.2.

4. Rearrange the scores from low to high. Use the formula $(N + 1)/2$ to find the median point. Count up from the bottom that number of values to locate the median score.

5. Make a list of 50 names. Using the table of random numbers (table A.1 in appendix A), randomly select 24 subjects, and then randomly assign them to two groups of 12 each.

6. Construct a stem-and-leaf display, using intervals of 10 scores (e.g., 10–19, 20–29, etc.) for the following 30 scores:

50	42	64	18	41	30	48	68	21	48
43	27	51	42	62	53	45	31	13	58
60	35	28	46	36	56	39	46	25	49

7. Consult equation 6.4 and construct a confidence interval for the mean of a sample ($M = 50$, $s = 6$, $n = 100$). Use the 95% confidence level. Interpret the results.

Statistical Issues in Research Planning and Evaluation

Statistics show that of those who contract the habit of eating, very few survive.—Wallace Irwin

■ ■ ■ ■ ■ ■ ■

To plan your own study or evaluate a study by someone else, you need to understand four concepts and their interrelationships: alpha, power, sample size, and effect size. In this chapter we present these concepts and show how to use them in research planning and evaluation.

Probability

Another concept that deals with statistical techniques is **probability,** which asks what are the odds that certain things will happen. You use probability in everyday events. What are the chances that it will rain? You hear from a weather report that probability of rain is 90%. You wonder whether this means that it will rain in 90% of the places or, more likely, that the chances are 90% that it will rain where you are, especially if you are planning to play golf or tennis.

A concept of probability related to statistics is called **equally likely events.** For example, if you roll a die, the chances of the numbers from 1 to 6 occurring are equally likely (i.e., 1 in 6, unless you are playing craps in Las Vegas). Another pertinent approach to probability involves **relative frequency.** To illustrate, suppose you toss a coin 100 times. You would expect heads 50 times and tails 50 times; the probability is one half, or .50. However, when you toss, you may get heads 48 times, or .48. This is the relative frequency. You might do this 10 times and never get .50, but the relative frequency would be distributed closely around .50, and you would still assume the probability as .50.

In a statistical test, you sample from a population of participants and events. You use probability statements to describe the confidence you place in the statistical findings. Frequently, you encounter a statistical test followed by a probability statement such as $p < .05$. This interpretation is that a difference or relationship of this size would be expected less than 5 times in 100 as a result of chance.

Alpha

In research, the test statistic is compared with a probability table for that statistic, which tells you what the chance occurrence is. The experimenter may establish an acceptable

probability

The odds that a certain event will occur.

equally likely events

A concept of probability in which the chances of one event occurring are the same as the chances of another event occurring.

relative frequency

A concept of probability concerning the comparative likelihood of two or more events occurring.

alpha (α)

A level of probability (of chance occurrence) set by the experimenter prior to the study; sometimes referred to as *level of significance*.

type I error

A rejection of the null hypothesis when the null hypothesis is true.

type II error

Acceptance of the null hypothesis when the null hypothesis is false.

truth table

A graphic representation of correct and incorrect decisions regarding type I and type II errors.

level of chance occurrence (called **alpha,** α) before the study. This level of chance occurrence can vary from low to high but can never be eliminated. For any given study, the probability of the findings being due to chance always exists, or, to quote Holten's Homily, "The only time to be positive is when you are positive you are wrong."

In behavioral research, alpha (probability of chance occurrence) is frequently set at .05 or .01 (the odds that the findings are due to chance are either 5 in 100 or 1 in 100). There is nothing magical about .05 or .01. They are used to control for a **type I error.** In a study, the experimenter may make two types of error. A type I error is to reject the null hypothesis when the null hypothesis is true. For example, a researcher concludes that there is a difference between two methods of training, but there really is not. A **type II error** is not to reject the null hypothesis when the null hypothesis is false. For example, a researcher concludes there is no difference between the two training methods, but there really is a difference. Figure 7.1 is called a **truth table,** which displays type I and type II errors. As you can see, to accept a true null hypothesis or reject a false one is the correct decision. You control for type I errors by setting alpha. For example, if alpha is set at .05, then, if 100 experiments are conducted, a true null hypothesis of no difference or no relationship would be rejected on only 5 occasions. Although the chances for error still exist, the experimenter has specified them exactly by establishing alpha before the study.

To some extent the issue is, If you had to make an error, which type of error would you be willing to make? The level of alpha reflects the type of error you are willing to make. In other words, is it more important that you avoid concluding that one training method is better than another when it really isn't (type I), or is it more important that you avoid the conclusion that one method is not any better than the other when it really is (type II)? For example, in a study of the effect of a cancer drug, the experimenter would not want to accept the null hypothesis of no effect if there were any chance that the drug worked. Thus, the experimenter might set alpha at .30 even though the odds of making a type I error would be inflated. The experimenter is making sure that the drug has every opportunity to show its effectiveness. On the other hand, setting alpha at a very low level, for example, .001, greatly decreases the odds of making a type I error and consequently makes it harder to detect a real difference (a type II error).

	H_0 true	H_0 false
Accept	Correct decision	Type II error (β)
Reject	Type I error (α)	Correct decision

Figure 7.1 Truth table for null hypothesis (H_0).

We cannot tell you where to set alpha; however, we can say that the levels .05 or .01 are widely accepted in the scientific community. If alpha is to be moved up or down, be sure to justify your decision, but consider this: "Surely God loves the .06 nearly as much as the .05" (Rosnow & Rosenthal, 1989).

Even when experimenters set alpha at a specific level (e.g., .05) before the research, they often report the probability of a chance occurrence for the specific effects of the study at the level it occurred (e.g., $p = .012$). Nothing is wrong with this procedure, as they are only demonstrating to what degree the level of probability exceeded the specified level.

It is debatable whether alpha should be specified before the research and the findings reported at the alpha that has been specified. Some argue that results are either significant at the specified alpha or they are not (not much is in between). It is like being pregnant—one either is or is not. Comparisons with alpha work the same way. It is established as a criterion, and the results either meet the criterion or they do not. Although experimenters sometimes report borderline significance (if alpha is .05), they label from .051 to .10 as borderline.

A more sound approach may be to report the exact level of probability (e.g., $p = .024$) associated with the test statistics (e.g., r, t) and then to estimate the meaningfulness of the difference or relationship. Using the statistical information (significance and meaningfulness), the researcher should interpret the findings within the theory and hypotheses that have been put forward. Rather than making the decision a statistical one, this approach places the responsibility of decision making where it belongs: on the researcher who has placed the study within a theoretical model and has considered related research. Much of the criticism of statistics has revolved around the blind application of statistical techniques to data without appropriate interpretation of the outcomes. For excellent articles on the use of common sense with statistical outcomes, read Cohen (1990) and Serlin (1987).

Beta

Although the magnitude of type I error is specified by alpha, you may also make a type II error, the magnitude of which is determined by **beta (β).** By looking at figure 7.2, you can see the overlap of the score distribution on the dependent variable for X (the sampling distribution if the null hypothesis is true) and Y (the sampling distribution if the null hypothesis is false). By specifying alpha, you indicate that the mean of Y (given a certain distribution) must be at a specified distance from the mean of X before the null hypothesis is rejected. However, if the mean of Y falls anywhere between the mean of X and the specified Y, you could be making a type II error (beta); that is, you do not reject the null hypothesis when, in fact, there is a true difference. As you can see, there is a relationship between alpha and beta; for example, as alpha is set increasingly smaller, beta becomes larger.

beta (β)

The magnitude of a type II error.

Meaningfulness (Effect Size)

In addition to reporting the significance of the findings, scholars need to be concerned about the **meaningfulness** of the outcomes of their research. The meaningfulness of a difference between two means can be estimated in many ways, but the one that has gained the most attention recently is **effect size** (suggested by Cohen, 1969; also called delta). You may be familiar with using effect size (ES) in meta-analysis (if not, you will be soon; see chapter 14 on research synthesis). The formula for ES is

$$ES = (M_1 - M_2)/s \qquad (7.1)$$

This formula subtracts the mean of one group (M_1) from the mean of a second group (M_2) and divides the difference by the standard deviation. That places the difference between the

meaningfulness

The importance or practical significance of an effect or relationship.

effect size

The standardized value that is the difference between the means divided by the standard deviation, also called *delta*.

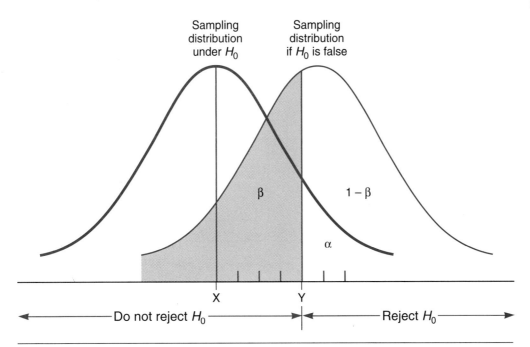

Figure 7.2   Regions under the normal curve corresponding to probabilities of making type I and type II errors.

From *Experimental Design: Procedures for the Behavioral Sciences*, by R.E. Kirk. Copyright © 1995. Reprinted with permission of Wadsworth, a division of Thomson Learning: www.thomsonrights.com. Fax 800-730-2215.

means in the common metric called *standard deviation units,* which can be compared to guidelines for behavioral research suggested by Cohen (1969): 0.2 or less is a small ES, about 0.5 is a moderate ES, and 0.8 or more is a large ES. For a more detailed discussion of the use of effect size in physical activity research, see Thomas, Salazar, and Landers (1991).

Many authors (e.g., Cohen, 1990; Serlin, 1987; Thomas, Salazar, & Landers, 1991) have indicated the need to report some estimate of meaningfulness with all tests of significance.

Power

Power is the probability of rejecting the null hypothesis when the null hypothesis is false (e.g., detecting a real difference), or the probability of making a correct decision. Power ranges from 0 to 1. The greater the power, the more likely you are to detect a real difference or relationship. Thus, power increases the odds of rejecting a false null hypothesis. Of course, to a degree, in behavioral research, *the null hypothesis is always false!*

What this statement reflects is that in behavioral research the means of two groups are never the same. Thus, if enough participants are obtained (one way of obtaining power), any two means can be declared significantly different. Remember what that means. If you do it again, you will get about the same answer. The more interesting questions in behavioral research are

- How large a difference is important in theory or practice? and,
- How many participants are needed to declare an important difference as significant?

Understanding the concept of power can answer these questions. If a researcher can identify the size of an important effect through previous research or even simply estimate an effect size (e.g., 0.5 is a moderate ES) and establish how much power is acceptable (e.g., a common estimate in the behavioral science is 0.8), then the size of the sample needed for a study can be estimated.

Figures 7.3 and 7.4 show the relationship among sample size (*y* axis), power (*x* axis), and effect size (ES) curves when alpha is either .05 or .01.

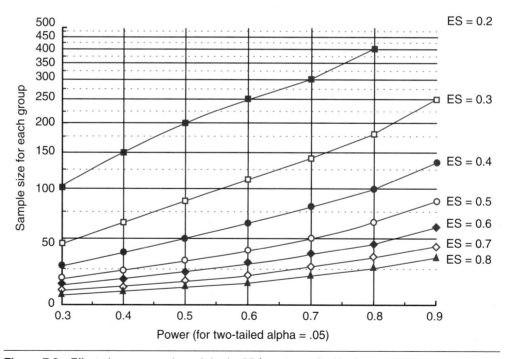

Figure 7.3   Effect size curves when alpha is .05 for a two-tailed test.

Reprinted, by permission, from C. Kraemer and S. Thiemann, 1987, *How many subjects?: Statistical power analysis in research* (Thousand Oaks, CA: Sage Publishing, Inc.), 111.

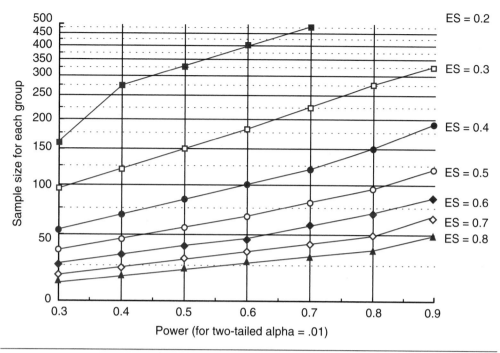

Figure 7.4   Effect size curves when alpha is .01 for a two-tailed test.
Reprinted, by permission, from C. Kraemer and S. Thiemann, 1987, *How many subjects?: Statistical power analysis in research* (Thousand Oaks, CA: Sage Publishing, Inc.), 112.

Consider the following example. An investigator plans a study that will have two randomly formed groups, but she does not know how many participants are needed for each group so as to detect a meaningful difference between treatments. However, there are several related studies, and the investigator has calculated an average ES = 0.7 (using equation 6.4) favoring the experimental group from the outcomes in these studies. The investigator decides to set alpha = .05 and wants to keep beta at four times the level of alpha (thus, beta = .20) because Cohen (1988) suggested that in the behavioral sciences, the seriousness of type I to type II error should be in a 4-to-1 ratio. Since power is 1 – beta (1.0 – 0.2 = 0.8), power is set at .8 (often recommended as an appropriate power in behavioral research, Green, 1991, 502). When this information is known—alpha, ES, and power—then the number of participants needed in each of the two groups can be estimated from figure 7.3. Read the 0.7 ES curve to where it passes the *x* axis (power) at .8. Then read across to the *y* axis (sample size), and note that 30 people would be needed in each group. Note how this relationship works: As the number of individuals in each group is reduced, power is reduced (at the same ES).

In figure 7.4 (alpha = .01), note that for the same level of power (.8) and ES (0.7), the number of people per group increases from 30 (from figure 7.3, alpha = .05) to 50. So if all else stays the same but a more stringent alpha is used (e.g., .05 to .01), a greater number of participants is required to detect a significant difference. Hence, with a more stringent (lower) alpha (e.g., .001) power is reduced, making it more difficult to detect a significant difference. A larger alpha such as .10 or .20 increases power; that is, it lessens the chance of making a type II error. Of course, it increases the chance of making a type I error (proclaiming that there is a difference when there is no difference). As stated before, the researcher has to decide which type error is more important to avoid. A practical guideline for applied research to follow is that if the costs (in money, time, etc.) are the same for both treatments, go for power: Find out whether there is a difference. However, if one treatment costs more than the other, avoid the type I error. You don't want to adopt a more expensive treatment (method, program, instrument) if it isn't *significantly* better.

The size of the sample is extremely influential on power. Power increases with increased *n*. Table 7.1 illustrates this using an ES of 0.5 and alpha at .05. With a very small number of participants per group, for example, *n* = 10, the power would be .20 (off the chart in

Table 7.1   Relationship of Sample Size and Power (ES = 0.5, alpha = .05)

n	Power
10	.20*
20	.30
50	.70
75	.85
100	.95*

*These values are off the chart in figure 7.3.

figure 7.3). This is only a 20% chance of detecting a real difference. Conversely, with a very large sample size, such as $n = 100$, the power is actually about .95, which is approaching certainty of finding a real difference.

Keep in mind the relationships of alpha, sample size, and ES in planning a study. If you have access to only a small number of participants, then you have to have a really large ES or use a larger alpha, or both. Don't just blindly specify the .05 alpha if detecting a real difference is the main issue. Use a higher one, such as .20 or even .30. This is very pertinent in pilot studies.

In chapter 9 we discuss some of the things you can do to increase the ES. Sometimes, however, in planning a study, you may not be able to determine an expected ES from the literature. This is where you can benefit from a pilot study to try to estimate the ES.

Our approach in chapter 9 is more post hoc, as it involves reporting the effect size and variance accounted for in significant findings and interpreting whether this is a meaningful effect. The a priori procedures described previously are more desirable but not always applicable. (For a more detailed discussion of important factors associated with establishing significance levels and determining power, see Franks & Huck, 1986, or Thomas, Lochbaum, et al., 1997.)

Finally, a summary of all this is well presented in the short poem, below, by Rosenthal (1991, 221).

To summarize, knowledge of four concepts is needed for planning and evaluating any quantitative study: alpha, power, sample size, and effect size.

- Alpha—establishes the acceptable magnitude of Type I error (chances of rejecting a true null hypothesis), the level of significance selected; typically an arbitrary value; often .05 or .01 in behavioral and biological areas of physical activity

- Power—the chances of rejecting a false null hypothesis; based on the magnitude of beta (chances of making a Type II error, accepting a false null hypothesis); beta is typically set at 4 × alpha, so if alpha is .05, then beta is .20; power is typically established as 1 − beta or .80 in the behavioral and biological areas of physical activity

- Sample size—the number of participants in the study being evaluated or planned

Achieving Power

I. The Problem

Oh, *F* is large and *p* is small
That's why we are walking tall.

What it means we need not mull
Just so we reject the null.

Or Chi-square large and *p* near nil
Results like that, they fill the bill.

What if meaning requires a poll?
Never mind, we're on a roll!

The message we have learned too well?
Significance! That rings the bell!

II. The Implications

The moral of our little tale?
That we mortals may be frail.
When we feel a *p* near zero
Makes us out to be a hero.

But tell us then is it too late?
Can we perhaps avoid our fate?
Replace that wish to null-reject
Report the *size* of the effect.

That may not insure our glory
But at least it tells a story
That is just the kind of yield
Needed to advance our field.

Reprinted, by permission, from R. Rosenthal, 1991, "Cumulating psychology: An appreciation of Donald T. Campbell," *Psychological Science* 2(213): 221.

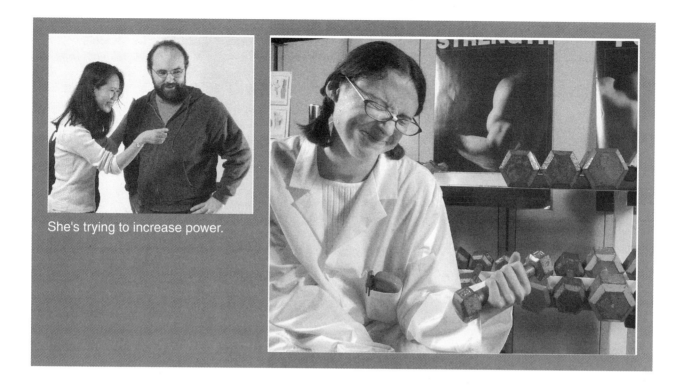

She's trying to increase power.

- Effect size—the outcome of a study typically expressed in standard deviation units (Thomas & Nelson, 2001):

$$ES = (M_1 - M_2) / SD \qquad (7.2)$$

r^2 is also a form of effect size or can be expressed in standard deviation units using the formula (Rosenthal, 1994):

$$ES = [r / \sqrt{(1 - r^2)}] \, [\sqrt{(df\,(n_1 + n_2)} / n_1\,n_2] \qquad (7.3)$$

Most often in evaluating published studies, all of the information to determine the four concepts is included, although everything might not be in final form (e.g., the effect size can be calculated from the means and standard deviations or converted from r).

Using Information in the Context of the Study

The item missing in these four concepts that is used to evaluate all research is the one of greatest interest: **context.** How do the findings from this study fit into the context of theory and practice? In fact, the question of greatest interest to the reader of research (which should make it of considerable interest to the researcher) is one that often goes unanswered: Are effect sizes for significant findings large enough to be meaningful when interpreted within the context of the study, or for the application of findings to other related samples, or for planning a related study?

In planning research, the experimenter often has either previous findings or pilot research from which appropriate effect sizes of important variables can be estimated. If the experimenter sets alpha at .05 and power at .80, then the question becomes, How many participants would be needed to detect an effect of a specific magnitude if the treatment works or the relationship is present? Charts and tables in books (see figure 7.3) as well as computer programs will estimate the number of participants needed, given that the experimenter has established or estimated the other three components. The Power Calculator, online at http://calculators.stat.ucla.edu/powercalc, will do a power analysis for you.

context

The interrelationships found in the 'real-world' setting. (See also chapter 12 on use of context.)

One of the major benefits of using effect sizes in estimating the meaningfulness of data is that they are not directly sensitive to sample size. Certainly effect sizes have some sensitivity to sample size because the size of the standard deviation is influenced by the sample size. Effect sizes are based on either the difference between the means (divided by the standard deviation) or the size of the correlation. Table 7.2 provides the magnitude of effect sizes as they reflect how two treatment groups overlap in their scores. Note that when an effect size is 0, the distribution of the scores of the two groups overlaps completely and the midpoints are the same (50% of participants' scores on either side of the mean). As effect sizes become larger, there is a net gain so that a greater percent of the participants' scores in one group (typically the experimental) exceeds the midpoint (50%) of the distribution of the other group's scores (control). Effect sizes are often interpreted by their absolute size. In table 7.2, a small effect size (often labeled 0.2 or less) suggests that the experimental group has a net gain (hopefully as a result of the treatments) of 8% over the control group; that is, 58% of the distribution of the experimental group's scores is beyond the midpoint (50%) of the control group's scores. A moderate effect size (often labeled about 0.5) shows a net gain of 19% over the midpoint of the control group's scores, whereas a large effect size (often labeled 0.8 and above) has a net gain of 29%. This is an excellent way of describing the influence of changes in an experimental group compared to a control group. That is, if the experimental and control groups begin at similar points on the dependent variable, what percent of experimental participants' scores are above the midpoint (50%) in the distribution of the control group's scores. Thus, the larger the effect size, the less the overlap between the distribution of scores in the two groups.

Finally, as we move into the next three chapters on statistics, remember that general linear model statistical analyses are only of value if the sample is neither too small nor too large. In very small samples not only can a single unusual value influence the results substantially, but within- and between-participant variation (the error variance) tends to be large and this makes the error term in tests of significance large, resulting in few significant

Table 7.2 Interpretation of Effect Size Using the Distribution From the Normal Curve

Effect size	Percent of experimental group difference above control group mean	Percent − 50
0.0	50	0
0.1	54	4
0.2	**58**	**8**
0.3	62	12
0.4	65	15
0.5	**69**	**19**
0.6	73	23
0.8	**79**	**29**
1.0	84	34
1.2	88	38
1.5	93	43
2.0	98	48

Adapted from McNamara 1994.

findings. On the other extreme of sample size, statistics has little value for very large samples because nearly any difference or relationship is significant. Recall that the error variance in significance tests is divided by the degrees of freedom, a calculation based on sample size. Thus, you will notice that large public health studies tend to discuss data analyses such as risk factors, percent of the sample with a characteristic (prevalence, incidence, mortality rate), and odds ratios rather than significance (see chapter 17 by B.E. Ainsworth and C.E. Matthews in this text).

In chapter 3 we discussed the research and null hypotheses. Cohen (1990, 1994) and Hagen (1997) have provided interesting insights into the value (or lack thereof) of the null hypothesis as a statistical test. However, in the behavioral and biological sciences of physical activity dealing mostly with human participants, at some level the null hypothesis is nearly always false (Cohen, 1990; but see Hagen, 1997, for a different view). Remember, conceptually the null hypothesis comes from the development of statistical tables that the researchers use to compare their test statistic (e.g., F ratios). These tables are based on population estimates of how often randomly selected samples might differ if there really is no difference. Thus, $p = .05$ means that two randomly selected samples of a certain size (n) will only differ by chance 5 times in 100. We use these tabled values for comparison with our calculated statistical values (e.g., F ratios). However, two groups of humans are hardly ever exactly alike on a measured variable, and if there are enough participants, p will be less than .05. Testing the null hypothesis is the wrong concept. The correct question is What is a meaningful difference (effect size) within the context of theory and application, and how many participants are required to reject a difference this large at $p < .05$ and power of $>.80$? Of course $p < .05$ and power greater than .80 are arbitrary values.

Remember, statistical routines are just that: computer routines into which numbers are input and that provide standardized output based on those numbers. Just as the numbers we assign to variables "don't know where they come from" (someone has developed the procedures for assigning values to certain levels of characteristics), statistical computer programs don't know where the numbers come from, nor do they care. You as the researcher must verify that the numbers are "good" ones. You are also the person who must verify that the numbers (data) meet the assumptions used in the development of the tables to which they are being compared (e.g., normal distribution of data for the F table, or the use of the χ^2 table for non-normal distributions). This is the very basis for evaluating data and making decisions between parametric and nonparametric statistical analyses (more on this in chapter 10).

Context is what matters with regard to meaningfulness. You must ask yourself, "Within the context of what I do, does an effect of this size matter?" The answer nearly always depends on who you are and what you are doing (and practically never on whether $p = .05$ or .01). Thus, having a significant (reliable) effect is a necessary, but not sufficient,

Context: The Key to Meaning

Here is a simple example of context. Suppose we told you that we have developed (and advertised on TV) a program of physical training, the "Runflex System," that will improve your speed in the 100-meter dash by 100 milliseconds. This system requires that you attach cables to your feet from flexible rods, lie on your back, and flex the rods on the Runflex apparatus by pulling your knees toward your chest. This system is guaranteed to work in 6 weeks or your money back. It has been tested by a major university laboratory in exercise physiology and does produce an average improvement of 100 milliseconds in speed in the 100-meter dash. Would you be willing to send us three easy payments of $99.95 for this system? Right now you are saying to yourself, "Are these people nuts? I don't even run 100-meter dashes." However, suppose you were the second fastest 100-meter sprinter in the world, and the fastest sprinter ran 100 milliseconds faster than you. Would you buy our Runflex apparatus now?

condition in statistics. To meet the criteria of being both necessary and sufficient, the effect must be significant and meaningful within the context of its use. Said another way:

- Estimates of significance are driven by sample size.
- Estimates of meaningfulness are driven by the size of the difference.
- Context is driven by how the findings will be used.

Quantitative researchers often do not provide appropriate contexts for their studies, especially the important findings. Some context may be provided in the introduction to papers when the rationale for the research is developed, but even then it is sometimes difficult to understand why a study is important and how it fits in the scheme of things. In the discussion section we often tend to "let the data speak for themselves" with a minimal attempt at explaining why the findings are important, how they support (or do not support) a theory, why the findings add to previous knowledge (or dispute it), what are logical subsequent research questions, and why practitioners should care about the outcomes.

Qualitative researchers are typically much stronger than quantitative researchers at placing findings in context (see Biddle et al., 2001, for a good discussion). Of course this is because context drives their research plan, data collection, data analysis, and interpretation of findings. We are not suggesting a switch to qualitative research by quantitative scholars; we have learned much, and much is yet to be learned from quantitative research. (In addition, we are too old to learn all the techniques.) However, considerable value exists in quantitative researchers learning to use context as the basis for planning and interpreting research, particularly important outcomes.

Reporting Statistical Data

A consistent dilemma among researchers, statisticians, and journal editors concerns the appropriate reporting of statistical information in published research papers. In recent years, some progress has been made that involves two issues in particular—always reporting some estimate of the size and meaningfulness of the finding along with the reliability or significance of the finding. Two organizations of importance to our field—the American Physiological Society (2004) and the American Psychological Association (1999)—have now published guidelines regarding these issues.

Following are summaries of general guidelines taken from these two sources:

- Information on how sample size was determined is always important. Indicate the information (e.g., effect sizes) used in the power analysis to estimate sample size. When the study is analyzed, confidence intervals are best used to describe the findings.

- Always report any complications that have occurred in the research, including missing data, attrition, and nonresponse and how these problems were handled in data analysis. "Before you compute any statistics, look at your data" *(American Psychologist)*. You should always screen your data (this is not tampering with data) to be sure the measurements make sense.

- Select minimally sufficient analyses. Using complicated methods of quantitative analyses may fit data appropriately and lead to useful conclusions, but many designs fit basic and simpler techniques. When they do, these should be the statistics of choice. Your job is not to impress your reader with your statistical knowledge and expertise but to appropriately analyze the research and present it so that a reasonably well-informed person can understand it.

- Report actual p values, or better yet, confidence intervals. Always report an estimate of the magnitude of the effect. If the measurement units (e.g., maximal oxygen consumption) have real meaning, then reporting them in the original metric, such as mean difference, is useful. Otherwise, standardized reporting such as effect size or r^2 is useful. In addition, placing these findings in practical and theoretical context adds much to the report.

- Control multiple comparisons through techniques such as Bonferroni.

- Always report variability using the standard deviation. Standard error characterizes the uncertainty associated with a population and is most useful in determining confidence intervals.
- Report the level of your data (e.g., how many decimal places) that is appropriate for scientific relevance.

Summary

In this chapter you have learned the interrelationships among alpha, power, sample size, and effect size. Appropriate use of this information is the most important aspect of planning your own study or evaluating someone else's. Placing this information in the context in which you plan the research or plan to use the results allows others to interpret and use your research findings appropriately.

✔ Check Your Understanding

1. Locate a data-based paper from a research journal. Answer the following questions about this paper:
    a. What probability levels did they use to test their hypotheses?
    b. What was the sample size(s) for the group(s)?
    c. What was the effect size for some aspect of the study?
    d. Estimate power by using a, b, and c.
2. Draw a figure (such as figure 7.2) with two normal curves (one for an experimental group and one for a control group) that reflects an effect size of 0.2. Do this again for an effect size of 0.5. Repeat again for an effect size of 0.8.

Relationships Among Variables

Statistics are like a bikini. What they reveal is suggestive, but what they conceal is vital.

—Aaron Levenstein

In chapter 6 we promised that after we had presented some basic information to help you understand statistical techniques, we would begin to explain in detail some specific techniques. We begin with correlation.

Correlation is a statistical technique used to determine the relationship between two or more variables. In this chapter we discuss several types of correlation, the reliability and meaningfulness of correlation coefficients, and the use of correlation for predictions, including partial and semipartial correlations and multiple-regression equations. Finally, we briefly overview multivariate forms of correlation: canonical correlation, factor analysis, and structural modeling.

correlation

A statistical technique used to determine the relationship between two or more variables.

What Correlational Research Investigates

Often a researcher is interested in the degree of relationship between, or correlation of, performances, such as the relationship between performances on a distance run and a **step test** as measures of cardiovascular fitness. Sometimes an investigator wishes to establish the relationship between traits and behavior, such as how personality characteristics relate to participation in high-risk recreational activities. Still other correlational research problems might involve relationships between anthropometric measurements, such as skinfold thicknesses and percentage fat as determined by underwater weighing. Here, the researcher may wish to predict percentage fat from the skinfold measurements.

Correlation may involve two variables, such as the relationship between height and weight. It may involve three or more variables, as when one investigates the relationship between a criterion (dependent variable) such as cardiovascular fitness and two or more predictor variables (independent variables) such as body weight, percentage fat, speed, muscular endurance, and so on. This is multiple correlation. Another technique, canonical correlation, establishes the relationships between two or more dependent variables and two or more independent variables. Factor analysis uses correlations among a number of variables to try to identify the underlying relationships or factors. Finally, structural modeling provides evidence for how variables may directly or indirectly influence other variables.

step test

Test used to measure cardiorespiratory fitness involving the measurement of pulse rate after stepping up and down on a bench.

coefficient of correlation

A quantitative value of the relationship between two or more variables that can range from .00 to 1.00 in either a positive or negative direction.

Understanding the Nature of Correlation

The **coefficient of correlation** is a quantitative value of the relationship between two or more variables. The correlation coefficient can range from .00 to 1.00 in either a positive or negative direction. Thus, perfect correlation is 1.00 (either +1.00 or –1.00), and no relationship at all is .00 (see below for examples of perfect correlations and examples of no correlation).

Correlation: Perfect and Not So Perfect

Quotes from anonymous famous people about perfect correlation (r = 1.00)

Smoking kills. If you are killed, you've lost a very important part of your life.

Traditionally, most of Australia's imports come from overseas.

It's bad luck to be superstitious.

Things are more like they are now than they ever were before.

The police are not here to create disorder but to preserve disorder.

The Internet is a great way to get on the net.

The President has kept all the promises he intends to keep.

China is a big country inhabited by many Chinese.

That low-down scoundrel deserves to be kicked to death by a jackass, and I'm just the one to do it.

It's like déjà vu all over again.

The loss of life will be irreplaceable.

Students' statements that are uncorrelated to teaching performance

Sometimes students write interesting things on the teaching evaluations for our research methods course. Below are a few of the statements that have an r = .00 to our teaching.

1. This class was a religious experience for me. I had to take it all on faith.
2. Textbook makes a satisfying "thud" when dropped on the floor.
3. Have you ever fallen asleep in one class and awakened in another?
4. Help! I've fallen asleep and can't wake up.
5. He teaches like Speedy Gonzalez on a caffeine high.
6. I'm learning by osmosis; I sleep with my head on the textbook.
7. Class is a great stress reliever—I was so confused I forgot who I was.
8. This class kept me out of trouble from 2:30–4:00 P.M. on Tuesday and Thursday.

---

positive correlation

A relationship between two variables in which a small value for one variable is associated with a small value for another variable, and a large value for one variable is associated with a large value for the other.

Positive Correlation

A **positive correlation** exists when a small value for one variable is associated with a small value for another variable and a large value for one variable is associated with a large value for another. Strength and body weight are positively correlated: Heavier people are generally stronger than lighter people. (The correlation is not perfect because some lighter people are stronger than some heavier people and weaker than some who weigh even less.)

Figure 8.1 is a graphic illustration of a perfect positive correlation. Notice that Tom's body weight is 1*s* above the mean, for he weighs 110 lb, and the mean is 90. Thus, he is 20 lb heavier than the mean, which is 1*s* (*s* = 20). His strength score is 250 lb, which is 50

She doesn't quite get the concept of *structural modeling.*

lb higher than the mean of 200. Because the standard deviation for strength is 50, he is 1*s* above the mean for strength, just as he was for body weight. A second boy, Bill, is 1*s* below the mean for weight and 1*s* below for strength. A third boy is exactly at the mean on both variables (he weighs 90 lb and has a strength score of 200). Joe is 0.50*s* above and Dick 0.50*s* below the means for weight and strength. Thus, when the scores are plotted, they form a perfectly straight diagonal line. This is perfect correlation (*r* = 1.00). The relative positions of the boys' pairs of scores are identical in the two distributions. In other words, each boy is the same relative distance from the mean of each set of scores. Common sense tells us that perfect correlation does not exist in human traits, abilities, and performances because of so-called people variability and other influences.

Figure 8.2 illustrates a more realistic relationship between body weight and strength (*r* = .67). (Fictional examples have been used for purposes of presentation in this and other chapters. These examples do not represent actual data.) When these 10 body weights and strength scores are plotted, they no longer constitute a straight line but do form a diagonal plot in the same lower-left-to-upper-right pattern as in figure 8.1.

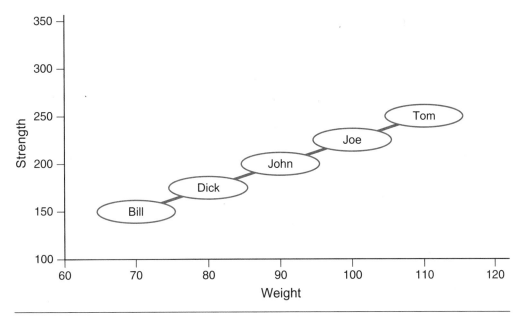

Figure 8.1   Perfect positive correlation (*r* = 1.00).

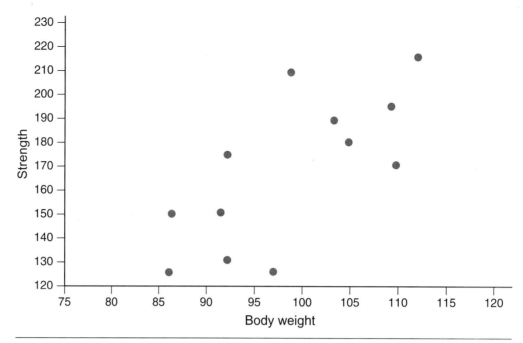

Figure 8.2   A more realistic correlation between body weight and strength (*r* = .67).

Negative Correlation

In figure 8.3, we have plotted the body weights and pull-up scores for the same 10 boys. A pull-up is performed by hoisting one's body weight from a hanging position until the chin is above the bar. For this test, body weight is somewhat of a liability; heavier people tend to do fewer pull-ups than do lighter people. As a result, a small number of pull-ups is associated with larger body weights and, conversely, a greater number of

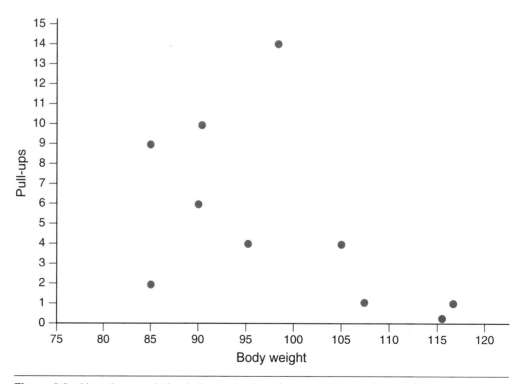

Figure 8.3   Negative correlation between body weight and pull-ups (*r* = −.54).

pull-ups with lesser body weights. This is a **negative correlation.** A perfect negative correlation would be a straight diagonal line at a 45° angle (the upper left corner of the graph to the lower right corner). Figure 8.3 depicts a negative correlation of a moderate degree ($r = -.54$). However, an upper-left-to-lower-right pattern is still apparent.

Patterns of Relationships

Figure 8.4 is a hypothetical example of four patterns of relationships between two variables. Figure 8.4*a* is a positive relationship as previously described. Figure 8.4*b* is a negative relationship, also previously described. When virtually no relationship exists between variables, the correlation is .00, as shown in figure 8.4*c*. This denotes independence between sets of scores. The plotted scores exhibit no discernible pattern at all. Finally, two variables may not have a linear relationship but can still be related, as in figure 8.4*d,* showing a curvilinear relationship. The interpretation of the reliability and meaningfulness of correlations is explained later in this chapter.

Correlation and Causation

At this point we must stress again that a correlation between two variables does not mean that one variable causes the other. In chapter 4 we used an example of a research hypothesis (albeit a dumb one) that achievement in math could be improved by buying a child larger shoes. This hypothesis resulted from a correlation between math achievement scores and shoe size in elementary school children (with age not controlled). This example helps to drive home the point that correlation does not mean causation.

negative correlation

A relationship between two variables in which a small value for the first variable is associated with a large value for the second variable, and a large value for the first variable is associated with a small value for the second variable.

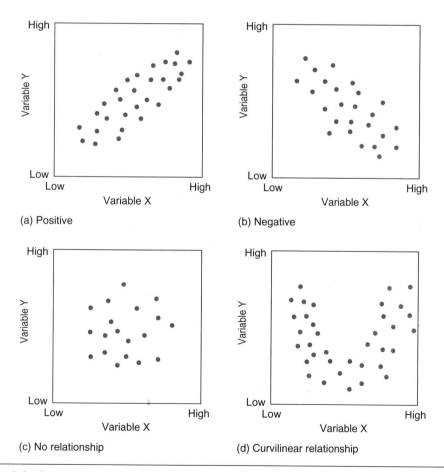

(a) Positive

(b) Negative

(c) No relationship

(d) Curvilinear relationship

Figure 8.4   Patterns of relationships.

In Ziv's 1988 study on the effectiveness of teaching and learning with humor, the experimental group participants were told a story to illustrate the fact that correlations do not show a causal effect. In the story, aliens from another planet, who were invisible to earthlings, decided to study the differences between fat and thin people. One alien observed the coffee drinking behavior of fat and thin people at a cafeteria over an extended period. The alien then analyzed the observations statistically and decided that there was a correlation between coffee-drinking behavior and body weight. Thin people tended to drink their coffee with sugar, whereas fat people mostly drank theirs with an artificial sweetener. The alien concluded from this study that sugar makes humans thin, whereas artificial sweeteners make humans fat.

We are not saying that one variable cannot be the cause of another. For cause and effect to occur, the two variables must be correlated. We are only saying that you can't infer this as a result of a correlation. Correlation is a necessary but not sufficient condition for causation. The only way that causation can be shown is with an experimental study in which an independent variable can be manipulated to bring about an effect.

Pearson Product Moment Correlation

Pearson product moment coefficient of correlation

The most commonly used method of computing correlation between two variables; also called *interclass correlation, simple correlation,* or *Pearson r.*

Several times in the preceding discussion we have used the symbol *r*. This symbol denotes the **Pearson product moment coefficient of correlation.** In this type of correlation, there is one criterion (or dependent) variable and one predictor (or independent) variable. Thus, every subject has two scores, such as body weight and strength. An important assumption for the use of *r* is that the relationship between the variables is expected to be linear, that is, a straight line is the best model of the relationship. When that is not true (e.g., figure 8.4*d*), *r* is an inappropriate way to analyze the data. If you were to apply the statistical formula to calculate *r* to the data in figure 8.4*d*, the outcome would be an *r* of about .00, no relationship. Yet we can clearly see from the plot that a relationship does exist; it is just not a linear one.

The computation of the correlation coefficient involves the relative distances of the scores from the two means of the distributions. The computations can be accomplished with a number of different formulas; we present just one. This formula is sometimes called the computer method because it involves operations similar to those performed by a computer. The formula appears large and imposing but actually consists of only three operations:

1. Sum each set of scores.
2. Square and sum each set of scores.
3. Multiply each pair of scores and obtain the cumulative sum of these products.

The formula is

$$r = \frac{N\Sigma XY - (\Sigma X)(\Sigma Y)}{\sqrt{N\Sigma X^2 - (\Sigma X)^2}\sqrt{N\Sigma Y^2 - (\Sigma Y)^2}} \qquad (8.1)$$

To illustrate the calculations involved, we use the body weight and pull-up scores from figure 8.3 and designate body weight as the *X* variable. In a correlation problem that simply determines the relationship between two variables, it does not matter which one is *X* and which is *Y*. If the investigator wants to predict one score from the other, then *Y* designates the criterion (dependent) variable (that which is being predicted) and *X* the predictor (independent) variable. In this example, pull-up performance would be predicted from body weight, as it would not make much sense to predict body weight from pull-up scores. Prediction equations are discussed later in this section.

The computations for the correlation between body weight and pull-ups are shown in example 8.1. Note that *N* refers to the number of paired scores, not the total number of scores.

For the $(\Sigma X)^2$ and the $(\Sigma Y)^2$ values, the sums of the raw scores for *X* and *Y* are then squared. Notice that these values are not the same as the ΣX^2 and ΣY^2 values, for which

Example 8.1

Known Values

Body weight X	X²	Number of pull-ups Y	Y²	Body weight × pull-ups XY
104	10,816	4	16	416
86	7,396	2	4	172
92	8,464	6	36	552
112	12,544	1	1	112
96	9,216	4	16	384
98	9,604	13	169	1,274
110	12,100	0	0	0
86	7,396	9	81	774
105	11,025	1	1	105
91	8,281	10	100	910

Number of paired scores: $N = 10$

Sum of body weight scores: $\Sigma X = 980$

Sum of pull-up scores: $\Sigma Y = 50$

Sum of body weight scores squared: $\Sigma X^2 = 96,842$

Sum of pull-ups squared: $\Sigma Y^2 = 424$

Sum of body weight × pull-ups: $\Sigma XY = 4,699$

Working It Out (equation 8.1)

$$r = \frac{(10)4,699 - (980)(50)}{\sqrt{10(96,842) - (980)^2}\sqrt{10(424) - (50)^2}}$$

$$r = \frac{46,990 - 49,000}{\sqrt{968,420 - 960,400}\sqrt{4,240 - 2,500}}$$

$$r = \frac{-2,010}{\sqrt{8,020}\sqrt{1,740}}$$

$$r = \frac{-2,010}{(89.6)(41.7)} = \frac{-2,010}{3,736.3} = -.54$$

X and Y are squared and then summed. The ΣXY (the sum of the cross products of the X and Y scores) determines the direction of the correlation, that is, whether it is positive or negative. In this example, a negative correlation was obtained because the first half of the numerator, $N\Sigma XY$, was smaller than the second half, $(\Sigma X)(\Sigma Y)$.

The negative correlation between body weight and pull-ups means that participants with greater body weight had lower pull-up performance, while those with lower body weight had higher pull-up performance. Sometimes a negative correlation coefficient results when the relationship is really positive. Confusing? The following examples should clarify the meaning. Suppose we were to correlate scores on the vertical jump and the 40-yard dash. Both performances are related to explosive power. People who score well on the vertical jump should also do well on the 40-yard dash because the two tests are measuring much the same thing; thus, the relationship between performances is positive. However, the vertical jump is scored by distance (in inches or centimeters), so a high score is good, and the 40-yard dash is scored in seconds, so a low number is good. Therefore, the correlation coefficient would be negative even though the relationship is positive.

The correlation between the distance a person could run in 12 minutes and heart rate after exercise would also be negative. This is because a greater distance covered is good while a lower heart rate after exercise is good. A person who has good cardiovascular endurance would have a high score on one test (the run) and a low score on the other.

What the Coefficient of Correlation Means

So far we have dealt with the direction of correlation (positive or negative) and the calculation of r. An obvious question that arises is, What does a coefficient of correlation mean in terms of being high or low, satisfactory or unsatisfactory? This seemingly simple question is not so simple to answer.

Interpreting the Reliability of r

significance

The reliability of or confidence in the likelihood of a statistic occurring again if the study were repeated.

First, there are several ways of interpreting r. One criterion is its reliability, or **significance.** Does it represent a real relationship? That is, if the study were repeated, what is the probability of getting a similar relationship? For this statistical criterion of significance, simply consult a table. In using the table, select the desired level of significance, such as the .05 level, and then find the appropriate degrees of freedom (df, which is based on the number of participants corrected for sample bias), which, for r, is equal to $N - 2$. Table A.3 in appendix A contains the necessary correlation coefficients for significance at the .05 and .01 levels. Refer to the example of the correlation between body weight and pull-ups ($r = -.54$). The degrees of freedom are $N - 2 = 10 - 2 = 8$ (remember, the variable N in correlation refers to the number of pairs of scores). When reading the table at 8 df, we see that a correlation of .632 is necessary for significance of a two-tailed test at the .05 level (and .765 at the .01 level). Therefore, we would have to conclude that our correlation of $-.54$ is not significant. (We explain whether to find the correlation values under the column for a one-tailed or two-tailed test in the section on interpreting t in chapter 9.)

Another glance at table A.3 reveals a couple of obvious facts. The correlation needed for significance decreases with increased numbers of participants (df). In our example, we had only 10 participants (or pairs of scores). However, if four more boys had been in the sample ($N = 14$), there would have been 12 df, and the correlation required for significance at the .05 level for 12 df is .532. Our correlation of $-.54$ would have met that test of significance. Very low correlation coefficients can be significant if you have a large sample of participants. At the .05 level, $r = .38$ is significant with 25 df, $r = .27$ is significant with 50 df, and $r = .195$ is significant with 100 df. In fact, with 1,000 df, a correlation of .08 is significant at the .01 level.

The second observation to note from the table is that a higher correlation is required for significance at the .01 level than at the .05 level. This should make sense. Remember, chapter 7 stated that the .05 level means that if 100 experiments were conducted, the null hypothesis (that there is no relationship) would be rejected incorrectly, just by chance, on 5 of the 100 occasions. At the .01 level, we would expect a relationship of this magnitude due to chance less than once in 100 experiments. Therefore, the test of significance at the .01 level is more stringent than at the .05 level, and so a higher correlation is required for significance at the .01 level. When you use most computer programs to calculate statistics like r, you are given the exact significance level in the printout. Using our example of body weight and pull-ups with $r = -.54$, the exact significance is $p = .109$.

Interpreting the Meaningfulness of r

The interpretation of a correlation for statistical significance is important, but because of the vast influence of sample size, this criterion is not always meaningful. As chapter

7 explained, statistics can answer two questions about data: Are the effects reliable? Are they meaningful?

The most commonly used criterion for interpreting the meaningfulness of the correlation coefficient is the **coefficient of determination** (r^2). In this method, the portion of common association of the factors that influence the two variables is determined. In other words, the coefficient of determination indicates the portion of the total variance in one measure that can be explained, or accounted for, by the variance in the other measure.

Figure 8.5 offers a visual depiction of this idea. Circle A represents the variance in one variable, while circle B represents the variance in a second variable. At the top is a relationship where r = .00 (no overlap between the two); thus, r^2 = .00. At the bottom of figure 8.5, r = .71; thus r^2 = .50 (considerable overlap between the two).

For example, the standing long jump and the vertical jump are common tests of explosive power. The tests are so commonly used that we tend to think of them as interchangeable, that is, as measuring the same thing. Yet correlations between the two tests usually range between .70 and .80. Hence, the coefficients of determination range from .49 ($.70^2$) to .64 ($.80^2$). Usually, the coefficient of determination is expressed as percentage of variation. Thus, $.70^2$ = .49 = 49% and $.80^2$ = .64 = 64%.

For a correlation of .70 between the standing long jump and the vertical jump, only about half (49%) of the variance (or influences) in one test is associated with the other. Both tests involve explosive force of the legs with some flexion and extension of the trunk and swinging of the arms. Both are influenced by body weight in that the participant must propel his or her body through space, both involve the ability to prepare psychologically and physiologically to generate explosive force, both involve **relative strength,** and so on. These are factors held in common in the two tests. If r = .80, then 64% of the performance in one test is associated with, or explained by, the factors involved in the performance of the other test.

But what about the unexplained variance ($1.0 - r^2$)? With a correlation of .70, there is 49% common (explained) variance and 51% ($1.00 - .70^2$) error (unexplained) variance. What are some factors unique to each test? We cannot explain this fully, but some of the factors could be that (a) the standing long jump requires that the body be propelled forward and upward, whereas the vertical jump is only upward; (b) the scoring of the vertical jump neutralizes one's height because standing reach is subtracted from jumping reach, but in the standing long jump, perhaps the taller person has some advantage; and (c) perhaps more skill (coordination) is involved in the vertical jump because the person must jump and turn and then touch the wall. This is not intended to be any sort of mechanical analysis of the two tests. These are simply suggestions of some possible factors of common association, or explained variance, and some factors that might be unexplained or unique to each test of explosive power.

When we use the coefficient of determination to interpret correlation coefficients, it becomes apparent that a rather substantial relationship is needed to account for a great amount of common variance. It takes a correlation of .71 to account for just half the variance in the other test, and a correlation of .90 accounts for only 81%. In some of the standardized tests used to predict academic success, correlations are generally quite low, often around .40. You can see by the coefficient of determination that a correlation of .40 accounts for only 16% of the factors contributing to academic success; therefore, there is a great deal of unexplained variance. Still, these measures are often used very rigorously as the criterion for admission to academic programs. Of course, the use of multiple predictors can greatly improve the estimate for success. The sizes of correlations can also be compared using the coefficient of determination. A correlation of .90 is not three times larger than a correlation of .30; it is nine times larger ($.30^2$ = .09, or 9%, and $.90^2$ = .81, or 81%).

Interpreting the correlation coefficient is further complicated because whether a correlation is "good" or "inadequate" depends on the purpose of the correlation. For example, if we are looking at the reliability (repeatability) of a test, a much higher correlation is needed than if we are determining simply whether there is a relationship between two variables. A correlation of .60 would not be acceptable for the relationship between two

coefficient of determination

The squared correlation coefficient; used in interpreting meaningfulness of correlations.

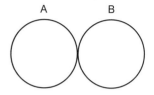

No relationship between A and B, r^2 = .00

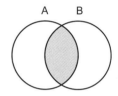

About 50% common variance between A and B, r^2 = .50

Figure 8.5 Example of shared variance between variables.

relative strength

The measure of the ability to exert maximum force in relation to a person's size.

similar versions of an exercise knowledge test, but a correlation of .60 between exercise knowledge and exercise behavior would be quite noteworthy.

Z Transformation of r

Fisher Z transformation

Method of approximating normality of a sampling distribution of linear relationship by transforming coefficients of correlation to Z values.

Occasionally, a researcher wants to determine the average of two or more correlations. It is statistically unsound to try to average the coefficients themselves because the sampling distribution of coefficients of correlation is not normally distributed. In fact, the higher the correlation, in either a positive or a negative direction, the more skewed the distribution becomes. The most satisfactory method of approximating normality of a sampling distribution of linear relationships is by transforming coefficients of correlation to Z values. This is often called the **Fisher Z transformation.** (This Z should not be confused with the z used to refer to the height of the ordinate in the area of the normal curve.)

The transformation procedure involves the use of natural logarithms. However, we need not use Fisher's formula to calculate the transformations (these conversions have been done for us in table A.4). We simply consult the table and locate the corresponding Z value for any particular correlation coefficient.

Suppose, for example, that we obtained correlations between maximal oxygen consumption and a distance run (e.g., an 8-min run-walk) on four groups of participants of different ages. We would like to combine these sample correlations to obtain a valid and reliable estimate of the relationship between these two measures of cardiorespiratory endurance. The data for the following steps are shown in table 8.1.

1. First, convert each correlation to a Z value using table A.4. For example, the correlation of .69 for the 13- and 14-year-olds has a corresponding Z value of 0.85; the next correlation of .85 for the 15- and 16-year-olds has a Z value of 1.26; and so on.

2. The Z values are then weighted by multiplying them by the degrees of freedom for each sample, which in this process is $N - 3$. So, for the 13- and 14-year-olds, the Z value of 0.85 is multiplied by 27 for a weighted Z value of 22.95. We do the same for the other three samples.

3. The weighted Z values are summed, and the mean weighted Z value is calculated by dividing by the sum of degrees of freedom ($N - 3$): 137.70/135 = 1.02.

4. The mean weighted Z value is converted back to a mean correlation by consulting table A.4 again. We see that the corresponding correlation for a Z value of 1.02 is .77.

Some authors declare that to average correlations by the Z-transformation technique, you must first establish that there are no significant differences among the four correlations. A comparison for differences could be made using a chi-square test of the weighted Z values (chi square is discussed in chapter 10). Other statisticians contend that averaging coefficients of correlation is permissible as long as the average correlation is not interpreted in terms of confidence intervals.

Table 8.1   Average of Correlation Coefficients by Use of the Z Transformation

Age group	N	r	Z	N – 3	Weighted Z
13 – 14	30	.69	0.85	27	22.95
15 – 16	44	.85	1.26	41	51.66
17 – 18	38	.70	0.87	35	30.45
19 – 20	35	.77	1.02	32	32.64
				135	137.70

The *Z* transformation is also used for statistical tests (such as those for the significance of the correlation coefficient) and for determining the significance of the difference between two correlation coefficients. After reading chapter 9, you may wish to consult a statistics text, such as that by Mattson (1981), for further discussion on the use of the *Z* transformation for these procedures.

Using Correlation for Prediction

We have stated several times that one purpose of correlation is prediction. College entrance examinations are used to predict success. Sometimes we try to predict a criterion such as percentage fat by the use of skinfold measurements or maximal oxygen consumption by a distance run. In studies of this type, the predictor variables (skinfold measurements) are less time-consuming, less expensive, and more feasible for mass testing than the criterion variable; thus, a **prediction,** or **regression, equation** is developed.

Prediction is based on correlation. The higher the relationship between two variables, the more accurately you can predict one from the other. If the correlation were perfect, you could predict with complete accuracy.

Working With Regression Equations

Of course, we do not encounter perfect relationships in the real world, but in introducing the concept of prediction (regression) equations, it is often advantageous to begin with a hypothetical example of a perfect relationship.

Verducci (1980) provided one of the best examples in introducing the regression equation concerning monthly salary and annual income. If there are no other sources of income, we can predict with complete accuracy the annual income of, for example, teachers simply by multiplying their monthly salaries by 12. Figure 8.6 illustrates this perfect relationship. By plotting the monthly salary (the *X*, or predictor variable), the predicted annual income (the *Y*, or criterion variable) can be obtained. Thus, if we know that another teacher (e.g.,

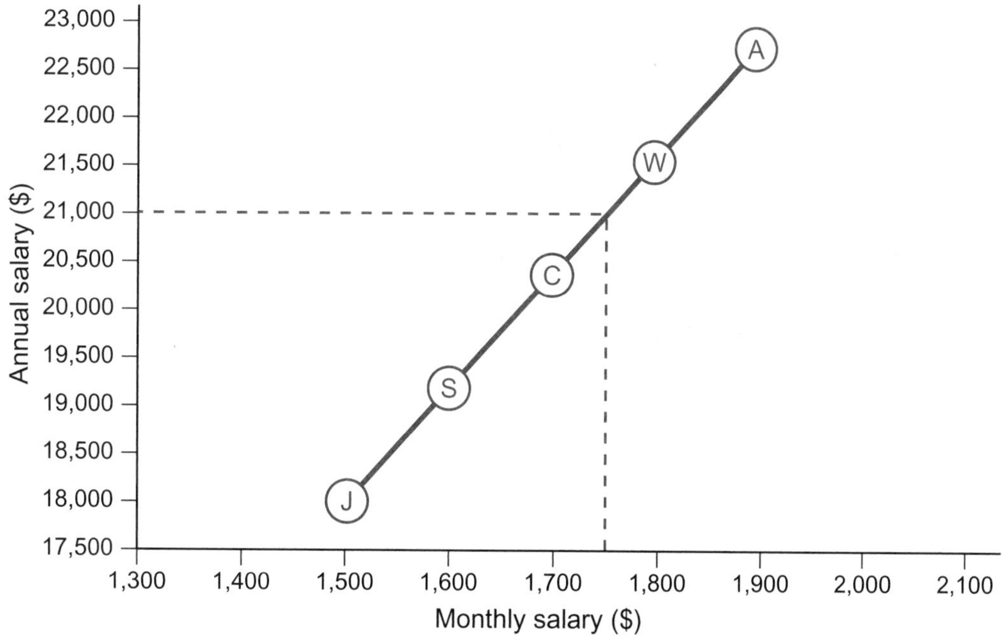

prediction equation

A formula to predict some criterion (e.g., some measure of performance) based on the relationship between the predictor variable(s) and the criterion; also called *regression equation.*

regression equation

See *prediction equation.*

Figure 8.6 Plotting monthly and annual salaries with perfect *r. Note:* Letters refer to teachers' initials.

abscissa

The horizontal, or x, axis of a graph.

ordinate

The vertical, or y, axis of a graph.

Ms. Brooks) earns a monthly salary of $1,750, we can easily plot this on the graph where $1,750 on the horizontal (x) axis (the **abscissa**) intercepts the vertical (y) axis (the **ordinate**) at $21,000. The equation for prediction (Y, predicted annual income) is thus $Y = 12X$. In the previous example, the teacher's monthly salary (X) is inserted in the formula: $Y = 12(1,750) = 21,000$.

Next, suppose that all teachers got an annual supplement of $1,000 for coaching or supervising cheerleaders or some other extracurricular activity. Now, the formula becomes $Y = 1,000 + 12X$. Ms. Brooks' annual income is predicted as follows: $Y = 1,000 + 12(1,750) = 22,000$. All teachers' annual incomes could be predicted in the same manner. This formula is the general formula for a straight line and is expressed as follows:

$$Y = a + bX \tag{8.2}$$

where Y = the predicted score, or criterion; a = the intercept; b = the slope of the regression line; and X = the predictor.

In this example, the b factor was ascertained by common sense because we know that there are 12 months in a year. The slope of the line (b) signifies the amount of change in Y that accompanies a change of 1 unit of X. Therefore, any X unit (monthly salary) is multiplied by 12 to obtain the Y value. In actual regression problems we do not intuitively know what b is, so we must calculate it by this formula:

$$b = r(s_Y/s_X) \tag{8.3}$$

where r = the correlation between X and Y, s_Y = the standard deviation of Y, and s_X = the standard deviation of X.

In our previous example, application of equation 8.3 uses these data:

X (monthly salary)	Y (annual income)
$M_X = 1,700$	$M_Y = 21,400$ (including $1,000 bonus)
$s_X = 141.42$	$s_Y = 1,697.06$
$r = 1.00$	

Therefore, $b = 1.00(1,697.06/141.42) = 12$. The a in the regression formula indicates the intercept of the regression line on the y axis. In other words, a is the value of Y when X is zero. On a graph, if you extend the regression line sufficiently, you can see where the regression line intercepts Y. The a is a constant because it is added to each of the calculated bX values. Once again, in our example we know that this constant is 1,000. In other words, this is the value of Y even if there were no monthly salary (X). But to calculate the a value, you must first calculate b. Then use the following formula:

$$a = M_Y - bM_X \tag{8.4}$$

where a = the constant (or intercept), M_Y = the mean of the Y scores, b = the slope of the regression line, and M_X = the mean of the X scores. In our example, $a = 21,400 - 12(1,700) = 1,000$. Then the final regression equation is $Y = a + bX$, or $Y = 1,000 + 12X$.

Next, let us use a more practical example in which the correlation is not 1.00. We can use the data from figure 8.2, where the correlation between body weight and dynamometer strength was .67. The means and standard deviations are as follows:

X (body weight)	Y (strength)
$M_X = 98.00$	$M_Y = 167.00$
$s_X = 9.44$	$s_Y = 33.52$
$r = .67$	

First, we calculate b as follows from equation 8.3:

$$b = r(s_Y/s_X) = .67(33.52/9.44) = 2.38$$

Then, a is calculated as follows from equation 8.4:

$$a = M_y - bM_x = 167 - 2.38(98) = -66.24$$

The regression formula (equation 8.2, $Y = a + bX$) becomes

$$Y = -66.24 + 2.38X$$

For any body weight, we can calculate the predicted strength score. For example, a boy weighing 100 lb (X) would have a predicted strength score $Y = -66.24 + (2.38)100 = 171.8$. The main difference between this example and that of monthly and annual salaries is that the latter had no error of prediction because the correlation was 1.00. When we predicted strength from body weight, however, the correlation was less than 1.00, so there is an error of prediction.

Calculating a Line of Best Fit

Before presenting the formula for calculating error of prediction, let us return to the derivation of the prediction formula. Figure 8.2 shows no straight line connecting the weight and strength scores, as there was in the hypothetical example in figure 8.1. Consequently, we calculate a **line of best fit** to predict Y from the X scores. To do this, we take a high X score (body weight) such as 110 and a low body weight such as 91 and apply the prediction formula. For a body weight of 110, we predict $Y = -66.24 + (2.38)110 = 195.6$. For a body weight of 91, we predict $Y = -66.24 + (2.38)91 = 150.3$.

Then we plot these two predicted values and connect them with a straight line. This line passes through the intersection of the X and Y means. Figure 8.7 shows this line of best fit. The 10 actual body weight and strength scores are also plotted. You can readily see that the scores do not fall on the straight line, as they did with the perfect correlation example. Note again our previous caution: This type of correlation is useful only if a straight line is the best fit to the data. This assumption of a straight line fitting the data applies to all of the statistical techniques we discuss in chapters 8, 9, and 10, as does the assumption that the data are normally distributed.

In constructing this line of best fit, we selected a high body weight (110) and a low body weight (91) and predicted their Y values. When we examine the actual Y values, we see that there is some error in prediction. The predicted strength score for the 110-lb boy was 195.6, yet the boy actually scored only 170, a difference of –25.6. The 91-lb boy was

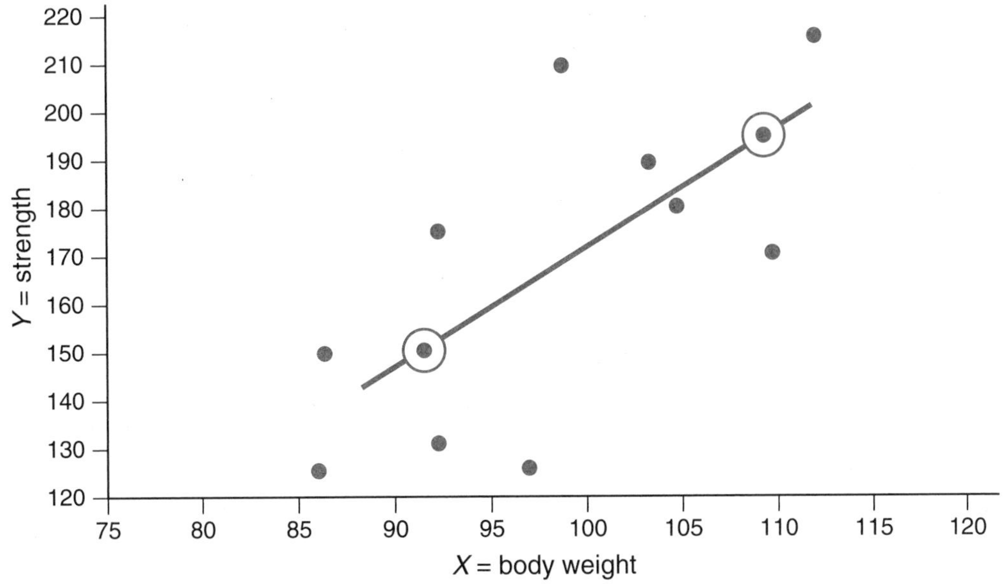

line of best fit

The calculated regression line that results in the smallest sum of squares of the vertical distances of every point from the line.

Figure 8.7   Regression line of best fit between body weight and predicted strength scores.

predicted to score 150.3 on the dynamometer, yet he scored 175, a difference of +24.7. These differences between predicted and actual Y scores represent the errors of prediction and are called **residual scores.** If we compute all the residual scores, the mean is zero, and the standard deviation is the **standard error of prediction,** or **standard error of estimate** ($s_{Y \cdot X}$).

A simpler way of obtaining the standard error of estimate is to use this formula:

$$s_{Y \bullet X} = s_Y \sqrt{1 - r^2}$$
$$s_{Y \bullet X} = 33.52\sqrt{1 - .67^2} = 24.9$$

(8.5)

The standard error of estimate is interpreted the same way as the standard deviation. In other words, the predicted value (strength) of a boy in our example, plus or minus the standard error of estimate, occurs approximately 68 times out of 100. Thus, we predict that a 104-lb boy will score 181.3 ± 24.9. To express it another way, the prediction range is between 156.4 and 206.2 lb 68 times out of 100 (or the chances are about 2 in 3).

The larger the correlation, the smaller the error of prediction. Also, the smaller the standard deviation of the criterion, the smaller the error. In the previous problem, if we had a correlation of .85, for example, the standard error of estimate would be only

$$s_{Y \bullet X} = 33.52\sqrt{1 - .85^2} = 17.6$$

The line of best fit is sometimes called the least-squares method. This means that the calculated regression line is one for which the sum of squares of the vertical distances of every point from the line is minimal. We will not develop this point here. Sum of squares is discussed in the next chapter.

Partial Correlation

The correlation between two variables is sometimes misleading and may be difficult to interpret when there is little or no correlation between the variables other than that caused by their common dependence on a third variable.

For example, many attributes increase regularly with age from 6 to 18 years, such as height, weight, strength, mental performance, vocabulary, reading skills, and so on. Over a wide age range, the correlation between any two of these measures will almost certainly be positive and will probably be high because of the common maturity factor with which they are highly correlated. In fact, the correlation may drop to zero if the variability caused by age differences is eliminated. We can control this age factor in one of two ways. We can select only children of the same age, or we can partial out (remove the influence of something or hold something constant) the effects of age statistically by holding it constant.

The symbol for partial correlation is $r_{12 \cdot 3}$, which means the correlation between variables 1 and 2 with variable 3 held constant (we could hold any number of variables constant, e.g., $r_{12 \cdot 345}$). The calculation of partial correlation among three variables is quite simple. Let us refer to the previous correlation of shoe size and achievement in mathematics. This is a good example of **spurious correlation,** which means that the correlation between the two variables is due to the common influence of another variable (age or maturing). When the effect of the third variable (age) is removed, the correlation between shoe size and achievement in mathematics diminishes or vanishes completely. We label the three variables as follows: 1 = math achievement, 2 = shoe size, and 3 = age. Then, $r_{12 \cdot 3}$ is the partial correlation between variables 1 and 2 with 3 held constant. Suppose the correlation coefficients between the three variables were $r_{12} = .80$; $r_{13} = .90$; and $r_{23} = .88$. The formula for $r_{12 \cdot 3}$ is

$$r_{12 \bullet 3} = \frac{r_{12} - r_{13}r_{23}}{\sqrt{1 - r_{13}^2}\sqrt{1 - r_{23}^2}}$$
$$= \frac{.80 - .90 \times .88}{\sqrt{1 - .90^2}\sqrt{1 - .88^2}}$$
$$= .039$$

(8.6)

Margin glossary

residual scores

The difference between the predicted and actual scores that represents the error of prediction.

standard error of prediction

The computation of the standard deviation of all of the residual scores of a population; the amount of error expected in a prediction. . Also called *standard error of estimate.*

standard error of estimate

See *standard error of prediction.*

spurious correlation

Relationship in which the correlation between two variables is due primarily to the common influence of another variable.

Thus, we see that correlation between math achievement and shoe size drops to about zero when age is held constant.

The primary value of partial correlation is to develop a multiple-regression equation with two or more predictor variables. In the selection process, when a new variable is "stepped in," its correlation with the criterion is determined with the effects of the preceding variable held constant. The size and the sign of a partial correlation may be different from the zero-order (two-variable) correlation between the same variables.

Uses of Semipartial Correlation

In the previous section on partial correlation, the effects of a third variable on the relationship between two other variables were eliminated by using the formula for partial correlation. In other words, in $r_{12 \cdot 3}$, the relationship of variable 3 to the correlation of variables 1 and 2 is removed. In some situations, the investigator may wish to remove the effects of a variable from only one of the variables being correlated. This is called **semipartial correlation.** The symbol is $r_{1(2 \cdot 3)}$, which indicates that the relationship between variables 1 and 2 is determined after the influence of variable 3 on variable 2 has been eliminated.

Suppose, for example, that a researcher is studying the relationship among perceived exertion (PE, a person's feelings of how hard he or she is working), heart rate (HR), and workload (WL). Obviously, WL is going to be correlated with HR. The researcher wants to investigate the relationship between PE and HR while controlling for WL. Regular partial correlation will show this relationship. However, regular partial correlation will remove the effects of WL on the relationship between PE and HR. But the researcher does not want to remove the effects of WL on the relationship of PE and HR, only to remove the effects of WL on HR. In other words, the main interest is in the net effect of HR on PE after the influence of WL has been removed. Thus, in semipartial correlation, the effect of WL is removed from HR but not from PE.

semipartial correlation

A technique in which just one variable is partialed out—effects removed—from two variables in a correlation.

Procedures for Multiple Regression

Multiple regression involves one dependent variable (usually a criterion of some sort) and two or more predictor variables (independent variables). The use of more than one predictor variable usually increases the accuracy of prediction. This should be self-evident. If you wished to predict basketball-playing ability, you would expect to get a more accurate prediction by using several basketball skills tests rather than by using only one.

The multiple correlation coefficient (R) indicates the relationship between the criterion and a weighted sum of the predictor variables. It follows then that R^2 represents the amount of the variance of the criterion that is explained or accounted for by the combined predictors. This is similar to the coefficient of determination (r^2), which was discussed earlier with regard to the common association between two variables. Now, however, we have the amount of association between one variable (the criterion) and a weighted combination of variables.

We wish to find the best combination of variables that gives the most accurate prediction of the criterion. Therefore, we are interested in knowing how much each of the predictors contributes to the total explained variance. Another way to say this is that we want to find the variables that most reduce the prediction errors. From a practical standpoint, in terms of time and effort involved in obtaining measures of the predictor variables, it is desirable to find the fewest number of predictors that account for most of the variance of the criterion. There are several selection procedures used for this purpose. (For additional information on multiple regression, see Cohen & Cohen [1983], and Pedhazur [1982].)

In deciding which predictors (or sets of predictors) to use, standard statistical packages offer several options. Table 8.2 includes the most common procedures. One option is to simply use all of the predictor variables. However, the fewer that can be used while still obtaining a good prediction, the more economical and valuable the prediction equation. Sometimes researchers specify the order and combinations of predictor variables. This is called hierarchical regression and should be based on theory and previous empirical

multiple regression

Model used for predicting a criterion from two or more independent, or predictor, variables.

Table 8.2 Selection Procedures in Multiple Regression

Procedure	How It Works
Full model	All prediction variables are included in the model.
Hierarchical	Prediction variables are entered as blocks (e.g., two together).
Forward	Predictor with highest correlation is entered first; then next highest is entered after accounting for relationship with first predictor.
Backward	All variables are entered; then the one with the least contribution is removed first, and then the one with the next least contribution.
Stepwise	A combination of forward and backward procedures in which checks are made at each step to see whether a previously entered predictor should be removed.
Maximum R squared	Program calculates the best variable combination.

evidence. The remaining four procedures listed in table 8.2 allow the program to select the order of predictor variables and how many are used based on statistical criteria. All are designed to end up with fewer predictor variables than the full model.

Other techniques that are very similar to multiple regression use these procedures. The difference is the variable being predicted. In multiple regression a continuous variable is predicted (e.g., monthly salary in the previous example). In *discriminant function analysis* a nominal variable is predicted (e.g., whether a child will continue to participate on a sports team or whether someone is using one of three types of exercise protocol). When using *logistic regression* a percentage is being predicted (e.g., the likelihood that a person will continue an exercise program). These techniques are being used in research in physical activity when the researcher wants to predict group membership or probability of an occurrence.

Multiple-Regression Prediction Equations

The prediction equation resulting from multiple regression is basically that of the two-variable regression model, $Y = a + bX$. The only difference is that there is more than one X variable; thus, the equation is

$$Y = a + b_1X_1 + b_2X_2 + \ldots + b_iX_i$$

We will not delve into the formula for the calculation of the a and the bs for the selected variables. As we said before, researchers undoubtedly use computers for multiple-regression problems. An example of a multiple-regression prediction formula follows. In this equation, a man's lean body weight (LBW) is being predicted from several anthropometric measures, including skinfold thicknesses, body-part circumferences, and muscle diameters. The following formula, developed by Behnke and Wilmore (1974), has a correlation of .958 and a standard error of estimate of 2.358, which is interpreted just as in the regression equation with only one predictor variable.

LBW = 10.138 + 0.9259 (wt) − 0.1881 (thigh skinfold) + 0.637 (bi-iliac diameter) + 0.4888 (neck circumference) − 0.5951 (abdominal circumference)

Some Problems Associated With Multiple Regression

The basic procedure to evaluate multiple regression is the same as in regression with only two variables: the size of the correlation. The higher the correlation, the more accurate the prediction. However, some other factors should be mentioned.

One limitation of prediction relates to generalizability. Regression equations developed with a particular sample often lose considerable accuracy when applied to others. This loss of accuracy in prediction is called **shrinkage.** The term **population specificity** also relates to this phenomenon. Shrinkage and the use of cross-validation to improve generalizability are further discussed in chapter 11. We need to recognize that the more accuracy researchers seek through selection procedures (forward, backward, stepwise, maximum R squared) that capitalize on specific characteristics of the sample, the more difficult it is to generalize to other populations. For example, a formula for predicting percentage body fat from skinfold measurements developed with adult men would lose a great deal of accuracy if used for adolescents. Thus, the researcher should select the sample carefully with regard to the population for which the results are to be generalized.

In prediction studies the number of participants in the sample should be sufficiently large. Usually, the larger the sample, the more likely that the sample will represent the population from which it is drawn. However, another problem with small samples in multiple-regression studies is that the correlation may be spuriously high. A direct relationship exists between the correlation and the ratio of the number of participants to the number of variables. In fact, the degree to which the expected value of R^2 exceeds zero when it is zero in the population depends on two things: the size of the sample (n) and the number of variables (k). More precisely, it is the ratio $(k - 1)/(n - 1)$. To illustrate, suppose you read a study in which the $R^2 = .90$. Impressive, right? However, the results would be meaningless if the study only had 40 participants and 30 variables, because we could expect an R^2 of .74 just on the basis of chance alone: $R^2 = (k - 1)/(n - 1) = (30 - 1)/(40 - 1) = .74$. You should be aware of this relationship between number of participants and number of variables when reading research that uses multiple regression. In the most extreme case, you can see that having the same number of variables as participants, $(k - 1)/(n - 1)$ yields an R^2 of 1.00! A participant-to-variable ratio of 10:1 or higher is often recommended.

Multivariate Forms of Correlation

We have been discussing forms of correlation that are typically labeled as univariate, meaning that there are one or more predictor variables (independent variables) but only one criterion variable (dependent variable). In this short section we introduce you to the multivariate forms of correlation. Our purpose is to provide enough information that you can read a paper using one of these procedures and understand what has been done.

Canonical Correlation

Canonical correlation is an extension of multiple correlation (several predictors and one criterion) to an analysis that has several predictors and several criteria, represented by the symbol R_C. In multiple correlation, a linear composite is formed of the predictor variables that maximally predict the single criterion variable. In canonical correlation two linear composites are formed, one of the predictor variables and one of the criterion variables. These two composites are formed to maximize the relationship between them, and this relationship is represented by R_C. R_C is interpreted just like R from multiple correlation; it represents the shared variance between the two linear composites of variables.

For example, McPherson and Thomas (1989) studied expert and novice tennis players at two age levels (10–11 and 12–13). They measured three predictor variables (tennis knowledge, tennis serve, and tennis ground strokes). They then videotaped the children playing tennis and coded the tapes for three criterion variables of tennis play: control (being in position), decisions (choice of what stroke to use and where to hit it), and execution (doing what was intended). They used canonical correlation to evaluate the relationship between the linear composites of the three predictor and three criterion variables. They reported a significant canonical correlation, $R_C = .79$, $F(6, 70) = 7.58$, $p < .01$ (the F ratio and the notation that describes it are discussed in chapter 9). When they evaluated the standardized canonical coefficients (like the standardized beta weights in multiple

shrinkage

Tendency for the validity to decrease when the prediction formula is used with a new sample.

population specificity

Phenomenon whereby a regression equation that was developed with a particular sample loses considerable accuracy when applied to others.

correlation) and multiple-regression follow-ups, they found that tennis knowledge was most closely related to good decisions during game play, while tennis ground strokes were most closely related to tennis execution during game play.

Factor Analysis

factor analysis

A statistical technique used to reduce a set of data by grouping similar variables into basic components (factors).

Many performance variables and characteristics are used to describe human behavior. Often it is useful to reduce a large set of performance and characteristic measures to a more manageable structure. We have already discussed the likelihood that two performance measures might to some extent assess the same underlying characteristic. This represents the degree to which they are correlated. **Factor analysis** is an approach to reducing a set of correlated measures to a smaller number of latent or hidden variables (Tinsley & Tinsley, 1987). Numerous procedures are grouped under the general topic of factor analysis. We do not discuss the various techniques in detail, but we do provide a general explanation that allows you to read and understand studies that use factor analysis. For a good explanation and example of factor analysis, read Tinsley and Tinsley (1987) and McDonald (1999).

Factor analysis is performed on data from a group of individuals on whom a series of measurements have been taken. The researcher usually wants to describe a reduced number of underlying constructs and possibly select the one or two best measures of each construct. Factor analysis begins by calculating the intercorrelations of all the measures used (the correlation between all possible pairs of variables; thus, if eight measures were used, the correlation would be determined between variables 1 and 2, variables 1 and 3, and so on for a total of 28 correlations). The goal of factor analysis is to discover the factors (underlying or hidden constructs) that best explain a group of measurements and describe the relation of each measure to the factor or underlying construct. There are two general types of factor analysis: exploratory, in which many variables are reduced to an underlying set, and confirmatory, which either supports or does not support a structure proposed from theory. Confirmatory factor analysis is the more useful technique.

For example, Marsh, Marco, and Aþçý (2002) examined whether the Physical Self-Description Questionnaire (PSDQ) had similar factor structures when used with cross-cultural samples—high school students in Australia, Spain, and Turkey. They translated the instrument into Spanish and Turkish and used back-translation techniques to assess that the translations produced a similar instrument in the new language. Children in each country then completed the instrument, and factor structures and fit statistics (such as root mean square error of approximation [RMSEA] and the relative noncentrality index [RNI]) were compared. The research supported the use of the PSDQ in a variety of cultural environments. As we note in chapter 11, confirmatory factor analysis is often used to validate scores from attitude instruments or other questionnaires, and this study extended that into multiple countries.

Structural Equation Modeling

linear structural relations (LISREL)

A statistical approach used to establish relationships and examine the structural equations model.

Path analysis and **linear structural relations (LISREL)** are structural, or causal, modeling techniques that are used to explain the way certain characteristics relate to one another and attempt to imply cause. You should remember from the discussion in chapter 4 and earlier in this chapter that cause and effect are not a statistical result but a logical one. That is, if the experimenter can argue theoretically that changing a certain characteristic should result in a specific change in behavior, and if the actual experiments (and statistical analysis) support this hypothesis, then a particular independent variable is often inferred to cause a dependent variable. Of course, this is true only if all other possible influences have been controlled.

The way variables influence one another is not always clear; for example, $X \rightarrow Y \rightarrow Z$, or $X \leftarrow Y \rightarrow Z$. In the first case, X influences Y, which in turn influences Z; however, in the second case, Y influences both X and Z. This is a very simple example. Path analysis and LISREL allow a more complex modeling of the way variables influence one another. But all you can say about these models is that they are either consistent or inconsistent with the data and hypotheses. Whether they imply cause and effect depends on other things (e.g., control of all other variables, careful treatments, logical hypotheses, valid theories). For a nice overview and practical explanation of structural equation modeling, read Schumacker and Lomax (2004).

A good way for you to gain a basic understanding of LISREL is through an example from the physical education literature. Greenockle, Lee, and Lomax (1990) were interested in describing the relationship between student characteristics and exercise behavior. Figure 8.8 shows the relationship they found in a sample of 10 existing high school physical education classes in which the teacher agreed to teach a specific physical fitness unit. The researchers measured four variables of the students' background: past activity level, perception of father's activity level, perception of mother's activity level, and knowledge of physical fitness. Two measures of attitude toward exercise were taken from a valid and reliable questionnaire. Two subjective norm measures, which indicated participants' perceptions of how others felt about the participants' exercise behavior, were obtained from the same questionnaire. One measure of intention to exercise was obtained. Five exercise behaviors were measured: jogging, fitness test score, percentage of participation, heart rate, and perceived exertion.

The technique of LISREL was used because it allows related measures to be grouped (as in factor analysis) into components (e.g., exercise behavior) and it shows the relationship among the components in terms of magnitude and direction. The estimates of each component are grouped together in a linear equation (using their factor loadings). Then a series of general linear equations (as in multiple regression) shows the relationships among the components. Model-fitting indexes are used to assess the nature of the fit. The results of the study by Greenockle et al. showed "the prediction of exercise behavior

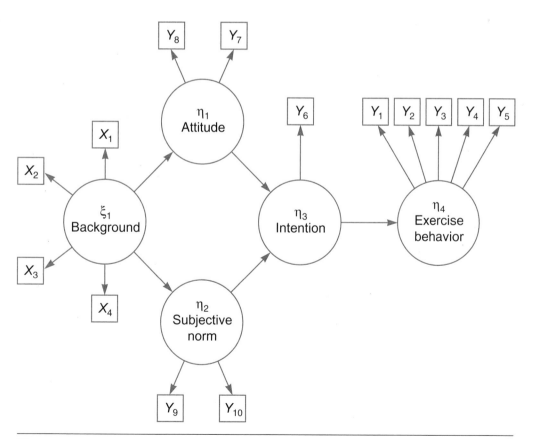

Figure 8.8 LISREL model.

by attitude and subjective norm to be significantly mediated by intention" (p. 59). You can see, then, that structural modeling techniques provide a way to evaluate the complex relationships that exist in real-world data. By establishing the directions of certain relationships, stronger inferences can be made about which characteristics are likely to cause other characteristics.

Summary

We have explored some statistical techniques for determining relationships among variables. The simplest type of correlation is the Pearson r correlation, which describes the relationship between two variables. We introduced linear regression, which can be used to predict one variable from another. Correlation is interpreted for significance (reliability) and for meaningfulness (r^2), which indicates the portion of the total variance in one measure that can be explained or accounted for by the other measure.

Partial correlation is a procedure in which a correlation between two variables is obtained while the influence of one or more other variables is held constant. Semipartial correlation removes the influence of a third variable on only one of the two variables being correlated. Partial correlation (or semipartial correlation) is used in multiple correlation and in developing multiple-regression formulas.

In multiple regression two or more predictor (independent) variables are used to predict the criterion variable. The most efficient weighted linear composite of predictor variables is determined through such techniques as forward selection, backward selection, stepwise selection, and maximum R squared.

Finally, we provided a brief overview of three multivariate correlational procedures: canonical correlation, factor analysis, and structural equation modeling. An example of the use of each technique was presented.

✔ Check Your Understanding

1. What is the correlation (r) between fat deposits on two different body sites? Variable X is the triceps skinfold measure and variable Y the suprailiac skinfold measure.

X	Y
16	9
17	12
17	10
15	9
14	8
11	6
11	5
12	5
13	6
14	5
4	1
7	4
12	7
7	1
10	3

2. Using table A.3, determine whether the correlation obtained in problem 1 is significant at the .05 level. What size would the correlation need to be for significance at the .01 level if you had 30 participants? What is the percentage of common variance between the two skinfold measurements in problem 1?

3. Determine the regression formula (equation 8.2) to predict maximal oxygen consumption ($\dot{V}O_2$max) from scores on a 12-min run. The information you need follows:

X (12-min run)	Y ($\dot{V}O_2$max)
M_X = 2,853 m	M_Y = 52.6 ml · kg · min–1
s_X = 305 m	s_Y = 6.3 ml · kg · min–1
r = .79	

4. Using the prediction formula developed in problem 3, what is the predicted $\dot{V}O_2$max for a participant who ran 2,954 m in 12 min? for a participant who ran 2,688 m in 12 min?

5. What is the standard error of estimate (equation 8.5) for the prediction formula in problem 3? How would you interpret the predicted $\dot{V}O_2$max for the participants in problem 4?

Differences Among Groups

CHAPTER

9

Remember, half the people you know

are below average!

■   ■   ■   ■   ■   ■   ■

Statistical techniques are used for describing and finding relationships among variables, as we discussed in chapters 6 and 8. They are also used to detect differences among groups. The latter techniques are most frequently used for data analysis in experimental and quasi-experimental research. They enable us to evaluate the effects of an independent (cause or treatment) or categorical (sex, age, race, etc.) variable on a dependent variable (effect, outcome). Remember, however, that the techniques described in this chapter are not used in isolation to establish cause and effect but only to evaluate the influence of the independent variable. Cause and effect are not established by statistics but by theory, logic, and the total nature of the experimental situation.

How Statistics Test Differences

In experimental research, the levels of the independent variable may be established by the experimenter. For example, the experiment might involve the investigation of the effects of intensity of training on cardiorespiratory endurance. Thus, intensity of training is the independent variable (or treatment factor), whereas a measure of cardiorespiratory endurance is the dependent variable. Intensity of training could have any number of levels. If it were evaluated as a percentage of maximal oxygen consumption ($\dot{V}O_2$max), then it could be 30%, 40%, 50%, and so forth. The investigator would choose the number and the intensity of levels. In a simple experiment, the independent variable might be two levels of intensity of training, for example, 40% and 70% of $\dot{V}O_2$max. The length of each session (e.g., 30 min), frequency (e.g., three times per week), and number of weeks of training (e.g., 12) are controlled (equal for both groups). The dependent variable could be the distance a person runs in 12 min.

The purpose of the statistical test is to evaluate the null hypothesis at a specific level of probability (e.g., $p < .05$). In other words, do the two levels of treatment differ significantly ($p < .05$) so that these differences are not attributable to a chance occurrence more than 5 times in 100? The statistical test is always of the null hypothesis. All that statistics can do is reject or fail to reject the null hypothesis. Statistics cannot accept the research hypothesis. Only logical reasoning, good experimental design, and appropriate theorizing can do so. Statistics can determine only whether the groups are different, not why they are different.

147

She's trying to solve the equation.

omega squared (ω²)

A method of interpreting the meaningfulness of the strength of the relationship between the independent and dependent variables; the proportion of total variance that is due to the treatments.

In using logical techniques to infer cause and effect after finding significant differences, you must be careful to consider all possibilities. For example, we might propose the theorem that all odd numbers are prime numbers (this example is from Ronen et al., cited in Scherr 1983, 146). You know that prime numbers are those that can be divided only by 1 and by themselves. Thus, 1 is a prime number, 3 is a prime number, 5 is a prime number, 7 is a prime number, and so on. Using the induction technique of reasoning, every odd number is a prime number. In Ronen et al.'s example, a very small number of levels of the independent variable (prime numbers) were sampled, and an error was made in inferring that all levels of the independent variable were the same.

When you use statistics that test differences among groups, you want to establish not only whether the groups are significantly different but also the strength of the association between the independent and dependent variables, or the size of the difference between two groups. The t and the F ratios are used throughout this chapter to determine whether groups are significantly different. **Omega squared (ω^2)** is used to estimate the degree of association (or percentage variance accounted for) between the independent and dependent variables. To some extent, ω^2 is similar to r^2 for the correlations presented in chapter 8; both represent the same idea, which is percentage variance accounted for. Another way of considering the meaningfulness of the differences is effect size (ES). Effect size (recall the discussion in chapter 7) is the standardized difference between two groups and is also used as an estimate of meaningfulness.

The uses of the t and the F distributions as presented in this chapter have four assumptions (in addition to the assumptions for parametric statistics presented in chapter 6; Kirk, 1982, 74, is a good book to read for more information):

- Observations are drawn from normally distributed populations.
- Observations represent random samples from populations.
- The numerator and denominator are estimates of the same population variance.
- The numerator and denominator of F (or t) ratios are independent.

Although t and F tests are robust to (only slightly influenced by) violations of these assumptions, the assumptions still are not trivial. You should be sensitive to their presence and to the fact that violations affect the probability levels that may be obtained in connection with t and F ratios.

Types of *t* Tests

We discuss three types of *t* tests: *t* test between a sample and a population mean, *t* test for independent groups, and *t* test for dependent groups.

t Test Between a Sample and a Population Mean

First, we may want to know whether a sample of students differs from a larger population. For example, suppose that for a standardized knowledge test on physical fitness, the mean is 76 for a large population of college freshmen. When tested, a fitness class ($n = 30$) that you are teaching has a mean of 81 and a standard deviation of 9. Does your class have significantly more knowledge about physical fitness than does the typical college freshman class?

The *t* test is a test of the null hypothesis, which states that there is no difference between the sample mean (M) and the population mean (μ), or $M - \mu = 0$. Equation 9.1 is the *t* test between a sample and a population mean:

$$t = \frac{M - \mu}{s_M / \sqrt{n}} \tag{9.1}$$

where s_M = the standard deviation for the sample mean and n = the number of observations in the sample. Example 9.1 shows this formula applied to the means and standard deviation of the example for the standardized fitness knowledge test.

Is the value of 3.14 significant? To find out you'll need to check table A.5 in appendix A. To use the table, first work out the degrees of freedom (*df*) for the *t* test. Degrees of freedom are based on the number of participants with a correction for bias:

$$df = n - 1 \tag{9.2}$$

In our example, the degrees of freedom are $30 - 1$, or 29. Degrees of freedom are used to enter a *t* table to determine whether the calculated *t* is as large as or greater than the table *t* value. Note that across the top of table A.5 are probability levels. We want to know whether our *t* value is significant at $p < .05$, the level of significance that we set. Read across to the .05 level. Now read down the left side (*df*) to the row for the degrees of freedom you calculated (*df* = 29). Read where the *df* = 29 row and the .05 column intersect. Is the calculated value (*t* = 3.05) larger than this value (2.045)? Yes, it is. So *t* is significant at $p < .05$. Thus, our sample class is reliably (significantly) different from the population average on the fitness test.

> ### Example 9.1
>
> **Known Values**
>
> Population: $N = 10{,}000$
>
> Fitness class: $n = 30$
>
> Sample mean: $M = 81$
>
> Population mean: $\mu = 76$
>
> Standard deviation: $s_M = 9$
>
> **Working It Out** (equation 9.1)
>
> $$t = \frac{81 - 76}{9 / \sqrt{32}} = \frac{5}{1.59} = 3.14$$

Independent *t* Test

The previous *t* test, applied to determine whether a sample differs from a population, is not used very frequently. The most frequently used *t* test determines whether two sample means differ reliably from each other. This is called an **independent *t* test.**

Suppose we return to our example at the beginning of this chapter: Do two groups, training at different levels of intensity (40% or 70% of $\dot{V}O_2$max, 30 min per day, 3 days per week for 12 weeks), differ from each other on a measure of cardiorespiratory endurance (12-min run)? Let us further assume that there were 30 participants who were randomly assigned to form two groups of 15 each.

Equation 9.3 is the *t*-test formula for two independent samples:

$$t = \frac{M_1 - M_2}{\sqrt{s_1^2 / n_1 + s_2^2 / n_2}} \tag{9.3}$$

independent *t* test

The most frequently used test to determine if two sample means differ reliably from each other.

Equation 9.4 is the version of the *t*-test formula most easily performed with a calculator:

$$t = \frac{M_1 - M_2}{\sqrt{\dfrac{\left[\Sigma X_1^2 - \left(\Sigma X_1\right)^2 / n_1 + \Sigma X_2^2 - \left(\Sigma X_2\right)^2 / n_2\right] \bullet \left(1 / n_1 + 1 / n_2\right)}{n_1 + n_2 - 2}}} \qquad (9.4)$$

The degrees of freedom for an independent *t* test are calculated as follows:

$$df = n_1 + n_2 - 2 \qquad (9.5)$$

In our example, $df = 15 + 15 - 2 = 28$ (or $N - 2 = 30 - 2 = 28$).

Example 9.2 shows how these formulas are applied to the intensity-of-training experiment. Thus, you can see that the 70% intensity of training allowed students to run reliably farther ($M = 3,004$ m) than did the 40% intensity of training ($M = 2,456$ m). If the 12-min run is a valid measure of cardiorespiratory endurance, if all other conditions were controlled, and if the theory, logic, and design of this experiment made sense, we can say that the cause of this increased level of cardiorespiratory endurance was the fact that the 70% group's training was more intense than that of the 40% group. Given that the theory and logic behind this experiment were sound, we might be able to say that the training resulted in a change in certain cardiorespiratory functions, allowing more work to occur in a specific time (i.e., more distance to be covered in 12 min).

The independent *t* test is a very common statistical technique in research studies. Sometimes it is the main technique used, and sometimes it is just one of several statistical tests in a study. Sachtleben and colleagues (1997) compared serum lipoprotein and apolipoprotein levels in a group of weight trainers who used anabolic steroids with those in a group of weight trainers who did not use steroids. A number of comparisons were made. Here is a sample of how the authors reported the results of the *t* tests: "The fasting HDL-C level in the nonusers group was significantly higher than the off-cycle users group, $t(21) = 4.66$, $p < .007$" (p. 112). This is how the result of a *t* test is reported in APA style. The number in parentheses (21) is the *df*, which is $N - 2$. So we know that there was a total of 23 people in the two groups. The computed *t* was 4.66, and this was significant at the established alpha level of $p = .007$. The authors used the Bonferroni procedure to adjust the alpha level. We discuss that procedure later in the chapter.

Checking for Homogeneity of Variance

All techniques for comparisons between groups assume that the variances (standard deviation squared) between the groups are equivalent. Although mild violations of this assumption do not present major problems, serious violations are more likely if group sizes are not approximately equal. Formulas given here and used in most computer programs

---

Example 9.2

Known Values

	Group 1 (70% $\dot{V}O_2$max)	Group 2 (40% $\dot{V}O_2$max)
Mean distance run:	$M_1 = 3,004$ m	$M_2 = 2,456$ m
Standard deviation:	$s_1 = 114$ m	$s_2 = 103$ m
Number of subjects:	$n_1 = 15$	$n_2 = 15$

Working It Out (equation 9.3)

$$t = \frac{3,004 - 2,456}{\sqrt{\dfrac{(114)^2}{15} + \dfrac{(103)^2}{15}}} = \frac{548}{39.67} = 13.81$$

allow unequal group sizes. However, the homogeneity assumption should be checked if group sizes are very different or even when variances are very different (these techniques are not presented here but are covered in basic statistical texts).

Estimating Meaningfulness of Treatments

How meaningful is the effect in our comparison of training intensities? Or, more simply, is the increase in cardiorespiratory endurance of running an additional 548 m (3,004 – 2,456) worth the additional work of training at 70% of $\dot{V}O_2$max as compared with 40% of $\dot{V}O_2$max? Given the total variation in running performance of the two groups, what we really want to know is how much of this variation is accounted for by (associated with) the difference in the two levels of the independent variable (70% vs. 40%).

Omega Squared

Omega squared (ω^2) is one way to estimate this variation. The following formula (Tolson, 1980) is used to calculate omega squared:

$$\omega^2 = (t^2 - 1)/(t^2 + n_1 + n_2 - 1) \tag{9.6}$$

Example 9.3 shows the application of this formula to our example.

We can conclude that $\omega^2 = .86$ means that 86% of the total variance in the scores for distance run can be accounted for by the difference in the two groups' levels of training. The remaining variance, 100% – 86% = 14%, is accounted for by other factors. The question now becomes, Is this a meaningful percentage of variance? No one can answer that except you. Are you willing to increase the intensity of training 30% to produce this effect? There is no statistical answer. Only theory, past research, and logic can answer that. The research has told you only what will happen if you increase the intensity of training from 40% to 70% of $\dot{V}O_2$max. This represents context as discussed in chapter 7.

Effect Size

Another way to estimate the degree to which the treatment influenced the outcome is by effect size (ES), the standardized difference between the means. Equation 9.7 (given previously as equation 7.1) is a way to estimate effect size (this concept was discussed in chapter 7 and is also used in meta-analysis, discussed in chapter 14):

$$ES = (M_1 - M_2)/s \tag{9.7}$$

where M_1 = the mean of one group or level of treatment, M_2 = the mean of a second group or level of treatment, and s = the standard deviation. The question is what standard deviation should be used. Considerable controversy exists over the answer to this question. Some statisticians think that if there is a control group, then its standard deviation should be used. If there is no control group, then the pooled standard deviation (equation 9.8) should be used. Some advocate the use of the pooled standard deviation on all occasions. Either can be defended; however, when there is no clear control group (as in the example used here), we recommend that you use the pooled standard deviation

$$s_p = \sqrt{\frac{s_1^2(n_1 - 1) + s_2^2(n_2 - 1)}{n_1 + n_2 - 2}} \tag{9.8}$$

where s_p = the pooled standard deviation, s_1^2 = the variance of group 1, s_2^2 = the variance of group 2, n_1 = the number of participants in group 1, and n_2 = the number of participants in group 2.

Effect size can be interpreted as follows: An ES of 0.8 or greater is large, an ES around 0.5 is moderate, and an ES of 0.2 or less is small. Thus, the ES calculated in example 9.4 of 5.0 is a large value and would typically be judged as a meaningful treatment effect. Sachtleben et al. (1997) used ES to report meaningfulness. For the t we quoted earlier, the ES was 1.2, which would be considered large.

Example 9.3

Known Values

Differences between groups: $t = 13.81$
Number of subjects in group 1: $n_1 = 15$
Number of subjects in group 2: $n_2 = 15$

Working It Out (equation 9.6)

$$\omega^2 = \frac{13.81^2 - 1}{13.81^2 + 15 + 15 - 1} = \frac{189.72}{219.72} = .86$$

Example 9.4

Known Values

	Group 1	Group 2
Mean distance run:	$M_1 = 3{,}004$ m	$M_2 = 2{,}456$ m
Standard deviation:	$s_1 = 114$ m	$s_2 = 103$ m
Number of subjects:	$n_1 = 15$	$n_2 = 15$

Working It Out (equations 9.7 and 9.8)

$$s_p = \sqrt{\frac{(114)^2(15-1) + (103)^2(15-1)}{15+15-2}} = 108.64$$

$$\text{ES} = \frac{3{,}004 - 2{,}456}{108.64} = 5.0$$

Dependent *t* Test

We have now considered use of the *t* test to evaluate whether a sample differs from a population and whether two independent samples differ from each other. A third application is called a **dependent *t* test,** which is used when the two groups of scores are related in some manner. Usually, the relationship takes one of two forms:

dependent *t* test

A test of the significance of differences between means of two sets of scores that are related, such as when the same participants are measured on two occasions.

- Two groups of participants are matched on one or more characteristics and thus are no longer independent.
- One group of participants is tested twice on the same variable, and the experimenter is interested in the change between the two tests.

The formula for a dependent *t* test is

$$t = \frac{M_{post} - M_{pre}}{\sqrt{\left[\left(s^2_{post} + s^2_{pre}\right) - \left(2r_{pp} \bullet s_{post} \bullet s_{pre}\right)\right] / (N-1)}} \tag{9.9}$$

Notice that the top part of this formula is the same as the independent *t*-test equation (9.3). In addition, the first term in the bottom (within the parentheses at the left) is also similar. However, an amount is subtracted from the bottom (error term) of the *t*-test formula. This term is read as "two times the correlation between the pre- and posttest (r_{pp}) times the standard deviation for the posttest times the standard deviation for the pretest." The independent *t* test (equation 9.3) assumes that the two groups of participants are independent. In this case, the same participants are tested twice (pretest and posttest). Thus, we adjust the error term of the *t* test downward (make it smaller) by taking into account the relationship (*r*) between the pretest and posttest adjusted by their standard deviations. The degrees of freedom for the dependent *t* test are

$$df = N - 1 \tag{9.10}$$

where N = the number of *paired* observations. Equation 9.9 is cumbersome to compute because the correlation (*r*) between the pre- and posttest must be calculated. Thus, the raw-score formula is much easier to use:

$$t = \frac{\Sigma D}{\sqrt{\left[N\Sigma D^2 - (\Sigma D)^2\right] / (N-1)}} \tag{9.11}$$

where D = the posttest minus the pretest for each participant and N = the number of *paired* observations. Let's work out an example. Ten dancers are given a jump-and-reach test (the difference between the height on a wall they can reach and touch while standing and how high they can jump and touch). Then they take part in 10 weeks of dance activity that involves leaps and jumps 3 days per week. The dancers are again given the jump-and-reach test after the 10 weeks. (This is not proposed to be a real experiment, just a simple example illustrating the statistical technique.) Our research hypothesis is that the 10 weeks of dance experience will improve jumping skills as reflected by a change in jump-and-reach scores. The null hypothesis (H_0) is that the difference between the pretest and posttest of jumping is not significantly different from zero, $H_0 = M_{post} - M_{pre} = 0$. Example 9.5 shows how to calculate the dependent t value.

The results indicate that the posttest mean (19.7 cm) was significantly better than the pretest mean (16.5 cm), $t(9) = 4.83$, $p < .05$. The null hypothesis can be rejected, and if everything else has been properly controlled in the experiment, we can conclude that the dance training produced a reliable increase in the height of jumping performance of 3.2 cm.

A standard formula is not available for calculating ω^2 for a dependent t test. But we could estimate the magnitude of the effect by dividing the average gain by the pretest mean, as in example 9.6.

The gain is 19.4% of the pretest, nearly a 20% improvement. Although this is a rather crude way of estimating the effect, it does suffice. On the other hand, we could estimate

Example 9.5

Known Values

Subject	Pretest score (cm)	Posttest score (cm)	Posttest – Pretest D	D²
1	12	16	4	16
2	15	21	6	36
3	13	15	2	4
4	20	22	2	4
5	21	21	0	0
6	19	23	4	16
7	14	16	2	4
8	17	18	1	1
9	16	22	6	36
10	18	23	5	25

Sum of pretest scores: Σ_{pre} = 165 cm

Sum of posttest scores: Σ_{post} = 197 cm

Sum of D: ΣD = 32

Sum of D^2: ΣD^2 = 142

Number of paired observations: N = 10

Posttest mean: M_{post} = 19.7

Pretest mean: M_{pre} = 16.5

Working It Out (equation 9.11)

$$t = \frac{32}{\sqrt{\dfrac{10(142) - (32)^2}{10 - 1}}} = \frac{32}{\sqrt{\dfrac{1,420 - 1,024}{9}}} = \frac{32}{6.63} = 4.83$$

Example 9.6

Known Values

Differences between post- and pretest means: $M_D = 3.2$

Mean of pretest scores: $M_{pre} = 16.5$

Working It Out

$$\text{Magnitude of increase} = \frac{M_D}{M_{pre}} = \frac{3.2}{16.5} = 0.194 = 19.4\%$$

ES for the pretest to posttest change by subtracting the pretest M from the posttest M and dividing by the pretest standard deviation. In our example this would be ES = 3.2/3.03 = 1.06, a large (and meaningful) effect size.

An example of a dependent t test in the literature is by Kokkonen, Nelson, and Cornwell (1998), who investigated the influence of acute stretching on maximal strength performance. The same participants were tested before and after stretching, which necessitated a dependent t test. One of the comparisons was reported thus: "The ST exercises altered sit-and-reach performance such that the poststretching sit-and-reach scores were significantly, $t(29) = 11.1$, $p < .05$, . . . increased 16% over the initial sit-and-reach scores" (p. 413).

Interpreting *t*

You have now learned to perform the calculations that determine differences among groups. What do the results mean? Are they significant? Does it matter whether they are or aren't? We answer these questions in this section by explaining the difference between one-tailed and two-tailed t tests and discussing the aspects of the t test that influence power in research.

One-Tailed Versus Two-Tailed *t* Tests

two-tailed *t* test

Test that assumes that the difference between the two means could favor either group.

one-tailed *t* test

Test that assumes that the difference between the two means lies in one direction only.

At this point, refer again to table A.5 in appendix A (the table of t values to which you compare the values calculated). Remember, you decide the alpha (probability) level (we have been using .05), calculate the degrees of freedom for t, and read the table t value at the intersection of the column (probability) and row (*df*). If your calculated t exceeds the table value, it is significant at the specified alpha and degrees of freedom. This table is used for a **two-tailed *t* test** because we assume that the difference between the two means could favor either mean. Sometimes we might hypothesize that group 1 will be better than group 2 or, at worst, no poorer than group 2. In this case, the test is a **one-tailed *t* test,** that is, it can go only one direction. Then, when looking at table A.5 for a one-tailed t test, the .05 level is located in the .10 column for the two-tailed test, and the value for the .01 level for the one-tailed test is in the .02 column for the two-tailed test. Generally in behavioral research, however, we are not so sure of our results that we can employ the one-tailed t table.

t Tests and Power in Research

In chapter 7, power was defined as the probability of rejecting the null hypothesis when the null hypothesis is false. To obtain power in research is very desirable, as the odds of rejecting a false null hypothesis are increased. The independent t test is used here to explain three ways to obtain power (in addition to setting the alpha level). These ways apply to all types of experimental research.

Consider the formula for the independent t test:

$$t = \frac{M_1 - M_2}{\sqrt{\dfrac{s_1^2}{n_1} + \dfrac{s_2^2}{n_2}}}$$

1
2
3

Note that we have placed numbers (1, 2, 3) beside the three horizontal levels of this formula. These three levels represent what can be manipulated to increase or decrease power.

The first level ($M_1 - M_2$) gives power if we can increase the difference between M_1 and M_2. You should see that if the second and third levels remain the same, a larger difference between the means increases the size of *t,* which increases the odds of rejecting the null hypothesis and thus increases power. How can the difference between the means be increased? The logical answer is by applying stronger, more concentrated treatments. One example would be to use a 12-week treatment instead of a 6-week treatment to give the treatment a better chance to show its effect. The 12-week treatment should move the means of the experimental and the control groups further apart than in the 6-week treatment. This results in less overlap in their distributions.

The second level of the independent *t* formula is s_1^2 and s_2^2, or the variances (s^2) for each of the two groups. Recall that the standard deviation represents the spread of the scores about the mean. If this spread becomes smaller (scores distributed more closely about the mean), the variance is also smaller. If the variance term is smaller and the first and the third levels remain the same, the denominator of the *t* test becomes smaller and *t* becomes larger, thus increasing the odds of rejecting the null hypothesis and increasing power. How can the standard deviation and thereby the variance be made smaller? The answer is to apply the treatments more consistently. The more consistently the treatments are applied to each participant, the more similar they become in response to the dependent variable. This groups the distribution more tightly around the means, thus reducing the standard deviation (and thereby the variance).

Finally, the third level (n_1, n_2) is the number of participants in each group. If n_1 and n_2 are increased and the first and second levels remain the same, the denominator becomes smaller (note *n* is divided into s^2) and *t* becomes larger, thus increasing the odds of rejecting the null hypothesis and obtaining power. Obviously, n_1 and n_2 can be increased by placing additional participants in each group.

Of course, power can also be influenced by varying the alpha (i.e., if alpha is set at .10 as opposed to .05, power is increased). But in doing this, we increase the risk of rejecting a true null hypothesis (i.e., we increase the chance of making a type I error).

Keep in mind that power can be influenced by your choice of design and the statistical techniques you use. We point this out as we discuss different statistical tests, but a simple example using tests we have already addressed makes the point. We indicated that a dependent *t* test is used when you have either matched pairs or the same participants being measured twice. This is because the independent *t* assumes that the groups are composed of different participants and that the error term is attributed to sampling error. The dependent *t* reduces this error. Another way of looking at this is that we increase power by using the dependent *t* when the situation warrants it. To illustrate, if you were to calculate an independent *t* test on the scores in example 9.5, you would find the *t* to be 2.32, whereas the dependent *t* was 4.83. So even with fewer degrees of freedom for the dependent *t,* we increased power by obtaining a higher *t* with that statistical test. We will see that we can also increase power by using techniques such as analysis of covariance and factorial analysis of variance.

In summary, power is desirable to obtain because it increases the odds of rejecting a false null hypothesis. Power can be obtained by using strong treatments, administering those treatments consistently, using as many participants as feasible, varying alpha, or using an appropriate research design and statistical analysis. Remember, however, that there is always a second question, even in the most powerful experiments. After the null hypothesis is rejected, the strength (meaningfulness) of the effects must be evaluated.

The *t* ratio has a numerator and a denominator. From a theoretical point of view, the numerator is regarded as **true variance,** or the real difference between the means. The denominator is considered **error variance,** or variation about the mean. Thus, the *t* ratio is

$$t = \frac{\text{true variance}}{\text{error variance}}$$

where true variance = $M_1 - M_2$ and error variance = $\sqrt{s_1^2 / n_1 + s_2^2 / n_2}$. If no real differences exist between the groups, true variance = error variance, or the ratio between the two is true variance/error variance = 1.0. When a significant *t* ratio is found, we are really saying that true variance exceeds error variance to a significant degree. The amount by which

true variance

The portion of the differences in scores that is (theoretically) real.

error variance

The portion of the scores that is attributed to participant variability.

the *t* ratio must exceed 1.0 for significance depends on the number of participants (*df*) and the alpha level established.

The estimate of the strength of the relationship (ω^2) between the independent and dependent variables is represented by the ratio of true variance to total variance.

$$\omega^2 = \frac{\text{true variance}}{\text{total variance}}$$

Thus, ω^2 represents the proportion of the total variance that is due to the treatments (true variance).

Effect size is also an estimate of the strength or meaningfulness of the group differences or treatments. Effect size places the difference between the means in standard-deviation units, $(M_1 - M_2)/s$. Figure 9.1 shows how the normal distribution of two groups differs for two different effect sizes, 0.5 and 1.0. In figure 9.1*a* the standardized difference between the group means is ES = 0.5; that is, the mean of group 2 falls 0.5 standard deviation to the right of the mean of group 1. (If the mean of group 2 fell to the left, the ES would be –0.5.) Figure 9.1*b* shows the distribution when ES = 1.0; that is, the mean of group 2 is 1 standard deviation to the right of group 1. Notice that there is less overlap between the two groups' distribution of scores in figure 9.1*b* than in figure 9.1*a*. Said another way, when scores on the dependent variable were grouped by the independent variable in figure 9.1*b*, the means were further apart and there was less distribution overlap than in figure 9.1*a*.

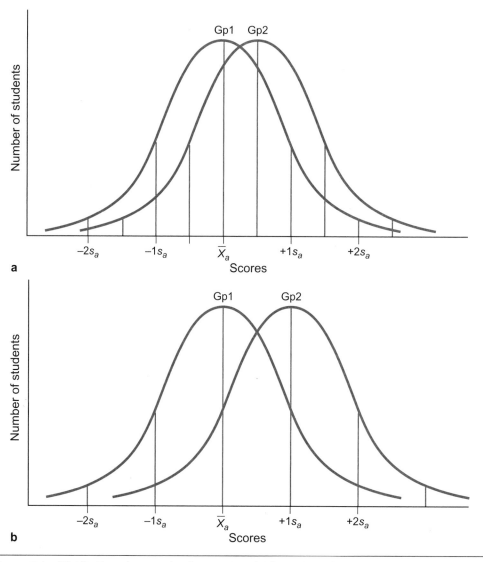

Figure 9.1   Distribution of scores for Groups 1 and 2 for which ES = 0.5 (a) and 1.0 (b).

Effect size is sometimes interpreted as percentile change attributed to the treatment. For example, in figure 9.1*b*, the treatment group (group 2) mean was 1.0*s* higher than the control group mean. If we consult table A.2 in appendix A, we see that a *z* of 1.00 (which is a distance of 1*s* above the mean) shows that only .1587 (16%) of the scores are higher than 1.0. In other words, the percentile rank for such a score is 84 (100 − 16). Consequently, in interpreting an effect size of 1.0, we can infer that the treatment improved average performance by 34 percentile points (i.e., treatment group = 84, control group = 50, 84 − 50 = 34).

Relationship of *t* and *r*

As we mentioned previously, our separation of statistical techniques into the two categories of relationships among variables (chapter 8) and differences among groups (this chapter) is artificial because both sets of techniques are based on the general linear model. A brief demonstration with *t* and *r* should make the point. However, the idea can be extended into the more sophisticated techniques discussed in chapter 8 and later in this chapter. Example 9.7 shows the relationship between *t* and *r*. Group 1 has a set of scores (dependent variable) for five people, as does group 2.

The point is that *r* represents the relationship between the independent and dependent variables (in fact, r^2 is a biased estimator of ω^2), and a *t* test can be applied to *r* (see step 4 in example 9.7) to evaluate the reliability (significance) of the relationship.

There are two sources of variance: true variance and error variance (true variance + error variance = total variance). The *t* test is the ratio of true variance to error variance, whereas *r* is the square root of the proportion of total variance accounted for by true variance. To get *t* from *r* only means manipulating the variance components in a slightly different way. This is because all parametric correlational and differences-among-groups techniques are based on the general linear model. This result can easily be shown to exist in the more advanced statistical techniques. (For a thorough treatment of this topic, see Pedhazur, 1982.)

You need to understand this basic concept because it is becoming increasingly common for researchers to use regression techniques to analyze experimental data. These data have traditionally been analyzed by techniques discussed in this chapter. We have, however, demonstrated that what is traditional is not required. What is important is that the data be appropriately analyzed to answer the following questions:

- Are the groups significantly different?
- Does the independent variable account for a meaningful proportion of the variance in the dependent variable?

Significance is always evaluated as the ratio of true variance to error variance, whereas the percentage of variance accounted for is always the ratio of true variance to total variance.

Analysis of Variance

Using *t* tests is a good way to determine differences between two groups. Often, though, experimenters work with more than two groups. A method of determining differences among them is needed in those cases. This section explains how analysis of variance is used to detect differences among two or more groups.

Simple Analysis of Variance

The concept of simple (sometimes called one-way but seldom considered simple by graduate students) **analysis of variance (ANOVA)** is an extension of the independent *t* test. In fact, *t* is just a special case of simple ANOVA in which there are two groups. Simple ANOVA allows the evaluation of the null hypothesis among two or more group means with the restriction that the groups represent levels of the same independent variable. In an earlier

analysis of variance (ANOVA)

Test that allows the evaluation of the null hypothesis between two or more group means.

Example 9.7

Known Values

Group 1			Group 2	
Participant	Dependent variable		Participant	Dependent variable
A	1		F	6
B	2		G	7
C	3		H	8
D	4		I	9
E	5		J	10

Sum of group 1: $\Sigma_1 = 15$
Mean of group 1: $M_1 = 3$
Standard deviation of group 1: $s_1 = 1.58$

Sum of group 2: $\Sigma_2 = 40$
Mean of group 2: $M_2 = 8$
Standard deviation of group 2: $s_2 = 1.58$

Working It Out

1. Conduct an independent t test (equation 9.3) and test for significance.

$$t = \frac{3 - 8}{\sqrt{\dfrac{1.58^2}{5} + \dfrac{1.58^2}{5}}} = 5.0^*$$

$$df = (n_1 + n_2) - 2 = 8$$
$$t(8) = 5.0, \, p < .05$$

* The sign is not important for this example.

2. Assign each participant a dummy code that stands for his or her group.

Group 1			Group 2	
Participant	Dummy code		Participant	Dummy code
A	1		F	0
B	1		G	0
C	1		H	0
D	1		I	0
E	1		J	0

3. Apply the correlation formula (equation 8.1). If we treat the dummy-coded variable as X and the dependent variable as Y and ignore group membership (10 participants with two variables), then the correlation formula used earlier can be applied to the data.

$$r = \frac{N\Sigma XY - (\Sigma X)(\Sigma Y)}{\sqrt{N\Sigma X^2 - (\Sigma X)^2}\sqrt{N\Sigma Y^2 - (\Sigma Y)^2}} =$$

$$\frac{10(15) - 5(55)}{\sqrt{10(5) - 25}\sqrt{10(385) - 3,025}} = \frac{125}{144} = .87^*$$

* The sign is not important for this example.

4. Apply a t test to r. In this example $t = 5.0$, the same as the t test done on the group means (within rounding error).

$$t = \sqrt{\frac{r^2}{(1 - r^2) / (N - 2)}} = \sqrt{\frac{.87^2}{(1 - .87^2) / (10 - 2)}} = \sqrt{\frac{.757}{.030}} = 5.0$$

example, we suggested that a t test was appropriate to test the difference between the means of two groups who trained at 40% and 70% of $\dot{V}O_2$max. This represents two levels (40% and 70%) of one independent variable (intensity of training). In fact, simple ANOVA and its test statistic, the F ratio, could just as easily have been used. But what would happen if there were more than two levels of the independent variable, for example, 40%, 60%, and 80% of $\dot{V}O_2$max? Simple ANOVA should be used in this situation to test the null hypothesis that the groups' average values on the 12-min run were not significantly different, $H_0 = M_1 = M_2 = M_3 = 0$.

Why not do a t test between the 40% and 60% groups, a second t test between the 40% and 80% groups, and a third t test between the 60% and 80% groups? The reason is that this would violate an assumption concerning the established alpha level (let it be $p = .05$). The .05 level means 1 chance in 20 of a difference due to sampling error, assuming that the groups of participants on which the statistical tests are made are from independent random samples. In our case, this is not true. Each group has been used in two comparisons (e.g., 40% vs. 60% and 40% vs. 80%) rather than only one. Thus, we have increased the chances of making a type I error (i.e., alpha is no longer .05). Making this type of comparison, in which the same group's mean is used more than once, is an example of increasing the experimentwise error rate (discussed later in this chapter). Simple ANOVA allows all three group means to be compared simultaneously, thus keeping alpha at the designated level of .05.

Calculating Simple ANOVA

Table 9.1 provides the formulas for calculating simple ANOVA and the F ratio. This method, the so-called *ABC method,* is simple:

- A = ΣX^2: Square each participant's score, sum these squared scores (regardless of which group the participant is in), and set the total equal to A.
- B = $(\Sigma X)^2/N$: Sum all participants' scores (regardless of group), square the sum, divide by the total number of participants, and set the answer equal to B.
- C = $(\Sigma X_1)^2/n_1 + (\Sigma X_2)^2/n_2 + \ldots + (\Sigma X_k)^2/n_k$. Sum all scores in group 1, square the sum, and divide by the number of participants in group 1; do the same for the scores in group 2, and so on for however many groups (k) there are. Then add all the group sums and set the answer to C.

Table 9.1   Formulas for Calculating Simple ANOVA

$$A = \Sigma X^2$$

$$B = \frac{(\Sigma X)^2}{N}$$

$$C = \frac{(\Sigma X_1)^2}{n_1} + \frac{(\Sigma X_2)^2}{n_2} + \ldots + \frac{(\Sigma X_k)^2}{n_k}$$

Summary table for ANOVA

Source	SS	df	MS	F
Between (true)	$C - B$	$k - 1$	$(C - B)/(k - 1)$	MS_B/MS_W
Within (error)	$A - C$	$N - k$	$(A - C)/(N - k)$	
Total	$A - B$	$N - 1$		

Note. X = a participant's score, N = total number of participants, n = number of participants in a group, k = number of groups, SS = sum of squares, df = degrees of freedom, MS = mean square.

sum of squares

A measure of variability of scores; the sum of the squared deviations from the mean of scores.

Next, fill in the summary table for ANOVA using A, B, and C. Thus, the between-groups (true variance) **sum of squares** (*SS*) is equal to C – B; the between-groups degrees of freedom (*df*) is the number of groups minus one ($k - 1$); the between-groups variance or mean square (MS_{B}) is the between-groups sum of squares divided by the between-groups degrees of freedom. The same is done within groups (error variance) and then for the total. The *F* ratio is $MS_{\mathrm{B}}/MS_{\mathrm{W}}$ (i.e., the ratio of true variance to error variance).

Example 9.8 shows how the formulas in table 9.1 are used. The scores for groups 1, 2, and 3 are the sums of two judges' skill ratings for a particular series of movements. The groups are randomly formed of 15 junior high school students. Group 1 was taught with videotape, and the teacher made individual corrections while the student viewed the videotape. Group 2 was taught with videotape, but the teacher made only general group corrections while students viewed the tape. The teacher taught group 3 without the benefit of videotape. From looking at the formulas in table 9.1, you should be able to see how each number in this example was calculated.

Example 9.8

Known Values

	Group 1		Group 2		Group 3	
	X	*X²*	*X*	*X²*	*X*	*X²*
	12	144	9	81	6	36
	10	100	7	49	7	49
	11	121	6	36	2	4
	7	49	9	81	3	9
	10	100	4	16	2	4
Σ	50	514	35	263	20	102
M	10	—	7	—	4	—

Working It Out

$$A = \Sigma X^2 = 514 + 263 + 102 = 879$$

$$B = \left(\Sigma X\right)^2 / N = \left(50 + 35 + 20\right)^2 / 15 = \left(105\right)^2 / 15 = 11,025 / 15 = 735$$

$$C = \left(\Sigma X_1\right)^2 / n_1 + \left(\Sigma X_2\right)^2 / n_2 + \left(\Sigma X_3\right)^2 / n_3 = \left(50\right)^2 / 5 + \left(35\right)^2 / 5 + \left(20\right)^2 / 5$$

$$= 2,500 / 5 + 1,225 / 5 + 400 / 5 = 825$$

Summary table for ANOVA

Source	SS	df	MS	F
Between	90	2	45.0	10.00*
Within	54	12	4.5	
Total	144	14		

*$p < .05$

Your main interest in example 9.8, after you make sure you understand how the numbers were obtained, is the *F* ratio of 10.00. Table A.6 in appendix A contains *F* values for the .05 and .01 levels of significance. Although the numbers in the table are obtained the same way as in the *t* table, you use the table in a slightly different way. The *F* ratio is obtained by dividing MS_{B} by MS_{W}. Note in example 9.8 that the term MS_{B} (the numerator of the *F* ratio) has 2 df ($k - 1 = 2$), and the term MS_w (the denominator of the F ratio) has 12 df ($N - k = 12$). Notice also that table A.6 has degrees of freedom across the top (numerator) and down the left-hand column (denominator). For our F of 10.00, read down the 2-df column to the 12-df row; there are two numbers where the row and the column intersect. The top number **(3.88)** in dark print is the table F value for the .05 level, whereas the bottom number (6.93) in light print is the table F value for the .01 level. If our alpha had been established as .05, you can see that our F value of 10.00 is larger than the table value of 3.88 at the .05 level (it is also larger than the value of 6.93 at the .01 level). So our F is significant and could be written in the text of an article as $F(2, 12) = 10.00$, $p < .05$ (read "F with 2 and 12 degrees of freedom equals 10.00 and is significant at less than the .05 level").

Buchowski, Darud, Chen, and Sun (1998) used a one-way (simple) ANOVA to compare work efficiency (WE) of aerobic instructors and noninstructors during an aerobic routine and found a significant difference, with instructors having the higher WE, $F(1, 22) = 13.01$, $p < .002$. The APA notation tells us that there were 2 groups with 24 total participants, because the degrees of freedom in the parentheses (1, 22) are for the between-groups variation ($k - 1$) first, then the within-groups variation ($N - k$). The calculated F was 13.01 and was significant at the $p < .002$ level.

Follow-Up Testing

In our example of ANOVA with three groups, we now know that significant differences exist among the three group means ($M_1 = 10$, $M_2 = 7$, and $M_3 = 4$). However, we do not know whether all three groups differ (e.g., whether groups 1 and 2 differ from group 3 but not from each other). So we next perform a follow-up test. One way to do this could be to use t tests between groups 1 and 2, groups 1 and 3, and groups 2 and 3. However, the same problem that we discussed earlier exists with alpha (type I error is increased). Several follow-up tests protect the experimentwise error rate (type I error, discussed later in this chapter). These methods include Scheffé, Tukey, Newman-Keuls, Duncan, and several others (see Toothaker, 1991, for both conceptual explanations and calculations of the various techniques). Each test is calculated in a slightly different way, but they all are conceptually similar to the t test in that they identify which pairs of groups differ from each other. The Scheffé method is the most conservative (which means it identifies fewer significant differences), followed by Tukey. Duncan is the most liberal, identifying more significant differences. Newman-Keuls falls between the extremes but has some other problems.

For our purpose, one example should suffice, so we explain the use of the Scheffé method of making multiple comparisons among means. Scheffé (1953) is the most widely recognized of all multiple-comparison techniques (Toothaker, 1991). Because we believe researchers should be conservative when using behavioral data, we generally recommend the Scheffé for follow-up testing. However, other multiple-comparison techniques are appropriate for various situations.

The Scheffé technique has a constant critical value for the follow-up comparison of all means when the F ratio from a simple ANOVA (or a main effect in a factorial ANOVA, discussed next) is significant. Scheffé controls type I error (alpha inflation) for any number of appropriate comparisons. Equation 9.12 is how the critical value (CV, required size of the difference) for significance is calculated using the Scheffé technique:

$$CV = \sqrt{(k-1)F_{\alpha;\, k-1,\, N-k}} \sqrt{2(MS_w / n)} \qquad (9.12)$$

where k = number of means to be compared, F_α = the table F ratio (from table A.6) for the selected alpha (e.g., .05) given dfs of $k - 1$ (for between groups) and $N - k$ (for within groups). For the second half of the equation, MS_w is obtained from the ANOVA summary table (SS_w/df_w), and n = the number in a single group.

Example 9.9 shows the Scheffé technique applied to the previous simple ANOVA shown in example 9.8. You find the F ratio in table A.6 at the intersection of the column $df = 2$ ($k - 1 = 3 - 1 = 2$) and the $df = 12$ row ($N - k = 15 - 3 = 12$) from the dfs in the summary ANOVA in table 9.1; the table F ratio is 3.88. Solving the left part of formula 9.12 in example 9.9, we find the Scheffé value for significance at the .05 level is 2.79. For the second part of formula 9.12, we obtain $MS_w = 4.5$ (from table 9.1) and the n for each group is 5. Thus, $\sqrt{2(4.5 / 5)} = 1.34$. Hence, formula 9.12 becomes

$$CV = (2.79)(1.34) = 3.74$$

This is the size of the difference between means that is needed in order to be significant at $p < .05$.

Example 9.9 arranges the means from highest (10) to lowest (4). The observed difference for each of the comparisons is computed; then each observed difference is compared with the needed difference (CV) for significance. By following these steps, you can see that only the comparison between groups 1 and 3 was significant at $p < .05$. We could conclude that the techniques used in group 1 were significantly better than those used in group 3 but not significantly better than those used in group 2. Also, the techniques used in group 2 were not significantly better than those used in group 3.

Example 9.9

Working It Out

1. Order the means from highest (10) to lowest (4).

2. Compute differences between all means.

Group 1 (M = 10)	Group 2 (M = 7)	Group 3 (M = 4)	Observed difference
10	7		3
10		4	6
	7	4	3

3. Compute critical value (difference needed) for significance at alpha = .05 (equation 9.12):

$$CV = \sqrt{(k-1)F_{\alpha;\ k-1,\ N-k}} \sqrt{2(MS_w / n)}$$
$$= \sqrt{(3-1)3.88} \sqrt{2(4.5 / 5)}$$
$$= (2.79)(1.34) = 3.74$$

4. Compare the CV of 3.74 with each observed difference to determine which difference or differences are significant. In this case, only the difference between groups 1 and 3 is as large as or larger than the CV of 3.74.

---

planned comparison

Comparison among groups that are planned before the experiment, rather than as a follow-up of a test such as ANOVA.

The researcher may also use **planned comparisons** to test for differences among groups. Planned comparisons are planned (testable hypotheses developed) before the experiment. Thus, an experimenter might postulate a test between two groups before the experiment because in theory this particular comparison is important and should be significant. However, the number of planned comparisons in an experiment should be small ($k - 1$, where k = number of groups or treatment levels of an independent variable) relative to the total number of possible comparisons.

Determining Meaningfulness of Results

Now that we know that F is significant for our example study and have followed it up to see which groups differ, we should answer our second question: What percentage of variance is accounted for by our treatments, or how meaningful are our results? One way to get a quick idea is to refer to table 9.1 and divide true variance by total variance: $(SS_{true}/SS_{total}) = 90/144 = .625$, or 62.5% of the variance is accounted for by the treatments.

Although this is sufficient for a quick estimate, it is biased. The more accurate way is to use the following formula from Tolson (1980),

$$\omega^2 = \frac{[F(k-1)] - (k-1)}{[F(k-1)] + (N-k) + 1} \qquad (9.13)$$

where F = the F ratio, k = the number of groups, and N = the total number of participants. If we do this with the data from example 9.8, we have

$$\omega^2 = \frac{[10.00(3-1)] - (3-1)}{10.00(3-1) + (15-3) + 1} = \frac{18}{33} = .545$$

Thus, ω^2 indicates that 54.5% of the total variance is accounted for by the treatments.

Putting our statistics together, we could say that the treatment was significant, $F(2, 12)$ = 10.00, $p < .05$, and accounted for a meaningful proportion of the variance, $\omega^2 = 54.5\%$. In addition, a follow-up Scheffé test indicates that group 1 had the best performance and was significantly different ($p < .05$) from group 3. However, group 1 was not significantly different from group 2, nor was group 2 significantly different from group 3.

Summarizing Simple ANOVA

One final point to recall before leaving our discussion of simple ANOVA is that t is a special case of F when there are only two levels of the independent variable (two groups). In fact, this relationship ($t^2 = F$) is exact, within rounding error (see example 9.10).

In other words, there is very little need for t, as F will handle two or more groups. However, t remains in use because it was developed first, being the simplest case of F. Also, t is convenient and can be computed by hand because the M and s are already calculated for each group to serve as descriptive statistics.

Factorial ANOVA

Up to this point, examples of two (t) or more levels (simple ANOVA) of one independent variable have been discussed. In fact, in our examples so far, all other independent variables have been controlled except the single independent variable to be manipulated and its effect on a dependent variable. This is called the law of the independent variable. But, in fact, we can manipulate more than one independent variable and statistically evaluate the effects on a dependent variable. This procedure is called **factorial ANOVA**, which is used when there is more than one factor, or independent variable. As you read the literature, you will see that there are many more studies that use factorial ANOVA than those that use one-way, or simple, ANOVA. Theoretically, a factorial ANOVA may have any number of factors (two or more) and any number of levels within a factor (two or more). However, we seldom encounter ANOVA with more than three or four factors. This is another good place to apply the KISS principle (Keep It Simple, Stupid).

factorial ANOVA

Analysis of variance in which there is more than one independent variable.

The Components of Factorial ANOVA

For the purpose of explaining, we consider the simplest form of the factorial ANOVA, which uses only two independent variables and only two levels of each variable. In this design, which is designated a 2 × 2 ANOVA, there are two main effects and one interaction. **Main effects** are tests of each independent variable when the other is disregarded (and controlled). Look at table 9.2 and note that the first independent variable (IV$_1$) has two levels, labeled A_1 and A_2. In our example this IV represents the intensity of training: high intensity and low intensity. The second independent variable (IV$_2$) represents the level of fitness of the participants: high fitness (B_1) and low fitness (B_2). The numbers in the cells represent mean attitude scores of each of the four groups toward the type of training (high intensity or low intensity). Thus, we have one group of high-fitness people who trained under a high-intensity program and one group of high-fitness people who trained with a low-intensity program. The low-fitness participants were also split into the two intensity programs.

main effects

Tests of each independent variable when all other independent variables are held constant.

Example 9.10

Known Values

	Control group			Experimental group		
	X	$(X - M)^2$	X^2	X	$(X - M)^2$	X^2
	2	.09	4	8	4	64
	4	2.89	16	7	1	49
	3	.49	9	5	1	25
	3	.49	9	4	4	16
	2	.09	4	7	1	49
	1	1.69	1	5	1	25
	1	1.69	1	6	0	36
Σ	16	7.43	44	42	12	264
M	2.3	—	—	6.0	—	—
s	1.11	—	—	1.41	—	—

Working It Out

1. Find t (equation 9.3).

$$t = \frac{M_1 - M_2}{\sqrt{(s_1^2 / n_1) + (s_2^2 / n_2)}} = \frac{2.3 - 6}{\sqrt{(1.11^2 / 7) + (1.41^2 / 7)}} = \frac{3.7}{0.68} = 5.44 *$$

* The sign is not important to report.

2. Compute the simple ANOVA and complete the summary table.

$$A = \Sigma X^2 = 44 + 264 = 308$$

$$B = (\Sigma X)^2 / N = (58)^2 / 14 = 240.29$$

$$C = (\Sigma X_1)^2 / n_1 + (\Sigma X_2)^2 / n_2 = (16)^2 / 7 + (42)^2 / 7 = 36.57 + 252 = 288.57$$

Summary table for ANOVA

Source	SS	df	MS	F
Between	48.28	1	48.28	29.80
Within	19.43	12	1.62	
Total	67.71	13		

$t^2 = F$, $(5.44)^2 = 29.6 \cong 29.8$ (within rounding error)

We can test the main effect of IV_1 (level of intensity) by comparing the row means (M_{A1} and M_{A2}), because IV_2 (B_1 and B_2)—fitness level of the participants—is equally represented at each level of A. Actually, a sum of squares with $A - 1$ df and mean squares for the between-group and within-group sources of variance are computed. Thus, we can compute an F for main effect A (intensity).

The same holds true for IV_2, fitness level, by comparing the column means (M_{B1} and M_{B2}). You can see that the two levels of A (IV_1), intensity of training, are equally represented in the two levels of B. Therefore, the main effect of the attitudes of high- and low-fitness participants toward exercise can be tested by the F ratio for B.

Table 9.2   Factorial (2 × 2) ANOVA Model

IV_2 (Fitness)

		B_1 (low fit)	B_2 (high fit)	
IV_1 (Intensity)	A_1 (high int)	$M = 10$	$M = 40$	$M_{A1} = 25$
	A_2 (low int)	$M = 30$	$M = 20$	$M_{A2} = 25$
		$M_{B1} = 20$	$M_{B2} = 30$	

Summary table for ANOVA

Source	SS*	df	MS	F
A (intensity)	—	$A - 1 = 2 - 1 = 1$	SS_A/df_A	MS_A/MS_{error}
B (fitness)	—	$B - 1 = 2 - 1 = 1$	SS_B/df_B	MS_B/MS_{error}
AB (Interaction)	—	$(A - 1)(B - 1) = 1$	SS_{AB}/df_{AB}	MS_{AB}/MS_{error}
Error	—	$(N - 1) - [(A - 1) + (B - 1) + (A - 1)(B - 1)]$	SS_{error}/df_{error}	
Total	—	$N - 1$		

* For simplicity, formulas for computing sums of square are not given here.

In a study of this type, the main interest usually lies in the interaction. We want to know whether attitude toward exercise programs of high or low intensity (factor A) depends on (is influenced by) the fitness level of the participants (factor B). This effect is tested by an F ratio for the interaction ($A \times B$) of the two IVs, which evaluates the four cell means: M_{A1B1}, M_{A1B2}, M_{A2B1}, and M_{A2B2}. Unless some special circumstance exists, interest in testing the main effects is usually limited by a significant interaction, which means that what happens in one independent variable depends on the level of the other. Thus, normally it makes little sense to evaluate main effects when the interaction is significant.

This particular factorial ANOVA is labeled a 2 (intensity of training) × 2 (level of fitness) ANOVA (read "2-by-2 ANOVA"). The true variance can be divided into three parts:

- True variance due to A (intensity of training)
- True variance due to B (level of fitness)
- True variance due to the interaction of A and B

Each of these true variance components is tested against (divided by) error variance to form the three F ratios for this ANOVA. Each of these Fs has its own set of degrees of freedom so that it can be checked for significance in the F table (table A.6 in appendix A).

Main effects (of more than two levels) can be followed up with a Scheffé test. However, no follow-up is required for an independent variable with only two levels. If the F is significant, you only have to see which has the higher M. If the F for the interaction is significant, meaning that attitude toward high- or low-intensity training depends on the fitness level of the participants, it will be reflected in a plot, as shown in figure 9.2. You can see in table 9.2 that there is no mean difference for the main effect of intensity of training (both A_1 and A_2 are the same, $M = 25$). However, the plot in figure 9.2 reflects a significant interaction, because the mean attitude scores of the low-fitness group toward the low-intensity training was higher than their attitude toward high-intensity training, while the opposite was shown for the high-fitness participants. They preferred the high-intensity program.

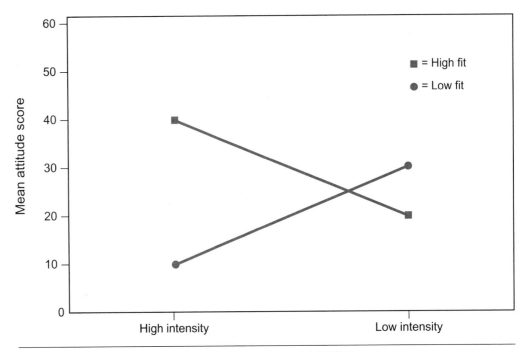

Figure 9.2   Plot of the interaction for a two-way factorial ANOVA.

This is an example of how power is increased by using a particular type of statistical test. If we had just used a *t* test or simple ANOVA, we would not have found any difference in attitude toward the two levels of intensity (both means were identical, *M* = 25). However, when we added another factor (fitness level) we were able to discern that there were indeed differences in attitude dependent on the level of fitness of the participants.

As you can see, all we have done to follow up the significant interaction is to describe the plot in figure 9.2 verbally. Considerable disagreement exists among researchers about how to follow up significant interactions. Some researchers use a multiple-comparison-of-means test (such as Scheffé) to contrast the interaction cell means. However, these multiple-comparison tests were developed for contrasting levels within an independent variable and not for cell means across two or more independent variables; thus, their use may be inappropriate. Other researchers use a follow-up called a test of simple main effects. If an interaction is to be followed up with a statistical test, simple main effects using the Scheffé procedure is the preferred technique.

Our preference for evaluating an interaction is to do as we have done: plot the interaction and describe it. This takes into account the true nature of an interaction; that is, what happens in one independent variable depends on the other. However, you are likely to encounter all these ways of testing interactions (and probably some others) as you read the research literature. Just remember that the researcher is trying to show you how the two or more independent variables interact. Also remember that follow-ups on main effects are usually unnecessary (or at least the interpretation must be qualified) when the interaction is significant.

Figure 9.3 shows a nonsignificant interaction. In this case, both groups preferred the same type of program over the

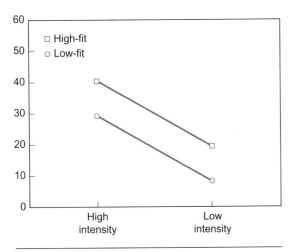

Figure 9.3   Plot of a nonsignificant interaction (parallel lines) where both fitness groups prefer low-intensity programs over high intensity.

other; hence, the lines are parallel. Significant interactions show deviations from parallel (as was shown in figure 9.2). The lines do not have to cross to reflect a significant interaction. Figure 9.4 shows a significant interaction in which the high-fitness group liked both forms of exercise equally, but there was a decided difference in preference in the low-fitness group, who preferred the low-intensity over the high-intensity program.

Determining Meaningfulness of Results

Having answered the first statistical question about a factorial ANOVA (are the effects significant?), we now turn to the meaningfulness of the effects, or what percentage of the variance in the dependent variable is accounted for by the independent variables and their interaction? The three formulas that follow (from Tolson, 1980) provide the test for each component of ANOVA:

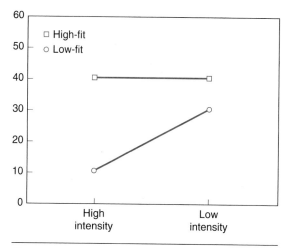

Figure 9.4 Plot of a significant interaction of no difference in high-fit preferences but significant difference in intensity preference among low-fit individuals.

$$\omega_A^2 = \frac{(p-1)F_A - (p-1)}{(p-1)F_A + (q-1)F_B + (p-1)(q-1)F_{AB} + (N-pq)+1}$$

$$\omega_B^2 = \frac{(q-1)F_B - (q-1)}{(p-1)F_A + (q-1)F_B + (p-1)(q-1)F_{AB} + (N-pq)+1}$$

$$\omega_{AB}^2 = \frac{(p-1)(q-1)F_{AB} - (p-1)(q-1)}{(p-1)F_A + (q-1)F_B + (p-1)(q-1)F_{AB} + (N-pq)+1}$$

(9.14)

where p = the number of levels of A and q = the number of levels of B. These formulas involve the proportion of true variance to total variance. Note that the total variance (the denominator) is the same for all three proportions.

Each ω^2 represents the percentage of variance accounted for by that component of the ANOVA model. The ω^2s can be summed to estimate the percentage of the total variance that is true variance. Of course, effect sizes can be calculated according to the previous equations (9.7 and 9.8) for any two means for which you want to compare main effects.

Summarizing Factorial ANOVA

Frequently, the levels of an independent variable are categorical (also called classified) rather than random. For example, we could do a simple ANOVA in which the levels (or groups) of the independent variable involve novice and expert tennis players, or the attitude toward intensity of training held by high-fitness participants as compared with low-fitness participants, as in our example in table 9.2. That is, we were interested in whether preference toward type of training programs affected high-fitness and low-fitness participants differently. Participants were randomly assigned to the levels of the first independent variable (intensity level of program), whereas people cannot be randomly assigned to the categorical variable of fitness level. Common categorical variables are sex and age. A study might look at the effects of the levels of a treatment (independent variable) to which male and female, or 6-, 9-, and 12-year-old, participants are randomly assigned. Clearly, the interest in this type of factorial ANOVA is the interaction: Does the effectiveness of treatment differ according to the sex (or age) of the participant?

Examples of factorial ANOVA are plentiful in the research literature. McPherson (1999) conducted a study in which she compared expert and novice tennis players from three different age groups on different performance skills. The data were analyzed with 3 (age levels) × 2 (expertise levels) factorial ANOVA. Omega squared (ω^2) was used to interpret meaningfulness. This is a sample description of the results of the factorial ANOVA: "Analysis of forceful game executions indicated significant main effects for age, $F(2, 46) = 7.6$,

$p < .002$, $\omega^2 = .10$, and expertise, $F(1, 46) = 93.0$, $p < .00001$, $\omega^2 = .48$. The interaction, $F(2, 46) = 0.9$, $p > .05$, was not significant" (p. 238). The significant main effects for age were followed up using the Scheffé method.

Repeated-Measures ANOVA

Much of the research in the study of physical activity involves studies that measure the same dependent variable more than once. For example, a study in sport psychology might investigate whether athletes' **state anxiety** (how nervous they feel at the time) differs before and after a game. Thus, a state-anxiety inventory would be given just preceding and immediately following the game. The question of interest is, Does state anxiety change significantly after the game? A dependent t test could be used to see whether significant changes occurred. This is the simplest case of **repeated-measures ANOVA.**

Another study might measure participants on the dependent variable on several occasions. Suppose we want to know whether the distance that children (between ages 6 and 8) throw a ball using an overhand pattern increases over time. We decide to measure the children's distance throw every 3 months during the 2 years of the study. Thus, each child's throwing distance was assessed nine times. We now have nine repeated measures (at the beginning, and in eight 3-month intervals) on the same children. We would use a simple ANOVA with repeated measures on a single factor. Basically, the repeated measures are used as nine levels of the independent variable, which is time (24 months).

The most frequent use of repeated measures involves a factorial ANOVA in which one or more of the factors (independent variables) are repeated measures. An example is an investigation of the effects of knowledge of results (KR) on skilled motor performance. There are three groups of participants (three levels of the independent variable, KR) who receive different types of KR (no KR, whether they were short or long of the target, and number of centimeters that they were short or long of the target). The task is to position a handle that slides back and forth on a track (called a linear slide) as close to the target as possible. But the participants are blindfolded, so they cannot see the target. They have only verbal KR to correct their estimates of where the target is on the track.

In this type of study, participants are usually given multiple trials (assume 30 trials in this example) so that the effects of the type of KR can be judged. The score on each trial is distance from the target in centimeters. This type of study is frequently analyzed as a two-way factorial ANOVA with repeated measures on the second factor. Thus, a 3 (levels of KR) × 30 (trials) ANOVA with distance error as the dependent variable is used to analyze the data. The first independent variable (level of KR) is a true one (three groups are randomly formed). The second independent variable (30 trials) is repeated measures. Sometimes this ANOVA is called a two-way factorial ANOVA with one between-subjects factor (levels of KR) and one within-subjects factor (30 repeated trials). Although an F ratio is calculated for each independent variable, the major focus is usually on the interaction. For example, do the groups change at different rates across the trials?

Advantages of Repeated-Measures ANOVA

Repeated-measures designs have three advantages (Pedhazur, 1982). First, they provide the experimenter the opportunity to control for individual differences among participants, probably the largest source of variation in most studies. In between-subject designs (completely randomized), the variation among people goes into the error term. Of course, this tends to reduce the F ratio unless it is offset by a large N. Remember, the error term of the F ratio is calculated by dividing the variation among participants by the degrees of freedom (based on the number of participants). In repeated-measures designs, variation from individual differences can be identified and separated from the error term, thereby reducing it and increasing power. As you can deduce from the first advantage, repeated-measures designs are more economical in that fewer people are required. Finally, repeated-measures designs allow the study of a phenomenon across time. This is particularly important in studies of change in, for example, learning, fatigue, forgetting, performance, and aging.

state anxiety

An immediate emotional state of apprehension and tension in response to a specific situation.

repeated-measures ANOVA

Analysis of scores for the same individuals on successive occasions, such as a series of test trials; also called *split-plot ANOVA* or *subject × trials ANOVA*.

Problems of Repeated-Measures ANOVA

Several problems adversely affect repeated-measures designs, including the following:

- Carryover effects. Treatments given earlier influence treatments given later.
- Practice effects. Participants get better at the task (dependent variables) as a result of repeated trials in addition to the treatment (also called the testing effect).
- Fatigue. Participants' performance is adversely influenced by fatigue (or boredom).
- Sensitization. Participants' awareness of the treatment is heightened because of repeated exposure.

Note that some of these problems may be the variables of interest in repeated-measures designs. Carryover effects may be of interest to a researcher of learning, whereas increased fatigue over trials may intrigue an exercise physiologist.

The tricky part of repeated-measures designs involves how to analyze the data statistically. We have already mentioned ANOVA models with repeated measures. Unfortunately, repeated-measures ANOVA has an assumption beyond the ones we have given for all previous techniques. This assumption is called **sphericity:** The repeated measures, "when transformed by a set of orthonormal weights, are uncorrelated with each other and have equal variances" (Schutz & Gessaroli, 1987, 134). If the design has a between-subjects factor, the pooled data (across all participants) must exhibit sphericity. How well the data meet these assumptions is best estimated by a statistic called epsilon (ϵ). Epsilon ranges from 1.0 (perfect sphericity, the assumption is met) to 0.0 (complete violation). An epsilon above .75 is desirable in repeated-measures experiments. Most of the widely used statistical packages (SPSS and SAS) have repeated-measures functions that provide both estimates of epsilon and tests that should be used to evaluate the F ratio for the repeated-measures factors in the designs. The failure to meet the sphericity assumption results in an increase in type I error; that is, the alpha level may be considerably larger than the researcher intended. Several statisticians (Davidson, 1972; Harris, 1985; Morrow & Frankiewicz, 1979) have suggested that multivariate techniques are more appropriate methods of analysis. However, two additional points are important (Pedhazur, 1982):

sphericity

An assumption that repeated measures are uncorrelated and have equal variance.

- When the assumptions are met, ANOVA with repeated measures is more powerful than the multivariate tests.
- If the number of participants is small, only the ANOVA repeated-measures test can be used.

If you are contemplating conducting a study that uses a repeated-measures design, an additional source of reading should be helpful to you. Schutz and Gessaroli (1987) provide a tutorial on the use of repeated measures for univariate and multivariate data that gives a sound rationale for making decisions as well as specific examples of how the data may be analyzed and evaluated.

Components of Repeated-Measures ANOVA

Table 9.3 provides the sources of variation in a one-way ANOVA with repeated measures. This kind of analysis is also called subject $\times$ trials ANOVA, within-subjects ANOVA, or two-way ANOVA with one subject per cell.

The values we used to arrive at the statistics in table 9.3 are the same as in example 9.8. Recall that in example 9.8, 15 participants were randomly assigned to treatment group 1, 2, or 3. Suppose that in a different study, five participants were each measured on three trials of a task, (or were given different treatments) and we coincidentally got the same 15 measurements as in example 9.8. In effect, rather than having 15 participants in groups 1, 2, and 3, we now have 5 participants who were measured on trials 1, 2, and 3 (or treatments 1, 2, and 3). The known values for table 9.3 would look identical to those in example 9.8, except that the columns could be labeled "Trial 1," "Trial 2," and "Trial 3" (or treatment 1, 2, and 3) instead of "Group 1," "Group 2," and "Group 3." We are using the same 15 values so that when table 9.3 and example 9.8 are compared, you can see why the repeated-measures design results in increased economy. Notice that the total sum of squares and degrees of freedom are the same. Also, the between-groups effect in example 9.8 and trials effect in table 9.3 are the same. However, the sum of squares for the within-group effect (error) in example 9.8 is divided into two components in the repeated-measures analysis in table 9.3. The residual effect (estimating error) has a sum of squares of 28 with 8 *df,* and the subjects effect has a sum of squares of 26 with 4 *df.* This results in an *F* ratio for the repeated-measures ANOVA (in table 9.3) that is larger than the *F* ratio for the simple ANOVA (in example 9.8), despite the repeated-measures ANOVA having only one third as many participants. The sum of squares for subjects in table 9.3 is not tested and simply represents the normal variation among subjects. Thus, across the three trials, the participants' mean performance decreased significantly (trial 1 = 10, trial 2 = 7, and trial 3 = 4). This analysis has all the strengths and weaknesses of repeated-measures designs previously discussed.

Table 9.3   Summary of Repeated-Measures ANOVA

Source	SS	df	MS	F
		Summary table for ANOVA		
Trials (T)	90	$(T-1)=2$	45.0	12.86*
Subjects (S)	26	$(S-1)=4$	3.5	
Residual (error)	28	$(S-1)(T-1)=8$		
Total	144	$N-1=14$		

* $p < .05$.

Note. $\epsilon = 1.00$; no adjustment of *df*s needed.

The epsilon estimate from this analysis is shown at the bottom of table 9.3. This was obtained by running a repeated-measures computer program from SPSS. If epsilon is less than 1.00, the degrees of freedom is adjusted according to the formula $df \times \epsilon$ = adjusted *df.* This is done for the degrees of freedom in both the numerator and the denominator of any *F* ratio that includes the repeated-measures factor.

Geisser/ Greenhouse correction

A conservative approach to the adjustment of the epsilon estimate in repeated-measures ANOVA that calculates adjusted degrees of freedom to find an *F* ratio to determine significance.

A very conservative approach to this adjustment (called **Geisser/Greenhouse correction**) can also be done as follows (Stamm and Safrit, 1975):

$$\theta = 1/(k-1) \qquad (9.15)$$

where *k* = the number of repeated measures. Then we multiply the degrees of freedom for trials by this value, $\theta(k-1)$, as well as the degrees of freedom for error, $\theta(n-1)(k-1)$. These adjusted degrees of freedom (to the nearest whole *df*) are used to look up the *F* ratio in the *F* table. If the *F* ratio is significant under the conservative test, then the effect likely is a real one. However, this procedure was advocated several years ago, and with the tests of repeated-measures designs now available, a computer program that provides the epsilon estimate is more appropriate.

It is probably safe to say that most ANOVA models in experimental research reported in national journals in our field use factorial ANOVA with repeated measures on one or more of the factors. An example of such a study is by M.R. Weiss and colleagues (1998), who examined the role of peer mastery and coping models on children's swimming skills, fear, and self-efficacy. If the sphericity assumption was violated ($\epsilon < .75$) the authors used a multivariate repeated-measures method. If sphericity was not violated, they used a 3 (peer model type) × 3 (assessment period) factorial ANOVA with repeated measures on the last factor.

Analysis of Covariance

Analysis of covariance (ANCOVA) is a combination of regression and ANOVA. The technique is used to adjust the dependent variable for some distractor variable (called the **covariate**), which is some variable that could affect the treatments.

Using ANCOVA

Suppose we want to evaluate the effects of a training program to develop leg power on the time required to run 50 m. However, we know that reaction time (RT) influences the 50-m-dash time because those who begin more quickly after the start signal have an advantage. We form two groups, measure RT in each group, train one with our power development program while the other serves as a control group, and measure each participant's time for running 50 m. This is a study in which ANCOVA might be used to analyze the data. There is an independent variable with two levels (power training and control), a dependent variable (50-m-dash time), and an important distractor variable, or covariate (RT).

Analysis of a covariance is a two-step process in which an adjustment is first made for the 50-m-dash score of each runner according to his or her RT. A correlation (r) is calculated between RT and the time in the 50-m dash. The resulting prediction equation, 50-m-dash time = $a + (b)$RT (the familiar formula $a + bX$), is used to calculate each runner's predicted 50-m-dash time (Y). The difference between the actual 50-m-dash time (Y) and the predicted time (Y') is called the residual ($Y - Y'$). A simple ANOVA is then calculated using each participant's residual score as the dependent variable (1 df for within-group sums of squares is lost because of the correlation). This allows an evaluation of 50-m-dash speed with RT controlled.

Analysis of covariance can be used in factorial situations and with more than one covariate. The results are evaluated as in ANOVA except one or more distractor variables are controlled. Also, ANCOVA is frequently used in situations in which a pretest is given, some treatment applied, and then a posttest given. The pretest is used as the covariate in this type of analysis. Note that in the preceding section on repeated-measures ANOVA, we indicated that this same situation could be analyzed by repeated measures. In addition, ANCOVA is used when comparing intact groups because the groups' performances (dependent variable) can be adjusted for distractor variables (covariates) on which they differ. If ANCOVA is warranted, it can increase the power of the F test (making it easier to detect a difference).

Limitations of ANCOVA

Although ANCOVA may seem to be the answer to many problems, it does have limitations. In particular, its use to adjust final performance for initial differences can result in misleading interpretations (Lord, 1969). In addition, if the correlations between the covariate and the dependent variable are not equal across the treatment groups, standard ANCOVA (there are nonstandard ANCOVA techniques) is inappropriate.

Turner and Martinek (1999) tested the validity of a technique called the "games for understanding" model in teaching field hockey. One group of participants was taught with this method, one group with a technique approach to instruction, and one group served as a control group. For some of the dependent variables (e.g., ball control and passing decision making), ANCOVA was deemed appropriate with the pretest as the covariate.

Experimentwise Error Rate

Sometimes researchers make several comparisons of different dependent variables using the same participants. Usually, a multivariate technique (discussed in the next section) is the appropriate solution. However, when dependent variables are combinations of other dependent variables (e.g., cardiac output is heart rate × stroke volume), a multivariate

analysis of covariance (ANCOVA)

A combination of regression and ANOVA that statistically adjusts the dependent variable for some distractor variable called the *covariate*.

covariate

A distractor variable that is statistically controlled in ANCOVA and MANCOVA.

model using all three dependent variables is inappropriate. (This book is not the appropriate place to explain why. For more detail, see Thomas, 1977.) Thus, an ANOVA among three groups might be calculated separately for each dependent measure (i.e., three ANOVAs). The problem is that this procedure results in increasing the alpha that has been established for the experiment. One of two solutions is appropriate for adjusting alpha. The first, called the Bonferroni technique, simply divides the alpha level by the number of comparisons to be made:

$$\alpha_{EW} = \alpha/c \qquad (9.16)$$

where α_{EW} = alpha corrected for the experimentwise error rate, α = alpha, and c = the number of comparisons. If, for example, α = .05 and c = 3, then the alpha for each comparison is .05/3 = .017. This means that the F ratio would have to reach an alpha of .017 to be declared significant.

In the article by Turner and Martinek (1999), the authors used the Bonferroni adjustment for establishing significance for each of their dependent variables. Sachtleben and colleagues (1997) made seven comparisons using the t test. The authors used the Bonferroni procedure to adjust the alpha level. With alpha at .05 the adjusted alpha was .05/7 = .007. Consequently, to be significant, the t tests had to reach $\alpha < .007$.

The second option is to leave the overall alpha at .05 but to calculate the upper limit that the alpha might be:

$$\alpha_{UL} = 1 - (1 - \alpha)^k \qquad (9.17)$$

where α_{UL} = alpha (upper limit) and k = the number of groups. Again, using the example with three comparisons and alpha at .05, $\alpha_{UL} = 1 - (1 - .05)^3 = .14$. Thus, the hypotheses are really being tested somewhere between an alpha of .05 if the dependent variables are perfectly correlated and .14 if they are independent. In instances where researchers make multiple comparisons using the same participants, they should either adjust alpha to the experimentwise rate or at least report the upper limit on their alpha.

Understanding Multivariate Techniques

Up to this point we have discussed experimental research examples involving one or more independent variables but only one dependent variable. Multivariate cases have one or more independent variables and two or more dependent variables. For example, it seems likely that when independent variables are manipulated, they influence more than one thing. The multivariate case allows more than one dependent variable. To use techniques that allow only one dependent variable (called univariate techniques) repeatedly when there are several dependent variables increases the experimentwise error rate (sometimes called probability pyramiding) in the same way as doing multiple t tests instead of simple ANOVA when there are more than two groups. However, there are instances where using univariate techniques in a study that has multiple dependent variables is acceptable or the only choice. For example, as with all research, theory should drive decision making. You might not want to include a theoretically important dependent variable in a multivariate analysis when less theoretically important dependent variables could disguise its importance. Also, a multivariate technique is sometimes just not feasible because the study has a small number of participants. Multivariate techniques often used in experimental studies are

- discriminant analysis,
- multivariate ANOVA (MANOVA) and special cases with repeated-measures designs, and
- multivariate ANCOVA (MANCOVA).

Remember that the general linear model still underlies all the techniques, and we are still attempting to learn two things: Are we evaluating something significant (reliable)? How meaningful are significant findings?

Discriminant Analysis

We use discriminant analysis when we have one independent variable (two or more levels) and two or more dependent variables. The technique combines multiple regression and simple ANOVA. In effect, discriminant analysis uses a combination of the dependent variables to predict or discriminate among the levels of the independent variable, which in this case is group membership. In the discussion of multiple regression in chapter 8, we used several predictor variables in a linear combination to predict a criterion variable. In essence, discriminant analysis does the same thing except that several dependent variables are used in a linear combination to predict the group to which a participant belongs. This prediction of group membership is the equivalent of discriminating among the groups (recall how t could be calculated by r). The same methods used in multiple regression to identify the important predictors are used in discriminant analysis. These include forward, backward, and stepwise selection techniques. For greater detail as well as a useful and practical description of discriminant analysis, read Betz (1987).

Forward, Backward, and Stepwise Selection

As mentioned in chapter 8, the forward selection technique enters the dependent variables in the order of their importance; that is, the dependent variable that contributes the most to separation of the groups (discriminates among or predicts group membership the best) is entered first. By correlation techniques, the effect of the first dependent variable on all others is removed, and the dependent variable that contributes the next greatest amount to separation of the groups is entered at step 2. This procedure continues until all dependent variables have been entered or until some criterion for stopping the process (established by the researcher) is met.

The backward selection procedure is similar except that all the dependent variables are entered and the one contributing the least to group separation is removed. This continues until the only variables remaining are those that contribute significantly to the separation of the groups.

The stepwise technique is similar to forward selection except that at each step all the dependent variables are evaluated to see whether each still contributes to group separation. If a dependent variable does not contribute, it is stepped out (removed) from the linear combination, just as in multiple regression.

An Example of Discriminant Analysis

Tew and Wood (1980) conducted a study in which varsity football players were classified into three groups: (a) offensive and defensive backs, (b) offensive and defensive linemen, and (c) linebackers and receivers. Data were collected for 28 athletes in each group ($N = 84$) on seven variables: 40-yard dash, 12-min run, shuttle run, vertical jump, standing long jump, bench press, and squat. Discriminant analysis was applied to determine how many of the seven dependent variables were needed to separate (predict) the three groups of football players.

Two criteria were set for the statistical computer program: to include the variable with the largest F ratio at each step and to stop when no remaining variable had an F significant at $p < .05$. The bench press was entered first because it had the largest univariate F. The 40-yard dash was selected next, followed by the vertical jump. At this point, none of the four remaining dependent variables had significant Fs, so the program provided the overall test of the linear composite of the dependent variables' (bench press, 40-yard dash, and vertical jump) ability to separate the three groupings of players. Squared canonical correlation, which is cumulative at each step, provides an estimation of meaningfulness: the percentage of variance accounted for (20% with the first variable, 34% with the first and second variables, and 35% with all three). The other four dependent variables were not selected because they failed to improve the ability to separate the three groups of players. This means that the characteristics underlying performance in the three dependent variables included were similar to the characteristics underlying performance in the four not included.

Discriminant analysis may be followed up with univariate techniques to determine which groups actually differed from one another on each of the dependent variables that are selected. There are several ways to approach this follow-up, but for simplicity, you could perform the Scheffé test among the three groups on the first dependent variable. Then you could use ANCOVA among the three groups on the second dependent variable using the first dependent variable as a covariate. This would provide adjusted means of the second dependent variable (means of the second dependent variable corrected for the first). A Scheffé test could be done among the adjusted means using the adjusted mean square for error. This procedure, called a *stepdown* F *technique,* is continued through each dependent variable using the previously stepped-in dependent variables as covariates. There are additional ways to follow up discriminant analysis.

Summarizing Discriminant Analysis

The study by Tew and Wood (1980) used discriminant analysis in a situation in which the three groups were intact (i.e., not randomly formed). This is a very common application of discriminant analysis. However, discriminant analysis does not overcome the need to form the groups randomly if determination of cause and effect is the purpose of the research.

Multivariate Analysis of Variance

From an intuitive point of view, multivariate analysis of variance (MANOVA) is a rather straightforward extension of ANOVA. The only difference is that the F tests for the independent variables and interactions are based on an optimal linear composite of several dependent variables. There is no need to consider simple MANOVA here, as this is discriminant analysis (one independent variable with two or more levels and two or more dependent variables). For a good practical discussion, definition, and example of MANOVA, see Haase and Ellis (1987).

Using MANOVA

The mathematics is complex for factorial MANOVA, but the idea is simple. An optimal combination (linear composite) of the dependent variables is made that maximally accounts for (predicts) the variance associated with the independent variables. The variance associated with each independent variable is then separated out (as in ANOVA), and each of the independent variables and interactions on the optimal linear composite is tested. The associated F and degrees of freedom for each test are interpreted in the same way as for ANOVA. There are several ways to obtain F in MANOVA: Wilks' lambda, Pillai's trace, Hotelling's trace, and Roy's greatest characteristic root. We point this out only because authors sometimes identify how the MANOVA Fs were obtained. For your purposes, these distinctions are not important: Just remember that the MANOVA F ratios are similar to ANOVA F ratios.

Once MANOVA has been used, a significant linear composite of dependent variables is identified that separates the levels of the independent variable or variables. Then, the important question usually is, Which of the dependent variables contribute significantly to this separation? One of many ways to answer this question is to use discriminant analysis and the stepdown F procedures discussed in the previous section as follow-up techniques. This works well for the main effects but not so well for interactions. Interactions in MANOVA are most frequently handled by calculating factorial ANOVA for each dependent variable, although this procedure fails to consider the relationships among the dependent variables.

An Example of MANOVA

As an example of the use and follow-up of MANOVA, consider an experiment reported by French and Thomas (1987). One aspect of this study was to evaluate the influence of two groupings of age level (8- to 10-year-olds and 11- to 12-year-olds) and level of expertise (expert and novice players within each age level) on basketball knowledge and performance. A basketball knowledge test and two basketball skill tests (shooting and dribbling) were given to all the children. Following is French and Thomas's description of these particular results:

A 2 × 2 (Age League × Expert/Novice) MANOVA was conducted on the scores of the knowledge test and both skill tests. The results of the MANOVA indicated significant main effects for age league, $F(3, 50) = 5.81$, $p < .01$, and expert/novice, $F(3, 50) = 28.01$, $p < .01$, but no significant interaction. These main effects were followed up by a stepdown procedure using a forward selection discriminant analysis. The alpha level used as a basis for stepping in variables was set at .05. The discriminant analysis for age league revealed that knowledge was stepped in first, $F(1, 54) = 8.31$, $p < .01$. Neither skill test was entered. Older children ($M = 79.5$) possessed more knowledge than younger children ($M = 64.9$). The discriminant analysis for expert/novice revealed shooting was stepped in first, $F(1, 54) = 61.40$, $p < .01$; knowledge second, $F(1, 53) = 5.51$, $p < .05$; and dribbling was not entered. Child experts ($M = 47.2$) performed significantly better than novices ($M = 25.7$) in shooting skills. The adjusted means for knowledge showed that child experts ($M = 77.1$) possessed more basketball knowledge than novices ($M = 64.2$). (p. 22)

If more than two levels had existed within an independent variable in the MANOVA (i.e., suppose three age-league levels had been used), then following the discriminant analysis, there would have been three means (one for each age-league level) to test among them for significance. It would be appropriate to use a Scheffé test for this follow-up, just as we did for ANOVA.

Multivariate Analysis of Covariance

Conceptually, multivariate analysis of covariance (MANCOVA) represents the same extension of ANCOVA that MANOVA did for ANOVA. In MANCOVA there are one or more independent variables, two or more dependent variables, and one or more covariates. Recall the earlier explanation of ANCOVA. A variable was used to adjust the dependent variable by correlation, and then ANOVA was applied to the adjusted dependent variable. In MANCOVA each dependent variable is adjusted for one or more covariates, then an optimal linear composite that best discriminates among the independent variables is formed from the adjusted dependent variables, just as in MANOVA. Follow-up procedures are the same as in MANOVA, except that the linear composite of adjusted dependent variables is used. This technique is seldom used and typically used incorrectly. Normally, there is no reason to adjust a dependent variable for a linear composite of covariates. We have purposely not provided an example of MANCOVA use from the physical activity literature. MANCOVA is a fine technique if you understand what it does and you specifically want to test for that. However, many researchers misunderstand MANCOVA and do not use it properly. Of course, the same is true of ANCOVA; for a good discussion of the issues, see A.C. Porter and Raudenbush (1987).

Repeated Measures With Multiple Dependent Variables

Rather frequently, experiments have multiple dependent variables that are measured more than once (over time). For example, in an exercise adherence study, both physiological and psychological variables might be measured once a week during a 15-week training program. If there were two training groups (different levels of training) and a control group, each with 15 participants, all of whom were measured every week (i.e., 15 times) on two psychological and three physiological measures, we would have a design that is 3 levels of exercise × 15 trials (3 × 15) for 5 dependent variables. This design offers several options for analysis.

We could do five 3 × 15 ANOVAs with repeated measures on the second factor. Here, we would follow the repeated-measures procedures described earlier in this chapter.

However, we would be inflating the alpha by doing multiple analyses on the same subjects. Of course, the alpha could be adjusted by the Bonferroni technique ($\alpha = .05/5 = .01$), but this fails to take into account the relationships among the dependent variables, which might be substantial and of considerable interest. An excellent tutorial on how to handle this problem is provided by Schutz and Gessaroli (1987). The following brief discussion is taken from their study, but if you are using this design and analysis, you should read their complete study and example.

There are two options for this analysis: multivariate mixed-model (MMM) analysis and doubly multivariate (DM) analysis. Which is used depends on the assumptions that your data meet. Using the study previously described (3 levels of exercise $\times$ 15 trials for 5 dependent variables), the MMM analysis treats the independent variable (levels of exercise) as a true multivariate case by forming a linear composite of the five dependent variables to discriminate among the levels of the independent variable. If this composite is significant, it can be followed up by the stepdown F procedures (for an alternative procedure, see Schutz & Gessaroli, 1987). For the repeated-measures factor (and interaction), a linear composite is formed for each trial, and the linear composite at each trial is treated as a regular repeated-measures analysis. This means that the sphericity assumption must be met as previously described and the epsilon can be used to test this assumption with the same standards previously described. The interpretation of the resulting F ratio for main effects for groups, trials, and the group $\times$ trial interaction are the same as in other designs. The question usually posed here is, Do the groups change at different rates across the trials on the linear composite of dependent variables? This is the preferred analysis if the sphericity assumption can be met because most authors believe it offers more power. However, this is a difficult assumption to meet, especially if there are more than two or three dependent variables measured on more than three to five trials.

The DM analysis does not require that the sphericity assumption be met. The analysis is the same for the independent variable of exercise. In the repeated-measures part of the analysis, however, a linear composite is formed not only of the dependent variables at each trial but also of the 15 trials (which themselves become a linear composite), thus the name *doubly multivariate*. The interpretations of the Fs for the two main effects and interaction remain essentially the same, but follow-ups become more complex.

McCullagh and Meyer (1997) compared four methods of providing information (physical practice with feedback, learning model with model feedback, correct model with model feedback, and learning model without model feedback) on learning correct form in the free-weight squat lift. There were two dependent variables (outcome and form) and five trials. A MANOVA with repeated measures was used to analyze the data. Univariate ANOVAs and post hoc comparisons were done as follow-ups for significant Fs.

Summary

This chapter presented techniques used in situations in which differences among groups are the focus of attention. These techniques range from the simplest situation of two levels of an independent variable with one dependent variable to complex multivariate techniques involving multiple independent variables, multiple dependent variables, and multiple trials. These techniques can be categorized as follows:

- A t test is used to determine how a group differs from a population, how two groups differ, how one group changes from one occasion to the next, and how several means differ (the Scheffé or some other multiple comparison test).
- ANOVA shows differences among the levels of one independent variable (simple ANOVA), among the levels of two or more independent variables (factorial ANOVA), and among levels of independent variables when there is a distractor variable or covariate (ANCOVA).

- Multivariate analysis of variance (MANOVA) is used when there is more than one dependent variable. Discriminant analysis is the simplest form of MANOVA (when there is only one independent variable and two or more dependent variables). More complex MANOVA involves two or more independent variables and two or more dependent variables.
- MANCOVA is an extension of MANOVA in which there is one or more covariates.
- Repeated-measures MANOVA is used when one or more of the independent variables are repeated measures.

Table 9.4 provides an overall look at the techniques presented in this chapter and chapter 8 (relationships among variables) and their appropriate use. Note that there are techniques for relationships among variables that parallel each technique for differences among groups. In fact, the relationship between t and r that was demonstrated earlier in this chapter holds, as ANOVA techniques are the equivalent of multiple regression. Each technique evaluates our two basic questions: Is the effect or relationship significant? Is the effect or relationship meaningful?

Table 9.4   Comparison of Statistical Techniques From Chapters 8 and 9

Description	Differences among groups	Relationships among variables
1 IV (2 levels) → 1 dv	Independent t test	
1 predictor → 1 criterion		Pearson r
2 or more IVs → 1 dv	Factorial ANOVA	
2 or more predictors → 1 criterion		Multiple regression
1 IV (2 or more levels) → 2 or more dvs	Discriminant analysis	
2 or more IVs → 2 or more dvs	MANOVA	
2 or more predictors → 2 or more criteria		Canonical correlation

IV = independent variable; dv = dependent variable.

Some ideas presented here are complex and may not be easily understood in one reading. We provide some suggested readings and problems in this chapter that should be helpful. If you do not feel confident in your understanding of this material, reread the chapter, do the problems, and consult some of the suggested readings. This is important because parts III and IV assume that you understand part II.

✔ Check Your Understanding

1. Critique the statistical part of a study that uses an independent t test. Calculate ω^2 for the t in this study.

2. Find a study that uses a multiple comparison test (Newman-Keuls, Duncan, or Scheffé) as a follow-up to ANOVA, and critique the use of this test.

3. A researcher wishes to compare the effectiveness of three different desk arrangements (rows, clusters, and circles) on classroom behavior (frequencies of observed on-task behaviors). Elementary school children ($N = 18$) were randomly assigned to the three desk-arrangement groups for instruction. The M, s, and n for each group are as follows:

	M	s	n
Rows	6.0	1.4	6
Clusters	6.8	1.9	6
Circles	9.5	1.0	6

Here is a partial summary of the ANOVA:

Source	SS	df	MS	F
Between	40.1			
Within	34.3			
Total	74.4			

a. Complete the table. Use table 9.1 and example 9.8 for help.

b. Test the F for significance at the .05 level using table A.6.

c. Compute ω^2 for meaningfulness.

d. Use the Scheffé method to determine where the differences are among the three groups (consult example 9.9).

e. Write a short paragraph interpreting the results of your analysis of data.

4. Critique the statistical part of a study that uses a two-way factorial ANOVA. Calculate ω^2 for each factor and the interaction.

5. Locate a study that used MANOVA in its data analysis. Identify the independent and dependent variables.

Nonparametric Techniques

He uses statistics as a drunken man uses lamp posts—for support rather than for illumination.
—Andrew Lang

In the preceding chapters, various parametric statistics were described. Recall that parametric statistics make assumptions about the normality and homogeneity of variance of the distribution. Another category of statistics is called nonparametric statistics (see the sidebar below for differences between parametric and nonparametric words). This category is also referred to as distribution-free statistics because no assumptions are made about the distribution of scores. Nonparametric statistics are versatile in that they can deal with ranked scores and categories. This can be a definite advantage when the investigator is dealing with variables that do not lend themselves to precise, interval-type or ratio-type data (which are more likely to meet parametric assumptions), such as categories of responses on questionnaires and various affective behavior rating instruments. Data from quantitative research are frequently numerical counts of events that can be effectively analyzed with nonparametric statistics.

The main drawback to nonparametric statistics that has often been voiced is that they are less powerful than parametric statistics. As you recall, power refers to the ability of a statistical test to reject a null hypothesis that is false. It should be pointed out that there is no agreement regarding the supposed power advantage of parametric tests (Harwell, 1990; Thomas, Nelson, & Thomas, 1999). Another drawback in the use of nonparametric

Parametric Versus Nonparametric Words

Parametric	Nonparametric	Description
Avoidable	uh-voy-duh-buhl	What the bullfighter tries to do
Counterfeiters	kown-ter-fit-ers	Workers who put together kitchen cabinets
Eclipse	e-klips	What an English barber does for a living
Misty	mis-tee	How golfers create divots
Paradox	par-o-docs	Two physicians
Relief	ree-leef	What trees do in the spring
Sudafed	sue-da-fed	Bring litigation against the government

That is definitely not a normal distribution.

tests has been the lack of statistical software for the more complex statistical tests, such as multivariate cases.

In chapter 6 we provided you with a set of procedures to judge the normality of data so that you can determine whether to use parametric or nonparametric approaches. These included looking at the distribution of the data (e.g., using a stem-and-leaf or histogram plot to evaluate whether the data fits a normal curve) and evaluating the skewness and kurtosis of the distribution. Deciding whether data meet the normality assumption or not is difficult (Thomas, Nelson, & Thomas, 1999). For example, how normal do data have to be to use parametric techniques? There are likely to be numerous occasions when a researcher's data do not meet the assumptions of normality. Micceri (1989), for example, maintains that much of the data in education and psychology are moderately or largely nonnormal, and thus nonparametric tests should be considered. Moreover, sometimes the only scores available are frequencies of occurrence or ranks (which often are not normally distributed), and the researcher should use nonparametric tests.

In this chapter, we present two categories of nonparametric analyses. First is chi square, which is used to analyze the frequency of responses that are in categories; for example:

- How many highly fit children are girls and how many are boys?
- How many former athletes or nonathletes participate in recreational sports after age 30?
- How many highly fit women regularly participate in swimming, running, or cycling?

Second, we present a standard approach to analyzing ranked data where the ranks are not normally distributed. To use any of the general linear model (GLM) techniques described in chapters 8 and 9, the data must be normally distributed, and the data points must fit a straight line. Examples of data that might not meet the normality assumption include these:

- Questionnaire responses of strongly agree, agree, neutral, disagree, and strongly disagree can be considered ranked data; the responses are ranked from 1 (strongly agree) to 5 (strongly disagree).
- Data such as time, velocity, acceleration, and counts (e.g., how many push-ups you can do) may not be normally distributed.

In these instances, use of parametric correlational and ANOVA procedures (including *t* tests) would be inappropriate because the assumptions for using the *r, t,* and *F* tables are not met.

In this chapter we provide a set of procedures for rank-order data that are parallel to the parametric procedures of correlation and ANOVA (including *t* tests). The ideas underlying these procedures are identical to the ones presented in chapters 8 and 9. The procedures use the same computer programs (e.g., SPSS, SAS) to analyze the ranked data, but instead of using the *r, t,* and *F* tables, one calculation is made (a statistic called *L*) that is then compared to a chi-square table, where the normality assumption is not required.

Chi Square: Testing the Observed Versus the Expected

Data are often sorted into categories, such as sex, age, grade level, treatment groups, or some other **nominal** (categorical) **measure.** A researcher is sometimes interested in evaluating whether the number of cases in each category is different from what would be expected on the basis of chance, some known source of information (such as census data), or some other rational hypothesis about the distribution of cases in the population. **Chi square** provides a statistical test of the significance of the discrepancy between the observed and the expected results.

The formula for chi square is

$$\chi^2 = \Sigma[(O - E)^2/E] \qquad (10.1)$$

where *O* = the observed frequency and *E* = the expected frequency. Thus, the expected frequency in each category—often labeled "cells" as in a table divided into four equal boxes—is subtracted from the observed (or obtained) frequency. This difference is then squared (which means all differences will be positive), and these values are divided by the expected frequency for their respective categories and then summed.

For example, tennis coach Roger Rabbitfoot believes in jinxes. He believes that one court at his university is definitely unlucky for his team. He has kept records of matches on four courts over the years and is convinced that his team has lost significantly more matches on court number 4 than on any other. Roger has set out to prove his point by comparing the number of losses on each of the four courts. The same number of matches has been played on each court, and his team has lost a total of 120 individual matches on these courts during this period. Theoretically, it would be expected that each court would have one fourth, or 30, of the losses. The observed frequencies and the expected frequencies are shown in example 10.1.

The resulting chi square is then interpreted for significance by consulting table A.7 in appendix A. Because there are four courts (or cells), the number of degrees of freedom (*df*) is *c* − 1 = 3. The researcher finds the critical value for 3 *df,* which is 7.82 for the .05 probability level. Roger's calculated value of 7.19 is less than this, so the null hypothesis that there were no differences among the four courts in the number of losses is not rejected. The observed differences could be attributed to chance. Roger probably still believes in his heart that he is right (called cardiac research).

In some cases, the expected frequencies for classifications can be obtained from existing sources of information, as in the following example. A new assistant professor, Nancy Niceperson, is assigned to teach the large sections of an introductory kinesiology course. After a few semesters, the department chairperson hears rumors that Dr. Niceperson is too lenient in her grading practices, so he uses chi square to compare the grades of the 240 students with the department's prescribed normal-curve grade distribution: 3.5% As and Fs, 24% Bs and Ds, and 45% Cs. If Dr. Niceperson were adhering to this normal distribution, she would be expected to have given 8 As and 8 Fs, (3.5% × 240), 58 Bs and 58 Ds (24% × 240), and 108 Cs (45% × 240). The observed frequencies (Dr. Niceperson's grades) and the expected frequencies (the department's mandated distribution) are compared using chi square in example 10.2.

nominal measure

Method of classifying data into categories, such as sex, age, grade level, or treatment groups.

chi square

A statistical test of the significance of the discrepancy between the observed and the expected results.

Example 10.1

Known Values

		Court number				
		1	2	3	4	Total
Observed number of losses:	$O =$	24	34	22	40	120
Expected number of losses:	$E =$	30	30	30	30	120

Working It Out (equation 10.1)

	Court number			
	1	2	3	4
$(O - E)$	−6	+4	−8	+10
$(O - E)^2$	36	16	64	100
$(O - E)^2/E$	1.20	0.53	2.13	3.33

$\chi^2 = \Sigma[(O - E)^2/E] = 7.19$

Example 10.2

Known Values

		Grade					
		A (3.5%)	B (24%)	C (45%)	D (24%)	F (3.5%)	Total
Observed number of grades given:	$O =$	21	75	114	28	2	240
Expected number of grades given:	$E =$	8	58	108	58	8	240

Working It Out (equation 10.1)

	Grade				
	A	B	C	D	F
$(O - E)$	13	17	6	−30	−6
$(O - E)^2$	169	289	36	900	36
$(O - E)^2/E$	21.13	4.98	0.33	15.52	4.50

$\chi^2 = \Sigma[(O - E)^2/E] = 46.46$

The chairperson, who is always fair, does not want to make a bad decision, so he decides to use the .01 level of probability. Table A.7 shows that for 4 df (there are five grades, or cells), a chi square of 13.28 is needed for significance at the .01 level. The obtained chi square of 46.46 exceeds the value, indicating a significant deviation from the expected grade distribution. Clearly, Dr. Niceperson's grades show too many As and Bs and too few Ds and Fs. The chairperson does the only fair thing by firing her on the spot.

The Contingency Table

Often a problem involves two or more categories of occurrences and two or more groups (a two-way classification). A common example is in the analysis of the results of questionnaires or attitude inventories in which there are several categories of responses (e.g., agree, no opinion, disagree) and two or more groups of respondents (e.g., exercise adherents and nonadherents). This type of two-way classification is called a **contingency table.**

To illustrate, suppose a group of athletes and a group of nonathletes respond to the following statement on a sportsmanship inventory: "A baseball player who traps a fly ball between the ground and his glove should tell the umpire that he did not catch it." In the previous examples of one-way classification, the expected frequencies were determined by some type of rational hypothesis or source of information. In a contingency table, the expected values are computed from the marginal totals. Example 10.3 shows the participants' responses.

A total of 144 respondents agreed with the statement. Because there are 450 people in all, then 144/450, or 32% of the total group, agreed with the statement. Thus, if there were no difference between athletes and nonathletes on sportsmanship (the null hypothesis) as reflected by this statement, then 32% of the athletes ($0.32 \times 200 = 64$) and 32% of the nonathletes ($0.32 \times 250 = 80$) would be the expected frequencies for these two cells.

contingency table

A two-way classification of occurrences and groups that is used for computing the significance of the differences between observed and expected scores.

Example 10.3

Known Values

Observed responses	Agree	No opinion	Disagree	Total
Athletes	30	46	124	200
Nonathletes	114	80	56	250
Total	144	126	180	450

Working It Out

1. Find the expected values (column total × row total)/N

Expected responses	Agree	No opinion	Disagree	Total
Athletes	$144 \times 200/450 = 64$	$126 \times 200/450 = 56$	$180 \times 200/450 = 80$	200
Nonathletes	$144 \times 250/450 = 80$	$126 \times 250/450 = 70$	$180 \times 250/450 = 100$	250
Total	144	126	180	450

2. Compute χ^2 (equation 10.1)

	$O - E$	$(O - E)^2$	$(O - E)^2/E$
Athletes agree	−34	1,156	18.06
Nonathletes agree	34	1,156	14.45
Athletes no opinion	−10	100	1.79
Nonathletes no opinion	10	100	1.43
Athletes disagree	44	1,936	24.20
Nonathletes disagree	−44	1,936	19.36
			$\chi^2 = 79.29$

A much faster method of calculating these expected frequencies is simply to multiply the column total by the row total for each cell and divide by the total number (N), as we've worked out in step 1 of example 10.3. Chi square is computed in the same manner as in the examples of one-way classification.

The degrees of freedom for a contingency table is $(r - 1)(c - 1)$, where r stands for rows and c for columns. Here, we have two rows and three columns, so $df = (2 - 1)(3 - 1) = 2$. As before, the investigator then looks up the table value for significance (in this study, the investigator had decided on the .01 level) and sees that a chi square of 9.21 is needed. The obtained chi square of 79.29 is clearly significant. This tells us that the null hypothesis is rejected: There is a significant difference between athletes and nonathletes in their responses to the statement. After inspecting the table, we could conclude that a significantly greater proportion of nonathletes agreed that the player should tell the umpire about trapping the ball and, conversely, that a higher proportion of athletes felt the player should not.

Restrictions in Using Chi Square

Although we indicated that nonparametric statistics do not require the same assumptions concerning the population as do parametric statistics, some restrictions apply in using this technique. The observations must be independent, and the categories mutually exclusive. By this we mean that an observation in any category should not be related to or dependent on other observations in other categories. You could, for example, ask 50 people about their activity preferences. If each gave three preferences, you would not be justified in using a total (N) of 150 because the preferences of any participant would likely be related and the chi square thus inflated. Moreover, an observation can be placed in only one category. The observed frequencies are exactly that: numbers of occurrences. Ratios and percentages are not appropriate. Another related point is that the total of the expected frequencies and the total of the observed frequencies for any classification must be the same. In example 10.3 notice that the total of the expected frequencies for athletes is the same as the total of the observed frequencies. The same is also seen for nonathletes and for the column totals.

Chi square is usually not applicable for small samples. The expected frequency for any cell should not be less than 1.0. Furthermore, some statisticians claim that no more than 20% of the cells can have expected values of less than 5. Opinions vary on this, however. Some say no cell should have less than 5, whereas other statisticians allow as many as 40% of the cells to have a frequency of less than 5. A common tactic in cases with several cells with expected frequencies less than 5 is to combine adjacent categories, thereby increasing the expected values.

Researchers generally agree that a 2 × 2 contingency table should have a so-called correction for continuity. This correction, usually referred to as **Yates' correction for continuity,** is to subtract 0.5 from the difference between the observed and the expected frequencies for each cell before it is squared:

$$\text{Corrected } \chi^2 = \Sigma[(O - E - 0.5)^2/E] \qquad (10.2)$$

Another limitation imposed on the 2 × 2 contingency table is that the total number (N) should be at least 20.

Finally, the expected distribution should be logical and established before the data are collected. In other words, the hypothesis (probability, equal occurrence, census data, etc.) precedes the analysis. Researchers are not allowed to look over the distribution and then conjure up an expected distribution that fits their hypothesis.

Yates' correction for continuity

Method of correcting a 2 × 2 contingency table by subtracting 0.5 from the difference between the observed and expected frequencies for each cell before it is squared.

contingency coefficient

Method of computing the relationship between dichotomous variables such as sex and race.

Contingency Coefficient

Several correlational techniques can be used in situations in which data are discrete (i.e., not continuous). You can compute the relationship between dichotomous variables such as sex and race by using a **contingency coefficient.** The test of significance is chi square. Recall the contingency table for use in detecting differences between groups or sets of data, described in the previous section. The contingency table can also be used to determine relationships. You can have any number of rows and columns in a contingency table. After the chi square is computed, the contingency coefficient C can be computed:

$$C = \sqrt{\chi^2 / \left(N + \chi^2\right)} \qquad (10.3)$$

If χ^2 is significant, C is also significant. The direction of the relationship is established by examining the data. There are several limitations concerning the ability of the contingency coefficient to estimate correlation. In general, one needs several categories and many observations to obtain a reasonable estimate.

Multivariate Contingency Tables: The Loglinear Model

Categorical data can be analyzed in combination with other variables. In other words, contingency tables can be studied in more than two dimensions. One can therefore identify associations among many variables, for example, the interrelationships among age, sex, skill level, and teaching method. This is similar to parametric multivariate analysis. However, with continuous quantitative data, variables are expressed as linear composites, whereas with categorical variables, the researcher deals with contributions to the expected frequencies within each cell of the multivariate contingency table. Any given cell represents the intersection of many marginal proportions.

Loglinear models are used to analyze multivariate contingency tables. Relative frequencies are transformed to logarithms, which are additive and similar to sum of squares in ANOVA. Main effects and interactions can be tested for significance. The probability of membership in a particular category can be predicted as a function of membership in other categories using a logistic regression equation based on the logarithmic odds of membership (called a **logit**).

The loglinear model has considerable potential application in research with categorical data (Schutz, 1989). This type of analysis is attracting much attention from researchers and theoreticians. The loglinear analysis provides a means for sophisticated study of the interrelationships among categorical variables. Readers who are interested in pursuing this topic are directed to the text by Kennedy (1983).

loglinear models

A system that analyzes multivariate contingency tables by transforming relative frequencies into logarithms.

logit

Probability of membership in a particular category as a function of membership in other categories in multivariate contingency tables.

Procedures for Rank-Order Data

Most books that report nonparametric procedures (including previous editions of this book) propose a series of rank-order techniques, including these:

- Mann-Whitney U test—Analogous to the parametric independent t test
- Wilcoxon matched-pairs signed-ranks test—Analogous to the parametric dependent t test
- Kruskal-Wallis ANOVA by ranks—Analogous to the parametric one-way ANOVA
- Friedman two-way ANOVA by ranks—Analogous to the parametric repeated-measures ANOVA
- Spearman rank-difference correlation—Analogous to the parametric Pearson r

You will still see these techniques used in the research literature, so you need to understand what each one does; our list shows you how they compare with parametric procedures. Eventually, as better methods become available, most of these techniques will no longer be used, but that takes time. Researchers who have been trained to use these procedures often do not keep up with current statistical methods; we acknowledge that this is difficult to do since researchers have their own research area to keep up with. One wag was heard to say that it takes 30 to 40 years for a new statistical approach to come into widespread use. This is because researchers never change the statistical procedures they are using, so you have to wait for them to retire and for a more recently trained group to begin using the new procedures.

> ## Using Computer Software: What the World Would Be Like if Microsoft Built Cars
>
> 1. Every time they repainted the lines on the road, you'd have to buy a new car.
> 2. Occasionally your car would just die on the highway for no reason; accept this, restart, and drive on.
> 3. Macintosh would make a car that was powered by the sun, twice as reliable, five times as fast, and twice as easy to drive, but it would only run on 5% of the roads.
> 4. The oil, engine, gas, and alternator warning lights would be replaced with a single "General Car Fault" warning light.
> 5. The airbag system would say "Are you sure?" before going off.

We can replace all the rank-order techniques previously listed with a standardized approach to rank-order data where the distribution of data is not normal (for greater detail, see Thomas, Nelson, & Thomas, 1999). The techniques parallel the parametric ones described in chapters 8 and 9. The procedures for all techniques involve these steps:

1. Change all the data values to ranks (most statistical software can do this).
2. Run the standard parametric statistical software (e.g., correlation, multiple correlation, ANOVA) on the ranked data. (See the sidebar above for some of the problems with software.)
3. Calculate the L statistic as the significance test, and compare it to the χ^2 table (table A.7).

The nonparametric test statistic (L) used to evaluate significance for all these procedures was developed by Puri and Sen (1969, 1985) and represents the test of the null hypothesis between X and Y where X can be groups or variables and Y is variables. This method can be used in place of all the techniques previously discussed and assumes that the data fit a straight line (as all the techniques in chapters 8 and 9 did) *but does not assume normal distribution of data.* L is very easy to calculate using the following formula:

$$L = (N - 1)r^2 \qquad\qquad (10.4)$$

where N = number of participants and r^2 = proportion of true variance = $SS_{regression}/SS_{total}$. The proportion of true variance in correlation is r^2 and in multiple correlation is R^2. In a t test, $r^2 = t^2/(t^2 + df)$, where $df = (n_1 + n_2) - 2$. In ANOVA, it is the sum of squares (SS) for the effect of interest (e.g., between groups in one-way ANOVA or for each factor in factorial ANOVA; each of these is the true variance) divided by the total sum of squares (these values are shown in the ANOVA printout from standard statistics programs).

When the L statistic is calculated for each significance test (either for correlation or ANOVA), it is compared to the chi-square table (table A.7 in appendix A) with degrees of freedom (df) = pq, where p = the number of df for the independent (or predictor) variable or variables and q = the number of df for the dependent (or criterion) variable or variables. Thus, for a correlation (r) between one predictor (abdomen skinfold) and one criterion (body weight), the pq df is $1 \times 1 = 1$. In a t test, $df = p$ (number of groups − 1) × q (number of dependent variables); thus, again the pq df = $1 \times 1 = 1$. Additional examples of how to calculate L and df are provided throughout the rest of this chapter.

Correlation

In this section we provide sample ranked data and how to calculate correlation and multiple correlation using standard correlational techniques. We then calculate the L and df for each example. In the simple correlation example, we also show how to decide whether to use parametric or nonparametric analysis, but we do not repeat this process for subsequent procedures (you can obtain this logic for each procedure by reading Thomas, Nelson, & Thomas, 1999).

Simple Correlation

We are interested in calculating the correlation between two skinfold measurements, biceps and forearm, that were collected on 157 participants. Descriptive data are given in figure 10.1, along with a histogram showing the data distribution with the normal curve overlaid. Recall from chapter 6 that skewness and kurtosis provide information about the nature of the distribution. Here they are given in the form of a z score, where 0.0 indicates a normal distribution and positive or negative values indicate specific variations from normality in the distribution. For the biceps skinfold, the skewness is above +1.0, indicating that the hump of the curve is shifted to the left. The kurtosis is nearly +2.0, indicating that the distribution is very peaked. For the forearm skinfold, the hump of the curve is also shifted left, and the curve is extremely peaked. Neither data set seems to be normal in its distribution. Although there is no set rule for how normal or nonnormal data must be, we do know that parametric techniques are not as resistant to nonnormality as once thought. In our view, these data are nonnormal enough to suggest use of rank-order procedures rather than parametric ones.

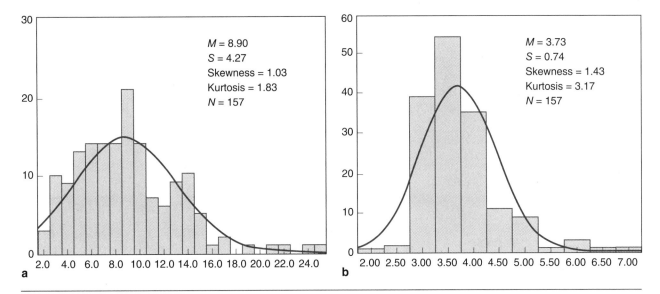

Figure 10.1   (a) Biceps, (b) forearm.

If we use a computer program for parametric correlation to determine the relationship between these two variables (using the original data), $r = .26$, and the test of significance is $F(1, 155) = 11.46$, $p < .001$. However, we are likely violating the assumption that data are normally distributed. If we rank subjects from the smallest biceps skinfold measurement (number 1) to the largest (number 157) and do the same for the forearm skinfold, we can then use the same computer program to correlate the ranked data; the correlation is actually larger, $r = .28$. If we square this value, we get .0784. If we then use equation 10.4 to calculate the significance test to replace F, we find that

$$L = (N - 1)r^2$$

$$= (157 - 1)(.28)^2 = (156)(.0784) = 12.23$$

This L has $df = 1$ because there is one X variable (biceps skinfold) and one Y variable (forearm skinfold), so $pq = 1 \times 1 = 1$. If we look this L up (table A.7) as a χ^2 with $df = 1$, we find a value of 10.83 for $p = .001$. Our value exceeds that, so it is significant at $p < .001$. Thus, we have now used a rank-order procedure that does not violate the assumption for normality, and our *r* and test of significance for the *r*, the *L* statistic, were both larger than the ones we would have obtained if we had used the parametric statistics. While this will not always be the case, this approach to nonparametric analysis does have good power compared with parametric procedures when data are not normal.

You can also calculate these same procedures by hand if you want to see exactly how they work. Return to chapter 8, where we calculate the correlation between body weight and pull-ups (example 8.1). Rank the body weight of each boy from 1 to 10, beginning with the highest body weight. Then do the same for pull-ups, beginning with the boy who did the most pull-ups. Then square and sum the values as we did in example 8.1, and calculate *r* using the formula provided. You will find the *r* for the rank-order procedures is .61 (note this is a positive correlation because we ranked pull-ups from most to least and weight from lowest to highest; if we had ranked weight from highest to lowest, the correlation would have been negative); this is higher than the *r* of −.54 for the parametric procedures.

Multiple Correlation

We can extend the exact same procedures to multiple correlation. In this example we use four skinfolds—abdomen, calf, subscapular, and thigh—to predict percentage of fat determined from hydrostatic weighting. (The data in this example are from Thomas, Keller, & Holbert, 1997. We thank these authors for allowing us to use their data.) All four skinfold measurements are changed to ranks, as are the percentage fat measurements. We then run the regular multiple-correlation program from SPSS using forward-stepping procedures on the ranked data. The correlations for all ranked pairs of variables (e.g., abdomen with calf skinfold, abdomen with subscapular skinfold, and so on) range from .45 to .74. These values are about the same as for the original (unranked) data, for which the correlations range from .41 to .78. Table 10.1 summarizes the multiple-regression results for the ranked data (note that all test statistics have been changed to *L*). The overall multiple correlation, *R*, was higher for the ranked data (.82) than for the original unranked data (.80).

Table 10.1 Summary of Forward Multiple Regression for the Ranked Data

Step	Variable	R	R²	β	df	L
1	Subscapular skinfold	.68	.46	.332	1	35.83*
2	Calf skinfold	.77	.60	.602	2	20.29*
3	Abdomen skinfold	.80	.64	.321	3	7.85*
4	Thigh skinfold	.82	.68	−.327	4	9.56*

*$p < .05$ for all.

$L(4) = 53.01$, $p < .001$, for the linear composite of predictors.

Differences Among Groups

We can apply the exact same logic and procedures when we want to test differences among groups. Here we provide examples of a *t* test, one-way ANOVA, and factorial ANOVA. If you want to extend these procedures to repeated-measures ANOVA and multivariate ANOVA, see Thomas, Nelson, & Thomas, 1999. In these examples we use data from Nelson, Yoon, & Nelson, 1991; we appreciate the use of their data.

t Test

Data are modified push-up scores for 90 boys and 90 girls in the upper elementary school grades. The distribution of data is positively skewed and somewhat peaked. Following are the means and standard deviations for the girls and boys and the *t* test comparing them:

$$\text{Boys} \quad M = 18.6 \quad s = 9.9$$

$$\text{Girls} \quad M = 12.7 \quad s = 9.8$$

$$t(178) = 4.00, p < .001$$

We changed the data to ranks from 1 to 180, disregarding whether the scores were for boys or girls. We then calculated *t* on the ranked data; it was 3.77. However, we need to change the *t* to *L* using equation 10.4, but first we have to calculate r^2. Following are those calculations:

$$r^2 = t^2/(t^2 + df) = 3.77^2/(3.77^2 + 178) = 14.21/(14.21 + 178) = .0739$$

$$L = (N - 1)r^2 = (179)(.0739) = 13.23$$

L is tested as a χ^2 with $df = pq$, where p = the number of groups − 1 (2 − 1 = 1) and q = the number of dependent variables (1), so 1 × 1 = 1 *df*. If we look in the chi-square table (table A.7), we see that our calculated value of 13.23 exceeds the table value of 10.83 at the .001 level; thus, our *L* is significant at $p < .001$.

You can practice this procedure by applying it to the data in example 9.8 in chapter 9. Use the data for groups 1 and 2, with 5 participants in each group (*N* = 10). Rank the scores from lowest to highest, disregarding whether the participant is in group 1 or 2. Then calculate the mean and standard deviation for each of the two groups and apply equation 9.3 to calculate *t*. The *t* value you find will be 2.79. Then obtain *r* from *t*:

$$r^2 = t^2/(t^2 + df) = 2.79^2/(2.79^2 + 8) = .49$$

We then calculate *L* using formula 10.4:

$$L = (N - 1)r^2 = (9)(.49) = 4.41$$

This *L* has 1 *df* [(groups − 1) × number of dependent variables]. With 1 *df* the critical χ^2 at the .05 level from table A.7 is 3.84. Since our value is larger than that, the two groups are significantly different.

One-Way ANOVA

The data from modified push-ups (Nelson, Yoon, & Nelson, 1991) was also evaluated by grade level: grades 4, 5, and 6. We use this as the example for one-way ANOVA. Grade level is the independent variable with three levels (this is really a categorical variable, but for ANOVA we treat it like an independent variable), and the modified push-up score is the dependent variable. Here are the descriptive data for each group:

$$\text{Grade 4} \quad M = 12.1 \quad s = 10.0$$

$$\text{Grade 5} \quad M = 16.3 \quad s = 9.6$$

$$\text{Grade 6} \quad M = 18.5 \quad s = 10.2$$

We then change the data to ranks, disregarding the grade level of the children, so we have 180 scores (number of children in all three grades) ranked from the highest modified push-up score (ranked number 1) to the lowest (ranked number 180). We then run the one-way ANOVA program on the ranked data, with the results shown in table 10.2. However, the printout shows *F* where we show *L* in table 10.2. To calculate the *L*, first calculate R^2 (the proportion of true variance in ANOVA) using

$$R^2 = SS_{between}/SS_{total}$$

$$= 31,646.61/485,412.00 = .065$$

Table 10.2   Summary of ANOVA Data for the Ranked Scores

Source	df	Sum of squares	Mean squares	L	p
Between groups	2	31,646.61	15,823.30	11.64	.01
Within groups (error)	177	453,765.39	2,563.65		
Total	179	485,412.00			

Then, calculate L as

$$L = (N - 1)R^2 = (179)(.065) = 11.64$$

The L has 2 df [(number of groups − 1) × number of dependent variables]. The value needed for significance at the .01 level with 2 df is 9.21 in table A.7, and our value exceeds that, so it is significant.

You can also do an example by hand if you return to example 9.8 in chapter 9 and use all three groups. As before, rank the scores from 1 to 15, disregarding which group the participants are in. Then work the one-way ANOVA as was done in example 9.8, except replace the original scores with the ranks. Calculate R^2 and L as previously described. You will find $R^2 = .657$ and $L(2) = 9.20$. The value for significance at the .02 level is 7.82 in table A.7, and the calculated value exceeds that, so it is significant.

Factorial ANOVA

Continuing with the same data set, we have a modified push-up score for girls and boys at each of three grade levels. Thus, we can rank the modified push-up scores from 1 to 180 as before and calculate a factorial ANOVA where one factor is sex (girls and boys) and the other is grade level (4, 5, and 6). This becomes a 2 (sex) × 3 (grade level) factorial ANOVA with the modified push-up score as the dependent variable.

These ranked data were analyzed using regular parametric statistical software. We found the outcome shown in table 10.3. We calculated R^2 using the same formula for each factor and for the interaction: $R^2_{grade} = SS_{grade}/SS_{total} = 31,646.61/485,412.00 = .065$; $R^2_{sex} = SS_{sex}/SS_{total} = 35,814.01/485,412.00 = .074$; and $R^2_{grade \times sex} = SS_{grade \times sex}/SS_{total} = 10,865.20/485,412.00 = .022$. We then used the R^2 to calculate L for each factor and for the interaction. In table A.7, the grade effect (with 2 df) is significant at the .005 level, the sex effect (with 1 df) is significant at the .001 level, but the interaction (with 2 df) is not significant.

Table 10.3   Summary of Factorial ANOVA Data for the Ranked Scores

Source	df	Sum of squares	Mean squares	L	p
Grade	2	31,646.61	15,823.30	11.64	.001
Sex	1	35,814.01	35,814.01	13.21	.001
Grade × sex	2	10,865.20	5,432.60	4.01	
Residual	174	407,086.18	2,339.58		
Total	179	485,412.00			

Extension to Repeated-Measures ANOVA and Multivariate ANOVA

While we do not provide examples here, the use of ranked data and the L statistic can be extended to ANOVA models with repeated measures and multivariate ANOVA. Conceptually, the ideas are identical, and you follow similar procedures to the ones used here, including ranking the data, using standard computer software, and changing the test statistic to L instead of F. If you need one of these techniques, see Thomas, Nelson, & Thomas (1999) for the exact steps to follow and how to interpret the results.

Summary

In this chapter we provided statistical tests for data that do not meet the assumptions of parametric data (chapters 8 and 9). Included are chi-square tests for frequency data and a set of procedures for ranked data that are parallel to parametric procedures. If distributions are not normal, the data are changed to ranks. The ranked data can be analyzed by standard SPSS and SAS statistical packages. One test, the L statistic, is calculated and compared with a chi-square table for significance. These procedures can be applied to all parametric linear models.

✔ Check Your Understanding

1. Compute chi square for the following contingency table of activity preferences of men ($n = 110$) and women ($n = 90$). Determine the significance of the chi square at the .01 level using table A.7. When calculating the expected frequencies, round off to whole numbers.

	Racquetball	Weight training	Aerobic dance
Men	35	45	30
Women	28	13	49

   a. Write a brief interpretation of your results.

   b. What would be the critical value for significance at the .05 level if you had five rows and four columns?

2. Critique the statistical part of a study that uses chi square in the analysis.

Measuring Research Variables

The average lifespan of a Major League baseball is 7 pitches.

■   ■   ■   ■   ■   ■   ■

A basic step in the scientific method of problem solving is the collection of data; therefore, an understanding of basic measurement theory is necessary. (We should point out that although measurement is discussed here as a research tool, measurement itself is an area of research.) In this chapter, we discuss fundamental criteria for judging the quality of measures used in collecting research data: validity and reliability. We explain different types of validity and different ways by which evidence of validity and reliability may be established. (The trustworthiness of qualitative research is discussed in chapter 19.) We conclude with some issues concerning the measurement of movement, the measurement of written responses used in paper-and-pencil instruments, the measurement of affective behavior, and the measurement of knowledge.

Validity

In gathering the data on which the results are based, we are also greatly concerned with the validity of the measurements we are using. If, for example, a study seeks to compare training methods for producing strength gains, the researcher must have scores that produce a valid measure of strength to evaluate the effects of the training methods. **Validity** of measurement indicates the degree to which the scores from the test, or instrument, measures what it is supposed to measure. Thus, validity refers to the soundness of the interpretation of scores from a test, the most important consideration in measurement.

There are different purposes for using certain measures. Consequently, there are different kinds of validity. We consider four basic types of validity: **logical, content, criterion,** and **construct.**

Although we list logical validity as a separate type of validity, the American Psychological Association and the American Educational Research Association consider logical validity to be a special case of content validity.

Logical Validity

Logical validity is sometimes referred to as **face validity,** although measurement experts dislike that term. Logical validity is claimed when the measure obviously involves the performance being measured. In other words, it means that the test is valid by definition.

validity

Degree to which a test or instrument measures what it purports to measure; can be categorized as *logical, content, criterion,* or *construct validity.*

logical validity

Degree to which a measure obviously involves the performance being measured; also known as *face validity.*

content validity

Degree to which a test (usually in educational settings) adequately samples what was covered in the course.

criterion validity

Degree to which scores on a test are related to some recognized standard or criterion.

construct validity

Degree to which a test measures a hypothetical construct; usually established by relating the test results to some behavior.

A **static balance** test that consists of balancing on one foot has logical validity. A speed-of-movement test, in which the person is timed in running a specified distance, must be considered to have logical validity. Occasionally, logical validity is used in research studies, but researchers prefer to have more objective evidence of the validity of measurement.

Content Validity

Content validity pertains largely to learning in educational settings. A test has content validity if it adequately samples what was covered in the course. As with logical validity, no statistical evidence can be supplied for content validity. The test maker should prepare a table of specifications (sometimes called a test blueprint) before making out the test. The topics and course objectives, as well as the relative degree of emphasis accorded to each, can then be keyed to a corresponding number of questions in each area.

A second form of content validity occurs with attitude instruments. Often a researcher wants evidence of independent verification that the items represent the categories for which they were written. When this happens, experts (in many cases 20 or more) are asked to assign each statement to one of the instrument categories. These categorizations are tallied across all experts and the percent who agreed with the original categorization are reported. Typically, the agreement of 80% to 85% would indicate the statement represents the content category.

Criterion Validity

Measurements used in research studies are frequently validated against some criterion. There are two main types of criterion validity: concurrent validity and predictive validity.

Concurrent Validity

Concurrent validity is a type of criterion validity that involves correlating an instrument with some criterion that is administered at about the same time (i.e., concurrently). The scores from many physical performance measures are validated in this manner. Popular criterion measures include an already validated or accepted measure, such as judges' ratings and laboratory techniques. Concurrent validity is usually employed when the researcher wishes to substitute a shorter, more easily administered test for a criterion that is more difficult to measure.

As an illustration, maximal oxygen consumption is regarded as the most valid measure of cardiorespiratory fitness. However, it requires a laboratory, expensive equipment, and considerable time for testing; furthermore, only one person can be tested at a time. Let us assume that a researcher, Douglas Bag, wishes to screen individuals by their fitness levels before assigning them to experimental treatments. Rather than using such an elaborate test as maximal oxygen consumption, Douglas determines it would be advantageous to use a stair-walking test he has devised. To determine whether it is a valid measure of cardiorespiratory fitness, he could administer both the maximal oxygen consumption test and the stair-walking test to a group of participants (from the same population as will be used in the study) and correlate the results of the two tests. If a satisfactory relationship exists, Doug can conclude that his stair-walking test is valid.

Written tests can also be validated in this way. For example, a researcher might wish to use a 10-item test of anxiety that can be given in a short period of time rather than a more lengthy version. Judges' ratings serve as criterion measures for some tests (sport skills and behavioral inventories are sometimes validated this way). A great amount of time and effort is required to secure competent judges, teach them to use the rating scale, test for agreement among judges, arrange for a sufficient number of trials, and so on. Consequently, judges' ratings cannot be used routinely to evaluate performances. The use of some skills tests would be more economical. Furthermore, the skills tests usually provide knowledge of results and measures of progress for the students. The skills tests could be initially validated, however, by giving the test and having judges rate the individuals on those skills. A validity coefficient can be obtained by correlating the scores on the skills tests with the judges' ratings.

Choosing the criterion is critical in the concurrent validity method. All the correlation can tell you is the degree of relationship between a measure and the criterion. If the criterion is inadequate, then the concurrent validity coefficient is of little consequence.

Predictive Validity

When the criterion is some later behavior, for example, entrance examinations used to predict later success, **predictive validity** is the major concern. Suppose that a physical education instructor wished to develop a test that could be given in beginning gymnastics classes to predict success in advanced classes. Students would take this test while they were in the beginner course. At the end of the advanced course, those test results would be correlated with the criterion of success (grades, ratings, etc.). In trying to predict a certain behavior, a researcher should try to ascertain whether there is a known "base rate" for that behavior. For example, someone might attempt to construct a test that would predict which female students might develop bulimia at a university. Suppose that the incidence of bulimia is 10% of the female population at that school. Knowing this, one could be correct 90% of the time in predicting that no one in the sample is bulimic. If the base rate is very low or very high, a predictive measure may have little practical value because the increase in predictability is negligible.

Chapters 8 and 16 discuss aspects of prediction in correlational research. Multiple regression is often used because several predictors are likely to have a greater validity coefficient than the correlation between any one test and the criterion. An example of this is the prediction of percentage fat from skinfold measurements. The criterion—percentage fat—can be measured by the underwater (hydrostatic) weighing technique. Several skinfold measures are then taken, and multiple regression is used to determine the best prediction equation. The researcher hopes to use the skinfold measures in the future if the prediction formula demonstrates an acceptable validity coefficient.

One limitation of such studies is that the validity tends to decrease when the prediction formula is used with a new sample. This tendency is called shrinkage. Common sense suggests that shrinkage is more likely when a small sample is used in the original study and particularly when the number of predictors is large. In fact, if the number of predictor variables is the same as N, you can achieve perfect prediction. The problem is that the correlations are unique to the sample, and when the prediction formula is applied to another sample (even similar to the first one), the relationship does not hold. Consequently, the validity coefficient decreases substantially (i.e., shrinkage occurs).

A technique recommended to help estimate shrinkage is **cross-validation.** In this technique, the same tests are given to a new sample from the same population to check whether the formula is accurate. For example, a researcher might administer the criterion measure and predictor tests to a sample of 200 people. Then, he or she would employ multiple regression on the results from 100 of them to develop a prediction formula. This formula is then applied to the other 100 individuals to see how accurately it predicts the criterion for them. Because the researcher has the actual criterion measures for these people, the amount of shrinkage can be ascertained by correlating (Pearson r) the predicted scores with the actual scores. A comparison of the R^2 from the multiple prediction with the r^2 between the actual and the predicted criteria yields an estimate of shrinkage.

Expectancy Tables

A problem of interpretation often arises when using a formula for prediction. A question frequently asked is, How large must a predictive validity coefficient be before it provides useful information? For example, correlation between attainment of advanced degrees and scores on the Graduate Record Examination (GRE) is typically quite small ($r < .40$). When we examine r^2 for meaningfulness, we see there is only 16% or less common variance, which is quite discouraging. However, when we look at the relationship from a different perspective, such as what percentage of students who score highly on the GRE attain a PhD and what percentage of people who score poorly attain the degree, we see that the

> **predictive validity**
>
> Degree to which scores of predictor variables can accurately predict criterion scores.

> **cross-validation**
>
> Technique to assess the accuracy of a prediction formula in which the formula is applied to a sample not used in developing the formula.

GRE does have good validity. We can get this type of perspective through the use of an **expectancy table.**

An expectancy table can be used to predict the probability of some performance—whether academic, job-related, or other performance. Expectancy tables are easily constructed. The table consists of a two-way grid containing the probability of a person with a particular assessment score attaining some criterion score. As an example, assume we have developed a test that purportedly measures sportsmanship, called the Jolly Good Show Inventory. We wish to see how well our inventory relates to ratings of sportsmanship by judges who have observed a group of 60 students at play (for hours and hours under various forms of competitive situations).

Our first step is to simply tally the number of students who fall in each cell. For example, there were three students who scored between 50 and 59 on the inventory. One of these was rated average, one good, and one excellent. Table 11.1a shows the results of this step. Next, we convert the cell frequencies to percentages of the total number of students in each row. For example, one of the three students scoring between 50 and 59 was rated average, thus the percentage is 1/3 = 33%. The percentages are shown in table 11.1b. That's all there is to it. We can see, for example, that of the nine students who scored in the 40–49 interval on the inventory, 22% were rated as excellent in sportsmanship. Further, none of the students with the top scores on the inventory were rated below average, and conversely, none of the students with the lowest inventory scores (0–9) were given ratings higher than poor. In this example, we are using an expectancy table to provide evidence of the concurrent validity of our inventory. Sometimes expectancy table information can be used to predict future success, such as grade-point average in college from an aptitude test given in high school. As with all situations involving criterion validity, the availability and relevance of the criterion and whether it can be measured reliably are key issues.

Table 11.1 Expectancy Table for Jolly Good Show Inventory Scores to Ratings of Sportsmanship

11.1a Frequency of Ratings for Each Predictor Score Level

Test score	Lousy	Poor	Average	Good	Excellent	Totals
50–59			1	1	1	3
40–49		1	2	4	2	9
30–39		4	8	2		14
20–29	2	5	7	5		19
10–19	3	6	2			11
0–9	3	1				4
						60

11.1b Expectancy Table After Converting Frequencies to Percentages

Test score	Lousy	Poor	Average	Good	Excellent
50–59			33	33	33
40–49		11	22	44	22
30–39		29	57	14	
20–29	11	26	37	26	
10–19	27	55	18		
0–9	75	25			

Construct Validity

Many human characteristics are not directly observable. Rather, they are hypothetical constructs that carry a number of associated meanings concerning how a person who possesses the traits to a high degree would behave differently from someone who possesses the traits to a low degree. Anxiety, intelligence, sportsmanship, creativity, and attitude are a few such hypothetical constructs. Because these traits are not directly observable, measurement poses a problem. Construct validity is the degree to which scores from a test measure a hypothetical construct and is usually established by relating the test results to some behavior. For example, certain behaviors are expected of someone with a high degree of sportsmanship. Such a person might be expected to compliment the opponent on shots made during a tennis match. For an indication of construct validity, a test maker could compare the number of times a person scoring high on a test of sportsmanship complimented the opponent with the number of times a person scoring lower on the test did so.

The **known group difference method** is sometimes used in establishing construct validity. For example, construct validity of a test of anaerobic power could be demonstrated by comparing test scores of sprinters and jumpers with those of distance runners. Sprinting and jumping require greater anaerobic power than distance running. Therefore, the tester could determine whether the test differentiates between the two kinds of track performers. If the sprinters and jumpers score significantly better than the distance runners, this would provide some evidence that the test measures anaerobic power.

An experimental approach is occasionally used in demonstrating construct validity. For example, a test of cardiovascular fitness might be assumed to have construct validity if it reflected gains in fitness following a conditioning program. Similarly, the originator of a motor skills test could demonstrate construct validity through its sensitivity in differentiating between groups of instructed and noninstructed children.

Correlation can also be used in establishing construct validity. Hypothesized structures or dimensions of the trait being tested are sometimes formulated and verified with factor analysis. The tester also uses correlation to examine relationships between constructs, for example, when it is hypothesized that someone with high scores on the test being developed (e.g., cardiovascular fitness) should also do well on some total physical fitness scale. Conversely, individuals with low scores on the cardiovascular test would be expected to do poorly on the fitness test.

Another example of using correlational techniques to assess construct validity is in the development of an attitude instrument. A discussion of the procedures for attitude research can be found in the paper by Silverman and Subramaniam (1999). Confirmatory factor analysis is used to see if the data from a pilot study fit the proposed model of constructs. If the data do not fit the model, additional pilot testing occurs until the model fits established standards. Examples of this technique with attitude measurement can be found in physical education and other areas (e.g., Keating & Silverman, 2004; Kulinna & Silverman, 1999; Subramaniam & Silverman, 2000).

Actually, all the other forms of validity we have discussed are used for evidence of construct-related validity. Indeed, it is usually necessary to use evidence from all the other forms to provide strong support for the validity of scores for a particular instrument and the use of its results.

Reliability

An integral part of validity is reliability, which pertains to the consistency, or repeatability, of a measure. A test cannot be considered valid if it is not reliable. In other words, if the test is not consistent—if you cannot depend on successive trials to yield the same results—then the test cannot be trusted. Of course, scores from a test can be reliable yet not valid, but it can never be valid if it is not reliable. For example, weighing yourself repeatedly on a broken scale would give reliable results but not valid ones. Test reliability is sometimes discussed in terms of **observed score, true score,** and **error score.**

known group difference method

Method used for establishing construct validity in which the test scores of groups that should differ on a trait or ability are compared.

observed score

In classical test theory, an obtained score which comprises a person's true score and error score.

true score

In classical test theory, the part of the observed score that represents the individual's real score and does not contain measurement error.

error score

In classical test theory, the part of an observed score that is attributed to measurement error.

A test score obtained by an individual is the observed score. It is not known whether this is a true assessment of this person's ability or performance. There may be measurement error pertaining to the test directions, the instrumentation, the scoring, or the emotional or physical state of the individual. Thus, an observed score theoretically consists of the person's true score and error score. Expressed in terms of score variance, the observed score variance consists of true score variance plus error variance. The goal of the tester is to remove error to yield the true score. The coefficient of reliability is the ratio of true score variance to observed score variance. Because true score variance is never known, it is estimated by subtracting error variance from observed score variance. Thus, the reliability coefficient reflects the degree to which the measurement is free of error variance.

Sources of Measurement Error

Measurement error can come from four sources: the participant, the testing, the scoring, and the instrumentation. Measurement error associated with the participant includes many factors, including mood, motivation, fatigue, health, fluctuations in memory and performance, previous practice, specific knowledge, and familiarity with the test items.

Errors in testing are those due to lack of clarity or completeness in the directions, how rigidly the instructions are followed, whether supplementary directions or motivation is applied, and so forth. Errors in scoring relate to the competence, experience, and dedication of the scorers and to the nature of the scoring itself. The extent to which the scorer is familiar with the behavior being tested and the test items can greatly affect scoring accuracy. Carelessness and inattention to detail produce measurement error. Measurement error due to instrumentation includes such obvious causes as inaccuracy and lack of calibration of mechanical and electronic equipment. It also refers to the inadequacy of a test to discriminate between abilities and to the difficulty of scoring some tests.

Expressing Reliability Through Correlation

The degree of reliability is expressed by a correlation coefficient, ranging from 0.00 to 1.00. The closer the coefficient is to 1.00, the less error variance it reflects and the more the true score is assessed. Reliability is established in several ways, which are summarized a little later in this chapter. The correlation technique used for computing the reliability coefficient differs from that used for establishing validity. Pearson r is often called **interclass correlation.** This coefficient is a bivariate statistic, meaning that it is used to correlate two different variables, as when determining validity by correlating judges' ratings with scores on a skill test. However, interclass correlation is not appropriate for establishing reliability because two values for the same variable are being correlated. When a test is given twice, the scores on the first test are correlated with the scores on the second test to determine their degree of consistency. Here, the two test scores are for the same variable, so interclass correlation should not be used. Rather, **intraclass correlation** is the appropriate statistical technique. This method uses ANOVA to obtain the reliability coefficient.

Interclass Correlation

There are three main weaknesses of Pearson r (interclass correlation) for reliability determination. The first is that, as mentioned previously, the Pearson r is a bivariate statistic, whereas reliability involves univariate measures. Second, the computations of Pearson r are limited to only two scores, X and Y. Often, however, more than two trials are given, and the tester is concerned with the reliability of multiple trials. For example, if a test specifies three trials, the researcher must either give three more trials and use the average or best score of each set of trials for the correlations or perhaps correlate the first trial with the second, the first with the third, and the second with the third. In the first case, extra trials must be given solely for reliability purposes; in the second case, meaningfulness suffers when several correlations among trials are shown. Finally, the interclass correlation does not provide a thorough examination of different sources of variability on multiple trials. For

interclass correlation

The most commonly used method of computing correlation between two variables; also called Pearson r or *Pearson product moment coefficient of correlation.*

intraclass correlation

An ANOVA technique used for estimating reliability of a measure.

example, changes in means and standard deviations from trial to trial cannot be assessed with the Pearson r method but can be analyzed with intraclass correlation.

Intraclass Correlation

Intraclass correlation provides estimates of systematic and error variance. For example, systematic differences among trials can be examined. The last trials may differ significantly from the first trials because of a learning phenomenon or fatigue effect (or both). If the tester recognizes this, then perhaps initial (or final) trials can be excluded or the point at which performance levels off can be used as the score. In other words, through ANOVA the tester can truly examine test performance from trial to trial and then select the most reliable testing schedule.

The procedures leading to the calculation of intraclass correlation (R) are the same as those of simple ANOVA with repeated measures, discussed in chapter 9. The components are subjects, trials, and residual sums of squares. An example of a summary ANOVA is shown in table 11.2. We use these data to compute R.

An F for trials determines whether there are any significant differences among the three trials. Refer to table A.6 in appendix A, and read down the 2-df column to the 8-df row. Our F of 5.94 is found to be greater than the table F of 4.46 at the .05 level of probability. Thus, there are significant differences among trials. At this point we should mention that there are different opinions about what should be done with trial differences (Baumgartner & Jackson, 1991; Johnson & Nelson, 1986; Safrit, 1976). Some test authorities argue that the test performance should be consistent from one trial to the next and that any trial-to-trial variance should be attributed simply to measurement error. If we decide to do this, the formula for R is

$$R = (MS_S - MS_E) / MS_S \tag{11.1}$$

in which MS_S is the mean squares for subjects (from table 11.2) and MS_E is the mean squares for error, which is computed as follows:

$$MS_E = \frac{SS \text{ for trials} + SS \text{ for residual}}{df \text{ for trials} + df \text{ for residual}}$$
$$= \frac{7.6 + 5.1}{2 + 8}$$
$$= 1.27$$

Thus, $R = (3.73 - 1.27)/3.73 = .66$.

Another way of dealing with significant trial differences is to discard the trials that are noticeably different from the others (Baumgartner & Jackson, 1991). Then a second ANOVA is calculated on the remaining trials, and another F test is computed. If F is not significant, R is calculated using equation 11.1, in which trial variance is considered measurement error. If F is still significant, additional trials are discarded, and another ANOVA is conducted. The purpose of this method is to find a measurement schedule that is free of trial differences (i.e., to find a nonsignificant F that yields the largest possible criterion score) and that is most reliable. This method is especially appealing when there is an apparent trend in trial differences, as when a learning phenomenon (or release of inhibitions) is evident from increased scores on a later trial or trials. For example, if five trials on a performance test yield mean scores of 15, 18, 23, 25, and 24, you might discard the first two trials and compute another analysis on the last three trials. Similarly, a fatigue effect may be evidenced by a decrease in scores on the final trials in some types of tests.

In table 11.2, note that the mean of trial 1 is considerably lower than for trials 2 and 3. Therefore, we discard the first trial and compute another ANOVA on trials 2 and 3 only (results are shown in table 11.3). The F for trials in the table is nonsignificant, so first we

Table 11.2 Summary of ANOVA for Reliability Estimation (3 Trials)

Source	SS	df	MS	F
Subjects	14.9	4	3.73	
Trials	7.6	2	3.80	5.94*
Residual	5.1	8	0.64	

*$p < .05$.

M (trial 1) = 2.4; M (trial 2) = 3.8; M (trial 3) = 4.0.

Table 11.3   Summary of ANOVA for Reliability With Trial 1 Discarded

Source	SS	df	MS	F
Subjects	11.4	4	2.85	
Trials	0.1	1	0.10	0.11*
Residual	3.4	4	0.85	

*$p > .05$.

intertester (interrater) reliability

The degree to which different testers can obtain the same scores on the same participants; also called *objectivity*.

stability

A coefficient of reliability measured by the test–retest method on different days.

test–retest method

Method of determining stability in which a test is given one day and then administered exactly as before a day or so later.

alternate-forms method

Method of establishing reliability involving the construction of two tests that both supposedly sample the same material; also called *parallel-form method* or the *equivalence method*.

parallel-form method

See *alternate-forms method*.

equivalence method

See *alternate-forms method*.

compute MS_E by combining the sums of squares for trials and residual and dividing by their respective degrees of freedom: $MS_E = (0.1 + 3.4)/(1 + 4) = 0.7$. Then we compute R with equation 11.1:

$$R = (2.85 - 0.7)/2.85 = .75$$

We see that R is considerably higher when we discard the first trial.

A third approach is simply to ignore trial-to-trial variance as measurement error. In this approach, trial-to-trial variance is not considered to be true score variance or error score variance. Consequently, R is notably higher than in the previous approaches because all trial-to-trial variance is removed. Although some measurement authorities advocate this approach, we do not recommend it because it does not seem to follow the theory that observed score variance equals true score variance plus error score variance. We believe that it should be of interest to the researcher or tester to ascertain whether there are trial-to-trial differences. Consequently, if significant differences are found, the tester can decide whether to eliminate some trials (as with a learning trend) or simply to consider trial-to-trial differences as measurement error.

Intertester reliability is determined by the same procedures we just outlined. Thus, the objectivity of judges or different testers is analyzed by intraclass R, and judge-to-judge variance is calculated in the same way as trial-to-trial variance. Of course, more complex ANOVA designs can be used in which trial-to-trial, day-to-day, and judge-to-judge sources of variance all can be identified. Baumgartner (1989) and Safrit (1976) discuss some models that can be used for establishing reliability.

Methods of Establishing Reliability

It is easier to establish reliability than validity. We first look at three types of coefficients of reliability: stability, alternate forms, and internal consistency.

Stability

The coefficient of **stability** is determined by the test–retest method on separate days. This method is used frequently with fitness and motor performance measures but less often with pencil-and-paper tests. This is one of the most severe tests of consistency because the errors associated with measurement are likely to be more pronounced when the two test administrations are separated by a day or more.

In the **test–retest method,** the test is given one day and then repeated a day or so later. The interval may be governed to some extent by how strenuous the test is and whether more than a day's rest is needed. Of course, the interval cannot be so long that actual changes in ability, maturation, or learning occur between the two test administrations.

Intraclass correlation should be used to compute the coefficient of stability of the scores on the two tests. Through ANOVA procedures, the tester can determine the amount of variance accounted for by the separate days of testing, test trial differences, participant differences, and error variance.

Alternate Forms

The **alternate-forms method** of establishing reliability involves the construction of two tests that supposedly sample the same material. This method is sometimes referred to as the **parallel-form method** or the **equivalence method.** The two tests are given to the same individuals. Ordinarily, some time elapses between the two administrations. The scores on the two tests are then correlated to obtain a reliability coefficient. The alternate-forms method is a widely used technique with standardized tests, such as those of achievement and scholastic aptitude. The method is rarely used with physical performance tests, probably because it is more difficult to construct two different sets of good physical test items than it is to write two sets of questions.

Some test experts maintain that, theoretically, the alternate-forms method is the preferred method for determining reliability. Any test is only a sample of test items from a universe of possible test items. Thus, the degree of relationship between two such samples should yield the best estimate of reliability.

Internal Consistency

Reliability coefficients can be obtained by several methods that are classified as **internal consistency** techniques. Some common methods are the same-day test–retest, the split-half method, the Kuder-Richardson method of rational equivalence, and the coefficient alpha technique.

The **same-day test–retest method** is used almost exclusively with physical performance tests because practice effects and recall tend to produce spuriously high correlations when this technique is used with written tests. The test–retest on the same day results in a higher reliability coefficient than does a test–retest on separate days. One would certainly expect more consistency of performance within the same day than on different days. Intraclass correlation is used to analyze trial-to-trial (internal) consistency.

The **split-half technique** has been widely used for written tests and occasionally in performance tests that require numerous trials. The test is divided in two, and the two halves are then correlated. A test could be divided into first and second halves, but this is usually not deemed satisfactory. Sometimes a person tires near the end of the test, and sometimes easier questions are placed in the first half. Usually, the odd-numbered questions are compared with the even-numbered ones. That is, the number of odd-numbered questions a person got correct is correlated with the number of correct answers on the even-numbered questions.

Because the correlation is between the two halves of the test, the reliability coefficient represents only half the total test; that is, behavior is sampled only half as thoroughly. Thus, a step-up procedure, the **Spearman-Brown prophecy formula,** is used to estimate the reliability for the entire test because the total test is based on twice the sample of behavior (twice the number of items). The formula is

$$\text{Corrected reliability coefficient} = \frac{2 \times \text{reliability for } 1/2 \text{ test}}{1.00 + \text{reliability for } 1/2 \text{ test}}$$

If, for example, the correlation between the even-numbered items and the odd-numbered items was .85, the corrected reliability coefficient would be

$$\frac{2 \times .85}{1.00 + .85} = \frac{1.70}{1.85} = .92$$

Another split-half method is the **Flanagan method,** which analyzes the variances of the halves of the test in relation to the total variance. No correlation or Spearman-Brown step-up procedure is involved.

In the **Kuder-Richardson (KR) method of rational equivalence,** one of two formulas, known as KR-20 and KR-21, is used for items scored dichotomously (e.g., right or wrong). Only one test administration is required, and no correlation is calculated. The resulting coefficient represents an average of all possible split-half reliability coefficients.

Highly regarded by many test experts, the KR-20 involves the proportions of students who answered each item correctly and incorrectly in relation to the total score variance. The KR-21 is a simplified, less accurate version of the KR-20.

The **coefficient alpha** technique is sometimes referred to as the **Cronbach alpha coefficient** (see Cronbach, 1951). It is a generalized reliability coefficient that is more versatile than other methods. One particularly desirable feature of coefficient alpha is that it can be used with items that have various point values, such as essay tests and attitude scales

internal consistency

An estimate of the reliability that represents the consistency of scores within a test.

same-day test-retest method

Method of establishing reliability in which a test is given twice to the same participants on the same day.

split-half technique

Method of testing reliability in which the test is divided in two, usually by making the odd-numbered items one half and the even-numbered items the other half, and the two halves are correlated.

Spearman-Brown prophecy formula

Equation developed to estimate the reliability for the entire test when the split-half technique is used to test reliability.

Flanagan method

A process for estimating reliability in which the test is split into two halves, and the variances of the halves of the test are analyzed in relation to the total variance of the test.

Kuder-Richardson (KR) method of rational equivalence

Formulas developed for estimating reliability of a test from a single test administration.

coefficient alpha

A technique used for estimating reliability of multiple-trial tests; also called *Cronbach alpha coefficient.*

Cronbach alpha coefficient

See *coefficient alpha.*

objectivity

See *intertester reliability.*

that have as possible answers "strongly agree," "agree," and so on. The method involves calculating variances of the parts of a test. The parts can be items, test halves, trials, or a series of tests such as quizzes. When the items are dichotomous (i.e., either right or wrong), the alpha coefficient results in the same reliability estimate as KR-20 (in fact, KR-20 is just a special case of coefficient alpha). When the parts are halves of the test, the results are the same as the Flanagan split-halves method. And when the parts are trials or tests, the results are the same as intraclass correlation. Coefficient alpha is probably the most commonly used method of estimating reliability for standardized tests.

Intertester Reliability (Objectivity)

A form of reliability that pertains to the testers is called intertester (or interrater) reliability, or, often, **objectivity.** This facet of reliability is the degree to which different testers can achieve the same scores on the same subjects.

Overall, most teachers and students prefer objective over subjective measures because so much depends on how valid and reliable the measures are. Objective measures are not automatically better than subjective ones, but they do yield quantitative scores that are more "visible" and that can be statistically handled more easily. In most research techniques, objective measurements are essential.

The degree of objectivity (intertester reliability) can be established by having more than one tester gather data. Then the scores are analyzed with intraclass correlation techniques to obtain an intertester reliability coefficient. It is possible to assess a number of sources of variance in one analysis, such as variance caused by testers, trials, days, participants, and error (for discussion of the calculations involved, see Safrit, 1976).

Some forms of research in physical activity involve the observation of certain behaviors in real-world settings, such as during physical education classes or sport participation. This involves the use or development of some sort of coding instrument. Most frequently, the instrument has a series of categories into which the various motor behaviors may be coded. The behaviors are observed using techniques described in chapter 16, such as event recording, time sampling, and duration recording.

In these types of scales, as with any measure, validity and reliability are important. However, it is usually much more difficult to obtain consistency in recording activities in a physical education class or sport than in recording error measurements from a laboratory task such as the linear slide. Consequently, observational researchers are concerned about coder consistency. Typically, coders are trained to a criterion level of reliability, and reliability is checked regularly throughout the project. A common way of estimating reliability among coders, often used in event recording and time sampling, is called **interobserver agreement (IOA),** which uses the following formula:

interobserver agreement (IOA)

Common way of estimating reliability among coders by using a formula that divides the number of agreements in behavior coding by the sum of the agreements and disagreements.

$$IOA = \frac{agreements}{(agreements + disagreements)}$$

Agreements are behaviors that are coded the same, whereas disagreements are behaviors coded differently. IOA is typically reported as the percentage of agreement.

Standard Error of Measurement

Previous chapters touched on the idea of standard error several times with regard to confidence intervals, the *t* and *F* tests, and interpreting significance levels. Chapter 8 also discussed standard error of prediction in correlational research. The standard error of measurement is an important concept in interpreting the results of measurement. Sometimes we get too carried away with the aura of scientific data collection and fail to realize that the possibility of measurement error always exists. For example, maximal oxygen consumption ($\dot{V}O_2max$) has been mentioned several times in this book as the most valid measure of cardiorespiratory fitness. Field tests are frequently validated

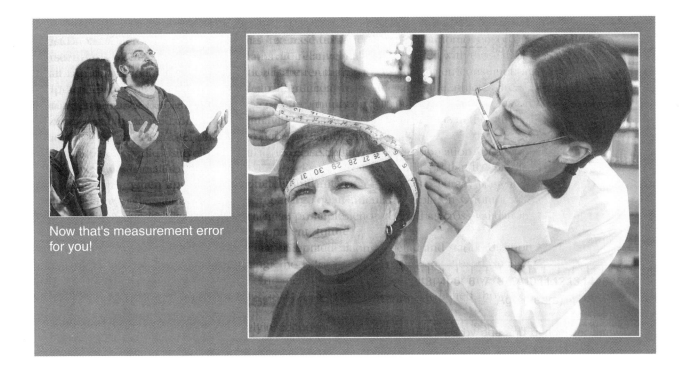

Now that's measurement error for you!

through correlation with $\dot{V}O_2$max. We must be careful, however, not to consider $\dot{V}O_2$max as a perfect test that is error free. Every test yields only observed scores; we can obtain only estimates of a person's true score. It is much better to think of test scores as falling within a range that contains the true score. The formula for the standard error of measurement ($S_{Y \cdot X}$) is

$$S_{Y \bullet X} = s\sqrt{1.00 - r} \qquad\qquad (11.2)$$

where s is the standard deviation of the scores and r is the reliability coefficient for the test. In the measurement of percentage body fat for female adults, assume that the standard deviation is 5.6% and the test–retest correlation is .83. The standard error of measurement would be

$$S_{Y \bullet X} = 5.6\sqrt{1.00 - .83} = 2.3\%$$

Assume further that a particular woman's measured percentage fat is 22.4%. We can use the standard error of measurement to estimate a range within which her true percentage fat probably falls.

Standard errors are assumed to be normally distributed and are interpreted in the same way as standard deviations. About two thirds (68.26%) of all test scores fall within plus or minus one standard error of measurement of their true scores. In other words, there is about a 68% chance that a person's true score will be found within a range of the obtained score plus or minus the standard error of measurement. In the example of the woman who had an obtained score of 22.4% fat, chances are 2 in 3 that her true percentage fat is 22.4% ± 2.3% (standard error of measurement), or somewhere between 20.1% and 24.7%.

We can be more confident if we multiply the standard error of measurement by 2 because about 95% (95.44%) of the time the true score can be found within a range of the observed score plus or minus two times the standard error of measurement. Thus, in the present example, we can be about 95% sure that the woman's true percentage fat is 22.4% ± 4.6% (±2.3% × 2), that is, between 17.8% and 27.0%.

From the formula we see that the standard error of measurement is governed by the variability of the test scores and the reliability of the test. If we had a higher reliability coefficient, the error of measurement would obviously decrease. In the present example, if the reliability coefficient was .95, the standard error of measurement would be only

1.3%. With the same reliability (.83) but with a smaller standard deviation, such as 4%, the standard error of measurement would be 1.6%.

Remember the idea of standard error of measurement when you are interpreting test scores. As indicated earlier, people sometimes have absolutely blind faith in some measurements, particularly if the measurements seem scientific. With percentage fat estimation from skinfold thicknesses, for example, we need to keep in mind that error is connected not only with the skinfold measurements but also with the criterion that these measurements are predicting, that is, the body density values obtained from underwater weighing and the percentage fat determined from them. Yet we have observed people accepting as gospel that they have a certain amount of fat because someone measured a few skinfolds. Newspapers have reported that some athletes have only 1% fat. This is impossible from a physiological standpoint. Moreover, obtaining a predicted negative percentage fat from regression equations is possible. Please do not misunderstand. We are not condemning skinfold measurements. We are simply trying to emphasize that all measurements are susceptible to errors and that common sense, coupled with knowledge of the concept of standard error of measurement, can help us better understand and interpret the results of measurements. We revisit standard error in chapter 15 when we discuss errors associated with surveys.

Using Standard Scores to Compare Performance

Direct comparisons of scores are not meaningful without some point of reference. Is a score of 46 cm on the vertical jump as good as a score of 25 push-ups? How can you compare centimeters and repetitions? If we know that the class mean for the vertical jump is 40 cm and that 20 is the mean for push-ups, we know that the performances of 46 cm and 25 push-ups are better than average, but how much better? Is one performance better than the other?

One way to compare the performances is to convert each score to a standard score. A standard score is a score expressed in terms of standard deviations from the mean. We now discuss how to determine standard scores by using z scores and T scales.

z Scores

z score

The basic standard score that converts raw scores to units of standard deviation in which the mean is zero and a standard deviation is 1.0.

The basic standard score is the **z score.** The z scale converts raw scores to units of standard deviation in which the mean is zero and a standard deviation is 1.0. The formula is

$$z = (X - M)/s \qquad (11.3)$$

Suppose that the mean and standard deviation for vertical-jump scores are 40 cm and 6 cm, respectively, and for push-ups 20 and 5, respectively. Thus, a score of 46 cm for the vertical jump is a z score of +1.00: $z = (46 - 40)/6 = 6/6 = 1.00$. A score of 25 push-ups is also a z score of +1.00: $z = (25 - 20)/5 = 5/5 = 1.00$.

We see that the person performed exactly the same on the two tests. Both performances were 1s above the mean. Similarly, scores of different students on the same test can be compared by z scores. A student jumping 37 cm has a z score of –0.5, a student who jumps 44 cm has a z score of 0.67, and so on. All standard scores are based on the z score. However, because z scores are expressed in decimals and can be positive or negative, they are not as easy to work with as some other scales.

T scale

Type of standard score that sets the mean at 50 and standard deviation at 10 to remove the decimal found in z scores and to make all scores positive.

T Scales

The **T scale** sets the mean at 50 and the standard deviation at 10. Hence, the formula is $T = 50 + 10z$. This removes the decimal and makes all the scores positive. A score 1s above the mean ($z = 1.0$) is a T score of 60. A score 1s below the mean ($z = -1.0$) is a T score of 40. Because more than 99% (99.73%) of the scores fall between ±3s, it is rare to have T

scores below 20 ($z = -3.0$) or above 80 ($z = +3.0$). Some standardized tests that use different transformations of means and standard deviations using the z-score distribution are shown in table 11.4.

The decision of which standard score to use depends on the nature of the research study and the extent of interpretation required for the test takers. In essence, then, it is a matter of choice in light of the use of the measures.

Table 11.4 Standardized Means and Standard Deviations of Well-Known Tests		
Scale	**M**	**s**
Graduate Record Examination	500	100_z
Stanford-Binet IQ	100	16_z
College Entrance Examination	500	100_z
National Teachers Examination	500	100_z
Wechsler IQ	100	15_z

Measuring Movement

Much of the research in physical activity obviously involves movement. The measurement of physical fitness has fascinated exercise physiologists and physical educators for years. The concept of health-related physical fitness is widely accepted as being represented by the components of cardiorespiratory endurance, muscular strength, muscular endurance, flexibility, and body composition. In addition, many research studies have examined other fitness parameters that are needed primarily for skilled performances, such as in athletics and dance. These components include power, speed, reaction time, agility, balance, kinesthetic perception, and coordination. Research in motor behavior generally deals with the acquisition and control of motor skills. Types of measurements include tests of basic movement patterns, sport skills, and controlled (often novel) laboratory tasks.

Research in biomechanics involves such measurements as high-speed cinematography, force transduction, and electromyography. Research in pedagogy typically involves observations of behavior in real-world settings, such as during physical education classes or sport participation. Videotaped observations are often used to allow more precise analyses.

The validity and reliability of all forms of measuring movement must be defended by the researchers. In some respects, measuring movement is simpler and more straightforward than measuring cognitive or affective behaviors. For example, the amount of force a person can apply in a given movement can be accurately and directly measured. The distance a person can jump can be assessed by a measuring tape, and everyone accepts that as a valid and reliable measure. On the other hand, cognitive and affective behaviors must usually be inferred from marks made on a sheet of paper. Nevertheless, the measurement of movement is rarely a simple manner of measuring distance jumped. It is usually complex and often difficult to standardize. Each type of measurement has its own methodological difficulties, which pose problems for the researcher with regard to validity and reliability. We do not attempt a more thorough description of measuring movement because much of this material is specific to your area of specialization (e.g., motor behavior, exercise, and wellness). The focus of your graduate program is on many of these issues.

Measuring Written Responses

The measurement of written (and oral) responses is part of the methodology of numerous research studies in physical education and exercise science. Research questions frequently deal with affective behavior, which includes attitudes, interests, emotional states, personality, and psychological traits.

Measuring Affective Behavior

Affective behavior includes attitudes, personality, anxiety, self-concept, social behavior, and sportsmanship. Many general measures of these constructs have been used over the years (e.g., *Cattell 16 Personality Factor (PF) Questionnaire*, Spielberger *State-Trait Anxiety Inventory*). Nelson (1989) presents an extensive discussion of the construction, strengths, and weaknesses of affective tests and scaling.

Attitude Inventories

A large number of attitude inventories have been developed. Undoubtedly, many test developers see a direct link between attitude and behavior. For example, if a person has a favorable attitude toward physical activities, that person will participate in such activities. Research, however, has seldom substantiated this link between attitude and behavior, although such a link seems logical.

Researchers usually try to locate an instrument that has already been validated and that is an accepted measure of attitude rather than constructing a new one. Finding a published test that closely pertains to the research topic and also that has scores validated with a similar sample is a problem. Another problem is that often the researcher wants to determine whether some treatment will bring about an attitude change. A source of invalidity discussed in chapter 18 is the reactive effects of testing. The pretest sensitizes the participant to the attitudes in question, and this sensitization rather than the treatment may promote a change.

Another problem (mentioned in chapter 15) inherent in any self-report inventory is whether the person is truthful. It is usually evident what a given response to an item in an attitude inventory indicates. For example, a person is asked to show his or her degree of agreement or disagreement to the statement "Regular exercise is an important part of our daily lives." The individual may perceive that the socially desirable response is to agree with this statement regardless of his or her true feelings. Some respondents deliberately distort their answers to appear "good" (or "bad"). Tests sometimes use so-called filler items to make the true purpose of the instrument less visible. For example, a test designer might include several unrelated items, such as "Going to the opera is a desirable social activity," to disguise the fact that the instrument is designed to measure attitude toward exercise. This is especially important when social desirability considerations may be biasing factors.

When a researcher seeks an attitude instrument to use in a study, validity and reliability must, of course, be primary considerations. Unfortunately, published attitude scales have not always been constructed scientifically, and limited information is provided about their validity and reliability. Reliability of methodology can be established easily. (We are not saying that attitude scales are easily made reliable.) Validity is usually the problem because of the failure to develop a satisfactory theoretical model for the attitude construct.

Personality Research

Many research studies in physical education and sports have attempted to explore the relationship between personality traits and various aspects of athletic performance. Interest in this topic can be attributed to several factors, including the great deal of public attention focused on athletics. Athletes are obviously "special" people with regard to physical characteristics. Beyond that, however, is the hypothesis that athletes have certain personality traits that distinguish them from nonathletes.

One area of investigation is to identify personality traits that might be uniquely characteristic of athletes in different sports. For example, do people who gravitate toward vigorous contact sports differ in personality from those individuals who prefer noncontact sports? That is, is there a "football type" or a "bowler type"? Are superior athletes different from average athletes in certain personality traits? Can participation in competitive sport modify one's personality structure? Moreover, do athletes within a sport differ on some traits that, if known, could point to different coaching strategies or perhaps be used to screen and predict which athletes will be "hard" to coach or will lack certain qualities associated with success?

People with a strong interest in sports, such as former coaches and players, have been greatly attracted to the study of personality and athletics. However, having a strong interest in athletics does not compensate for a lack of preparation and experience in psychological evaluation. Unfortunately, there have been cases in which personality measuring instruments have been misused. Recently, sport psychologists have recognized the value of sport-specific measures of various behaviors and perceptions. Much of the impetus for this approach can be traced to Martens' *Sport Competition Anxiety Test* (Martens, 1977).

Some sport-specific measures relate to group cohesiveness (Carron, Widmeyer, & Brawley, 1985), intrinsic and extrinsic motivation (Weiss, Bredemeier, & Shewchuck, 1985), confidence (Vealey, 1986), and sport achievement orientation (Gill & Deeter, 1988). The rationale for the sport-specific approach is that general measures of achievement, anxiety, motivation, self-concept, social behavior, and fair play do not have high validity for sport situations. Recent studies in test development indicate a commitment by sport psychologists to develop measures of affective behavior that are multidimensional and specific to the competitive environment.

Scales for Measuring Affective Behavior

In measuring affective behavior, a variety of scales are used to quantify the responses. Two of the most commonly used are the Likert-type scale and the semantic differential scale.

The Likert-Type Scale

The **Likert-type scale** is referred to in chapter 15 in connection with survey research techniques. It is usually a five- or seven-point scale with assumed equal intervals between points. The Likert-type scale is used to assess the degree of agreement or disagreement with statements and is widely used in attitude inventories. An example of a Likert-type item follows:

> **I** prefer quiet recreational activities such as chess, cards, or checkers rather than activities such as running, tennis, or basketball.

Strongly agree   Agree   Undecided   Disagree   Strongly disagree

A principal advantage of scaled responses such as the Likert-type is that they permit a wider choice of expression than responses such as "always" or "never," or "yes" or "no." The five, seven, or more intervals may help increase the reliability of the instrument. (For more comprehensive information concerning the Likert-type, semantic differential, and other scales, see Edwards, 1957, and Nunnaly, 1978.)

The Semantic Differential Scale

The **semantic differential scale** employs bipolar adjectives at each end of a seven-point scale. The respondent is asked to make judgments about certain concepts. The scale is based on the importance of language in reflecting a person's feelings. A sample of a semantic differential item follows:

The Coach

7   6   5   4   3   2   1

1. creative ←—————————→ unoriginal

2. supportive ←—————————→ critical

3. fair ←—————————→ unfair

The 1–7 scale between adjectives is scored, with 7 being the most positive judgment. Factor analysis studies have consistently identified the same three dimensions that are assessed by the semantic differential technique: evaluation (e.g., fair–unfair), potency (e.g., powerful–feeble), and activity (e.g., dynamic–static).

Likert-type scale

Type of closed question that requires the participant to respond by choosing one of several scaled responses; the intervals between items are assumed to be equal.

semantic differential scale

Scale used to measure affective behavior in which the respondent is asked to make judgments about certain concepts by choosing one of seven intervals between bipolar adjectives.

Rating Scales

rating scale

A measure of behavior that involves a subjective evaluation based on a checklist of criteria.

rating of perceived exertion (RPE)

Self-rating scale developed by G.A. Borg (1962) to measure an individual's perceived effort during exercise.

Rating scales are sometimes used in research to evaluate performance. For example, in a study that compares different strategies in teaching diving, the dependent variable (diving skill) would most likely be derived from expert ratings because diving does not lend itself to objective skill tests. Thus, after the experimental treatments have been applied, people knowledgeable in diving rate all the participants on their diving skills. To do this in a systematic and structured manner, the raters need to have some kind of scale with which to assess skill levels in different parts or phases of the performance.

A self-rating scale concerning an individual's perceived efforts during exercise that has been widely used in research is the **rating of perceived exertion (RPE)** scale by Borg (1962). The underlying rationale for the scale is to combine and integrate the many physiological indicators of exertion into a whole, or gestalt, of subjective feeling of physical effort. This feeling of perceived exertion was quantified by Borg into a scale with numbers ranging from 6 to 20, reflecting a range of exertion from "very, very light" to "very, very hard."

Types of Rating Scales

There are different kinds of rating scales. Some scales use numerical ratings, some use checklists, some have verbal cues associated with numerical ratings, some require forced choices, and still others use rankings. Some scales are simple, whereas others are rather complex. Whatever the degree of complexity, however, practice in using the scale is imperative.

When more than one judge is asked to rate performances, some common standards must be set. Training sessions with videotaped performances of people of different ability levels are helpful in establishing standard frames of reference before judging the actual performances. Intertester agreement and reliability were discussed earlier in this chapter.

Rating Errors

leniency

Tendency for observers to be overly generous in rating.

central tendency errors

Inclination of the rater to give an inordinate number of ratings in the middle of the scale, avoiding the extremes of the scale.

halo effect

A threat to internal validity wherein raters allow previous impressions or knowledge about a certain individual to influence all ratings of that individual's behaviors.

proximity error

Inclination of a rater to consider behaviors to be more nearly the same when they are listed close together on a scale than when they are separated by some distance.

Despite efforts to make ratings as objective as possible, there are inherent pitfalls in the process. Recognized errors in rating include leniency, central tendency, the halo effect, proximity, observer bias, and observer expectation.

Leniency is the tendency for observers to be overly generous in their ratings. This error is less likely to occur in research than in evaluating peers (e.g., co-workers). Thorough training of raters is the best means of reducing leniency.

Central tendency errors result from the inclination of the rater to give an inordinate number of ratings in the middle of the scale, avoiding the extremes of the scale. Several reasons are attributed to this. Sometimes it may be due in part to ego needs or status. For example, the judge is acting in the role of an expert and, perhaps unconsciously, may grade good performers as average to suggest that he or she is accustomed to seeing better performances. Sometimes, errors of central tendency are due to the observer's wanting to "leave room" for better future performances. A common complaint in large gymnastics, diving, and skating competitions is that the performers scheduled early in the meet are scored lower for comparable performances than athletes scheduled later in the competition. Another central tendency error is the inclination to avoid assigning very low scores and is likely due to a judge's reluctance to be too harsh. This is, of course, a form of leniency.

The **halo effect** is the commonly observed tendency for a rater to allow previous impressions or knowledge about a certain individual to influence all ratings of that individual's behaviors. For example, knowing that an individual excels in one or more activities, a judge may rate that person highly on all other activities. The halo effect perhaps is not the most appropriate term because negative impressions of a person tend to lead to lower ratings in subsequent performances.

Proximity errors are often the result of overly detailed rating scales, insufficient familiarity with the rating criteria, or both. Proximity errors are manifested when the rater tends to rate behaviors that are listed close together as more nearly the same than when the behaviors are separated on the scale. For example, if the qualities "active" and "friendly" are listed side by side on the scale, proximity errors result if raters evaluate performers as more similar on those characteristics than if the two qualities were listed several lines apart on the rating scale. Of course, if the rater does not have adequate knowledge about all facets of the behavior, he or she may not be able to distinguish between different behaviors that logically should be placed close together on the scale. Thus, the different phases of behavior are rated the same.

Observer bias errors vary with the judge's own characteristics and prejudices. For example, a person who has a low regard for movement education may also tend to rate students from such a program too low. Racial, sexual, and philosophic biases are potential sources of rating errors. Observer bias errors are directional in that they produce errors that are consistently too high or too low.

Observer expectation errors can operate in various ways, often stemming from other sources of errors such as the halo effect and observer bias. Observer expectations can contaminate the ratings in that a person who expects certain behaviors is inclined to see evidence of those behaviors and interpret observations in the "expected" direction.

Research has demonstrated the powerful phenomenon of expectation in classroom situations in which teachers are told that some children are gifted or slow learners. The teachers tend to treat the pupils accordingly, giving more attention and patience to the "gifted" and less time and attention to the "slow learners."

In the research setting, potential observer expectation errors are likely when the observer knows what the experimental hypotheses are and is thus inclined to watch for these outcomes more closely than if the observer were unaware of the expected outcomes. The double-blind experimental technique described in chapter 18 is useful for controlling expectation errors. In the double-blind method, the observers do not know which individuals received which treatments. They also should not know which performances they are shown are the pretest and which are the posttest.

In summary, rating errors are always potentially present. The researcher must recognize and strive to eliminate or reduce them. One way to minimize rating errors is to define the behavior to be rated as objectively as possible. In other words, avoid having the observer make many value judgments. Another suggestion is to keep observers ignorant of the hypotheses and which participants received which treatments. Bias and expectation can be reduced if the observer is given no information about the participants' achievements, intelligence, social status, and other characteristics. The most important precaution the researcher can take is to train the observers adequately to achieve high levels of accuracy and interrater reliability.

observer bias error

Inclination of a rater to be influenced by his or her own characteristics and prejudices.

observer expectation error

Inclination of a rater to see evidence of certain expected behaviors and interpret observations in the expected direction.

Measuring Knowledge

Obviously, the measurement of knowledge is a fundamental part of the educational aspects of physical education, exercise science, and sport science. However, the construction of knowledge tests is relevant for research purposes as well. The procedures for establishing the validity and reliability of most pencil-and-paper measuring instruments used in research are similar to the procedures previously discussed. However, we must also determine whether individual items' difficulty and ability to discriminate between levels of ability are functioning in the desired manner; that is, we must perform item analysis. (For a more thorough discussion of measurement methodology for knowledge tests, see Mood, 1989.)

Analyzing Test Items

The purpose of **item analysis** is to determine which test items are suitable and which need to be rewritten or discarded. Two important facets of item analysis are determining the difficulty of the test items and determining their power to discriminate between different levels of achievement.

Item Difficulty

Analysis of **item difficulty** is usually accomplished easily. You simply divide the number of people who correctly answered the item by the total number of people who responded to the item. For example, if 80 people answered an item and 60 answered it correctly,

item analysis

Process in analyzing knowledge tests in which the suitability of test items' difficulty and discrimination is evaluated.

item difficulty

Analysis of the difficulty of each test item in a knowledge test, determined by dividing the number of people who correctly answered the item by the total number of people who responded to the item.

the item would have a difficulty index of .75 (60/80). The more difficult the item, the lower its difficulty index. For example, if only 8 of 80 answered an item correctly, the difficulty index is 8/80, or .10. Most test authorities recommend eliminating questions with difficulty indices below .10 or above .90. The best questions are those that have difficulty indices around .50. Occasionally, a test maker may wish to set a specific difficulty index for screening purposes. For example, if only the top 30% of a group of applicants are to be chosen, this could be accomplished by using questions with difficulty indices of .30. Questions that everyone answers correctly or that everyone misses provide no information about people differences in norm-referenced measurement scales.

Item Discrimination

item discrimination

The degree to which a test item discriminates between people who did well on the entire test and those who did poorly; also called *index of discrimination.*

index of discrimination

See *item discrimination.*

Item discrimination, or the degree to which test items discriminate between people who did well on the entire test and those who did poorly, is an important consideration in analyzing norm-referenced test items. There are many ways to compute an **index of discrimination.** The simplest way is to divide the completed tests into a high-scoring group and a low-scoring group and then use the following formula:

$$\text{Index of discrimination} = (n_H - n_L)/n$$

where n_H is the number of high scorers who answered the item correctly, n_L is the number of low scorers who answered the item correctly, and n is the total number in either the high or the low group. To illustrate, if we have 30 in the high group and 30 in the low group and if 20 of the high scorers answered an item correctly and 10 of the low scorers answered it correctly, the index of discrimination would be $(20 - 10)/30 = 10/30 = .33$.

Various percentages of high and low scorers, such as the upper and lower 25%, 30%, or 33%, are used in determining discrimination indices. The Flanagan method uses the upper and the lower 27%. The proportion of each group answering each item correctly is calculated; then a table of normalized biserial coefficients is consulted to obtain the item reliability coefficient. Thus, item reliability is the relationship between responses to each item and total performance on the test.

If approximately the same proportion of high scorers answer an item correctly as did the low scorers, the item is not discriminating. Most test makers strive for an index of discrimination of .20 or higher for each item. Obviously, a negative index of discrimination would be unacceptable. In fact, when this happens, the question needs to be examined closely to see whether something in the wording is throwing off the high scorers.

Item Response Theory

classical test theory (CTT)

A measurement theory built on the concept that observed scores are composed of a true score and an error score.

Most of the information concerning the validity, reliability, and item analysis presented thus far pertains to what is called **classical test theory (CTT).** There have been some radical changes recently in the study of the measurement of cognitive and affective behaviors. The advance that has received the most attention in the educational and psychological literature is **item response theory (IRT).**

Characteristics of Item Response Theory

item response theory (IRT)

A theory that focuses on the characteristics of the test item and the examinee's response to the item as a means of determining the examinee's ability; also called *latent trait theory.*

latent trait theory

See *item response theory.*

In CTT, inferences are made about items and student abilities from *total* test score information, whereas item response theory (IRT), as the name implies, attempts to estimate an examinee's ability on the basis of his or her responses to test items. CTT requires only a few assumptions about the observed scores and the true scores of individuals on a test. Group statistics regarding the total score on a particular test for the total group being examined are used to make generalizations to an equivalent test and population. The estimate of error is assumed to be the same for all individuals.

IRT is based on stronger assumptions than CTT. The two major assumptions are unidimensionality and local independence. Unidimensionality means that a single ability or trait is being measured. This ability is not directly measurable, so IRT is sometimes referred to as **latent trait theory.** For an introduction to IRT and some of its applications, see

Spray (1989). According to Spray, the real advantage of IRT is that the measurement of an examinee's ability from responses to test items is not limited to a particular test. Rather, it can be measured by any collection of test items that are considered to be measuring the same trait.

In CTT, item difficulty is measured as a function of the total group. In IRT, item difficulty is fixed and can be assessed relative to an examinee's ability level. Thus, the probability of an examinee with a particular ability level making a correct response to an item can be mathematically described by an **item characteristic curve (ICC).** The ICC is a nonlinear regression for any item that increases from left to right, indicating an increase in the probability of a correct response with increased ability, or latent trait. **Parameter invariance** means that the item difficulty remains constant regardless of the group of examinees. The item's discriminating power is indicated by the steepness of the curve. The ICC can be used to analyze the difficulty of the item, the discriminating power of the item, and a so-called guessing parameter.

Application of Item Response Theory

Space limitations (and lack of knowledge by the authors) prohibit a detailed description and discussion of IRT. It is not a simple concept, and complex computer programs are required. Large sample sizes are needed for item calibration and ability estimates. The IRT model has been the subject of intense research in psychology and education for several years. It definitely has potential application for assessment problems in physical education, exercise science, and sport science.

Spray (1989) described several ways that IRT can be used: item banking, adaptive testing, mastery testing, attitude assessment, and psychomotor assessment. **Item banking** is the creation of large pools of test items that can be used for constructing tests that have certain characteristics concerning the precision of estimating latent ability. **Adaptive testing,** sometimes called tailored testing, refers to selecting items that best fit (are neither too difficult nor too easy for) the ability level of each individual. This function must be done on a computer by using items drawn from an item bank. Some of the widely used standardized tests such as the Graduate Record Examination use adaptive testing.

In criterion-referenced measurement, tests are constructed that use a cutoff score to show the proportion of items that should be answered correctly to represent mastery of the subject matter. Item response theory can be used to select the optimal number of items that yield the most precise indication of mastery for an examinee. In other words, individuals of different ability levels would require different numbers of items.

Item response theory has considerable potential application for assessing attitudes and other affective behaviors. Models of IRT have been proposed that estimate the attitude or trait parameter of each respondent on an interval scale regardless of the ordinal nature of the scale. A score for each trait level is available for each category of each item. Changes in attitude or other traits over time can also be assessed with IRT models. To date, few affective measuring instruments have been constructed using IRT procedures. In our field, Tew (1988) used IRT methods in the construction of a sport-specific test of mental imagery.

The potential use of IRT for psychomotor assessment has been postulated (Spray, 1987) but has not yet been implemented to any extent. Motor performance tests are different from written tests in terms of the numbers of items and trials. Also, some assumptions (particularly local independence) of IRT are not easily accommodated in psychomotor testing. Preliminary research on the application of IRT to motor performance has been conducted by Safrit, Cohen, and Costa (1989). More research will undoubtedly be done on the application of IRT models to our field.

Summary

In this chapter we discussed the concepts of validity and reliability of measurements and how they apply to research. Criterion validity (which includes both concurrent and predictive validity) and construct validity are two of the most popular methods of validating measures

item characteristic curve (ICC)

Nonlinear regression curve for a test item that increases from left to right, indicating an increase in the probability of a correct response with increased ability, or latent trait.

parameter invariance

A postulate in item response theory that the item difficulty remains constant regardless of different populations of examinees and that examinees' abilities should not change when a different set of test items is administered.

item banking

The creation of large pools of test items that can be used for constructing tests that have certain characteristics concerning the precision of estimating latent ability.

adaptive testing

Selecting test items that best fit the ability level of each individual; also called *tailored testing.*

used in research studies. One problem often identified regarding predictive measures is population or situation specificity. The value of using expectancy tables was discussed.

The topic of test reliability has prompted hundreds of studies and innumerable discussions among test theorists and researchers. The reason for this interest is that a measure that does not yield consistent results cannot be valid. Classical test theory views reliability in terms of observed scores, true scores, and error scores, and the coefficient of reliability reflects the degree to which the measure is free of error variance. The rationale for using intraclass R instead of r for reliability and the computational procedures for intraclass R were presented. Various methods of estimating reliability were mentioned, such as stability, alternate forms, and internal consistency. An understanding of the concept of standard error of measurement is vital. Any test score should be viewed as only an estimate of an individual's true score that probably falls within a range of scores.

Standard scores allow direct comparisons of scores on different tests using different types of scoring. The basic standard score is the z score, which interprets any score in terms of standard deviations from the mean.

Research studies in physical activity frequently use instruments involving written responses to measure knowledge or affective attributes such as attitudes, interests, emotional states, and psychological characteristics. More recently, sport- and exercise-specific affective measures have been developed.

Responses to items on these scales often use the Likert-type scale (a five- or seven-point scale ranging from "strongly agree" to "strongly disagree") and the semantic differential scale (a seven-point scale anchored at the extremes by bipolar adjectives). When using scales like these, researchers must be aware of specific problems such as rating errors, leniency, central tendency, halo effects, proximity errors, observer bias errors, and observer expectation errors.

Knowledge testing requires item analysis techniques to ensure objective responses. The difficulty and discrimination of each test item must be established to determine the quality of the knowledge test.

Item response theory (IRT, also called latent trait theory) differs from classical test theory. It allows item difficulty to be fixed so that the participant's ability level can be determined. Item response theory is useful in tailoring tests for adaptive testing, mastery testing, attitude assessment, and psychomotor assessment.

✔ Check Your Understanding

1. Briefly describe two ways that evidence of construct validity could be shown for either a motor performance test (such as throwing), a test of power, or a test of manipulative skill. How could criterion validity be shown in your example?

2. Find a study in a research journal that used a questionnaire (e.g., an attitude inventory). Describe how the author reported the reliability of the instrument scores. What is another technique that could have been used to establish reliability?

3. A girl received a score of 78 on a test for which the reliability is .85 and the standard deviation is 8. Interpret her score in terms of the range in which her true score would be expected to fall with 95% confidence.

PART III

Types of Research

*If you have always done it that way,
it is probably wrong.—Charles Kettering*

■   ■   ■   ■   ■   ■   ■

Research may be divided into five basic categories: analytical, descriptive, experimental, qualitative, and creative (the last category includes art and dance but is not presented in this book). Chapter 12 discusses historical research, which is a type of analytical research that answers questions through the use of past knowledge and events. Substantial changes have occurred in recent years in the reporting of historical research, particularly regarding the integration of the events of interest with other related events. In this chapter, Nancy Struna provides an excellent overview of historical research methods. Note that this chapter uses University of Chicago humanities style for notes because that is the style most commonly used in reporting historical research.

Chapter 13 describes another type of analytical research that can be applied to physical activity: philosophic methods. Scott Kretchmar does a fine job of explaining and grouping philosophic research methods and uses various examples to identify the strengths and weaknesses of these methods. Chapter 14 presents research synthesis (specifically meta-analysis), another type of analytical research that focuses on the shortcomings of the typical literature analysis. Meta-analysis is a more useful solution in analyzing a large body of literature. Although several other types of research are considered analytical, historical research, philosophic research, and meta-analysis are the most common and useful in the study of physical activity.

Chapter 15 discusses descriptive research and focuses on survey techniques. Then in chapter 16, other descriptive techniques, such as case studies, correlational studies, developmental studies, and observational studies, are presented. In general, descriptive research shows relationships among people, events, and performances as they currently exist.

Chapter 17, written by Barbara Ainsworth and Chuck Matthews, provides methods in a relatively new type of descriptive research in our field: exercise epidemiology. Often this approach uses large databases from various sources on exercise and health behaviors.

Chapter 18 introduces experimental research, which deals with future events or the establishment of cause and effect. Which independent variables can be manipulated to create change in the future in a certain dependent variable? After reviewing how difficult cause and effect are to establish, we divide the chapter according to the strengths of various designs: preexperimental, true experimental, and quasi-experimental. Our purpose is to demonstrate which designs and principles are best suited for controlling the various sources of invalidity that threaten experimental research.

Finally, chapter 19 presents qualitative research techniques, which are increasingly used in the study of physical activity. The assumptions underlying qualitative research differ from types of research that adhere to the traditional scientific method. This does not mean that qualitative research is not science (i.e., systematic inquiry); however, the techniques for acquiring and analyzing knowledge differ from the typical steps in the scientific method. This chapter discusses differences in the quantitative and qualitative paradigms, procedures used in qualitative research, interpretation of qualitative data, and theory construction.

If you are a producer or a consumer of research in physical activity, you must understand accepted techniques for solving problems systematically. The following eight chapters attempt to provide the basic underpinnings of research planning. Many research types reported here are closely associated with the appropriate statistical analyses previously presented. We refer to the appropriate statistics as we discuss the types of research. Use the knowledge gained from the previous section to understand this application. You should learn about the relationship between the correct type of design and the appropriate statistical analysis as soon as possible.

Historical Research in Physical Activity

Nancy L. Struna
University of Maryland

*Often it does seem a pity that Noah and his party
did not miss the boat.—Mark Twain*

History is the systematic examination and explanation of change, or the lack of it, over time in human affairs. In our field, the phrase *human affairs* means virtually anything having to do with human movement and the body. Indeed, the subfield that scholars commonly refer to as the "history of sport" focuses on systematic examinations and explanations of an array of practices, including sports, health, the body, sports medicine, exercise, recreation, and leisure.

In pursuing the evidence and the meaningfulness of change or persistence in the behaviors and attitudes of humans toward sport, health, leisure, and so on, historians operate similar to, but not exactly like, scientists. First, they do much of their research in distinctive laboratories: libraries, archives, and historical societies. Second, in these laboratories they work to identify and then to analyze patterns in the "evidence," which is our equivalent to scientists' "data." Third, as good scientists do, good historians expect to construct meaningful generalizations from the historical evidence, or data. Generalizations are large, synthetic statements that offer a historian's sense of multiple pieces of evidence, or data. They present one's interpretation of the data, the information from and about a set of experiences in or across time. Fourth, many sport historians draw on theory, as this chapter discusses more fully in the next section. Some scholars also test theory against their evidence and use theory as a basis for generalization. Finally, sport historians are, like scientists, attuned to scholarship in other disciplines, and one result is that the history of sport is increasingly interdisciplinary. Indeed, in particular departments, history of sport is comfortably located in larger scholarly categories such as sport studies and cultural studies.

Research Paradigms

Good history and good science share one other characteristic: a meaningful paradigm, or framework. A paradigm is an intellectual device that contains a scholar's beliefs and assumptions about the world, the past, and the evidence; her or his conceptions, or definitions, of theory and data; and the questions he or she pursues. A paradigm operates, in very simple terms, like a picture frame. It contains and shapes particular elements and guides the researcher. It also separates one's beliefs, assumptions, theories, and views from alternative ones. A historian who adheres to one paradigm will not ask the same questions as someone who adheres to another paradigm (or framework, perspective, or school of thought), nor will these two historians end up telling the same "story."[1]

When you read the research literature today, you will undoubtedly see a number of paradigms. One early framework drew from modernization theory and was effectively employed by Allen Guttmann in his classic *From Ritual to Record: The Nature of Modern Sports.*[2] The modernization paradigm, which could also be called a theory and has a predictive component, maintains that "progressive increases in per capita productivity set in motion" a whole series of changes common to a range of human institutions and activity.[3] After defining the characteristics of modern sports, Guttmann proceeded to find evidence of these characteristics in mid-nineteenth-century records. Then, to explain how and why these particular forms of sport emerged as they did in the nineteenth century, he examined processes and changes in the cities where modern sports emerged and saw evidence of the increases in productivity as described by the modernization framework (or theory). In the end, he argued, the existence *in time* of particular economic "drivers" (the per capita increases) *and* the emergence of modern sports was not coincidental; the former accounted for the appearance of the latter.

Especially in the last decade, the modernization paradigm has ceased to be as popular and intellectually powerful as it once was. In part, this is so because sport historians have found evidence of things that the modernization theory or framework does not consider or explain. For example, one of the characteristics of modern sports, specialization, appeared earlier than Guttmann acknowledged, in the mid-eighteenth century.[4] Consequently, sport historians have turned away from the modernization framework toward other frameworks.

A second highly influential paradigm draws from British and European social and cultural historical traditions. Although it lacks a simple title, this framework might be understood as the "human agency" paradigm, because it elevates people to the status of makers of their history rather than viewing them in more reactive ways (to conditions) as modernization theory does. One powerful early work in this vein was Elliott Gorn's *The Manly Art: Bare-Knuckle Prize Fighting in America,* which suggested that traditional forms of bare-knuckle prize fighting expressed the values and relations of working-class men in New York City and that this sport was one of the few venues in which they could exert their agency.[5] In the last decade and a half, a number of historians have reshaped and focused this larger paradigm, with the effect that there are now several offshoot paradigms that emphasize human agency and social processes. These include cultural hegemony and social or cultural constructionism and owe much to the British-born field of cultural studies.[6] Much of the recent scholarship on the gendering and racializing of sport, on the historical experiences of women and of people of color, on institutional hegemony, and on the social and cultural contextualizing of sport and leisure employs these paradigms.[7]

Since the 1990s, another European-developed framework has penetrated sport history scholarship in the United States: postmodernism. Few North American sport studies scholars have adopted all of the tenets of postmodernism or its linguistic source, poststructuralism. Poststructuralism and postmodernism developed in the second half of the twentieth century, contributed to an array of scholars such as Roland Barthes, Pierre Bourdieu, Jacques Derrida, Frederic Jameson, and Michel Foucault. At the risk of overreduction, poststructuralism (often referred to as the method) and postmodernism (the aesthetic) reject positivism, absolute objectivity, and master narratives (or single, universal stories and explanations), among other things. They emphasize the significance of language (through which humans represent thoughts and ideas and thus not only make sense of things but also make things), the fragmentation and unpredictability of social life and cultural forms, and power (knowledge is power) and power relations (those who control knowledge have power).[8]

In much of today's sport history (and sport studies, more generally), the influence of the poststructuralist/postmodern paradigm is clear. Few scholars assume that master narratives are adequate or possible; rather, particular sports or people will have particular histories. *Discourse* and *deconstruction* are commonly used words in historical analyses, with the former referring to language and often meaning a "conversation" and the latter serving as a method of analyzing language. Questions about power and power relations, which also draw from materialist paradigms, abound.

Few sport historians today would probably say that they "belong" to this or that paradigm. Most of us blend elements from various frameworks, but we do work at defining our assumptions, views of evidence, and questions. And all of this is work! Defining an integrated and logically consistent paradigm is an arduous task, and students who are just beginning to think about historical research will probably not frame their research as completely as they would like. Most historians readily admit that it takes them a long time to achieve a clear and consistent framework, and this is one of the reasons they maintain that they become good historians only in the second half of their careers. So I introduce paradigms in part to help students think about the things that historians think about and in part to help them realize not only that beliefs and assumptions do bear on the doing of history but also that every scholarly methodology (which is derived from a paradigm) is also an ideology.

Lines of Inquiry and Topics

For most students, the real beginning of "doing history" involves choosing a topic, and about the only limit on possible topics for historical research today is one's imagination! Your interest will, and should, likely be the trigger. What do you want to do? What do you want to know? What interests you? These are questions that all students should ask themselves.[9]

Historians of sport, exercise, and leisure in the United States have made great strides in exploring a variety of topics, but we have only really "broken the surface" on most topics. The particular history of baseball, for example, has an extensive literature, but many dimensions of baseball experiences in the past remain un- or underexplored. These topics include local teams, the experiences of women and minorities, struggles between professional players and owners, and the international transmission and adaptation of the game. Other organized sports such as professional football, basketball, and hockey remain to be even more fully explored, as do the many sports about which few histories have been written: rodeo, water and winter sports, and more recent forms such as in-line skating. There are also volumes yet to be written about collegiate athletics beyond the Ivy League, industrial recreation programs, a broad range of sport business enterprises, and sport policies in both the public and the private sectors.

Research is also needed on the history of health, leisure, and public recreation. We have only just begun to construct histories of sports medicine, the use of drugs to enhance human physical performance, and health programs. Indeed, little is known about the latter beyond the nineteenth century. Scholars have also begun to turn, or return, to historical studies of leisure experiences and to examine the experiences, both in and out of sport activities, that people who have not received adequate attention—women, European ethnic groups, African Americans, Asian Americans, Native Americans, and Hispanics—have had and found meaningful. Histories of the relationships between the media and sport organizations, the relationship between government agencies and sports groups, sports spectatorship, gender relations and the gendering of sports, and health interests and programs, among other things, are also waiting for interested and well-prepared students of history.

The history of sport and physical activity is also a global enterprise, and it is expanding in subject matter and depth of understanding even as you read this. Importantly, as well, this global spread is making both extra-U.S. and comparative historical topics more significant than was the case even a decade ago. There are some fine books about sports and leisure in Africa, South America, and Asia—but not enough.[10] The same statement applies to Australia and New Zealand, which have become hotbeds of sport and leisure history. Comparative historical research that examines two or more countries, regions, or peoples simultaneously, on the other hand, has received little attention, especially from historians in the United States. Consequently, this is a wide-open enterprise that may be just right for a curious and ambitious student!

Secondary Sources

Where does a student go to find out whether anyone has investigated any of these topics or what historians think we know about a given topic? To the secondary sources, or the existing literature, which is the student's real beginning point, as it is for all researchers. The secondary literature includes the books, articles, and other media (such as films or tape recordings) that are histories, in contrast to the *primary* or firsthand sources that are the actual evidence.

Today historians of sport, leisure, kinesiology, physical education, and exercise science have a vast array of secondary literature upon which to draw. In fact, we actually have several literatures. First, of course, is the *direct* literature, which consists of monographs, anthologies, journal articles, and films about the history of sport, health, and so on.[11] There is also an ever-widening body of *ancillary* literature. This literature derives from numerous fields, such as history, anthropology, sociology, the various "area" studies (including American studies, African American studies, Asian American studies, Native American studies), women's studies, public policy, political science, and management, among others. Individual topics may require a student to examine the literatures in medicine, law, and other fields also.

Students may not know precisely what all of their secondary literatures are until they have begun to "dig" for background reading. People generally start their study of history with an interest area, some thing or person or era about which or whom they wish to know more—the amorphous *topic*. An adviser will undoubtedly suggest some books and articles for a student to read and understand in order to begin to realize what is already known about this topic. Graduate advisers will also encourage students to do precisely what they do to locate the literatures: use the computer searches to proceed systematically through various literature banks. Not too many years ago, students searched for secondary literature through the various published indexes, such as *America: History and Life, Education Index,* or *Social Science Abstracts.* Today, however, students can define a string of key words and search multiple online databases. A student may also communicate via e-mail with working historians about topics they have investigated and secondary sources they find important.

What does one take from this secondary literature? The simple answer is *as much as possible.* At one level, students read as much as possible on a topic so that they know what other scholars do and do not know about that topic. At another level, they read the literature to understand both the broader society that affected the subject and the frameworks and theories that will bear on the research and their questions and interpretations. Finally, students read the literature to uncover sources of evidence. In historical research, evidence is discussed in the text, but the actual identification and location of that evidence usually appears in the notes and appendixes. Consequently, historians often study the notes as long and hard as they do the body of an article or book.

History students' eventual success or failure often hinges on how thoroughly they are grounded in the literature. This literature will assist them in shaping the framework, or paradigm, that directs and constrains the research. It will guide them to some sources of data and ways of interrogating the evidence. And perhaps most important, at least as they begin to investigate the past, the literature will help students refine and transform the original, amorphous topic into a researchable question.

Questions

All research is about answering questions, and on this point history is exactly like science. But one issue critical to the life of the historian may raise the stake of the question in historical research: the issue of objectivity. There are three parts to this issue. One involves what historians really do: Do they recapture or replicate the past, or do they admit that nothing can be known for certain and what they must do is to "tell a story" about the past? Not too many years ago, most historians positioned themselves on one side or the other of this particular debate.[12] Either they were objective, or they were not. The *nots* were known as *subjectivists,* or relativists. Today, few of us see the choices in such stark terms because we see a third possibility: We are attempting to make sense of some fragment of the past.

A second part of the objectivity issue involves what one believes about the evidence. Do you believe that there is evidence about the past that is "real" (it exists independently of your mind) that anyone else can make sense of? To this question, a doctrinaire post-structuralist would say no, which may be why few of us are doctrinaire poststructuralists. Instead, most historians seem to believe that there is evidence of real people doing things that were real to them, of institutions, of processes, and so on, about which we can get some information.

The third dimension of the objectivity issue involves one's approach to research. Do you start your research already knowing the answer to your question (you want to prove something) and therefore ignore anything that does not support your argument? In other words, are you trying to prove a hypothesis, or a guess? Or, are you willing to explore all the evidence, to look for additional evidence, and to anticipate both evidence and patterns that you did not expect? Most working historians today would probably admit that they will be more or less objective, or *relatively objective,* on all three matters. They would also agree that beginning with questions helps them operate with more rather than less objectivity—or, in simple terms, as an open-minded detective.

How does a student develop a *good* beginning question, and what criteria can one apply to distinguish a good question from a poor one? On the general matter of questions, the French historian Henri-Irénée Marrou offered some sage advice: "There is an unlimited number of different questions to which the documents can provide answers." However, he cautioned, one must ask each question "properly."[13] And that *proper* questioning of the past begins with the secondary sources. At the very least, a good question is one that is grounded in and derived from the literature. The *unknown* it focuses on is clear, as is the significance of the question. But a good question is something more, as well: It is also answerable.

Let us work through an example that moves us from the vague topic to the good beginning question(s). The topic is the women's high school athletic association in a state, say Louisiana. In the beginning the student knows only that the organization first appeared about 1925. The topic itself suggests the background reading, because every word in that topic has a literature: social histories of the 1920s (and subsequent decades), women's and gender histories, histories of organized athletics, histories of schools and education, and Louisiana history. From all of this, the student learns nothing about the particular athletic organization but much about the issues concerning competition for girls and women, about occupational and social expectations and roles for women, about other organizations that women formed, and about the changing curriculum, population, and context of high schools in Louisiana.

The logical transformation of topic to questions can now occur. Who were the organizers, and what goals did they have? What activities and what kinds of competitive formats did they incorporate? When did they introduce different or additional sports? What were the relationships of athletic competitions to other physical education, health, and recreational programs? What was the structure of the association, and what were the relationships of the association to other governing structures within the school and community? What were the community discourses in which athletics were meaningful? Questions such as these will be precursors to some *bigger* questions: How and why did the association change over time? What accounted for the change(s)? In what ways did the planning and actual operation of the association affect power relations between supervisors and participants? How did the association affect the broader student and school life, including gender, race, and class relations? What were other consequences of the association in and over time, and what was the significance (meaningfulness) of the association for women's athletics and for women more generally in Louisiana?

Designing the Research

Once a student has replaced the vague topic with specific beginning questions, he or she can begin to construct a research design. This is not a phrase historians frequently use, but we do *design* our research, even if such designs lack the formulaic precision found in experimental designs. A design in historical research is precisely what it is in scientific research: a systematic, even hierarchical, layout of the questions and a plan for answering them. One needs to know what questions to ask, in what order to ask them, what evidence one needs to answer the questions, what evidence is actually available, and how to treat the evidence (precise qualitative procedures, quantitative procedures, or both).

Not too many years ago, historians also used the terms *descriptive history* and *analytic history* in talking about research design. In fact, earlier editions of this book did precisely that. However, historians have come to realize that neither these categories nor those distinguishing "narrative" and "analytic" history are particularly useful, in part because differences between them are matters of degree rather than kind. Narrative history emphasizes a "story," but writing this story requires both description and explanation, as well as synthesis. Analytic history, on the other hand, tends to stress the explanation for some phenomenon or practice. Analytic histories also usually employ all three intellectual processes—description, analysis, and synthesis—and may focus on a theme, as opposed to a full "story."[14]

Working With the Evidence

One thing that has not changed in recent times in the "doing" of history is the importance of the evidence. Only from the *evidence*—or, in scientists' terms, the *data*—can the historian construct answers to the questions that she or he has posed. Moreover, the evidence that one identifies and uses will affect the increasingly specific, but invariably more complicated, questions that one asks. Located in *primary sources,* or firsthand accounts, historical evidence is a major determinant in the research process.[15]

What is historical evidence, or data? In *Truth in History,* Oscar Handlin offered a simple and clear definition: Evidence is "everything made or recalled."[16] Each "thing" is a piece of the past, and such pieces come in many forms—artifacts such as equipment and clothes; photographs and, increasingly, movies and digital material; oral tales and reminiscences; numerical evidence such as box scores, prize money, and census accounts; and literary materials such as letters, laws, manuscripts, and newspapers. The evidence that a student will use depends, of course, on the questions that he or she is asking. The questions also determine whether a student uses qualitative or quantitative, or both, forms of and means of treating the evidence.[17] However, about one thing there is no "depending": There can be no answers to questions without evidence. Just as a physiologist must have raw data about heart rate to answer questions about cardiovascular condition or the psychologist must have records of brain activity to study hemispheric dominance, the historian must have information created and constructed by subjects in the past.

Locating Primary Sources

Before a student can actually begin to work with evidence, he or she needs to identify the sources that may contain the evidence. To locate where historical sources exist—usually in distinct kinds of laboratories such as archives, libraries, or privately run collections—the student can proceed in a number of ways. If one is working on a particular sport, recreational facility, curricular program, or any other project that is locale specific, one may go directly to the catalogs or indexes in the holdings of the local historical society (state or town), to the college or university archives, to a hall of fame, or to an organization's archives and library. Increasingly, organizations are digitizing and making their holdings available via the Internet or on CD-ROM. Students will also find that many newspapers, censuses, and government records have indexes that aid in locating potential information, and there are many published topical bibliographies and catalogs of sources.

Particularly for students who are interested in recent history, other forms of historical evidence consist of oral and visual sources. Unlike many of us, some historians have the luxury of working with people who are very much alive and well, or they are examining movements and events that have been filmed or recorded in other ways. These contemporary "subjects" have lived through experiences and processes and can provide firsthand accounts. Although many of the suggestions this chapter makes about doing history are applicable to oral and visual history projects, these kinds of projects have particular issues that are not addressed in this chapter. Interested students should search out books and articles on oral and visual history research methods available in most university libraries.

Historical Criticism

Once we have located some primary sources, even though the identification of all the sources will probably not yet be complete, we begin to criticize the sources. This is an evaluative process that has two steps. The first step determines the form of the material. In this step we ask whether a given artifact or document is really a source of evidence about the past. Called **external criticism,** this phase of the evaluation establishes the *authenticity* of the source. We have only to think about Watergate or the Iraqi prisoner mistreatment affairs to realize why the historian must establish a source as an authentic witness to the past. For any number of reasons, people in the past—who were the makers of historical evidence—could, and have been known to, forge documents. More common than outright forgeries, however, are the unintended errors that resulted when time or another person intervened between the actual occurrence of an event and the recording of it. So historians must either know how to establish that a piece of evidence is really a record of what it purports to be—through a technique such as carbon-14 dating for very old artifacts or through textual and stylistic analysis for documents—or they must rely on experts to perform these tests. The simple point, of course, is that a piece of evidence must be the genuine article![18]

The second kind of test that historians apply to firsthand records is called **internal criticism.** This examination deals with the nature of the source. It asks specifically whether a now-determined genuine artifact or document is *credible*. Internal criticism thus involves matters of consistency and accuracy; at one level, internal criticism is comparable to the

external criticism

Establishing the authenticity of the source.

internal criticism

Establishing that the source is credible.

That's her idea of historical research.

scientist's efforts to establish validity.[19] Just as the scientist must establish the adequacy and accuracy of a test, instrument, or construct, so the historian must determine both whether an observation or some other form of record is believable and the context and perspective of the record. Historians who use quantitative data and sampling techniques will actually use the same techniques to establish validity as biophysical scientists do. Investigators who rely on more traditional historical data sources, especially literary documents, cannot approach the mathematical precision of validity coefficients, but they do ask appropriate questions systematically. In fact, there are rules for a student to apply in the process of internally criticizing a document. One is the "rule of context," which maintains that a word must be understood in relation to the words that precede and follow it—and not according to the historian's own contemporary usage. Reading sentences and paragraphs requires the same sequence and connotation guidelines. A second guideline is the "rule of perspective," which encourages the student to ask who (or what agency or group) left the record, what the source's relationship to an event or group was, and even how the source collected the information. Finally, there is the "rule of omission or free editing," which holds that most historical sources—whether official records, newspaper accounts, diaries, or formal agreements—are not accounts of complete scenes. The minutes of an NCAA meeting, a photograph of a home run, a diary entry by someone such as Albert Spalding or Senda Berenson, and even a census or map record some things and leave out others. Diarists may omit things that happened or words that were spoken in the course of a day, and a meeting or newspaper report may use different words, leave out entire sections of a report, or even purposefully misrepresent conversations and actions.[20] Thus, because sources may leave information out, whether intentionally or not, the historian needs to use more than one source to view any event or program or person in the past—in much the same way as a biomechanist will use three cameras to view one movement from different angles.

Reading the Evidence

Locating and criticizing historical sources are important tasks that all historians perform over and over again. They are not, however, ends in themselves. A newspaper, for example, is a source of information about the past, just as an EMG record is a source of information about muscle contractions. Like the EMG firing pattern, the newspaper text or picture requires a "reading." In short, a student needs to know what information the newspaper, or any historical source, contains.

To accomplish this, the student will ask a relatively simple question of each piece of information: What is this evidence evidence of? Consider a practical example—a health and physical education curriculum from, say, 1950. Most students in our fields are familiar with curricula; they lay out goals, activities, and learning outcomes. But about what or whom can a curriculum really inform us? Is it evidence of what went on in a classroom or gymnasium? Is it evidence of all of the thoughts about health and physical education that the writers may have held? Is it evidence of the state of health and physical education in the country? The answer, of course, is "none of the above." Insofar as it was written in behavioral terms, the health and physical education curriculum is nothing more, or less, than evidence of a set of expected behaviors. Moreover, a curriculum is evidence of behaviors *expected* by the writers rather than of those for whom the curriculum was written.

Students will see that all historical sources provide evidence that is narrow in scope once they ask what the evidence is evidence of. Typical newspaper accounts of a ball game, for example, provide us with a portion of the reporter's observations, rather than a comprehensive picture of all that occurred in the contest. The same is true of diaries and letters. Rule books provide information about what behaviors should and should not occur, but they do not tell us what behaviors actually did or did not occur. Likewise, numerical evidence can inform us of only certain things. A salary figure, for instance, informs us only of dollars paid to a player, not what the value of the athlete was, and salaries across time are not even easily comparable until they have been adjusted for inflation. The number of fans at an event tells us how many people were present, rather than whether the event was popular or even important in any large sense.[21]

This process of determining what the evidence "is evidence of" thus really has two goals. The more obvious of the two is knowing what information we *can* obtain from any given piece. Less obvious but no less important is knowing what information we *cannot* obtain from a source. A student must come to grips with both sides of the coin to determine whether the evidence is *adequate and appropriate*. If one finds that the evidence meets both criteria, then one can proceed to answer the research questions. If it is not, however, one has a big problem and cannot proceed with the original questions. This situation might have occurred with our health education curriculum—had we expected it to be evidence of behaviors in the classroom. It is also a situation that can, and does, occur when historians ask questions about beliefs, attitudes, values, processes, and conditions. Many of our common forms of evidence, whether literary or numerical, provide data about what people did or said rather than about what they believed. Consequently, unless historians make some hard-to-establish assumptions about the relationships between what people saw or did and what they believed, they will find that most historical sources do not provide appropriate information about a person's or a society's beliefs and attitudes.

Several options are available for resolving this dilemma of having the wrong (or inadequate) evidence for the questions one wants to asks. First, students can change the questions. Second, they can retain the questions and search for other, appropriate sources of evidence. Finally, they can *translate* what is evidence of one thing, usually behavior, into evidence of another thing, such as beliefs or other concepts. This is an inferential process that requires the careful construction of *indicators* and the testing of relationships between the "thing" that we cannot observe or find direct evidence of and the "thing" that we think will indicate or implicate it. It is also a process about which historians can learn much from their peers in science, who frequently investigate what they cannot directly observe. Physiologists, for example, examine cardiovascular fitness, a condition that cannot be directly observed or measured. To do so they have developed and validated an inferred indicator: expired air. In studying aggression, psychologists have proceeded similarly. They cannot observe and measure aggression directly, but they can observe and measure an inferred indicator: violent acts.

Constructs and inferential relationships are important to historians who are looking to examine concepts, values, or attitudes that cannot be directly observed. Social class is such a concept. It is also a complex of behaviors and attitudes about which no one piece of evidence is direct evidence. Historical sources do, however, provide numerous and appropriate indicators from which evidence of class can be inferred when taken all together: occupation, income, goods purchased and used, club membership, and so on. Indicators and inferences are also essential to the study of beliefs and values, as in the case of research on the beliefs of members of health clubs. A student may suspect that particular beliefs—in good health, fitness, even beauty—may have influenced, or perhaps even motivated, people to join a given club. The hypothesis is not directly testable, however, because of the nature of the evidence. All of the records of the club and even those of the members—daily logs, accounts, equipment used and purchased, exercise regimens, even diaries—are readable only as evidence of what the members did and not what they believed. Again, the student must infer a relationship between the observed and the unobservable and, of course, *test it*.

In part because of the critical role of inferences in historical description and explanation, other information about a person, an event, or a process becomes essential to the historian. This other information will often be in sources that may at first seem tangential to a student's major sources. In the previous example of the research about beliefs of health club members, one might not think that evidence about members' occupations, the other activities and groups in which they participated, their eating habits and medical histories, and their neighborhoods, among other things, would be particularly important. All of these are important, of course. Only with this large body of evidence can a student test the inferences about beliefs that he or she has produced from the health club behavior data. If the student projected a belief in fitness as one of the factors, perhaps even a "motive" for a subject's membership in the health club, he or she must "test" that belief in other behavioral situations, the evidence for which lies in this other information. If no such evidence is forthcoming, or if there is even counterevidence (e.g., if the subject drank heavily, ate high-cholesterol food, or did not regularly see a doctor), the student will have to look for a "new" belief within the health club behaviors. In effect, this seemingly tangential evidence did not bear out the inferred relationship between behavior and belief.[22]

Context

Seemingly tangential evidence is important in historical research for another reason as well. It is a part of the *context* in which a historical subject lived or a historical process occurred and in which sport, recreation, or health experiences and processes need to be placed if they are to be understood. Context refers to the "ensemble" of data about a person, event, or period. Rather than "background" material, in other words, context means the network of facts and meanings in which a given piece of historical evidence existed. The term is so important in historical research that one scholar has defined the field with the term: History is the "discipline of context."[23]

Context figures in many tasks that the historian performs. Students may recall the "rule of context" discussed in the section on internal criticism. They may also recognize the allusions to context in the paragraphs on indicators and inference. But it is in the reading of evidence that the importance of context becomes most obvious, as one extended example should reveal. The source here is a typical literary document, the *American Turf Register and Sporting Magazine,* which was an urban journal published in the 1820s and 1830s. The particular article, "The Great Foot Race" (on pages 225-227), is an account of a pedestrian race held at the Union race track on Long Island in June 1835. Beyond the kind of race and its site, the piece contains information about the distance covered (10 miles), the numbers of competitors (nine), some descriptive information about the runners (name, age, weight, occupation, dress), the order in which they ran, the prize money ($1,300), and the times for each mile achieved by the winner.[24]

From this evidence, we can construct a description of the event. Nine men, ranging in age from 22 to 33 and of various manual occupations, competed against one another over a distance of 10 miles on what was normally a horse track on a Friday in June 1835. The winner, Henry Stannard, covered the distance in the prescribed time (just under an hour) and won the purse of $1,300 in the presence of a crowd estimated at between 16,000 and 20,000. With any certainty, this is all the information we can "read" in this account.

The rest of the information in the article is in bits and pieces and generally in the form of judgments and suggestions by the reporter. For example, the first sentence of the piece introduces the event as the "great trial of human capabilities." We cannot accept this statement at face value; the phrase may have been a literary device employed by the writer to interest readers. The report also mentions that a 10th runner, Francis Smith, had appeared but did not run because he had not entered his name "by a certain day." Such a "reason" may or may not be true; the runner was a black man who may have been excluded on the grounds of race. The report also indicates that another man, John Cox Stevens, was particularly active in the race; he even rode around the track with Stannard. But Stevens' exact role remains unclear. Finally, we cannot be certain that the purse represented the total of Stannard's winnings, or that he was the only one who "won" money. The reporter also noted that betting, even by the runners, was a part of the scene.

So beyond the earlier "bare-bones" and not very telling description of this event, we cannot proceed from this piece. As a consequence, many fairly simple questions must go unanswered. Was pedestrianism common? Was the $1,300 purse typical or "a lot"? Was this really a "Great Race"? Was Smith prevented from running because of his late entry or his race? Was the crowd large? Answers to these questions hinge on a student's willingness to read the evidence in relation to, and in conjunction with, other pieces of evidence. The first question, for example, requires data on other races—in fact, a counting of them and a numerical comparison to other sport and public events in this period. The second question can only be answered by a comparison of this purse to other purses during that period and to the annual incomes of similar working men. The question of crowd size requires a major shift in the evidential base, from qualitative to quantitative evidence. Descriptive statistics from both sport and nonsport crowds are essential; inferential statistics would

The Great Foot Race

The great trial of human capabilities, in going ten miles within the hour, for $1,000, to which $300 was added, took place on Friday, on the Union Course, Long Island; and we are pleased to state, that the feat was accomplished twelve seconds within the time, by a native born and bred American farmer, Henry Stannard, of Killingworth, Connecticut. Two others went the ten miles—one a Prussian, in a half a minute over; the other an Irishman, in one minute and three quarters over the time.

As early as nine o'clock, many hundreds had crossed the river to witness the race, and from that time until near two, the road between Brooklyn and the course presented a continuous line, (and in many places a double line) of carriages of all descriptions, from the humble sand cart to the splendid barouche and four; and by two o'clock, it is computed that there were at least from sixteen to twenty thousand persons on the course. The day, though fine, being windy, delayed the start until nineteen minutes before two, when nine candidates appeared in front of the stand, dressed in various colors, and started at the sound of a drum.

The following are the names, &c. of the competitors, in the order in which they entered themselves:

Henry Stannard, a farmer, aged twenty-four years, born in Killingworth, Connecticut. He is six feet one inch in height, and weighed one hundred and sixty-five pounds. He was dressed in black silk pantaloons, white shirt, no jacket, vest, or cap, black leather belt and flesh colored slippers.

Charles R. Wall, a brewer, aged eighteen years, born in Brooklyn. His height was five feet ten and a half inches, and he weighted one hundred and forty-nine pounds.

Henry Sutton, a house painter, aged twenty-three years, born in Rahway, New Jersey. Height five feet seven inches; weight one hundred and thirty-three pounds. He wore a yellow shirt and cap, buff breeches, white stockings and red slippers.

George W. Glauer, rope-maker, aged twenty-seven, born in Elberfeldt, Prussia. Height five feet six and a half inches; weight one hundred and forty-five pounds. He had on an elegant dress of white silk, with a pink stripe and cap to match; pink slippers and red belt.

Isaac S. Downes, a basket-maker, aged twenty-seven, born at Brookhaven, Suffolk county. Height five feet five and a half inches; weight one hundred and fifty pounds. He was dressed in a white shirt, white pantaloons, blue stripe, blue belt, no shoes or stockings.

John Mallard, a farmer, aged thirty-three, born at Exeter, Otsego Co., New York. Height five feet seven and a half inches; weight one hundred and thirty pounds. Dress, blue calico, no cap, shoes, or stockings.

William Vermilyea, shoemaker, aged twenty-two years, born in New York. Height five feet ten and a half inches; weight one hundred and fifty pounds. Dressed in green calico, with black belt; no shoes or stockings.

Patrick Mahony, a porter, aged thirty-three, born in Kenmar county, Kerry, Ireland. Height five feet six inches. Weight one hundred and thirty pounds. Dress, a green gauze shirt, blue stripe calico breeches, blue belt, white stockings, and black slippers.

John M'Gargy, a butcher, aged twenty-six, born at Harlaem. Height five feet ten inches. Weight one hundred and sixty pounds. Dressed in shirt, pink stripe calico trousers, no shoes or stockings.

There was a tenth candidate, a black man, named Francis Smith, aged twenty-five, born in Manchester, Virginia. Mr. Stevens was willing that this man should run; but as he had not complied with the regulation requiring his name to be entered by a certain day, he was excluded from contesting the race.

The men all started well, and kept together for the first mile, except Mahony, who headed the others several yards, and Mallard, who fell behind after the first half mile. At the end of the second mile, one gave in; at the end of the fourth mile, two more gave up; in the fifth, a fourth man fell; at the end of the fifth mile, a fifth man gave in; during the eighth mile, Downes, one of the fastest, and decidedly the handsomest runner, hurt

(continued)

225

The Great Foot Race   *(continued)*

his foot, and gave in at the termination of that mile, leaving but three competitors, who all held out the distance.

The following is the order in which each man came up to the judges' stand at the close of each mile.

Miles

	1st.	2d.	3d.	4th.	5th.	6th.	7th.	8th.	9th.	10th.
Stannard	3	4	3	3	3	2	2	1	1	1
Glauer	2	2	1	1	2	3	3	3	2	2
Mahony	1	1	5	5	5	4	4	4	3	3
Downes	5	3	2	2	1	1	1	2	gave in.	
M'Gargy	6	7	7	7	4	gave in.				
Wall	4	5	4	4	gave in.					
Sutton	8	8	6	6	gave in.					
Mallard	9	9	8	8	fell and gave in.					
Vermilyea	7	6	gave in.							

The following is the time in which each mile was performed by Stannard, the winner. Mahony, the Irishman, did the first mile in five minutes twenty-four seconds.

	Min.	Sec.
1st mile	5	36
2d mile	5	45
3d mile	5	58
4th mile	6	25
5th mile	6	2
6th mile	6	3
7th mile	6	1
8th mile	6	3
9th mile	5	57
10th mile	5	54
	59	44

The betting on the ground both before and after starting, was pretty even, and large sums were staked both for and against time. Downes was undoubtedly the general favorite; and was well known in the neighborhood; he did the eight miles in forty-eight and a half minutes; he had been well trained under his father, who in his thirty-ninth year, performed seventeen miles in one hour and forty-five minutes; accomplishing the first twelve and a half miles in one hour and fifteen minutes.

Mallard was known to be an excellent runner; he had performed sixteen miles in one hour and forty-nine minutes, stopping during the time to change his shoes. He was not sober when he started, and he fell in the fifth mile.

The German had performed the distance between New York and Harlaem, and returned thence (twelve miles) in seventy minutes; his friends were very sanguine of his success. He betted nearly $300 that he would win the prize. He was within the time until the sixth mile, and he performed the ten miles in one hour and twenty-seven seconds. He was four seconds behind time in the eighth mile. Part of the distance he carried a pocket handkerchief in his mouth.

Mahony, the Irishman, had undergone no training whatever; he left his porter's cart in Water street, went over to the course, ran the first mile in less than five and a half minutes; at the end of the sixth mile he was one minute and a quarter behind; at the end of the eighth mile two minutes behind; at the ninth he was three minutes behind, and he performed the ten miles in sixty-one and three quarter minutes. On the 25th of last month, this man ran eight miles in forty-one minutes fifty-six seconds. M'Gargy was out of condition; but he did the five miles in thirty-two and a half minutes. Vermilyea was very thin and in a wreched state of health; he travelled thirty-eight minutes on foot, on Tuesday last, to be here in time to enter, and the next day performed eight miles in forty-six minutes; he is an excellent runner, but gave in at the end of the second mile from a pain in the side; he was also thrown down by a man crossing the course in the first mile. Wall and Sutton ran remarkably well, but gave in at the end of the fourth mile for want of training.

Stannard, the winner, we understand, has been in good training for a month. He is a powerful stalwart young man, and did not seem at all fatigued at the termination of the race. He was greatly indebted to Mr. Stevens, for his success; Mr. S. rode round the course with him the whole distance, and kept cheering him on, and cautioning him against over-exertion in the early part of the race; at the end of the sixth mile, he made him stop and take a little brandy and water, after which his foot was on the mile mark just as the thirty-six minutes were expired; and as the trumpet sounded he jumped forward gracefully, and cheerfully exclaimed "Her am I to time;" and he was within the time every mile. After the race was over, he mounted a horse and rode round the course in search of Mr. Richard Jackson, who held his overcoat. He called up to the stand and his success (and the reward of $1,300) was announced to him, and he was invited to dine with the Club; in which he replied in a short speech thanking Mr. Stevens, and the gentlemen of the Club for the attention shewn to the runners generally throughout the task. After this, it was announced by Mr. King, the President of the Jockey Club, that the German and the Irishman, who had both performed the ten miles, though not within the time, would receive $200 each.

We are happy to state that none of the men seemed to feel any inconvenience from their exertions; every thing went off remarkably satisfactory, nor did we hear of the slightest accident the whole day. After the foot race was over, a purse of $300, two mile heats, for all ages, was run for by the following horses, and decided as under:

	1st.	2d.
Tarquin	1	1
Post Boy	2	3
Columbia Taylor	3	dist.
Rival	4	2
Ajax	5	dist.
Sir Alfred	6	d'rn.

The first heat was performed in three minutes forty-seven seconds—the second in three minutes fifty seconds.

During the running of this match, a written paper was handed to Mr. King, stating that two native Americans were willing to attempt to walk five hundred miles without eating or drinking, as soon as a purse of $500 should be made up.

The day was remarkably fine, but the wind blew very strongly on the course, and considering the vast amount of money (in bets, &c.) at stake, Mr. Stevens felt uncertain at first how to act, and decided to postpone the race; but the general opinion and desire seem to be against any postponement, and he yielded to this. The result on this account was most fortunate. The race was won handsomely; although when it wanted but twenty eight seconds to the hour, bets at five to three were offered, and taken, that the task would not be accomplished. It is certain that if the wind had not been so high, Stannard would have performed the ten miles in fifty-seven minutes.

American Turf Register and Sporting Magazine 6 (June 1835): 518–520.

be more informative. The remaining questions, as well, demand additional data on other sports, events, and social practices of the times.

The significance of this practice of reading the evidence in its full and appropriate context should not be underestimated. Divorced from the network of facts and meanings in which any given piece of evidence about a sport, a health practice, or a recreational program originally existed, the evidence becomes drab and lifeless. To picture such a situation, consider a ball. Unless a student places a ball in the context of a game played by real live people who hit and throw and catch it, it exists only as a round object. Taken out of its context, a ball becomes just another bit of rubber, cloth packing, or even animal bladder. Considering your own life as a graduate student might even make the point a bit more dramatically. Taken out of the context of graduate school, in a particular place at a particular time in your life, much of what you do (such as reading this chapter) would make absolutely no sense to anyone, even to you! So like you, if historians are to achieve their goal of making sense of the past, they must place their subjects in context.

Summary

We now have some of the components for the making of a good history. From the secondary literature, we have derived our research questions: significant ones, *telling* ones, answerable ones. We have laid out the questions in a logical, hierarchical order and identified the evidence needed to answer them. We have located and criticized the sources, and we have spent hours reading and putting the evidence in context. We have, in effect, been both scientist and detective. So, are we finished?

Not quite! Historians are not finished until they *make sense* of whatever piece of the past they have investigated. This is clearly a matter of constructing a case, a theme, or, perhaps more pointedly, an argument. With everything done so far—the questions raised, the evidence read and interrogated—the historian now assumes the guise of both architect and prosecuting attorney.

The construction of a history demands an answer to one final question: So what happened here? Another way of asking this is, So what does all of this mean? Regardless of how one poses it, however, this question needs an answer before a student can begin to write. Like the architect who would not presume to build a house before knowing what the house will look like, the historian would not presume to write the story before knowing its content and meaning.

How does one construct the answer to the big question, So what does all of this mean? I do not believe that there is any single or simple answer to this question, nor is there any intellectual formula. An analogy of the prosecuting attorney may be helpful, however. Like the attorney, the historian is committed to making a case (or interpretation). This goal, in turn, requires laying out the evidence and the interpretations of various pieces of evidence in a logical fashion, which leads to the climax, the "point." When we talk about laying out the evidence, historians usually mean discussing and analyzing the primary material (even in descriptive studies) and pointing out and discussing the patterns of experiences about which particular sources are revealing. We also lay out the evidence, and our interpretations of its meanings, in a sequence that leads to the "end" of the story. No prosecuting attorney would take a case to court without being clear about the sequence of events and which evidence to produce when. So it is with the historian's story, or case. Good historians never "list" sentences within paragraphs or paragraphs in an article or chapter. Each sentence and each paragraph has a logical "fit" in a sequence of sentences and paragraphs.

What determines this sequence? The historian's mind, of course, and the "big" story or argument that he or she wishes to offer, which in turn is the answer to the question(s) asked. And this "story," in turn, moves us to the task of writing, which many people, working scholars and students alike, view as the hardest task of doing history. When I hear my students or colleagues complain about how hard writing is, I usually respond with a short and cryptic "nuts!" It's not the writing that's hard—it's the *thinking*. Too often, many of us try to write before we have labored through all the thinking that good history requires. That thinking must include the answer to the question "so what happened here?" In other words, students need to know the story line, the interpretation, or the argument before sitting down to write. They must, in other words, work out precisely what they are going to write and in what sequence to present the evidence and conclusions so that they lead the reader through the parts to the "big" conclusion.

There are undoubtedly as many ways to work through the "end of the story" and determine the best sequencing of evidence and subarguments as there are historians. A couple of suggestions, however, may be in order. Some people work well with topical outlines, in part because each part of such an outline is really a code for a more extensive thought that they retain in memory. For other people, however, topical outlines are insufficient to prevent the "problem" they have with writing. Single words or phrases are often vague; they fall short of the complete answers that historical questions require. Students in history who find themselves in this second category would be better off spending their time drafting complete sentences or even paragraphs that answer specific questions. In effect, they may find the writing of a history easier after they have already done a lot of thinking (and writing) and can honestly say "this is how the story ends" or "this is the point," as well as "this is how I am going to develop it."

The writing of history deserves a more extended treatment than that presented in this chapter. As a skill that, like most others, takes practice, it actually warrants several volumes of material. But that is not in the purview of this book or this chapter. There are many chapters and books that students can consult about the writing of history.[25] There is not, however, such an extensive literature on historical research especially oriented to graduate students in kinesiology, sport studies, exercise science, and physical education. Hopefully this chapter provides that orientation. Perhaps only time—and good history—will tell.

✔ Check Your Understanding

1. Locate and read a research study that employs the historical method.

   a. Identify the primary sources.

   b. Critique the methods of criticism (external and internal) of the sources that were described in the study.

Notes

[1] A classic discussion of paradigms appears in Thomas Kuhn, *The Structure of Scientific Revolutions* (Chicago: University of Chicago Press, 1962). About the implications of Kuhn's work for history and the social sciences, see Barry Barnes, *T.S. Kuhn and Social Science* (New York: Columbia University Press, 1982); David Hollinger, "T.S. Kuhn's Theory of Science and Its Implications for History," *American Historical Review* 78 (April 1973):370-93.

[2] Allen Guttmann, *From Ritual to Record: The Nature of Modern Sports* (New York: Columbia University Press, 1978).

[3] Joyce Appleby, "Modernization Theory and the Formation of Modern Social Theories in England and America," *Comparative Studies in Society and History* 20 (1978):261.

[4] Nancy L. Struna, *People of Prowess: Sport, Leisure, and Labor in Early Anglo-America* (Urbana: University of Illinois Press, 1996), esp. ch. 6.

[5] Elliott Gorn, *The Manly Art: Bare-Knuckle Prize Fighting in America* (Ithaca, NY: Cornell University Press, 1986).

[6] Richard Johnson, "What Is Cultural Studies Anyway?," *Social Text* (Winter 1986/87):38-80; Stuart Hall, "Cultural Studies: Two Paradigms," in Nicholas B. Dirks, Geoff Eley, and Sherry B. Ortner, eds., *Culture/Power/History: A Reader in Contemporary Social Theory* (Princeton, NJ: Princeton University Press, 1994), 520-38.

[7] See, for example, Troy D. Paino, "Hoosiers in a Different Light: Forces of Change Versus the Power of Nostalgia," *Journal of Sport History* 28 (Spring 2001):63-80; Patricia Vertinsky, "The Erotic Gaze, Violence and 'Booters with Hooters,'" *Journal of Sport History* 29 (Fall 2002):387-94; Michael Ezra, "Main Bout, Inc., Black Economic Power, and Professional Boxing: The Cancelled Muhammad Ali/Ernie Terrell Fight, *Journal of Sport History* 29 (Fall 2002): 413-38; Stephen W. Pope, *Patriotic Games: Sporting Traditions in the American Imagination 1876-1926* (Urbana: University of Illinois Press, 1997); Struna, *People of Prowess: Sport, Leisure, and Labor in Early Anglo-America* (Urbana: University of Illinois Press, 1996).

[8] Roland Barthes, *Elements of Semiology* (New York: Hill & Wang, 1964); Pierre Bourdieu, "The Forms of Capital," in John Richardson, ed., *Handbook of Research for the Sociology of Education* (New York: Greenwood Press, 1986), 241-58; Jacques Derrida, *Of Grammatology* (1967; Baltimore: Johns Hopkins University Press, 1974); Fredric Jameson, *Postmodernism, or, The Cultural Logic of Late Capitalism* (Durham, NC: Duke University Press, 1991); Michel Foucault, *History of Sexuality, Vol. 1: An Introduction* (1979; New York: Vintage Books, 1990). See also Madan Sarup, *An Introductory Guide to Post-Structuralism & Postmodernism* (Athens: University of Georgia Press, 1993). Some classic (all nonsport) essays exist in Keith Jenkins, ed., *The Postmodern History Reader* (London: Routledge, 1997).

[9] One of the most useful research methods books for students working in the humanities, to which sport history is akin, is Wayne C. Booth, Gregory G. Colomb, and Joseph M. Williams, *The Craft of Research* (Chicago: University of Chicago Press, 1995).

[10] Rather than list the books, which are numerous, I refer students to an essay that considers international publications, as well as books about U.S. sport history. See Nancy L. Struna, "Social History," in Eric Dunning and Donald Sabo, eds., *Handbook of Sport and Society* (Beverley Hills, CA: Sage Publications, 2000), 187-203.

[11] Reviews of the literature on the history of sport, leisure, and health can be very helpful to students and working historians alike. None seems to have been completed since the late 1990s, however. Good prior ones include Larry R. Gerlach, "Not Quite Ready for Prime Time: Baseball History, 1983-1993," *Journal of Sport History* 21(Summer 1994):103-37; Roberta J. Park, "A Decade of the Body. Researching and Writing About the History of Health, Fitness, Exercise and Sport, 1983-1993," *Journal of Sport History* 21 (Spring 1994):59-82; Steven A. Riess, "From Pitch to Putt: Sport and Class in Anglo-American Sport," *Journal of Sport History* 21 (Summer 1994):138-84; Jeffry T. Sammons, "Race and Sport: A Critical, Historical Examination," *Journal of Sport History* 21 (Winter 1994):203-78; Patricia Vertinsky, "Gender Relations, Women's History and Sport History. A Decade of Changing Enquiry, 1983-1993," *Journal of Sport History* 21 (Spring 1994):1-58; Stephen Hardy, "Sport in Urbanizing America. A Historical Review," *Journal of Urban History* 23 (September 1997):675-708. For two relatively recent histories of sport history overviews, see Struna, "Social History," which has an international content, and "Sport History," in John Massengale and Richard Swanson, eds., *History of Exercise and Sport Science* (Champaign, IL: Human Kinetics, 1996), 143-80.

[12] This debate commenced over a century ago with the introduction of "scientific history" by the German scholar, Leopold Von Ranke. He maintained that objective history, history "written as it really happened," was both possible and necessary. Wilhelm Dilthey just as clearly stated the opposing position, the subjectivist or interpretivist position, which was predicated on his view that the "historical world was a text to be deciphered." Later relativists raised the level of the debate by establishing a clear role for the present; the past had meanings for and could be used to solve problems in the present. This objective versus subjective issue bears on questions about theory in history and on the nature of evidence. See also John R. Hall, "Temporality, Social Action, and the Problem of Quantification in Historical Analysis," *Historical Methods* 17 (Fall 1984):206-18; Joyce Appleby, Lynn Hunt, and Margaret Jacob, *Telling the Truth About History* (New York: W. W. Norton, 1994).

[13] Henri-Irénée Marrou, *The Meaning of History,* trans. Robert J. Olsen (Baltimore: Helicon, 1967), 76-7.

[14] Description and analysis are most commonly presented in methodology textbooks as categories of historical research. See, for example, Robert Shafer, *A Guide to Historical Method,* 3rd ed. (Homewood, IL: Dorsey Press, 1980). To be sure, narrative history does not mean the same thing as description, but some historians do use the terms almost interchangeably. Narration refers to a mode of historical thought, specifically the "recounting" of something in the past. Narration will, then, take place in the creation of a descriptive history. Both the narrative and explanatory modes of thought may also appear in analytic histories. See Robert F. Atkinson, *Knowledge and Explanation in History* (Ithaca, NY: Cornell University Press, 1978); Maurice Mandelbaum, *The Anatomy of Historical Knowledge* (Baltimore: John Hopkins University Press, 1977); Dale H. Porter, *The Emergence of the Past: A Theory of Historical Explanation* (Chicago: University of Chicago Press, 1981); Allan Megill, "Recounting the Past: 'Description,' Explanation, and Narrative in Historiography," *American Historical Review* 94 (June 1989): 627-53. Some historians also describe synthetic history, which builds primarily on existing histories (secondary sources) to produce overarching generalizations. See Thomas Bender, "Wholes and Parts: The Need for Synthesis in American History," *Journal of American History* 73 (June 1986):120-36.

[15] E.P. Thompson, *The Poverty of Theory and Other Essays* (NY: Monthly Review Press, 1978), 27-28. In Thompson's exact words: "A historian is entitled . . . to make a provisional assumption. . .: that the evidence which he handles has a 'real' determinant existence independent of its existence within the forms of thought, that this evidence is witness to a real historical process, and that this process (or some approximate understanding of it) is the object of historical knowledge." However, he continued, "The historical evidence is there . . . not to disclose its own meaning but to be interrogated" (pp. 28-29).

[16] Oscar Handlin, *Truth in History* (Cambridge: Harvard University Press, 1979), 120. See also Stephen R. Humphrey, "The Historian, His Documents, and Elementary Modes of Historical Thought," *History and Theory* 19 (1980):1-20.

[17] Quantitative evidence, numerical evidence, is slowly but surely finding its way into historical research in kinesiology, exercise science, and physical education. Much more needs to be used, particularly in cases in which historians want to establish conditions, extent of change, and categories such as social class. The most helpful journal about quantitative techniques in history is *Historical Methods*. A useful methods text is Konrad H. Jarausch and Kenneth A. Hardy, *Quantitative Methods for Historians: A Guide to Research, Data and Statistics* (Chapel Hill: University of North Carolina Press, 1991).

[18] One classic case of a document of questionable authenticity that had a long-standing impact on the history of a sport was that of a diagram of a baseball field, presumably authored by Abner Doubleday when he was a schoolboy in 1839. Abner Graves, who claimed to have been a classmate of Doubleday's and an eyewitness to the design episode, included the diagram in a letter to the Mills Commission, which had been set up to establish the "origin" of baseball in 1907. The commission accepted the diagram and Graves' testimony as "proof" that Doubleday had "invented" the game (see Shafer, *Historical Method,* 127-47; Handlin, *Truth in History,* 111-24). For a marvelous, and humorous, discussion of a suspect document and the problems it created, see John D. Milligan, "The Treatment of an Historical Source," *History and Theory* 18 (1979):177-96.

[19] Shafer, *Historical Method,* 149-70; Handlin, *Truth in History,* 124-44.

[20] Shafer, *Historical Method,* 150-158.

[21] Handlin, *Truth in History,* 165-226.

[22] Adrian Wilson, "Inferring Attitudes From Behavior," *Historical Methods* 14 (Summer 1981):143-44.

[23] E.P. Thompson, "Anthropology and the Discipline of Historical Context," *Midland History* 3 (Spring 1972):41-55.

[24] "The Great Foot Race," *American Turf Register and Sporting Magazine* 6 (June 1835):518-20.

[25] Henry W. Fowler and F.G. Fowler, *The King's English* (Oxford: Oxford University Press, 3rd ed., 2003); Savoie Lottinville, *The Rhetoric of History* (Norman: University of Oklahoma Press, 1976).

Philosophic Research in Physical Activity

R. Scott Kretchmar
Pennsylvania State University

*Being a philosopher, I have a problem
for every solution.—Robert Zend*

◼ ◼ ◼ ◼ ◼ ◼ ◼

Philosophic research is considered by some in our field to be a contradiction in terms. This is due to the rise of empirical science, historic relationships of kinesiology to the medical profession, and widespread contemporary doubts about the validity of reflective, reason-based procedures (Kretchmar, 1997, 2005). Some believe that philosophy involves little more than sharing opinions, albeit while using impressively long sentences and incomprehensible words. Philosophers themselves do not agree on the utility of their methods and the validity of their results. Yet with all the accomplishments of science added to traditional antiphilosophic biases of some kinesiologists, and in spite of a degree of disarray within the field of philosophy, nonempirical analysis and informed speculation have not disappeared from the research landscape. If anything, recognition of the need for philosophic insights related to movement activity may have grown over the past two decades (Fahlberg & Fahlberg, 1994; Glassford, 1987; *ICSSPE Bulletin,* no. 27, Fall, 1999; Kretchmar, 2005; Lawson, 1993; Sheets-Johnstone, 1999).

Reasons for this increased interest are tied to the purposes of philosophy and the potentially complementary relationship between philosophic inquiry and empirical study. However, the best way to appreciate the irreplaceable nature of philosophy and its potential significance in affecting human lives is to get involved with it—to experiment with philosophic thinking yourself and enter into a philosophic debate. Consequently, this chapter includes an examination of the purposes of philosophic research and draws attention to the similarities and differences among philosophy, qualitative research, and science. This is followed by descriptions of five general methods and brief analyses of each method's strengths and weaknesses. This preliminary material is designed to prepare you to read an article by Gardner (1989) and participate in a contemporary sports ethics debate that is presented in the "Check Your Understanding" section at the end of this chapter.

Identifying the Purposes of Philosophic Research

A fundamental goal of philosophy is to examine reality by using reflective procedures, not the empirical tools of science. Accordingly, philosophers and scientists differ not so much on what they look at but rather how they study it. Both kinds of researchers are interested, for instance, in understanding exercise. Empirical scientists, however, approach this phenomenon by looking through microscopes at such things as muscle tissue, by collecting respiratory or blood pressure data, and by employing statistical procedures to determine the strengths of possible causal relationships. Philosophers, on the other hand, reflect on exercise and use such things as ideas and ideals, meanings, lived experience, values, logical relationships, and reasons in attempting to shed some light on this phenomenon.

This distinctive approach allows philosophers to answer questions that empirical methodologies cannot. For instance, after data on exercise have been gathered and analyzed by chemists, physiologists, sociologists, psychologists, and historians, questions still remain about the human meanings and values associated with this phenomenon. Why should we exercise? To live longer? To live better? Both? If both, which is more important, the presence of more life per se or the existence of a certain quality of life? On what criteria would we determine quality of life, and if we can identify some of them, what are the in-principle or logical relationships between exercise and those criteria of good living?

Philosophic research is needed not because empirical methodologies are ineffective but because they are, on their own, incomplete. For example, it has long been a hope of science to uncover the various mechanisms that govern natural processes and human behavior. If this can be done, future events and behavior could be predicted and some of them might even be controlled.

Considerable progress has been made along these lines. In the realm of exercise, a better empirical understanding of various physiological mechanisms, coupled with more sophisticated information on exercise as a physiological stressor, has helped us to predict and control the outcomes of an active lifestyle. But human exercise is a very complex event, and this prediction and control has been tenuous at best. This is particularly so when prediction involves a complete, thinking human being in the natural world rather

Let me guess . . . philosophic research?

than an isolated physiological system in a controlled laboratory setting. Some would say that this lack of predictability is due to the current state of empirical science and that it is only a matter of time before a far more impressive degree of control is achieved. But many philosophers would argue that complete prediction is, in principle, not possible no matter how sophisticated the methods of empirical science become (Merleau-Ponty, 1964; Midgley, 1994; Ridley, 2003; Sheets-Johnstone, 1999; Wallace, 2000).

The reason the predictability and control of exercise are so tenuous is that exercise is an electrical-chemical-biomechanical-meaningful event. To be sure, linear or chaotic electrical and chemical reactions help us understand exercise, but only to a degree. And linear or dynamical biomechanical rules shed light on various mechanisms of exercise, but again, only to a degree. Inescapably (because it is people who exercise, not mere machines) there is also meaning, poignancy, interest, hope, lived experience, and idiosyncratic perception in the exercise event. These things play a role in behavior, and they cannot be fully appreciated by looking at them electrically, chemically, biomechanically, or through any other strictly empirical window. There may be no getting around it. Reflective techniques are needed to measure and analyze meanings and values as they are encountered by actual, live people. Once again, empirical methodologies, on their own, produce research findings that are incomplete.

Whether philosophers are analyzing the nature of concrete things such as chairs and footballs or intangible things such as determination and fair play, they must bring their subject matter to mind reflectively. Consequently, their data, when it is being described, logically analyzed, or otherwise worked on, must be present to consciousness. It must reside in what behaviorists popularly termed "the black box." This metaphorical box was colored black because it was thought to be impenetrable.

Skinner recommended that the CNS (central, or conceptual, nervous system) be ignored because it could never be understood well enough to be of use. Many contemporary psychologists acknowledge the importance of cognition, but they still try to reduce experience to physics, chemistry, physiology, or computer theory (Dennett, 1991; Hamlyn, 1990). Even though they dare to venture inside the black box, they often refuse to allow meanings to be just meanings! As did many of their empirical predecessors, they refuse to take lived experience and its ideas seriously.

Among those researchers who take consciousness seriously are those who use qualitative methodologies. (See chapter 19 in this volume.) However, qualitative researchers and philosophers deal with the subjective domain in very different ways. Typically, qualitative scholars rely on some form of systematic data gathering—observation, interviews, conversations, interpretations of cultural artifacts. Although philosophers are not prohibited from using such information, they typically do not gather data in these ways, nor do they usually have any reason to do so. They can begin their research with an idea, an exemplar, a theoretical problem abstracted from experience.

It is true that both qualitative researchers and philosophers make judgments, evaluations, or interpretations that cannot be demonstrated, proven, or even reduced to a level of statistical probability. However, qualitative work aims at making claims about what an actual group of people, for instance, really thought or believed. Philosophers are less interested in making such attributions, particularly in claiming they know what went through someone else's head at any certain time and place. Philosophers of physical activity argue about what might be thought about exercise, what should be thought about the value of health, what makes sense when comparing exercise to sport, what contents could be in aesthetic experiences of "flow" in weightlifting, what reasons can lie behind good ethical behavior related to taking drugs in sport, and what human intelligence, in general, is capable of negotiating.

Philosophy, however, should not be considered simply an alternative or complement to science. Increasingly, scholars from both measurement and nonmeasurement domains are finding common ground where the findings of all levels of research—from the micro to the macro, from particle physics to subjective meaning—are needed to provide a more complete picture of human nature and behavior. Philosophers and ethicists, for instance, are increasingly relying on research from genetics, biology, and evolution to more fully explain the nature of human intellection and our ambiguous capabilities as moral creatures (see, e.g., Midgley, 1994; Singer, 1995; and Wilson, 1993).

Conversely, contemporary mathematicians, computer scientists, and neuro-physiologists more regularly refer to meaning when trying to explain how the brain is wired and how its capabilities exceed those of even the most sophisticated "smart machines." Some are inclined to conclude that meanings and other abstractions are realities that we must factor into our understanding of how "meat" (the brain) produces nonphysical ideas (meanings), and in turn, how the brain and our genes are modified by those ideas. (See, e.g., Pinker, 1997, 2002; Ridley, 2003; and Wallace, 2000.)

Philosophy's overall mission, it can be concluded, is to examine reality through reflective techniques, typically in the absence of systematic empirical data gathering and usually without an interest in making claims about what specific people actually thought, believed, or felt in real-life circumstances. In addition, philosophic research both complements the findings of science and interacts with them. Philosophers are now invited to join interdisciplinary teams that attempt to unravel the mysteries of human existence that lie at subcellular levels, thoughtful levels, and everywhere in between.

This very general characterization of philosophic research must now be fleshed out in some detail. Accordingly, it should be useful to identify several subdivisions or branches of reflective inquiry and describe the work accomplished by each one.

- Metaphysics
- Axiology
- Epistemology
- Poetry

metaphysics

A branch of philosophy that addresses the nature of things.

axiology

A branch of philosophy that addresses the value of things.

epistemology

A branch of philosophy that addresses securing knowledge and understanding its status.

poetry

A branch of philosophy that deconstructs truth claims.

In **metaphysics** philosophers attempt to analyze the *nature of things*. The scope of this work runs the gamut from simple distinctions (e.g., telling sport from dance) to more complex and controversial theses (e.g., describing the nature of excellence in competitive events). It also may portray categories as cleanly defined with sharp edges, or alternately as less distinct . . . perhaps with fuzzy or blurry lines of demarcation.

Work in a second branch of philosophy called **axiology** focuses on the *value of things*—on achievements, acquisitions, or states of affairs such as fitness, health, knowledge, and excellence (theories of nonmoral value), on human behavior such as the breaking or bending of game rules (theories about ethics), and finally on art and beauty such as the qualities of certain routines in gymnastics (aesthetic research). The debate at the end of this chapter falls into the second subdomain of axiology—that of ethics.

A third branch of philosophy, **epistemology,** is about *securing knowledge* (How do we come to know things?) and *understanding its status*. (How authoritative is knowledge and by what criteria can we determine its credentials?) Research is conducted on logic (e.g., determining the clarity and force of any apparently fixed relationships between cooperation and competition in sport), on sense perception (e.g., describing the importance of one's perspective in the lived experience of a dance, say, as a performer in contrast to a spectator), and on the status of reason itself (e.g., deciding if we can trust what appear to be reasonable arguments for allowing or disallowing the use of performance-enhancing drugs in sport).

A final branch of philosophy has emerged in the postmodern era, one that some are calling **poetry** (Rosen, 1989). You may encounter this work under the heading of deconstruction and often in the context of political philosophy. Paradoxically, poetry is an antiphilosophic movement designed to liberate people from the claims of traditional metaphysics, ethics, and epistemology while hinting at alternatives. The liberating aspect of this research draws attention to the possible contamination of reasons given for traditional claims about, for example, the nature and value of physical activity. The hinting function tentatively suggests ways that understandings and values of exercise might be expanded. It could be, for instance, that some notions of health and the ideal human body are promulgated by dominant cultures under the guise of truth, and that broader understandings are warranted. It is possible that some definitions of femininity have had unnecessarily restrictive effects on participation in sport by women and girls and that more liberating understandings are needed.

This brief discussion has produced five reasons for doing philosophic research. The overarching rationale points to the fact that philosophers look at reality with different tools than those used by empirical scientists. They reflect rather than actually measure and count. As a result they see aspects of things, actions, and behavior that are inaccessible to those who work only with real images and data from measurements. As researchers who reflect on reality, and similarly to those who use qualitative methods, they take meaning or cognition itself seriously and are reluctant to reduce lived experience to any of its electrical, chemical, and biological antecedents or concomitants. However, unlike these qualitative researchers, they do not typically collect data (or record observations) in any systematic way and are usually not interested in making claims about actual instances of subjective activity. A recent rise of interest in philosophy, as noted previously, may stem from a growing understanding that any failure to take meaning seriously will make it difficult if not impossible to understand fully (let alone predict and control) whole-person behavior. Moreover, many modern notions of brain functioning (e.g., Searle, 1980)—and even some that rely on the metaphor of mental computing (e.g., Pinker, 1997)—require references to, and depend on the substantial reality of, meaning!

Four additional purposes of philosophic research include the need to understand the nature of things and the differences between them (metaphysics); the highest values of life, proper ways of behaving, and the qualities of art and beauty (axiology, including ethics and aesthetics); the avenues by which humans know things and the foundations on which such knowledge rests (epistemology); and the dangers of building grand philosophic systems combined with tentative suggestions about what should guide life in their stead (poetry).

Now you are ready to consider the tools or methods needed to accomplish these purposes. Space is not sufficient here to survey the full range of techniques employed in philosophy, nor is it possible to review all the strengths and weaknesses of these diverse approaches. In fact, it is far more important to use them yourself than to merely read or talk about them. Consequently, this discussion is designed only to prepare you to engage the exercise at the end of the chapter with some foreknowledge.

Locating a Research Problem

Philosophic problems are not difficult to find. They are literally everywhere. You cannot wake up in the morning without being confronted with potential philosophic issues. What is worth doing today? What is the good life? How should I go about my work or play? Why should I be ethical—particularly if everyone else is cutting corners and there is little chance of getting caught?

If you are breathing, conscious, and have even a modicum of curiosity, philosophic questions will find you! You will probably not even need to go out looking for them (Kretchmar, 2005).

Difficulties come, however, in getting a handle on these questions and problems. Because they usually do not involve physical things that can be turned around in your hand or repositioned under a microscope, they need to be manipulated reflectively. And just as you cannot look at too much under a microscope at one time, you cannot reflect on too much at any one moment—particularly if it is vague or poorly defined.

Good philosophic research therefore begins with definitions, descriptions, clarifications, and disclaimers, particularly when the object of analysis is too large or excessively vague. In the article highlighted in the appendix at the end of the chapter, you will see that the author began in just this way. Gardner notes that performance-enhancing drugs might be prohibited for any number of reasons, but he wants to examine only one of them in depth—the issue of unfair advantages. In addition, many different kinds of drugs exist, and their effects vary considerably. This too needs to be cleared up before substantial progress can be made. Consequently, Gardner provides definitions, descriptions, and disclaimers at the start of his article to clarify the core issue he will investigate.

This process can be done well or poorly, and some philosophic research is scuttled before it even gets started. If clarifications and definitions are too aggressive, you may get the sense that researchers are begging the question. That is, they want you to assume too much; they provide too few arguments for their position. On the other hand, if this preliminary clearing of the field is not done aggressively enough, you may be able to discount key findings because the researchers have not thought about extenuating circumstances or, worse yet, because they have not clearly defined what they are talking about.

Analyzing a Research Problem

Philosophers, much like researchers in other fields, do not always agree on the best techniques for analyzing reality, and their disagreements can be very serious ones. For this reason this chapter introduces several methods, and you will need to choose the techniques that seem most useful to you.

In the broadest terms, philosophic methods can be divided into those that are ambitious and those that are more limited in scope. These two positions reflect conflicting judgments about the twin poles of all reflection—(1) the thinking or reflecting itself (the research act) and (2) the thing thought about or reflected on (the research object).

Ambitious philosophers are impressed by human powers of reasoning and believe that common biases and potentially limiting perspectives of thinking (e.g., from history, socialization, language, religion, and sense perception) can be controlled. Philosophic acts, in other words, can be objective and dispassionate or at least partially so. According to this view, logic, faithful description, and even careful speculation can lead to accurate conclusions with universal appeal. Some researchers in this optimistic school of thought (e.g., Husserl, 1962) believe that philosophic thinking can approach, or perhaps even surpass, the precision of science.

Pessimistic researchers, on the other hand, doubt the powers of reason, see thinking as contaminated, and regard most traditional philosophic conclusions as limited, useless, or even harmful. Some of these researchers regard philosophic work as witting or unwitting rationalizations for unjust economic systems, unfair gender relationships, or individually enslaving value systems. All thinking, according to this view, proceeds from historical-, political-, gender- and language-skewed perspectives, and no methods exist for effectively eliminating or controlling these biases.

Dramatically different judgments have also been made regarding the second half of reflection—the object thought about or reflected on. The more ambitious school of philosophy sees the world as composed of discrete classes of items—for instance, chairs, tables, automobiles, games, play, dance, good ethics, bad ethics, and so on. Each of these things has a nature that distinguishes it from other things around it. If biases in the process of reflecting can be avoided, these and other categories of things that populate the world can be described over time ever more faithfully and completely. In other words, philosophers discover (more than create), and genuine philosophic progress is regarded as possible.

The more cautious philosophers, on the other hand, challenge hypotheses about the neat packaging of reality into separate classes of items. Objects in the world, according to this view, are individuals; there are no essential natures of things—no neat lines that can be drawn between abstract categories such as "chairness" and "tableness," games and play, or good moral behavior and its counterpart. Rather, human beings create their world; they draw lines of distinction to suit themselves and then redraw them when old ideas no longer work. Philosophic progress makes little sense for these researchers, although it is still far better to understand that philosophic classes are arbitrarily defined parts of a continuous reality of individuals than to invent comfortable or convenient fictions.

Optimism and pessimism about the range and significance of philosophy fall across a number of techniques—induction, deduction, description, speculation, and critical reasoning. These tools, in other words, can be used aggressively with high hopes or more timidly with lesser expectations. However, the methods of inductive and deductive thinking and the strategies employed by descriptive and speculative philosophers are, generally speaking, representative of the more ambitious school of philosophic research. Critical and poetic techniques tend to be used by philosophers who distrust such ambition.

Inductive Reasoning

Inductive reasoning is thought that moves from a limited number of specific observations to general conclusions about the thing or class of thing that was observed. It relies on intelligent discernment to identify common elements or similarities in the specific samples.

inductive reasoning

See page 28.

If we were doing some metaphysics on the nature of competition, we might select a number of events that fall safely within this realm—say, contests in baseball, racquetball, swimming, spelling, and Trivial Pursuit. We want to reason inductively from these five particular exemplars of competition to general statements about them—statements that accurately describe these events, and especially, all (genuinely competitive) past and future activities that were not included in the sample.

Can we say anything about the number of parties or sides that must be involved in competition? These five examples suggest that at least two sides must be present for competition to take place.

Can we say anything about the nature of the activity? These five examples suggest that participants all face tests or problems. That is, what they are doing is not easy. It also appears that the two or more parties are facing the same kind of problem—for instance, hitting and catching baseballs in rule-defined ways, recalling "trivia" under a time limit in certain specified categories, and so on.

Can we say anything about the commitments, if any, made by the two or more competing sides? The exemplars would indicate that each party intends to solve its rule-defined problems in a superior fashion to any other teams or players in the game. For example, baseball players intend to score more runs than their opponents do; Trivial Pursuit players intend to fill their disk with wedges before anyone else does.

In short, the inductive process has provoked a number of claims about the nature of competition—that competition requires two or more parties, that these sides face problems, that these problems are common to all, and that each side must intend to solve their problems better than the other team(s) or player(s) do. If these statements are valid, we now know something explicit about competition in general, something perhaps about which we were mistaken before, sensed only implicitly, or simply had never thought about.

Inductive reasoning has its assets and liabilities. On the positive side, this method places something concrete in front of the philosopher. It is a procedure that is not mystical, that uses data that everyone can see or reflect on. Also, because exemplars are limited in number, inductive reasoning is manageable. Although the inductive process might go on indefinitely (how would we ever know that we had finished?), the process of examining a finite number of items for common features or common elements is not technically difficult even though it can be done with differing degrees of success.

Inductive thinking is a commonsense approach. In daily life we seem to deal with classes of things quite naturally and without difficulty (Pinker, 1999). In spite of the fact that no two instances of competition are exactly alike, we have no difficulty identifying or producing the conditions for it. Clearly we deal easily and regularly with things such as competition on an abstract level. By examining particulars inductively, we can better appreciate these generalizations or abstractions that have heretofore remained in the background.

Finally, this technique is useful in distinguishing what is essential to the makeup of some phenomenon and what is accidental or unnecessary. Suppose we had a group of exemplars that included four team sports and one individual activity. Focusing only on the team sports at first, we might hypothesize that competition requires that groups of players face other groups of players. However, when we come to the individual sport, we see that the team-versus-team phenomenon cannot be an essential feature of competition. We can then conclude that it is an option or possibility, not a requirement. We have made some progress in weeding out peripheral characteristics of our theme from central ones.

Inductive reasoning is not foolproof. Three particular pitfalls can be mentioned here. The first involves the procedure by which the original list of exemplars is chosen. If this list is faulty (for instance, if it is incomplete or includes one or more incorrect examples of the thing in question), it can produce incorrect conclusions. As we already saw, had we begun our analysis of competition with a list that included only team sports, we might have concluded inaccurately that competition requires groups of people to face off against one another.

Additionally, induction is based on certain assumptions. First, we assume that future exemplars will not contradict the generalizations drawn from past ones. Also, we assume a certain stability in the world. The examination of examples of competition, for instance, suggests that valid tests are a prerequisite. That is how the world works today. And we believe that is how the world will work tomorrow. If we are wrong about this, our generalized conclusion about relationships between valid tests and competition may be wrong too.

The third problem involves the process of seeing abstractions accurately and insightfully. This is a skill that can be done with different degrees of finesse and insight. For instance, some of the concepts of the nature of competition identified earlier may seem relatively obvious and thus not particularly helpful. In truth there is nothing inherent in the inductive process that guarantees that the abstractions generated will be good ones, that is, concepts that tell us important and heretofore unnoticed things about competition or whatever else is being examined.

Deductive Reasoning

deductive reasoning

See page 29.

Deductive reasoning is a companion technique of inductive reasoning, and many philosophers adroitly and spontaneously intermix the two. Deduction requires intellectual movement in the opposite direction from induction. Whereas inductive reasoning has you working from particulars to general abstractions, deduction has you start with general claims to see what particulars follow.

These general claims or premises are of two sorts. The first are statements of fact and are often phrased, "Because such and such is true, then it follows that . . . ." But premises can also be hypothetical: "If such and such is true, then it follows that . . . ." We will take a look at a deductive line of reasoning that uses hypothetically stated premises based on our inductively generated concepts about the nature of competition.

If competition requires two or more parties to take a test, and

If these two or more parties must face the same test, and

If these two or more parties must be committed to passing the common test in a superior fashion, and

If taking and facing tests and being committed to superior performances are acts that can be undertaken only by conscious beings, and

If mountains are not conscious beings, then

It follows that humans, as conscious beings, cannot compete against mountains. That is, people can neither defeat mountains nor lose to them in contests.

This conclusion may be of some interest because certain successful climbers have been quoted as saying things like, "I defeated the mountain." For many of us, such competitive-sounding statements make at least some sense. However, if the previous line of deductive reasoning is valid, we now know that such loose claims cannot refer to victories in contests. Their meaning has to lie elsewhere, perhaps in the direction of passing a very severe or dangerous test that was provided, in part, by the mountain. Once again, reflective philosophic procedures (deductive ones, in this case) allowed us to clear up things about which we may have been mistaken, sensed only implicitly, or never thought about.

An important strength of deduction is the promise it holds for obtaining sure or certain knowledge. Some deductions necessarily follow from their premises. As long as the premises are sound in these cases, the conclusions must also be true. Our previous example may come close to providing this kind of forceful conclusion. If our premises about competition, conscious beings, and mountains are valid, then it makes no sense to talk literally of people competing against, defeating, or losing to mountains. As mentioned, one might still speak metaphorically of a contest with a mountain, and this might carry some meaning having to do with struggle, danger, success, and failure. However, the fact remains that claims about winning against mountains are literally nonsense.

A second advantage of deduction is its ability to permit speculations on the basis of unproven premises. We do not need to await confirmation of uncertain information. We can assume that it is accurate, and then proceed to see what follows. Researchers who take this tact, of course, need to be frank about the tentative status of their premises. If later research confirms them, their deductions will stand. If premises are eventually shown to be inaccurate, then new deductions will need to be drawn. You will see later, in the appendix, that the validity of certain premises is absolutely crucial when Gardner's arguments are framed deductively.

Deduction also embodies some weaknesses. First, uncontested premises are difficult to find. Because of this fact alone, most deductive conclusions need to be held tentatively. Second, errors can be made in deduction itself even if the premises are allowed to stand. For instance, some people make errant ethical claims about competition based on faulty deductions. Their reasoning may go something like this:

> If competition requires two or more parties to face the same test with the intent of performing better than the other(s), and
>
> If competition produces (save for occasional ties) scores that symbolize better and lesser performances, and
>
> If at least one party must be identified with the lesser performance (i.e., as having lost),
>
> Then it follows that competition harms one or more parties and is, at least for that reason, morally suspect.

Although there may be some vague persuasive force to this line of reasoning, the deductive conclusions here do not necessarily follow from the premises. Something crucial is missing, and that "something" is related to important assumptions about losing, harm, and morality. Minimally, two more premises are needed—a statement to the effect that losing necessarily brings harm to people, and a claim that the magnitude or quality of such harm is sufficient to raise moral concerns.

Descriptive Reasoning

Some philosophers prefer to conduct their research in less roundabout ways than those provided by inductive and deductive logic. They do not line up a series of particulars and ask themselves what they all have in common (induction), nor do they begin with a set of givens or premises to see what follows (deduction). They simply *describe* what they see reflectively when they examine an object. This may strike you as disarmingly simple and, in a sense, it is.

However, the processes of looking carefully and describing accurately require considerable skill, and like all skills they can be done well or poorly. Because human beings are inattentive, tend to jump to conclusions, fail to notice details, take things for granted, and often confuse parts and wholes, accurate, insightful, and careful description is not particularly easy. When it is done well, it can be extremely useful and enlightening.

As we have already seen with our very cursory analysis of competition, even though most of us have experienced it countless times in a variety of settings, we still may not have thought much about it. We can identify competition when we encounter it, but we may be at a loss to describe it much beyond a superficial level. Imagine, however, the important things still to be discovered about competition, things that cannot be examined with the tools of empirical research but must be seen reflectively and then described.

Competition is often understood to involve disagreement, aggression, controlled fighting, and striving for something that only one person or team can have but that both or all parties desire. Certainly there is more than a little truth to these descriptions, but if you are to be a good philosopher, you must look more closely. Does competition involve cooperation? If cooperation is present, is it accidentally or necessarily there? Does it exist, for instance, only in ethically praiseworthy examples of sport or in all instances of competition? Does cooperation play any crucial role in allowing competition to happen? For instance, can two or more sides in a game face the same test without cooperating in some meaningful sense?

Initially, competition seems to be associated with attempts to put other people down, to show their inferiority, perhaps even to "blow them out" or otherwise embarrass them. But might competition be compatible with friendship or even love? Philosophers would need to look more carefully and describe faithfully what they see to determine whether competition has a structure that permits opposing sides to uplift one another. If this is possible, in what precise ways does competition allow this to happen? For instance, if one side in a contest provides an interesting, challenging test for the other side, could that be seen as a helpful service? Is this act of providing a challenging test compatible conceptually with what friends might do for one another? When competition goes well, say when it is mutually uplifting, what pushes it in this direction rather than in a negative one? Is there any sense in which losing is not a putdown? If so, what precisely is that sense? Is it possible, for example, to play well and still lose? What exactly does it mean to play well if it does not mean "to win"? The philosopher must look skillfully for concepts that effectively explain this. Optimistic philosophers believe that, if this research is done well, conclusions will coincide with findings produced independently by others who wonder about the same issues.

Descriptive methodologies place a great deal of confidence in the capacity of reason to portray reality accurately. It is important to recognize that this confidence resides in different traits and skills of human intelligence depending on the level of description attempted. If description is intended to portray an actual lived experience—with its real perceptions, feelings, ups, and downs—then there must be a strong reliance on good memory, honesty, attention to detail, an unwillingness to embellish facts, and so on. This level of description (similar to that used in qualitative methodologies) is in effect making claims about what someone really experienced, and such portrayals can vary between fact and fiction.

A second level of descriptive work is disinterested in the actual subjective experience itself, such as a peak experience in basketball that occurred last week. At this level, the concern is with the nature of things in principle. This level of description places confidence in the power of reason to notice important differences.

The danger with this methodology is not poor memory, dishonesty, or tendencies to exaggerate actual events but rather with internalized biases that contaminate reason's ability to see clearly. For instance, your religious upbringing might affect the way you observe the world. You may not be able to describe acts of cheating dispassionately and objectively. This, of course, could influence your descriptions of its effect on contests and the values that are available there.

Speculative Reasoning

Speculation can be thought of as an extension of descriptive philosophy. As evidence and argumentation diminish for analyses of either actual subjective events (level 1 description) or the nature of things in principle (level 2 portrayals), the room for speculation increases.

This does not mean that speculation requires no argumentation whatsoever. Once again, speculating can be done well or poorly and, although speculative conclusions cannot be proven or demonstrated, they can vary in their degree of plausibility.

When looking reflectively at competition, we could argue, for instance, that a mutual quest for excellence (e.g., Simon, 2004; Weiss, 1969) is the highest value of contesting; it may, in other words, be the feature that best turns competition from an activity that is, on balance, humanly harmful to one that is helpful or uplifting. Or we might claim that this highest value is a cooperative search for knowledge (Fraleigh, 1984) or a shared experience of play (Hyland, 1990). On the other hand, the best forms of sport might focus on fair play (Loland, 2002) or the development of virtues such as trust and honesty (Clifford & Feezell, 1997; McNamee, 1998) or courageous experimentation with new and sometimes unusual forms of competition (Roberts, 1998). Although many have argued for each of these descriptions of what is good about competition, none of them can be regarded as conclusive. Moreover, no amount of future research will likely change this situation. These value claims are plausible, perhaps even attractive, and additional arguments may be advanced at some future time to increase their plausibility or attractiveness, but no value can be shown conclusively to be the highest one, even though one of them may well be. These axiological claims must therefore be identified as products of speculative philosophy.

The advantage of speculative philosophy is that it broaches subjects that have human significance and have markedly changed human lives. Wars, for instance, have been fought for a way of life that is grounded on the perceived superiority of some values over others—say, democracy over some form of totalitarianism. Accordingly, given the apparent fact that people mold their lives around values, it may well be important to do research on intangibles, even if knockdown arguments are not available.

The research principles here would be twofold. First, it is better to know a little (even if it is inconclusive or incomplete) than to know nothing at all. And second, one can be unable to prove a claim and still be right about it. Some speculate that a spiritual reality of some sort exists, and they provide evidence that, they believe, points in that direction. Others disagree and present their own arguments to the effect that nothing is real but matter and void. Neither side can present knockdown proof. Yet somebody is still right. Or perhaps, in some hard-to-imagine way, both of them are right.

Disadvantages also exist. Narrow, selfish, manipulative, and even mean philosophies can masquerade as legitimate speculative statements about reality. Many contemporary philosophers, as you saw, have made this claim about much previous research. Either purposely or unintentionally, some people have spewed forth groundless opinions, biases, and self-serving and status-quo-preserving visions of life in the name of philosophy. Even arguments that are intended to show the plausibility and attractiveness of certain claims may only play to the biased predispositions of the audience. For instance, claims about the superiority of sport when conceptualized as a mutual quest for excellence may seem persuasive in a capitalistic, achievement-oriented culture in which almost everyone is raised to be "all that they can be." In another society, such arguments for excellence might seem odd or misplaced.

Critical and Poetic Reasoning

As you have already seen, critical and poetic reasoning is based on skeptical attitudes toward the power of reason (a position called "relativism") and doubts about the organization of reality (a stance frequently identified as "nominalism"). The rise of empirical science had much to do with the former concerns, those related to philosophic acts of thinking or reflecting. Thought processes were found to be influenced by chemicals, genes, language, and the socializing influence of family, state, and religion, to name only a few factors. Thinking, and its philosophic products, was therefore seen as dependent on these, and a host of other, potential influences. Reason, in short, was no longer thought to be an objective, independent, cross-cultural, and a-temporal accessor of truth.

Related to the demise of reason is the judgment that reality is not as neatly organized as metaphysicians claimed it is. Abstract categories that were once thought to accurately depict reality came under attack—even categories such as "competitive behavior," "games," "play," "exercise," and "sport." These were described as convenient fictions that failed to represent anything that was truly "out there" in the world.

Given this brief introduction to contemporary critical and poetic philosophy and its doubts about the power of reason and the validity of philosophic categories, you might wonder what philosophers thusly persuaded do. What research agenda could there possibly be for those who seem to have eliminated the only means of access to reflective truth (a trust in the powers of reason) and the only subject matter that would be worth scrutinizing (reality as objective, stable, and distinctive)?

Actually, critical and poetic philosophers undertake at least two important research activities. Criticism is a debunking activity intended to show the errors of traditional philosophy. As noted, this is often termed "deconstruction." Analysis attempts to reveal the futility of all metaphysical and ethical system building and points out where claims have been faulty and beliefs misplaced.

Texts are often used by deconstructionists to show where, for instance, groundless value claims have been made. One method is to show where unwarranted shifts in language and meaning have occurred—for example, the movement from "is" claims to "ought" recommendations. Here is an example from our field:

Claim #1. Exercise *is* effective in promoting both a vigorous life and a longer existence.

Claim #2. Costs of many exercises *are* minimal, and some *can* even *be* fun.

Recommendation: People *ought* to exercise.

The concern here is that facts ("is" statements) about vigor, long life, and costs still leave unanswered questions about the desirability of seeking those things (the "ought" conclusion). It makes sense to grant that exercise promotes longer life, but still ask whether we would want to pursue that. Some additional evidence is needed to warrant any ought recommendation. To many deconstructionists, therefore, the recommendation about exercise is groundless.

Another method is to examine texts for possible political, economic, gender, racial, historical, or other biases (see, e.g., Fernandez-Balboa, 1997, and Kirk, 1992). Curriculum recommendations about exercise, competition, and play, for example, can be debunked for their provincialism or self-serving chauvinism. The strategy typically employed is one of showing the plausibility of alternate interpretations of certain biased philosophic conclusions that have been misleadingly presented as objectively valid. Sometimes possible cause-and-effect relationships are implied between conceptually reinforcing socioeconomic forces, on the one hand, and prevailing philosophic "truths" on the other (see, e.g., Gruneau, 1983). Big-time college sport, for instance, has been depicted by some metaphysicians as objectively valuable, perhaps even as uplifting as art. Yet this could be nothing more than a rationalization for certain economic imperatives related to profit and control.

Although critical and poetic philosophic work is largely destructive in nature, it can still be valuable. It is undoubtedly better to be suspicious of grand philosophic claims and know their shortcomings than to be blindly supportive of them. Even though this research generally does not replace supposedly defective claims with better ones, it suggests that it is important to realize that we do not (and possibly cannot) know certain things.

A constructive side to this type of philosophic research exists, however. In this essay it is called *poetry* because it cannot (without contradicting itself) make use of the reasoned argumentation and system building that it so vigorously criticizes. It is suggestive rather than explicit; it is tentative rather than sure of itself; it points to generally superior directions for living rather than laying them out systematically.

Nietzsche (1967) suggested that life should be lived without resorting to crutches of various sorts (e.g., blind religious belief or neatly packaged visions of the world) but by relying on a type of courageous freedom. Although Nietzsche did not carry out any extended discussion of relationships between game playing and his "will to power," his views lend themselves to poetic commentaries on the similarities between life and games.

Hurdles that we face in games and life, it could be said, are essentially arbitrary, and when we win in either domain, we haven't gained or accomplished anything significant. Nevertheless, we can still choose to play, compete, and create new solutions *as if* the outcome mattered. In the face of absurdity, there is a degree of nobility in this persistence.

If this analogy is meaningful, poetry has done its job. However, because reason itself is held suspect, and because reality is believed not to be neatly packaged in categories, Nietzsche and other philosophers doing this kind of work would not want to make too much of the logic implicit in this analogy; nor would they want to claim that games are essentially different from other activities. It is enough to make a poetic suggestion and move on.

A strength of this methodology revolves around its purposes—its skepticism toward reason and its attempt to prevent people from relying on fictitious crutches and other false beliefs. It is also, for many, a scientifically palatable philosophy because it treats thinking as a very natural, in-the-world process and describes life's purposes without resorting to myth, absolute values, mysteries, the divine, or special meanings.

A recurring difficulty is the tendency of critical and poetic philosophy to practice what it opposes. It may use reason and thereby build systems to show that reason is irredeemably corrupt and that systems are nothing but false human constructs. This paradoxical enterprise requires that the products of this methodology be suggestive rather than definitive, poetic rather than systematic (Rosen, 1989). This involves some difficult tightrope walking, and for some philosophers, these high-wire routines of ironic debunking make philosophy less than it can and should be.

Summary

In some ways, doing philosophy well is like performing a motor action skillfully. Within broad boundaries of orthodoxy, alternative strategies and styles work for different people. Although the differences in philosophic methodology described here are not trivial, it is nevertheless more important to begin thinking philosophically in some manner than to wait for consensus on a perfect methodology. Consequently, you would be better off getting on with developing an ability to use some of them than fretting long and hard over which one to choose, thereby delaying your introduction to any of them.

Note to readers: After working on the section "Check Your Understanding," you are encouraged to read Gardner's article, "On Performance-Enhancing Substances and the Unfair Advantage Argument." This essay is an example of good philosophic research, albeit in the very difficult area of ethics. Reading this article and considering how different methodologies can be applied to it will give you a feel for the persuasive tug of philosophic arguments and perhaps help you to generate new arguments yourself.

A good class setting for such an experience is a formal debate. Following the appendix is a brief chronological bibliography to help debaters trace historical arguments on the ethics of taking performance-enhancing substances.

Such an in-class experience will allow you to judge for yourself the matter of taking performance-enhancing drugs. Has Gardner shown clearly that taking anabolic steroids does not give the performer an unfair advantage over opponents or the game itself?

✔ Check Your Understanding

These activities will help you learn to use the tools or methods needed to accomplish the purposes of philosophic research.

Locating a Research Problem

Suppose that one day you heard someone say that exercise was pure drudgery. They continued, "It is impossible to play while exercising." Your curiosity has been piqued. A philosophic problem has found you, but it is a large and unwieldy one. How will you phrase your research question? And then how will you define exercise and play so you know what you are researching?

Inductive Reasoning

What other generalities or abstractions can you draw from the five examples on pages 244-246? Can you say anything about the nature of the parties involved? For example, must they be human? What can you say about the nature of the problem faced in competition? Must it require motor skill? Is the function of motor skill in relationship to solving the game problem the same for each example of competition? If you do not like the five examples provided, how would you go about selecting others?

Deductive Reasoning

What additional deductions might you draw from the set of premises provided in this section? Based on the first three premises, what can you say about competition in which one party is playing to win and the other is out for some exercise? Or one in which two parties are taking the same test at the same time (e.g., bowling), but they do not know about one another (one is bowling in California and the other in Ohio)? Or one in which two teams are playing the same game (e.g., baseball), but one side has resorted to using illegal pitches?

Descriptive Reasoning

Have you ever had a peak experience in sport—a time when your performance was qualitatively unique, delightful, and unforgettable? Can you remember some of the characteristics of that event and describe them clearly and accurately? If you were not concerned with your actual experience, could you (in principle) describe features of a peak experience? Could you say what makes movement experiences extraordinary?

Speculative Reasoning

How would you rank the three values of excellence, knowledge, and play in terms of their significance for achieving a good life? What additional values in competitive activity might be added to this list, and what arguments can be forwarded in their defense (see, e.g., Kretchmar, 2005)?

Critical and Poetic Reasoning

Are claims about the value of competitiveness for survival in life gender biased from a uniquely male perspective, or are they rationally defensible for all human life? What about arguments to the effect that life itself is inherently competitive and that sport therefore is needed to prepare effectively for it?

The Ethics of Taking Drugs to Enhance Sporting Performance

(Note: Another version of this exercise appears in Kretchmar, 2005.)

1. Read Roger Gardner's essay "On Performance-Enhancing Substances and the Unfair Advantage Argument," *Journal of the Philosophy of Sport,* 1989, *XVI,* 59-73. Also in Kretchmar, R. Scott, (2005). *Practical Philosophy of Sport and Physical Activity* (2nd ed., pp. 269–85). Champaign, IL: Human Kinetics; and in Morgan, William J., and Meier, Klaus V. (Eds.), *Philosophic Inquiry in Sport* (2nd ed., pp. 222-31). Champaign, IL: Human Kinetics.

2. Notice Gardner's thesis. He claims that no compelling arguments have been advanced for banning performance-enhancing drugs—at least not on the grounds that they provide an unfair advantage. Therefore, he suggests, current rules that prohibit the use of these substances may not be justified. Consequently, you will be looking for arguments that support this minority position—arguments to the effect that rules against drug usage are arbitrary, inconsistent, or otherwise unjustified.

3. Locating the Problem (Introduction and section titled, "What Is an Unfair Advantage?"). Gardner has a large and unwieldy issue in tow—the ethics of drug use in sport. Because this is so complex and multifaceted, he must take considerable pains to reduce the problem to a manageable size and to clarify exactly what he will be claiming. To clearly evaluate his argument about unfair advantages and drugs, you will need to understand his stipulations about the following:

   • What counts as a performance-enhancing drug?

   • What if certain drugs really do not enhance performance? What are you to assume in this regard?

   • What about arguments besides unfair advantage that have been advanced to ban drugs, such as harm, coercion, and unnaturalness? Are you to consider these or not?

- Aren't unfair advantages fairly commonplace in sport, and aren't they tolerated without moral condemnation? If so, the issue cannot be one of unfair advantage per se, but rather how the advantage was gained.
- If unfair advantages are gained in ways that are morally unacceptable, wouldn't you expect rules to prohibit such means, and wouldn't that stand as a justification for such rules?

At the end of this section, Gardner believes he has cleared away the brush and is ready to meet the arguments head on. You should, for instance, no longer be distracted by the argument that taking anabolic steroids provides an unfair advantage simply because it involves breaking a rule that others follow. Gardner agrees that this is unfair, but he then reminds you that the larger issue is one of justifying that rule. Why is the rule there in the first place? Is taking drugs a morally objectionable means for gaining an advantage? *That* is the key question!

4. Analyzing the Problem (the section "An Advantage Over Other Athletes"). Gardner argues that a variety of complaints about using drugs because they offer an improper or somehow unsporting way of gaining an advantage over an opponent are not persuasive. Gardner can show this by using an *inductive* train of thought in reference to the claim that the drugs provide an unfair advantage because they are *not equally accessible to all competitors*.

Sample A: An athlete who has benefited competitively from having wealthy parents and the best coaching money can buy

Sample B: An athlete who performs better because she lives in a country with good scientific and technological support

Sample C: A skier who performs exceptionally well because he was born in Norway

What is common to these cases? Gardner's answer: They all involve unfair advantage because these athletes have *unequal access* to performance-favoring conditions—access that other athletes do not enjoy. However, we also notice that these advantages are morally acceptable, even if they are not ideal. Such athletes are allowed to compete, no one discounts their victories, and no one holds them morally blameworthy for benefiting from the advantages available to them.

Gardner can then, from these particulars and others, abstract his broader claim relative to unequal access to drugs such as anabolic steroids. If unequal access to performance-enhancing factors is common and unobjectionable in sport, why would we single out unequal access to anabolic steroids as a morally objectionable way to show superiority? Do you agree with Gardner that the unequal access argument is not conclusive?

A further argument in the section titled "An Advantage Over Other Athletes" can be framed in *descriptive* terms. Some people who favor the banning of drugs argue that *advantages from using anabolic steroids are not earned*. Gains from "popping pills" require no training, no sacrifice, no effort. Sport should reward those who work harder, not those who have better pharmacists.

Gardner's descriptive counteranalysis could go something like this: Although sport has much to do with measuring merit, effort is not the only way to show merit. In fact, diverse sports require different levels of effort. Training for baseball is not the same as training for football or distance running. Likewise different athletes in the same sport, depending on their dispositions and genetic endowments, may find success with considerably less effort than their peers. But just because they did not have to work as hard as someone else, we do not discount their accomplishments.

Gardner wants you to agree with his descriptive conclusions. High levels of effort expenditure are not necessary conditions for the attribution of athletic merit. Do you concur then that arguments for banning performance-enhancing drugs because they reduce (or serve as a substitute for) effort are not conclusive?

(continued)

5. Analyzing the Problem (the section "An Advantage Over the Sport"). In this part of his essay, Garner considers claims that *the use of drugs may compromise the integrity of sport,* that is, provide an avenue for solving sport problems that end up ruining the activity. One argument he uses could be framed in *deductive* terms.

Premise 1: All new technologies or methods that make a sport test too easy should be prohibited.

Premise 2: The use of square-grooved golf clubs makes the (golfing) sport test too easy, but the use of steroids does not make sport tests too easy.

Conclusion: The use of square-grooved golf clubs should be prohibited; the use of steroids should not be prohibited.

Once again, Gardner believes he has provided persuasive arguments in favor of his case. If you disagree, you would need to identify weakness in his premises, look for premises that are missing, or argue that his conclusion is not warranted.

Summary. Through the use of inductive, descriptive, and deductive arguments, Gardner believes he has provided evidence that supports his contention that the unfair advantage argument for prohibiting drugs in sport is inconclusive. If you think his arguments are weak, or that evidence on the opposite side is stronger, you have an obligation to show the superiority of your view. Remember, Gardner has limited his discussion to only one facet of the issue. Drug taking could be justifiably banned for other reasons (e.g., harm to oneself or coercion of others to take dangerous drugs) even if Gardner is right about the issue of unfair advantage.

Those who use critical and poetic or speculative techniques would be among those who would challenge Gardner. *Critical and poetic philosophers* might attempt to deconstruct the edifice built by Gardner. How can he place so much emphasis on what people commonly think is an acceptable advantage—as if this provides some insight into the way the moral universe is constructed? The fact is, they might argue, people do not object to privilege and technological advantages in sport today simply because this expresses their current tastes. Tomorrow this could all change, and then Gardner's argument would crumble.

Speculative philosophers might argue that there is a sporting ideal that would have you (and all of us) working to eliminate unfair advantages—as much as possible. Why, they might ask, should you put up with advantages of birth, wealth, and genetic endowment when these can be controlled, at least to a degree? Why not work toward and promote a better version of sport, one that levels the playing field more than the current version does? If access to steroids is uneven (say, only the wealthy can afford them), or if biological reactions to steroids are extremely diverse (favoring some and not helping others), then why not take them out of play? Why not work toward the ideal of making competition genuinely fair?

Selected Bibliography on Ethics of Performance Enhancement (Chronologically Arranged)

Brown, M. (1980). Drugs, ethics, and sport. *Journal of the Philosophy of Sport, 7,* 15–23.

Simon, R. (1985). Good competition and drug-enhanced performance. *Journal of the Philosophy of Sport, 11,* 6–13.

Gardner, R. (1989). On performance-enhancing substances and the unfair advantage argument. *Journal of the Philosophy of Sport, 16,* 59–73.

Schneider, A. (1993–1994). Why Olympic athletes should avoid the use and seek the elimination of performance-enhancing substances and practices from the Olympic Games. *Journal of the Philosophy of Sport, 20–21,* 64–81.

Burke, M., & T. Roberts. (1997). Drugs in sport: An issue of morality or sentimentality? *Journal of the Philosophy of Sport, 24,* 99–113.

Kretchmar, S. (Fall, 1999). The ethics of performance-enhancing substances in sport. *Bulletin: International Council of Sport Science and Physical Education, 27,* 19–21.

Holowchak, A. (2002). Ergogenic aids and the limits of human performance in sport: Ethical issues and aesthetic considerations. *Journal of the Philosophy of Sport, 29,* 75–86.

Simon, R. (2005). *Fair play: The ethics of sport* (2nd ed., chapter 4). Boulder, CO: Westview.

Research Synthesis (Meta-Analysis)

Get your facts first, and then you can distort them as much as you please.—Mark Twain

■ ■ ■ ■ ■ ■ ■

An analysis of the literature is a part of all types of research. The scholar is always aware of past events and how they influence current research. Sometimes, however, the literature review stands by itself as a research paper, one that involves the analysis, evaluation, and integration of the published literature. A term used to describe this is *research synthesis*. As mentioned in chapter 1, many journals consist entirely of literature review papers, and nearly all research journals publish review papers occasionally.

All the procedures discussed in detail in chapter 2 apply to research synthesis. The difference is that the purpose here is to use the literature for empirical and theoretical conclusions rather than to document the need for a particular research problem. A good research synthesis results in several tangible conclusions and should spark interest in future directions for research. Sometimes a research synthesis leads to a revision to or the proposal of a theory. The point is that a research synthesis is not simply a summary of the related literature; it is a logical type of research that leads to valid conclusions, hypothesis evaluations, and the revision and proposal of theory.

The approach to a research synthesis is like the approach to any other type of research. The researcher must clearly specify the procedures that are to be followed. Unfortunately, the literature review paper seldom specifies the procedures the author used. Thus, the basis for the many decisions made about individual papers is usually unknown to the reader. Of course, this makes an objective evaluation of a literature review nearly impossible. Questions that are important yet usually unanswered in the typical review of literature include the following:

• How thorough was the literature search? Did it include a computer search and hand search? In a computer search, what descriptors were used? Which journals were searched? Were theses and dissertations searched and included?

• On what basis were studies included or excluded from the written review? Were theses and dissertations arbitrarily excluded? Did the author make decisions about inclusion or exclusion based on the perceived internal validity of the research, on sample size, on research design, or on appropriate statistical analysis?

• How did the author arrive at a particular conclusion? Was it based on the number of studies supporting or refuting the conclusion (called *vote counting*)? Were these studies weighted differentially according to sample size, meaningfulness of the results, quality of the journal, and internal validity of the study?

Many more questions could be asked about the decisions made in the typical literature review paper, for good research involves a systematic method of problem solving. However, in most literature reviews, the author's systematic method remains unknown to the reader, thus prohibiting an objective evaluation of these decisions.

In recent years, several attempts have been made to solve the problems associated with literature reviews. The most notable of these attempts was made in a paper by Glass (1976) and followed up with a book by Glass et al. (1981) who proposed a technique called **meta-analysis.** The major purpose of this chapter is to present an overview of meta-analysis and the procedures developed for use in meta-analysis (but for a more general approach to research synthesis, see Cooper & Hedges, 1994).

meta-analysis

A technique of literature review that contains a definitive methodology and quantifies the results of various studies to a standard metric that allows the use of statistical techniques as a means of analysis.

Using Meta-Analysis to Synthesize Research

Since Glass introduced meta-analysis in 1976, thousands of meta-analyses have been published, especially in the social, behavioral, and medical sciences. At a national conference in 1986 sponsored by the National Institutes of Health, Workshop on Methodological Issues in Overviews of Randomized Clinical Trials, participants focused considerable attention on meta-analysis for the health-related and medical research. Handbook of Research Synthesis, edited by Cooper and Hedges (1994), is still the most useful source for planning and conducting meta-analytical reviews.

Purpose of Meta-Analysis

Like any research procedure, meta-analysis involves the selection of an important problem to address. However, meta-analysis involves two steps lacking in the typical literature review paper. First, a definitive methodology is reported concerning the decisions in a literature analysis. Second, the results of various studies are quantified to a standard metric called *effect size* (ES) that allows the use of statistical techniques as a means of analysis. Here are the steps in a meta-analysis:

1. Identify a problem.
2. Perform a literature search by specified means.
3. Review identified studies to determine inclusion or exclusion.
4. Carefully read and evaluate to identify and code important study characteristics.
5. Calculate effect size.
6. Apply appropriate statistical techniques.
7. Report all these steps and the outcomes in a review paper.

Of course, one of the major problems in a literature review paper is the number of studies that must be considered. To some extent, analyzing all these studies is like trying to make sense of all the data points collected on a group of participants. However, within a study, statistical techniques are used to reduce the data to make them understandable. The procedures of meta-analysis are similar. The findings within individual studies are considered the data points to use in a statistical analysis of the findings of many studies.

How, then, can findings based on different designs, data collection techniques, dependent variables, and statistical analyses be compared? Glass (1976) addressed this issue by using the estimate effect size (ES, or the symbol Δ). Note that we discussed this general concept in chapters 7 and 9 as a way of judging the meaningfulness of group differences. ES is determined by the following formula:

$$ES = (M_E - M_C)/s_C \qquad (14.1)$$

where M_E is the mean of the experimental group, M_C is the mean of the control group, and s_C is the standard deviation of the control group. Note that this formula places the difference between the experimental and control groups in control group standard deviation units. For example, if $M_E = 15$, $M_C = 12$, and $s_C = 5$, then ES = (15 – 12)/5 = 0.60. The experimental group's performance exceeded the control group's performance by 0.60 standard deviations. If this were done across several studies addressing a common problem, the findings of the studies would be in a common metric, ES (Δ), which could be compared. The mean and standard deviations of ES can be calculated from several studies. This allows a statement about the average ES of a particular type of treatment.

Suppose we want to know whether a particular treatment affects males and females differently. In searching the literature, we find 15 studies on males comparing the treatment effects and 12 studies on females. We calculate an ES for each of the 27 studies and the mean (and standard deviation) of the ES for males ($n = 15$) and females ($n = 12$). If the ESs were distributed normally, an independent t test could then be used to see whether the average ES differed for males and females. If the t values were significant and the average ES for the females were greater, we could conclude that the treatment had more effect on females than on males. Cooper and Hedges (1994) provide considerable detail on the methods of meta-analysis, including literature search strategies, ways to calculate ES from the statistics reported in various studies, suggestions for how and what to code from studies, and examples of the use of meta-analysis.

Certainly, meta-analysis is not the answer to all the problems associated with research synthesis. But Glass has provided an objective way to evaluate the literature. Advances (Hedges, 1981, 1982a, 1982b; Hedges & Olkin, 1980) in studying the statistical properties of ES have contributed considerably to the appropriate use of meta-analysis. The text by Hedges and Olkin (1985) provides a complete accounting of the procedures and statistical analyses appropriate for meta-analysis. (Other books have appeared on meta-analysis advocating slightly different procedures, e.g., Hunter & Schmidt, 1990.) A tutorial by Thomas and French (1986) is an overview of Hedges' and colleagues' techniques and includes examples from the study of physical activity. Some examples of meta-analyses that have appeared in the physical activity literature follow.

Examples of Meta-Analysis

Meta-analyses have been published since 1976, including many about physical activity. A brief overview of some of these studies shows the value of meta-analysis.

In a meta-analysis on the effects of perceptual-motor training for improving academic, cognitive, or perceptual-motor performance, Kavale and Mattson (1983) reported a distinct lack of success for perceptual-motor training in the 180 studies included. The largest ES they found was 0.198, and it was associated with the 83 studies rated low in internal validity. In fact, in studies with high internal validity, the trained participants did worse. Thus, perceptual-motor training does not appear useful for any type of outcome (academic, cognitive, or perceptual-motor) for any type of participant (normal, educable mentally retarded, trainable mentally retarded, slow learning, culturally disadvantaged, learning disabled, reading disabled, or motor disabled) at any age level (preschool, kindergarten, primary grades, middle grades, junior high school, or high school).

Sparling (1980) reported a meta-analysis of ES differences between males and females for maximal oxygen uptake ($\dot{V}O_2$max). One of the most interesting findings was that when ES was averaged across studies and corrections were made for the differing body compositions of males and females, ES was reduced. When $\dot{V}O_2$max was expressed as liters per minute, 66% of the variance was explained by knowing the participant's sex; when $\dot{V}O_2$max was expressed in milliliters per minute per kilogram of body weight, the explained variance was reduced to 49%; and when $\dot{V}O_2$max was expressed as milliliters per minute per kilogram of fat-free weight, the explained variance was reduced to 35%. Thus, the advantage of males over females in $\dot{V}O_2$max is reduced when corrected for body weight and is reduced even more when corrected for fat-free weight.

In a meta-analysis on the effects of exercise on blood lipids and lipoproteins, Tran, Weltman, Glass, and Mood (1983) reported an analysis of 66 training studies involving a total of 2,925 participants. They found significant relationships between training and beneficial changes in blood lipids and lipoproteins: "Initial levels, age, length of training, intensity, $\dot{V}O_2$max, body weight, and percent fat have been shown . . . to interact with exercise and serum lipids and lipoprotein changes" (p. 400).

Payne and Morrow (1993) reported a meta-analysis of the effects of exercise on children's $\dot{V}O_2$max. From 28 studies yielding 70 effect sizes, they compared cross-sectional (trained vs. untrained children) and pre- to posttraining designs. Effect sizes were large in cross-sectional studies (0.94 ± 1.00) but small in training studies (0.35 ± 0.82). In training studies, the average improvement in $\dot{V}O_2$max was only 2 ml · kg · min⁻¹. Effect size for the training studies was not influenced by sex, training protocol, or the way participants were tested on $\dot{V}O_2$max.

Feltz and Landers (1983) reported a meta-analysis of the effects of mental practice on motor skill learning and performance. From 60 studies they calculated an average ES of 0.48, less than half of a standard deviation unit. They concluded that mentally practicing a motor skill is slightly better than not practicing one at all.

In studying gender differences in motor performance across childhood and adolescence, Thomas and French (1985) reported findings from 64 studies based on 31,444 participants. They found motor performance differences to be related to age in 12 of the 20 motor tasks. These 12 tasks followed four general types of curves across age. Figure 14.1 shows one typical type of curve for three tasks (long jump, shuttle run, and grip strength). The differences are moderate (about 0.5 to 0.75 standard deviation units) before puberty but then increase dramatically during and following puberty (over 1.5 standard deviation units). They concluded that the differences before puberty were likely to be induced by environmental factors (differential treatment by parents, teachers, coaches, and peers) but that they represented an interaction of biology and environment beginning at puberty. Note the difference between the curves in figures 14.1 and 14.2. The effect size for throwing performance is 1.5 standard deviation units at ages 3–4 and increases constantly across childhood and adolescence until the differences are 3.5 standard deviation units by age 18. Thomas and French (1985) suggested that differences that are so large so early in life might have some basis in biology and in the influence of cultural treatment and expectations for girls and boys.

Carron, Hausenblas, and Mack (1996) reported a meta-analysis of social influence on exercise. They found 87 studies containing 224 ESs based on 49,948 participants. The outcomes indicated that social influence has a positive but generally small to moderate relationship (ESs ranged from 0.2 to 0.5) to exercise. However, four variables had a moderate to large relationship (0.5 to 0.8): "family support and attitudes about exercise, task cohesion and adherence behavior, important others and attitudes about exercise, and family support and compliance behavior" (p. 1).

We use these examples to illustrate the value of the meta-analysis approach. In the perceptual-motor meta-analysis, the controversial issue of the value of this type of training appears resolved: no benefit. In the meta-analysis of gender differences in $\dot{V}O_2$max, the large differences appear to be accounted for mainly by differences in body composition rather than any differences in underlying

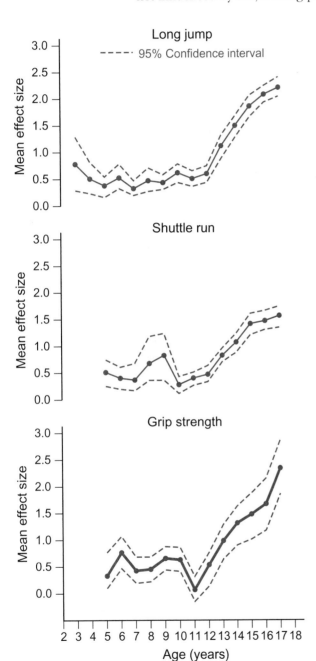

Figure 14.1 ES for three motor performance tasks. *Note.* Dotted lines are confidence intervals.

Reprinted from J.R. Thomas and K.E. French, 1985, "Gender differences across age in motor performance: A Meta-analysis," *Psychological Bulletin* 98(2): 260-282. Copyright ©1985 by the American Psychological Association. Reprinted with permission.

mechanisms. Also, exercise appears to have a positive effect on cholesterol and its components. Exercise training seems to benefit children, but the gains are relatively small by adult standards. Mental practice is better than no practice, but not by much. Gender differences in motor performance before puberty appear to be mainly environmentally induced, but throwing may be a skill for which biology plays a greater role before puberty. Finally, social influence was positively related to exercise, but most effects were small to moderate.

Meta-analysis, when applied appropriately and interpreted carefully, offers a means of reducing a large quantity of studies to underlying principles. These principles can become the bases for program development, future research, and theory testing, as well as various practical applications, such as practice and training.

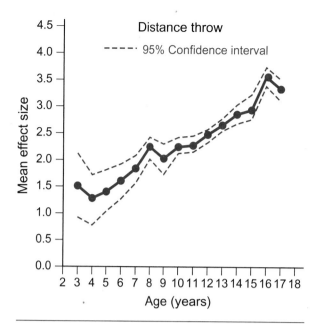

Figure 14.2 ES by age and gender for throwing for distance. *Note.* Dotted lines are confidence intervals.

Reprinted from J.R. Thomas and K.E. French, 1985, "Gender differences across age in motor performance: A Meta-analysis," *Psychological Bulletin* 98(2): 260-282. Copyright ©1985 by the American Psychological Association. Reprinted with permission.

Methodological Considerations

Although meta-analysis has been a widely used technique in the last 25 years, considerable controversy exists concerning the validity of meta-analysis (e.g., Carlberg et al., 1984; Slavin, 1984a, 1984b). Among the most severe criticisms of meta-analysis is that it combines findings from studies representing different measurement scales, methodologies, and experimental designs. This "mixing of apples and oranges" was compounded by the results of early meta-analyses using the methods of Glass (1977), which tended to reveal few differences in ESs, even between studies in which internal validity and methodological control clearly varied. Hedges (1981, 1982a, 1982b) and Hedges and Olkin (1983, 1985) extended the original work of Glass (1977) and proposed a new set of techniques and statistical tests specifically designed to address the following questions and criticisms of meta-analysis:

- How should the process of deciding the variables to code and the systematic coding be organized?
- What should be used as the standard deviation when calculating an ES?
- Because sample ESs are biased estimators of the population of ESs, how can this bias be corrected?
- Should ESs be weighted for their sample size?
- Are all ESs in a sample from the same population of ESs? This is the "apples and oranges" issue: Is the sample of ESs homogeneous?
- What are appropriate statistical tests for analyzing ESs?
- If a sample of ESs includes outliers, how can they be identified?

In the following sections, we address each of these questions, summarize the theoretical basis for the statistical procedures introduced by Hedges (1981, 1982a, 1982b) and Hedges and Olkin (1983, 1985), and suggest applications of these techniques in the study of physical activity. Rosenthal (1994) has an excellent chapter (16) that provides all of the formulae for conducting a meta-analysis when data are normally distributed.

Deciding What to Code

One of the most difficult yet most important tasks in doing a meta-analysis is choosing variables to code and developing a scheme for coding them. Stock (1994) provided a useful discussion of this topic. In particular, the meta-analyst must realize the trade-off between the number and importance of items to code and the time required to code. Of course,

She's looking for the drawer effect.

the more information that is coded about a study, the more time is required to develop a coding scheme and actually do the coding. However, omitting potentially important items to code also creates major problems because the scholar then has to go back through all studies to pick up these omitted items.

The best way to be successful in selecting items to code and a coding scheme is to know the theoretical and empirical literature on which you are doing the meta-analysis and to understand meta-analysis procedures. Stock (1994) suggested the following two important considerations for planning the coding for a meta-analysis.

The meta-analyst is looking for items to code that are likely to influence (or be related to) the effect sizes that will be calculated. For example, when Thomas and French (1985) calculated effect sizes for gender differences in motor performance, they thought that children's ages would be related to gender differences. Specifically, they hypothesized that gender differences would become greater as children grew. Once important items to code are selected, the meta-analyst needs an organizational scheme for the items. At the end of this chapter (pp. 258–267) is a sample of a coding form and code book developed by Sheri Parks from the School of Human Movement Studies at the University of Queensland, Brisbane, Australia, for her dissertation review on sport expertise. We appreciate her permission to use the materials. Parks' coding scheme used these categories (with several subparts of each): descriptive information about the paper, type of sport/area of expertise, information about the task, type of measures used for each task, and information about the participants.

Stock (1994) provides a generalized example of a coding form. An important issue here is how to code variables, or how to assign numbers to characteristics to be coded. Sometimes specific information is maintained; for example, if the average age of participants from each study is important, this average age can be entered as the code. In other instances, variables are grouped into categories for coding; for example, categories of level of expertise could be coded as 1 = internationally ranked athletic performers, 2 = nationally ranked athletic performers, 3 = college or university athletic performers, 4 = high school athletic performers. Other important issues include the training of coders and establishing and maintaining intra- and intercoder reliability.

Choosing the Standard Deviation for the Effect Size

Originally, Glass et al. (1981) proposed the use of the control group's standard deviation as the most appropriate measure of group variability. From an intuitive perspective, the control group's variability represents the "normal" variation in an untreated population. The control group's standard deviation also has the advantage of assigning equal ESs to equal treatment means when a study contains two or more experimental groups that have heterogeneous variances. Therefore, the control group's standard deviation serves as a common standard metric from which treatment differences can be compared.

In most cases, estimates of group variance are homogeneous. Hedges (1981) argued that a pooled estimate of the variance provides a more precise estimate of the population variance. (The square root of the pooled estimate of the variance is the pooled estimate of the standard deviation.) One advantage of pooling variances is an increase in the degrees of freedom associated with the estimate of variance. The pooled estimate Hedges (1981) suggested is given in equation 14.2. Note that the variance of each group is weighted by the sample size of that group in a way that is similar to the procedure used in t tests with unequal n's. The formula for the pooled estimate of standard deviation (s_p) is

$$s_p = \sqrt{\frac{(N_E - 1)s_E^2 + (N_C - 1)s_C^2}{N_E + N_C - 2}}$$

(14.2)

where N_E is the sample size of the experimental group (group 1), N_C is the sample size of the control group (group 2), s_E^2 is the variance of the experimental group (group 1), and s_C^2 is the variance of the control group (group 2).

Many studies involve tests of effects between or among categorical variables (e.g., race and gender) where there is no control condition. Early meta-analyses used an average of the standard deviations of the groups compared (Glass & Smith, 1979; Hyde, 1981; Smith, 1980). Hedges and Olkin (1985) suggested using the weighted pooled standard deviation (s_p, equation 14.2) as the estimate of the standard deviation (for an example using gender differences in motor performance, see Thomas & French, 1985).

There are cases in which the variance for the groups is heterogeneous. Glass et al. (1981) showed that the ESs are biased when the group variances are unequal. Because parametric tests are based on the assumption of equal variances, ESs calculated from t and ANOVA may be biased if the group variances are unequal.

Researchers should evaluate whether heterogeneous variances are a common phenomenon in the area of research for the meta-analysis. We have encountered unequal variances when comparing the motor performance of children of different age levels (French & Thomas, 1985). We offer these suggestions: When the ES compares an experimental group and a control group and the group variances are unequal, use the standard deviation of the control group to calculate the ES in all studies. If the ES compares two groups (such as different groups by age or gender) in which there is no clear control condition, we believe the weighted pooled estimate (equation 14.2) suggested by Hedges (1981) is the best choice.

Calculating Effect Sizes for Within-Subjects Designs

Often a researcher wants to calculate an effect size for a within-subjects design (or repeated-measures effect), usually between a pretest and posttest for a treatment effect (i.e., experimental group). The appropriate formula for this effect size is to use the pretest standard deviation in the denominator of equation 14.1 (Looney, Feltz, & VanVleet, 1994). This represents the best source of untreated variance against which to standardize differences between the pre- and posttest means.

Using Effect Size As an Estimator of Treatment Effects

Hedges (1981) provided a theoretical and structural model for the use of ES as an estimator of treatment effects. An individual ES may be viewed as a sample statistic that estimates the population of possible treatment effects within a given experiment.

Hedges (1981) also demonstrated that ESs are positively biased in small samples; however, the bias is 20% or less when the sample size exceeds 20. A virtually unbiased estimate of ES can be obtained by multiplying the ES by the correction factor given in the following formula:

$$c = 1 - \frac{3}{4m - 9} \qquad (14.3)$$

where $m = N_E + N_C - 2$ when a pooled estimate is used as the standard deviation, $m = N_1 + N_2 - 2$ if a pooled estimate is used in a categorical model, or $m = N_C - 1$ when the control-group standard deviation (or pretest standard deviation in the case of a within-subjects ES) is used.

Each ES should be corrected before averaging or further analysis. If ESs are not corrected before averaging, the average of even a large number of ESs remains biased and simply estimates an incorrect value more precisely (Hedges, 1981). Early meta-analyses in exercise and sport (Feltz & Landers, 1983; Sparling, 1980) did not correct each ES for small sample bias. The ESs reported in these studies are most likely slight overestimates.

Although the individual ES estimate can be corrected for bias, the variability associated with the estimate remains a function of the sample size. Hedges (1981) showed that the variance of an individual ES can be directly calculated from the following formula:

$$\mathrm{var}\!\left(\mathrm{ES}_i\right) = \frac{N_E + N_C}{N_E N_C} + \frac{\mathrm{ES}_i^2}{2\!\left(N_E + N_C\right)} \qquad (14.4)$$

where N_E is the sample size of the experimental group (group 1), N_C is the sample size of the control group (group 2), and ES_i is the estimate of ES. Note that the variability associated with the sample statistic, or ES, is a function of the value of the ES and the sample size. An ES based on a large sample has a smaller variance than an ES based on a small sample. Therefore, ESs based on large samples are better estimates of the population parameter of treatment effects. Hedges (1981) and Hedges and Olkin (1985) suggested that each ES should be weighted by the reciprocal of its variance, that is 1/variance (as estimated from equation 14.4). Therefore, ESs that are more precise receive more weight in each analysis.

Testing Homogeneity

Meta-analysis has been criticized for mixing apples and oranges, or combining studies with different measurement scales, designs, and methodologies. No appropriate test existed to determine whether all the ESs were estimating the same population treatment effect until Hedges (1982b) introduced his test for homogeneity. The homogeneity statistic, H, is specifically designed to test the null hypothesis, H_0: $\mathrm{ES}_1 = \mathrm{ES}_2 = \ldots = \mathrm{ES}_i$. This is equivalent to saying that all ESs tested come from the same population of ESs.

The statistic H is the weighted sum of squared deviations of ESs from the overall weighted mean. The contribution of each ES to the overall mean is weighted by the reciprocal of its variance (equation 14.4). Effect sizes with smaller variances receive more weight in the calculation of the overall mean. Under the null hypothesis, H has a chi-square distribution with $N - 1$ degrees of freedom, where N equals the number of ESs.

When the null hypothesis is not rejected, all ESs are similar and represent a similar measure of treatment effectiveness. In this case, the researcher should report that the homogeneity statistic indicates that ESs are homogeneous and use the weighted mean ES with confidence intervals for interpretation. If the null hypothesis is rejected, ESs are not homogeneous and do not represent a similar measure of treatment effectiveness, or grouping. Two methods have been proposed by Hedges (1982b) and Hedges and Olkin (1983) to examine explanatory models for ESs. We look at these models next.

Analysis of Variance and Weighted Regression

The first method to fit an explanatory model to ES data is analogous to ANOVA, in which the sum of squares for the total statistic is partitioned into sum of squares between (or among groups of) ESs (H_B) and sum of squares within groups of ESs (H_W). Each sum of squares can be tested as a chi square with $k - 1$ df for H_B and $N - k - 1$ df for H_W (where N is the number of ESs and k is the number of groups). Therefore, a test can be conducted for between- or among-group differences (H_B), and a test can be conducted to determine whether all ESs within a group are homogeneous (H_W). Further discussion of the categorical model is presented in Hedges (1982b) and Hedges and Olkin (1985).

The second method proposed by Hedges and Olkin (1983) to fit an explanatory model to ES data is a weighted regression technique. We have chosen to present a more detailed discussion of the regression techniques for several reasons. First, one of the meta-analyses discussed previously in exercise and sport used regression techniques to analyze ES (Thomas & French, 1985). Second, it is common for a continuous variable to influence ES. Third, often more than one study characteristic influences ES. This is especially true when there are many ESs. The regression procedures can accommodate a larger number of variables in the analysis without inflating the alpha level of the statistical test. Conducting many tests using the categorical or ANOVA-like procedure results in alpha inflation, so that the researcher would need to report the experimentwise error rate or adjust the alpha level using the Bonferroni technique (see chapter 9). Fourth, categorical variables can easily be dummy- or effect-coded and entered into the regression procedures. For example, published versus unpublished papers can be dummy-coded (1 or 0) and entered into the regression. Thus, it is not necessary to conduct a separate analysis for categorical variables.

Each ES is weighted by the reciprocal of its variance in the weighted regression technique. Most standard statistical packages (e.g., SAS, SPSS, BIMED) have an option to perform weighted regression. Thus, the computations can be done easily on most computers.

The sum of squares total for the regression is equivalent to the homogeneity statistic, H. It is partitioned into sum of squares regression and sum of squares error. The **sum of squares** for regression yields a test of the variance due to the predictor variables (H_R). The H_R is tested as a chi square with $df = p$, where p equals the number of predictor variables. The sum of squares for error (H_E) provides a test of model specifications, or whether the ESs are homogeneous when the variance caused by the predictors is removed. The H_E is tested as a chi square with $df = N - p - 1$, where N equals the number of ESs and p equals the number of predictor variables. The test of model specification evaluates the deviation of the ESs from the regression model. A nonsignificant test of model specification indicates that the ESs do not deviate substantially from the regression model, whereas a significant test for model specification shows that one or more ESs deviate substantially from the regression model. Ideally, the researcher wants the H_R for regression to be significant and the H_E for model specification to be nonsignificant.

When the test for model specifications is significant, one or more ESs do not follow the specified regression. Often, the ESs that do not follow the same pattern may suggest other characteristics that can be added to the model (e.g., published vs. unpublished studies). Moreover, some ESs may represent outliers (unrepresentative scores). In either case, the use of outlier techniques is helpful in identifying these ESs.

sum of squares

A measure of variability of scores; the sum of the squared deviations from the mean of scores.

Testing for Outliers

Hedges and Olkin (1985) outlined procedures to identify outliers in categorical and regression models. We use an example from a regression model (for more information concerning the categorical model, see chapter 7 in Hedges & Olkin, 1985).

Outliers in regression models can be identified by examining the residuals of the regression equation. The absolute values of the residuals are standardized to z scores by subtracting the mean and dividing by the standard deviation. Effect sizes with standardized residuals larger than 2 are often examined as potential outliers, as they fall outside 95% of the distribution.

According to Hedges and Olkin (1985), each ES makes a contribution to the regression model. The outliers identified by standardized residuals may vary, depending on which ESs are in the model at the time. Hedges and Olkin (1985) suggested computing the standardized residuals multiple times with a different ES deleted from the model each time. For example, if you have 10 ESs, standardized residuals would be computed 10 times with a different ES deleted from the model each time. However, some of our preliminary work suggested that if more than 20 ESs are included in the regression model, going through all possible combinations of deleting ESs from the model may be of limited value. Simply calculate the residuals with all the ESs in the model, standardize the residuals, and evaluate whether ESs with a z score larger than 2 might be outliers.

Alternatively, if the ESs are not normally distributed (as Hedges & Olkin, 1985, reported is often the case), the rank-order procedures described in chapter 10 (and developed in greater detail by Thomas, Nelson, & Thomas, 1999) are very effective if ESs are changed to ranks prior to statistical analysis.

Accounting for Publication Bias

Journals have a tendency to accept papers that report significant findings. Thus, there may be some number of studies on any topic (including one on which a meta-analysis is being done) that have not been submitted to or published in scholarly journals. This problem has been labeled the "file-drawer" effect.

An important question in a meta-analysis is how many of these studies are sitting around in file drawers. The meta-analyst should find every study possible, but how many studies have been done that have not been reported? Hedges and Olkin (1985) suggested a technique for estimating how many unpublished studies with no significant effect on the variable of interest would be needed to reduce the mean ES to nonsignificance.

$$K_0 = [K(d_{ES\ mean} - d_{ES\ trivial})]/d_{ES\ trivial} \qquad (14.5)$$

where K_0 is number of studies needed to produce a trivial ES, K is the number of studies in the meta-analysis, $d_{ES\ mean}$ is mean of all the ESs in the meta-analysis, and $d_{ES\ trivial}$ is estimate of a trivial ES.

For example, if there were 75 studies in the meta-analysis (K), if the mean ES ($d_{ES\ mean}$) was 0.73, and if a trivial effect size ($d_{ES\ trivial}$) were established as 0.15, to estimate the number of "file-drawer" studies needed to produce a trivial ES (K_0), we solve equation 14.5:

$$K_0 = [75(0.73 - 0.15)]/0.15 = 290$$

Thus, there would need to be an inordinate number (290) of unpublished studies on this topic to reduce the average ES of 0.73 in the study to a trivial ES of 0.15.

Other Considerations

Sometimes all the information needed to calculate an ES is not available in a research report. Rosenthal (1994) provides an excellent discussion, formulas, and examples of how to estimate effect sizes from other statistics that are often provided in individual studies. However, on occasion, only the statement that two groups do not differ significantly is reported. What should the meta-analyst do? To drop the study is to bias the results of the meta-analysis toward significant differences. Thomas and French (1985) inserted a *zero* ES, using the rationale that no significant differences means that the differences did not reliably differ from zero. However, Thomas and French (1985) had few zero ESs. Using many zero ESs may cause problems in the analysis of homogeneity by making the studies look overly consistent. Hedges (1984) suggested a correction factor that adjusts for this problem. Regardless of the solution selected, the meta-analyst must cope with this issue and select a logical solution.

Presenting Effect Size Data

Graphic displays are very useful when presenting effect size data. According to Light, Singer, and Willett (1994), graphs enable the reader to see important characteristics of relationships in the effect size data. In particular, graphic displays of effect size data are useful for showing sources of variation in effect size displaying the center, spread and shape of the distribution—this includes stem-and-leaf plots, schematic plot (box and whiskers) and funnel graphs (effect sizes plotted against sample sizes). Graphic displays of effect size data are also useful for sorting effect sizes by important study characteristics—this might include treatment intensity, age, or gender. These displays should also reflect the variability of effect sizes for sorted characteristics (e.g., 95% confidence intervals, see figure 14.2 for an example).

Light, Singer, and Willett (1994, 451-452) list five valuable concepts for graphic displays of meta-analytic data (of course these concepts apply equally well to studies reporting original data):

1. *The spirit of good graphics can even be captured by a table of raw effect sizes, provided they are presented in a purposeful way* (e.g., by magnitude or important study characteristics).

2. *Graphic displays should be used to emphasize the big picture* (e.g., make the main points clear).

3. *Graphics are especially valuable for communicating findings from a meta-analysis when each individual study outcome is accompanied by a confidence interval* (this not only demonstrates variability of the effect sizes of interest but also whether or not they are different from the null value).

4. *Graphic displays should encourage the eye to compare different pieces of information.*

5. *Stem-and-leaf plots and schematic plots are especially efficient for presenting the big picture that emerges in a research synthesis.*

Summary

A research synthesis is not just a summary of studies arranged in some kind of sequence. It should be structured, analytical, and critical and lead to specific conclusions. The meta-analysis is a definitive methodology that quantifies results and allows statistical techniques to be used as a means of analysis.

The meta-analysis explicitly details how the search was done, the sources, the choices made regarding inclusion or exclusion, the coding of study characteristics, and the analytical procedures used. The basis of a meta-analysis is the effect size, which transforms differences between experimental and control groups' performances to a common metric, which is expressed in standard deviation units.

The meta-analysis technique has been widely used in recent years in many fields, including the study of physical activity. Meta-analysis offers a means of reducing large quantities of studies to underlying principles. Refinements in technique continue to be made regarding the choice of the standard deviation to be used in calculating an effect size, the weighting of effect sizes for sample bias, the test for outliers, and the statistical procedures that can be used for the analysis.

✔ Check Your Understanding

Select a possible problem (e.g., gender differences in running speed, effects of weight training on males and females, or influence of training at 60% vs. 80% of $\dot{V}O_2max$). Find five studies that compare the characteristics you select (e.g., males and females). Make sure each study has the means and standard deviations for the variables of interest. Calculate the ES for each study and correct it for bias. Calculate the average ES for the five studies and their pooled standard deviations. Interpret this finding.

Sample Coding Form for a Meta-Analysis of Sport Expertise

This section provides a detailed account of the procedures used to code the 92 characteristics outlined for each paper in Sheri Parks' meta-analysis in the area of expertise in sport.

Step 1 Choosing Appropriate Papers

This review is interested in any and all research that investigates expertise in sport using some form of between-group (expert–novice) comparison and addressing it using a motor behavior approach.

Each paper must provide the means and standard deviations for each of the performance measures listed so comparisons on the basis of effect sizes can be calculated and compared with other like papers. In other words, all papers and research in this review must be of a database format.

Step 2 Providing Descriptive Information About Each Paper

Column 1: Article number

Each paper chosen for inclusion in analysis should be given a unique number code, and an up-to-date list of these should be kept in appendix 1, complete list of articles used in this analysis.

Column 2: Year completed

Column 3: Is the research (1) published or (0) unpublished?

Column 4: Source information

A. If published, is it from a

(2) refereed journal?

(1) nonrefereed journal?

(0) conference proceedings?

B. If unpublished, is it from a

(2) PhD dissertation?

(1) master's thesis?

(0) unpublished manuscript?

Column 5: If published, where did the research originate?

(4) Sport science journal

(3) Movement science journal

(2) Mainstream psychology journal

(1) Other source

(0) Not applicable

Column 6: If published, name of journal

Keep a list of the journals that have been coded (1 to 1,000), where 0 denotes "not applicable," as in the case of unpublished work. This list should be kept up-to-date and included in appendix 2.

Column 7: Affiliation

This column includes a coding for where the specific research was done (i.e., which university, which lab, which particular group of researchers). A list of the affiliations should be coded (1 to 1,000) and included in appendix 3.

Column 8: Internal validity ranked on a scale of

(2) Strong

(1) Moderate

(0) Weak

This is developed from the following 6-point checklist. Refer to appendix 4 for further explanation of the rating scale for internal validity used in this analysis.

1. Did this experiment include a control group? Y/N

2. Did the experiment use randomization techniques to control for the following threats to internal validity?

(2a) History and maturation—by random placement or matched pairing of participants into each of a control and experimental group? Y/N

(2b) Order effects/learning—by random ordering/presentation of the tasks? Y/N

3. Did the experiment establish appropriate instrumentation (or test reliability/psychometrics)? Y/N

4. Was the participant retention/dropout rate limited? Y/N

5. Were the participants unaware of the purposes and intentions of this experiment? Y/N

Total yes 5–6 = strong
Total yes 2–4 = moderate
Total yes 0–1 = weak

Step 3 Providing Information About the Type of Sport/Area of Expertise

Column 9: Type of sport (area of expertise) studied

Keep a list of the sports/activities that have been coded (1 to 1,000) in appendix 5.

Column 10: Classification of sport/activity

Which cell from Higgin's 8 cell model does each sport/activity fit?
Nature of environment (open vs. closed) × Body stability × Body transport
Refer to appendix 6 for further explanation of the categorizing of sports scale used in this analysis.

Column 11: Is it (1) a team sport or (0) an individual/dual sport?

Column 12: Type of motor skills involved in sport or activity

(2) Gross motor skills

(1) Fine motor skills

(0) A combination of both gross and fine motor skills

Step 4 Providing Information About the Task(s) Performed

Column 13: How many tasks are involved in study?

This involves counting the number of experimental tasks identified in the experiment. Many experiments keep the same participants and have them undergo a series of different tasks. This is acceptable, as long as the same participants are not "reused," so to speak, in completely different research sources, as previously stated in notes for column 4.

Column 14: Type of task

Itemize all tasks in studies, and try to group like tasks within a similar range of numbers; for example, all perceptual tasks are identified by numbers 1–10, all decision-making tasks should be identified by numbers 15–25, and so on. A list of these coded tasks should be included in appendix 7.

Column 15: Levels/conditions of the independent measure for each task

The independent measure can be defined as the variable of interest, or the part of the experiment that the researcher is manipulating (i.e., blocks of practice, occlusion condition, etc.). This can also be referred to as the experimental or treatment variable.

Column 16: Levels/conditions of the dependent measure for each task

The dependent measure can be defined as the particular performance measures used in

(continued)

the experiment (i.e., reaction time, error scores, etc.). This can also be considered the effect of the independent variable or the yield of the experiment.

Column 17: Scoring direction for task, where

(1) Higher score is better, lower is worse (i.e., performance scores)

(0) Lower score is better, higher score is worse (i.e., error scores)

Step 5 Providing Information About the Type of Measures Used in Each Task

Column 18: Ecological validity ranked on a scale of

(2) Strong (total ≥ 7)

(1) Moderate (total 4–6)

(0) Weak (total ≤ 3)

This is developed from the following nine-point checklist. Refer to appendix 8 for further explanation of the rating scale for ecological validity used in this analysis.

1. Is the *action* involved in performing the task . . . that of the real-world sporting situation?

   (a) Exactly the same, +3

   (b) Somewhat the same, +2

   (c) Nothing like, +1

   (d) Not applicable, 0

2. Is the *perceptual display* offered in performing the task . . . that of the real-world sporting situation?

   (a) Exactly the same, +3

   (b) Somewhat the same, +2

   (c) Nothing like, +1

   (d) Not applicable, 0

3. Are the *time constraints* involved in performing the task (e.g., preparation time, decision time, performance time, etc.) . . . that of the real-world sporting situation?

   (a) Exactly the same, +3

   (b) Somewhat the same, +2

   (c) Nothing like, +1

   (d) Not applicable, 0

Column 19: Validity of measures ranked on a scale of

(4) Strong

(3) Moderate

(2) Weak

(1) Not mentioned or can't be interpreted on the basis of the information provided in the source

(0) If column 20 reliability of measures is rated as weak or not mentioned in study (0), rate validity as 0 in this column.

Column 20: Reliability of measures ranked on a scale of

(2) Strong

(1) Moderate

(0) Weak

Column 21: Were the statistics and statistical procedures/analyses used appropriate, as rated on the following scale:

    (2) Appropriate

    (1) All right/adequate

    (0) Not appropriate

Column 22: Is it a (1) lab-based or (0) field-based assessment/report?

Column 23: Is the action (1) natural or (0) contrived?

Column 24: Does the task duplicate normal time constraints? (1) Yes or (0) No

Column 25: If a visual display is used, is it

    (2) Static

    (1) Dynamic

    (0) Not applicable

Column 26: If a visual display is used, are the stimuli

    (2) Task specific

    (1) Alphanumeric

    (0) Not applicable

Column 27: If a visual display is used, how well does it mimic the natural situation across all modalities?

    (3) Strong

    (2) Moderate

    (1) Weak

    (0) Not applicable

Step 6 Providing Information About the Experimental Group of Participants

Column 28: Group number code, where groups can be one of the following:

    (6) Competitive at an international or Olympic level

    (5) Competitive at a national level

    (4) Competitive at a state/regional level

    (3) Participating in the sport at a recreational intramural/local level

    (2) Competitive in sports other than the one of interest

    (1) Nonathletes

Column 29: Quality of definition of the group (expert, elite, skilled, experienced, etc.)

    (2) Well defined

    (1) Not enough information provided

    (0) Not defined

Column 30: Confidence rating in performance selection

    (2) High

    (1) Average

    (0) Low

(continued)

This column attempts to address the following questions: Are these groups the best at this level? Are they only moderately competitive at this level, or do they lose first round at the Olympics, for example?

Column 31: How was this level determined?

(3) Previous performance record

(2) Coach's determination

(1) Other criteria

(0) Not mentioned

Column 32: Total number of participants in group

Column 33: Gender of participants

(3) All males

(2) All females

(1) Mixed

(0) Not stated

Column 34: Mean age of sample used in group

(5) > 30

(4) 20–29

(3) 16–19

(2) 13–15

(1) < 12

(0) Not stated

Column 35: Mean number of years' experience with sport

(5) > 10

(4) 7–10

(3) 4–6

(2) 2–3

(1) < 2

(0) Not stated

Column 36: Status of participants

(2) Paid

(1) Unpaid

(0) Not stated

Column 37: Previous experience with task

(2) Familiar

(1) New/novel

(0) Not stated

Column 38: Mean group performance score on task

Column 39: Group standard deviation on task

Step 7 Providing Information About the Control Group of Participants

Column 40: Group number code, where groups can be one of the following:

    (6) Competitive at an international or Olympic level

    (5) Competitive at a national level

    (4) Competitive at a state/regional level

    (3) Participating in the sport at a recreational intramural/local level

    (2) Competitive in sports other than the one of interest

    (1) Nonathletes

Column 41: Quality of definition of the group (expert, elite, skilled, experienced, etc.)

    (2) Well defined

    (1) Not enough information provided

    (0) Not defined

Column 42: Confidence rating in performance selection

    (2) High

    (1) Average

    (0) Low

Column 43: How was this level determined?

    (3) Previous performance record

    (2) Coach's determination

    (1) Other criteria

    (0) Not mentioned

Column 44: Total number of participants in group

Column 45: Gender of participants

    (3) All males

    (2) All females

    (1) Mixed

    (0) Not stated

Column 46: Mean age of sample used in group

    (5) > 30

    (4) 20–29

    (3) 16–19

    (2) 13–15

    (1) < 12

    (0) Not stated

Column 47: Mean number of years' experience with sport

    (5) > 10

    (4) 7–10

    (3) 4–6

    (2) 2–3

    (1) < 2

    (0) Not stated

(continued)

Column 48: Status of participants

(2) Paid

(1) Unpaid

(0) Not stated

Column 49: Previous experience with task

(2) Familiar

(1) New/novel

(0) Not stated

Column 50: Mean group performance score on task

Column 51: Group standard deviation on task

Step 8 Providing Information About the Intermediate Group of Participants

Column 52: Group number code, where groups can be one of the following:

(6) Competitive at an international or Olympic level

(5) Competitive at a national level

(4) Competitive at a state/regional level

(3) Participating in the sport at a recreational intramural/local level

(2) Competitive in sports other than the one of interest

(1) Nonathletes

Column 53: Quality of definition of the group (expert, elite, skilled, experienced, etc.)

(2) Well defined

(1) Not enough information provided

(0) Not defined

Column 54: Confidence rating in performance selection

(2) High

(1) Average

(0) Low

Column 55: How was this level determined?

(3) Previous performance record

(2) Coach's determination

(1) Other criteria

(0) Not mentioned

Column 56: Total number of participants in group

Column 57: Gender of participants

(3) All males

(2) All females

(1) Mixed

(0) Not stated

Column 58: Mean age of sample used in group

    (5) > 30

    (4) 20–29

    (3) 16–19

    (2) 13–15

    (1) < 12

    (0) Not stated

Column 59: Mean number of years' experience with sport

    (5) > 10

    (4) 7–10

    (3) 4–6

    (2) 2–3

    (1) < 2

    (0) Not stated

Column 60: Status of participants

    (2) Paid

    (1) Unpaid

    (0) Not stated

Column 61: Previous experience with task

    (2) Familiar

    (1) New/novel

    (0) Not stated

Column 62: Mean group performance score on task

Column 63: Group standard deviation on task

Step 9 Providing Information About Any Other Group of Participants Included in Study

Column 64: Group number code, where groups can be one of the following:

    (6) Competitive at an international or Olympic level

    (5) Competitive at a national level

    (4) Competitive at a state/regional level

    (3) Participating in the sport at a recreational intramural/local level

    (2) Competitive in sports other than the one of interest

    (1) Nonathletes

Column 65: Quality of definition of the group (expert, elite, skilled, experienced, etc.)

    (2) Well defined

    (1) Not enough information provided

    (0) Not defined

Column 66: Confidence rating in performance selection

    (2) High

    (1) Average

    (0) Low

(continued)

Column 67: How was this level determined?

(3) Previous performance record

(2) Coach's determination

(1) Other criteria

(0) Not mentioned

Column 68: Total number of participants in group

Column 69: Gender of participants:

(3) All males

(2) All females

(1) Mixed

(0) Not stated

Column 70: Mean age of sample used in group

(5) > 30

(4) 20–29

(3) 16–19

(2) 13–15

(1) < 12

(0) Not stated

Column 71: Mean number of years' experience with sport

(5) > 10

(4) 7–10

(3) 4–6

(2) 2–3

(1) < 2

(0) Not stated

Column 72: Status of participants

(2) Paid

(1) Unpaid

(0) Not stated

Column 73: Previous experience with task

(2) Familiar

(1) New/novel

(0) Not stated

Column 74: Mean group performance score on task

Column 75: Group standard deviation on task

Step 10 Calculating Effect Sizes and Standard Deviations

Refer to appendix 12 for further explanation of the calculations used in this analysis.

Column 76: Experimental versus control: effect size corrected for bias

Column 77: Experimental versus control: variance of effect size

Column 78: Experimental versus control: weighted effect size (inverse variance × corrected ES)

Column 79: Experimental versus intermediate: effect size corrected for bias

Column 80: Experimental versus intermediate: variance of effect size

Column 81: Experimental versus intermediate: weighted effect size (inverse variance × corrected ES)

Column 82: Experimental versus other: effect size corrected for bias

Column 83: Experimental versus other: variance of effect size

Column 84: Experimental versus other: weighted effect size (inverse variance × corrected ES)

Column 85: Control versus intermediate: effect size corrected for bias

Column 86: Control versus intermediate: variance of effect size

Column 87: Control versus intermediate: weighted effect size (inverse variance × corrected ES)

Column 88: Control versus other: effect size corrected for bias

Column 89: Control versus other: variance of effect size

Column 90: Control versus other: weighted effect size (inverse variance × corrected ES)

Column 91: Intermediate versus other: effect size corrected for bias

Column 92: Intermediate versus other: variance of effect size

Column 93: Intermediate versus other: weighted effect size (inverse variance × corrected ES)

Our thanks to Sheri Parks from the Department of Human Movement Studies, University of Queensland, Brisbane, Australia, for the use of her coding booklet.

The Survey

*Most surveys represent the average opinion
of people who don't know.*

■　■　■　■　■　■　■

Descriptive research is a study of status and is widely used in education and the behavioral sciences. Its value is based on the premise that problems can be solved and practices improved through objective and thorough description. The most common descriptive research method is the **survey.** The survey is generally broad in scope. The researcher usually seeks to determine present practices (or opinions) of a specified population. The survey is used in education, psychology, sociology, and physical activity. The questionnaire, the Delphi method, the personal interview, and the normative survey are the main types of surveys.

The Questionnaire

The **questionnaire** and the **interview** are essentially the same except for the method of questioning. Questionnaires are usually answered in writing, whereas interviews are usually conducted orally. The procedures for developing questionnaire and interview items are similar. Consequently, much of the discussion regarding the steps in the construction of the questionnaire also pertains to the interview.

Researchers use the questionnaire to obtain information by asking participants to respond to questions rather than by observing their behavior. The obvious limitation of the questionnaire is that the results consist simply of what people say they do or what they say they believe or like or dislike. However, certain information can be obtained only in this manner, so it is imperative that the questionnaire be planned and prepared carefully to ensure the most valid results. There are eight steps in the survey research process, and we discuss each in this section.

Determining the Objectives

This step may seem too obvious to mention, yet countless questionnaires have been prepared without clearly defined objectives. In fact, poor planning may account for the low esteem in which survey research is sometimes held. The investigator must have a clear understanding of what information is needed and how each item will be analyzed. As with any research, the analysis is determined in the planning phase of the study, not after the data have been gathered.

The researcher must decide on the questionnaire's specific purposes: What information is wanted? How will the responses be analyzed? Will they merely be described by listing the percentages of participants who responded in certain ways, or will the responses of one group be compared with those of another?

survey

Technique of descriptive research that seeks to determine present practices or opinions of a specified population; can take the form of a questionnaire, interview, or normative survey.

questionnaire

Type of paper-and-pencil survey used in descriptive research in which information is obtained by asking participants to respond to questions rather than by observing their behavior.

interview

Survey technique similar to the questionnaire except that participants are questioned and respond orally rather than in writing.

One of the most common mistakes in constructing a questionnaire is not specifying the variables to be analyzed. In some cases, when investigators fail to list the variables, they ask questions unrelated to the objectives. In other cases, the investigator forgets to ask pertinent questions.

Delimiting the Sample

Most researchers who use questionnaires have in mind a specific population to be sampled. Obviously, the participants selected must be the ones who have the answers to the questions. In other words, the investigator must know who can supply what information. If information about policy decisions is desired, the respondents should be those involved in making such decisions.

Sometimes the source used in selecting the sample is inadequate. For example, some professional associations are made up of teachers, administrators, professors, and other allied professional workers. Thus, such an association's membership is not a good choice for participants for a study that is geared only for public school teachers. Unless there is some screening mechanism regarding place of employment, many incomplete questionnaires will be returned because the questions were not applicable.

The representativeness of the sample is an important consideration. Stratified random sampling, as discussed in chapter 6, is sometimes used. In a questionnaire surveying the recreational preferences of a university student body, the sample should reflect the proportion of students at the different class levels. Thus, if 35% of the students are freshmen, 30% are sophomores, 20% are juniors, and 15% are seniors, then the sample should be selected using those percentages. Similarly, if a researcher is studying school program offerings and 60% of the schools in the state have fewer than 200 students enrolled, then 60% of the sample should be from such schools.

The selection of the sample should be based on the variables specified to be studied. This affects the generalizability of the results. If an investigator specifies that the study deals with just one sex, one educational level, or one institution, then the population is narrowly defined, and it may be easy to select a representative sample of that specific population. However, the generalizations that can be made from the results are also restricted to that specific population. On the other hand, if the researcher is aiming the questionnaire at all of a specific population (e.g., all athletic directors or all fitness instructors), then the generalizability is enhanced, but the sampling procedures become more difficult. The representativeness of the sample is more important than its size.

Sampling Error

Frequently, you read in the newspapers about a national poll about some national issue or whether somebody should resign or whether some defendant received a fair trial. Generally, there is a statement about the sample size and sampling error, such as, "The poll involved 1,022 Americans, and the results were subject to sampling error of plus or minus 3.1 percentage points." Ever wonder how they arrive at that error? We discussed standard error in earlier chapters. Standard error of sampling can be expressed as

$$\text{SE} = \sqrt{\text{variance} / n} \tag{15.1}$$

where SE is the standard error of sampling and n is the size of the sample. We know that variance can be defined as the sum of the squared deviations from the sample mean divided by n. In this case, however, variance is calculated as a function of proportions, that is, the percentage of a sample that has a certain characteristic or gives a certain response. Variance of a proportion is calculated as $p(1 - p)$, where p is the proportion that has a characteristic or responds a certain way and $1 - p$ is the proportion that does not or responds differently. These proportions could be of many things, such as males and females, voters and nonvoters, smokers and nonsmokers, or those who believe one way versus those who believe differently, as is common in surveys. So the formula can be written

$$SE = \sqrt{p(1 - p) / n} \qquad (15.2)$$

If we do not know the proportions (or expected proportions), $p = .50$ is used. This proportion results in the largest error. Let's go back to our example of the survey of 1,022 Americans with the 3.1% sampling error. Using equation 15.2, we have

$$SE = \sqrt{.50(1 - .50) / 1022} = .0156, \text{ or } 1.6\%$$

However, remember that standard errors are interpreted as standard deviations in that ±1 SE includes roughly 68% of the population; to say it another way, we are only 68% confident that our sample is representative. So most polls use a 95% confidence interval, which we know is ±1.96 SE. Hence, our finalized formula should be

$$\begin{aligned} SE &= 1.96\sqrt{p(1 - p) / n} \\ &= 1.96\sqrt{.50(1 - .50) / 1022} \\ &= 3.1\% \end{aligned} \qquad (15.3)$$

Sample Size

The size of the sample needed is an important consideration from two standpoints: (1) for adequately representing a population and (2) for practical considerations of time and cost. Certain formulas can be used to determine the adequacy of sample size (Vockell, 1983, 111–118). These formulas involve probability levels and the amount of sampling error deemed acceptable. The practical considerations of time and cost need attention in the planning phase of the study. Surprisingly, students often ignore these considerations until they are forced to sit down with a calculator and tabulate the costs of printing, initial mailing (which includes self-addressed, stamped return envelopes), follow-ups, scoring, and data analysis. Sometimes the costs are so substantial that a sponsoring agency or grant must be found to subsidize the study, or else the project must be narrowed or abandoned entirely. Time is also important with regard to the availability of participants, possible seasonal influences, and various deadlines. The reader is referred to Fink (2003) and Fowler (2002) for discussions on survey sampling, including the calculation of sample size.

Constructing the Questionnaire

The notion that constructing a questionnaire is easy is a fallacy. Questions are not simply "thought up." Anyone who prepares a questionnaire and asks someone to read it soon discovers that it is not such an easy task after all. Those questions that were so clear and concise to the writer may be confusing and ambiguous to the respondent.

One of the most valuable guidelines for writing questions is to continually ask yourself what specific objective this question is measuring. Then ask how you are going to analyze the response. While you are writing questions, it is a good idea to prepare a blank table that includes the categories of responses, comparisons, and other breakdowns of data analysis so that you can readily determine exactly how each item will be handled and how each will contribute to the objectives of the study. Next, you must select the format for the questions, some examples of which follow.

Open-Ended Questions

Open-ended questions, such as "How do you like your job?" or "What aspects of your job do you like?" may be the easiest for the investigator to write. Such questions allow the respondent considerable latitude to express feelings and to expand on ideas. However, several drawbacks to open-ended questions usually make them less desirable than **closed questions.** For example, most respondents do not like open-ended questions. For that matter, most people do not like questionnaires because they feel that they are encroachments on their time. Also, open-ended questions require more time to answer than closed

open-ended question

Category of question in questionnaires and interviews that allows the respondent considerable latitude to express feelings and to expand on ideas.

closed question

Category of question found in questionnaires or interviews that requires a specific response and that often takes the form of rankings, scaled items, or categorical responses.

questions. Another drawback is limited control over the nature of the response: The respondent often rambles and strays from the question. Also, such responses are difficult to synthesize and to group into categories for interpretation. While open-ended items can yield valuable information, they are hard to analyze by any means other than simple description.

Sometimes open-ended questions are used to construct closed questions. Student course evaluations are often developed by having students list all the things they like and dislike about a course. From such lists, closed questions are constructed by categorizing the open-ended responses.

Closed Questions

Closed questions come in a variety of forms. A few of the more commonly used closed questions are rankings, scaled items (some measurement scales were covered in chapter 11), and categorical responses.

ranking

Type of closed question that forces the respondent to place responses in a rank order according to some criterion.

A **ranking** forces the respondent to place responses in a rank order according to some criterion. As a result, value judgments are made, and the rankings can be summed and analyzed quantitatively. Here is an example of a rank-order response question:

Rank the following activities with regard to how you like to spend leisure time. Use numbers 1–5, with 1 being the most preferred and 5 the least preferred.

___ Reading

___ Watching television

___ Arts and crafts

___ Vigorous sports such as tennis and racquetball

___ Mild exercise activity such as walking

scaled items

Type of closed question that requires participants to indicate the strength of their agreement or disagreement with some statement or the relative frequency of some behavior.

Scaled items are one of the most commonly used types of closed questions. Participants are asked to indicate the strength of their agreement or disagreement with some statement or to cite the relative frequency of some behavior, as in the following example:

Indicate the frequency with which you are involved in committee meetings and assignments during the academic year.

Rarely   Some   Often   Frequently

A Likert-type scale is a scale with three to nine responses, in which the intervals between responses are assumed to be equal:

In a required physical education program, students should be required to take at least one dance class.

Strongly agree   Agree   Undecided   Disagree   Strongly disagree

The difference between "strongly agree" and "agree" is considered equivalent to the difference between "disagree" and "strongly disagree." Different response words can be used in scaled responses, such as "excellent," "good," "fair," "poor," and "very poor"; "very important," "important," "not very important," and "of no importance"; and so on.

categorical response

Type of closed question that offers the participant only two responses, such as yes or no.

Categorical responses offer the respondent only two choices. Usually, the responses are yes and no or "agree" and "disagree." An obvious limitation of categorical responses is the lack of other options such as "sometimes" or "undecided." Categorical responses do not require as much time to administer as scaled responses but also do not provide as much information about the respondent's degree of agreement or frequency of the specified behavior.

Sometimes questions in questionnaires are keyed to the responses of other items. For example, a question might ask whether the respondent's institution offers a doctoral degree.

If the answer is yes, then he or she would be directed to answer subsequent questions about the doctoral program. If the respondent answers no, he or she is directed either to stop or to move on to the next section.

Borg and Gall (1989) offered the following rules for the construction of questionnaire items:

- The items must be clearly worded so that the items mean the same to all respondents. Avoid words that have no precise meaning, such as "usually," "most," and "generally."

- Use short questions rather than long questions because they are easier to understand.

- Do not use items that have two or more separate ideas in the same question, for example, "Although everyone should learn how to swim, passing a swimming test should not be a requirement for graduation from college." This item cannot really be answered by a person who agrees with the first part of the sentence but not the last, or vice versa. Another example is, "Does your department require an entrance examination for master's and doctoral students?" This is confusing because the department may have such an examination for doctoral students but not for master's students, or vice versa. If the response choices are only yes and no, a no response would indicate no examination for either program, and a yes response would mean exams for both. This should be divided into two questions.

- Avoid using negative items, such as "Physical education should not be taught by coaches." Negative items are often confusing, and the negative word is sometimes overlooked, causing the individual to answer in exactly the opposite way from what he or she intended.

- Avoid technical language and jargon. Attempt to achieve clarity and the same meaning for everyone.

- Be careful that you do not bias the answer or lead the respondent to answer in a certain way.

Sometimes questions can be stated in such a way that the person knows what is the "right" response. For example, you would know what response was expected if you encountered the question, "Because teachers work so hard, shouldn't they receive higher salaries?" The same advice applies to the problem of threatening questions. If the respondent perceives certain items to be threatening, he or she will probably not return the questionnaire. A questionnaire for teachers on grading practices, for example, may be viewed as threatening because poor grading practices would indicate that the teachers are not doing a good job. Thus, a teacher who feels threatened either may not return the questionnaire or may give responses that seem to be "right answers."

Results of experiments on surveys show that very minor changes in wording, or even in ordering the alternatives, can cause differences in responses. At a seminar at Washington State University, Don Dillman, a recognized authority on the survey method, reported results of experiments on the effects of question form and survey mode on survey results. As an illustration of some of the variations studied, when students were asked how many hours they studied per day, 70% said more than 2.5 hours, but when the same question was asked another way, only 23% indicated more than 2.5 hours.

Appearance and Design

Finally, the entire appearance and format of the questionnaire can have a significant bearing on the return rate. Questionnaires that appear to be poorly organized and prepared are likely to be "filed" in the wastebasket. Remember, many people have negative attitudes toward questionnaires, so anything that the researcher can do to overcome this negative attitude enhances the likelihood that the questionnaire will be answered. Some suggestions are merely cosmetic, such as the use of colored paper or an artistic design. Even little things, such as having dotted lines from the questionnaire item to the response options or grouping related items together, may pay big dividends.

The questionnaire should provide the name and address of the investigator. It is especially important that the instructions for answering the questions are clear and complete and that examples are provided for any items that are anticipated to be difficult to understand.

The first few questions should be easy to answer; the respondent is more likely to start answering easy questions and is also more apt to complete the questionnaire once he or she is committed. A poor strategy is to begin with questions that require considerable thought or time to gather information. Every effort should be made to make difficult questions as convenient to answer as possible. For example, questions that ask for enrollment figures, size of faculty, number of graduate assistants, and so on can often ask for responses in ranges (e.g., 1–10, 11–20, 21–30). First, you will probably group the responses for analysis purposes anyway. Second, the respondent can often answer range-type questions without having to consult the records or at least can supply the answers more quickly. An even more basic question to ask yourself is, Do I really need that information? or Does it pertain to my objectives? Unfortunately, some investigators simply ask for information off the top of their heads with no consideration for the time required to supply the answer or whether the information is relevant.

Generally, shorter questionnaires have higher response rates and more validity than longer ones. It is not a matter of more data, but better data (Punch, 2003). According to W.R. Borg and Gall (1989), an analysis of 98 questionnaire studies showed that, on average, each page added to a questionnaire reduced the number of returns by 0.5%. Because many people are prejudiced against questionnaires, the cover letter (discussed later) and the size and appearance of the questionnaire are crucial. A lengthy questionnaire requiring voluminous information will very likely be put aside until later if not discarded immediately.

Conducting the Pilot Study

A pilot study is recommended for any type of research but is imperative for a survey. Actually, the designer of a questionnaire may be well advised to do two pilot studies. The first trial run consists simply of asking a few colleagues or acquaintances to read over the questionnaire. These people can provide valuable critiques about the questionnaire format, content, expression and importance of items, and whether questions should be added or deleted.

After revising the questionnaire in accordance with the criticisms obtained in the first trial run, respondents who are a part of the intended population are selected for the second pilot study. The questionnaire is administered, and the results are subjected to item analysis (discussed in chapter 11). In some questionnaires, correlations may be determined between scores on each item and the whole test to see whether the items are measuring what they are intended to measure. Responses are always examined to determine whether the items seem clear and appropriate. First, questions that are answered the same way by all respondents need to be evaluated; they probably lack discrimination. Unexpected responses may indicate that the questions are poorly worded. Some rewording and other changes might also be necessary if the participants, who might be sensitive to some questions, do not respond to them. Furthermore, the pilot study determines whether the instructions are adequate. Another value of the pilot study relates to determining the length of the survey. The researcher should record how long it takes the average participant to complete the survey.

A trial run of the analysis of results should always be accomplished in the pilot study. The researcher can see whether the items can be analyzed in a meaningful way and then ascertain whether some changes are warranted for easier analysis. This is one of the most profitable outcomes of the pilot study. Of course, if substantial changes are mandated by the results of the pilot study, another pilot study is recommended to determine whether the questionnaire is ready for initial mailing.

cover letter

The letter attached to a survey that explains the purposes and importance of the survey.

Writing the Cover Letter

Unquestionably, the success of the initial mailing depends largely on the effectiveness of the **cover letter** that accompanies the questionnaire. If the cover letter explains the purposes and importance of the survey in a succinct and professional manner (and if the purposes are worthy of study), the respondent will likely become interested in the problem and will be inclined to cooperate.

An effective cover letter should also assure respondents that their privacy and anonymity will be maintained. Furthermore, the cover letter should make an appeal for the respondent's cooperation. Some subtle flattery about the respondent's professional status and the importance of his or her response may be desirable. This must be done tactfully, however, and only when appropriate. In this situation, the person's name and address should appear on the cover letter. The letters should look as if they were individually typed, even when a word processor is used. It is insulting to try to convince people that they have been handpicked for their expertise and valued opinions when the letters are addressed "Dear Occupant."

If the survey is endorsed by recognized agencies, associations, or institutions, specify this in the cover letter. If possible, use the organization's or institution's stationery. Respondents will be much more cooperative if some respected person or organization is supporting the study. Also acknowledge whether financial support is being given and by whom. Identify yourself by name and position. If the study is part of your thesis or dissertation, give your advisor's name. Sometimes it is advantageous to have the department chairperson or the dean of the college sign the letter. To increase the response rate, contact individuals by letter, card, or telephone, asking for their participation in the survey. Provided that the purposes of the study are worthwhile, offering the respondent a summary of the results is usually effective at this time. Before making this offer, give ample consideration to how much work and money such a summary will entail. However, if you do offer to provide a summary, be sure that you follow through on your promise.

Because a questionnaire imposes on a person's time, strategies involving rewards and incentives are sometimes used in an effort to amuse or involve the respondent. Some questionnaires include money (perhaps a dollar) as a token of appreciation. This may appeal to a person's integrity and evoke cooperation. Then again, you may just be out a dollar. The disadvantage is that inflation plays havoc with the effects of such rewards. A quarter may have been effective years ago, whereas one or more dollars may now be required to elicit the same "sense of guilt" for not responding. Research results concerning the effectiveness of monetary incentives have been mixed. Denton and Tsai (1991) compared several sizes of monetary incentives and a raffle for a professional journal and failed to find significant benefits.

The cover letter should request that the questionnaire be returned by a certain date. (This information should also be specified on the questionnaire.) When establishing the date for the return of data, consider such things as the respondents' schedules, responsibilities, vacations, and so forth. Be reasonable in the time you allow the person to respond; however, it is advisable not to give the respondent too much time because of the tendency to put a questionnaire aside and later forget it. One week is ample time for the respondent to answer (in addition to the mailing time).

The appearance of the cover letter is just as important as the appearance of the questionnaire. Grammatical errors, misspelled words, and improper spacing and format give the respondent the impression that the author does not attach much importance to details and that the study will probably be poorly done.

Researchers have tried a number of subtle and tactful approaches to establish rapport with the respondent. Some have tried a very solemn appeal to the monumental importance of the survey, some have attempted casual humor, and others have tried a folksy approach. Obviously, the success of any approach depends on the skill of the writer and the receptiveness of the reader. Any of these attempts can backfire.

A number of years ago, one of this book's authors (Nelson) received via his department head a note to which a letter like this was attached:

Dear Dr. _____:

It has been said: "There are two kinds of information, knowing it or knowing where to find it." I asked three men you know, Doctors Eeney, Meeney, and Moe, a certain question, and each said: "Sorry. I don't know his name." But they came back with this helpful suggestion: write to you.

Now for the question: What is the name of the person in your department or school who is in charge of your graduate program in physical education? . . .

Cordially yours,

Harry Homespun

The department head answered the letter, naming Nelson as head of the graduate program; shortly thereafter, this letter arrived:

Dear Dr. Nelson:

Your good work in administering a graduate program is well known through-out our profession. Even though there are many knotty problems, everyone wants to upgrade the resources for graduate studies. But how? Because of your scholarly approach, my associates Doctors Eeney, Meeney, and Moe suggested that I write to you.

It has been said: "If you want to get something important done, ask a busy person." My associates tell me you are busier than a bird dog in tall grass. They also say that you have a high regard for excellence. . . . If you could find it convenient to return your checked copy on or before May 1, you might help reduce the cultural lag.

Cordially yours,

Harry Homespun

We have, of course, omitted the main parts of the letters concerning what was being studied and what the instructions specified, but these parts were generally well done.

About a year later, another questionnaire arrived from the same university with essentially the same folksy approach, including the bird dog in tall grass analogy. It is safe to assume that these two graduate students had the same research methods course at the same institution. Most likely, in the discussion of the survey method, some examples of cover letters were given with different approaches. Professors Eeney, Meeney, and Moe would undoubtedly be embarrassed if they knew that the two students had written such nearly identical letters.

In most respects, the cover letters contained the essential information, and the topics were worthwhile. The authors were simply too heavy-handed in their attempts at a down-home approach, and their efforts to flatter the respondent regarding his "well-known" expertise lacked subtlety.

Page 277 shows an example of a cover letter for a questionnaire that deals with a potentially threatening topic. The letter effectively explains to the respondent how confidentiality will be ensured. The second cover letter (page 278) focuses on the importance of the topic and on the agencies endorsing the study. It also tactfully appeals to the respondent's ego. Bourque and Fielder (2003) offer helpful suggestions regarding what should be included in cover letters.

Sample Cover Letter #1

Assuring confidentiality

Dear _____

Your participation in a national survey of perceived leader behavior in physical education is needed. As a doctoral student in physical education at the University of Georgia, I am conducting this study to compare and contrast male and female faculty members' perceptions of their male and female department heads' leader behavior. This institution was randomly selected to take part in this research project. Your department head has already been contacted and expressed willingness to take part in the study. I am now asking randomly selected faculty members from your institution to become involved. Your name was one of those selected to ask to participate by answering the enclosed questionnaires.

Participation will require approximately 15 minutes of your time to answer both questionnaires that will be used and to fill out an information sheet. The questionnaires and instructions as to how they are to be completed have been included with this letter in the hopes you will agree to be a participant. There is no evaluation intended or implied with this study. The instruments used describe the perceived leader behavior of the administrator. The analysis of the data will be utilized in group mean scores. Following the completion of the survey and the statistical analysis of the data, I will gladly send you a summary of the findings. All data will be dealt with confidentially and no institution or individual taking part in the study will be identified.

Hopefully, you will find time in your busy schedule to participate in this study. Thank you for your time and participation. I look forward to your early response.

Sincerely,

Kay A. Johnson

Clifford G. Lewis
Major Professor

Enclosures: 5

Reprinted courtesy of C.G. Lewis.

Sending the Questionnaire

The investigator needs to carefully consider the best time for the initial mailing. Such considerations include holidays, vacations, and especially busy times of the year for the respondents. A self-addressed, stamped envelope should be included. It is almost an insult to expect respondents not only to answer your questionnaire but also to provide an envelope and a stamp to mail it back to you.

Other matters regarding the mailing of the questionnaire, such as establishing the date to be returned, have been covered in previous sections. The initial mailing represents a substantial cost to the sender. Securing a sponsor to underwrite or defray expenses and using bulk-mailing services are important considerations.

Lately, researchers have been utilizing e-mail, fax correspondence, and the Internet to conduct surveys. Generally, the results in terms of return rate have been good. Naturally, these methods limit the participant pool to people with access to these kinds of technology.

Sample Cover Letter #2

Justifying the topic

Dear _____

The responsibilities that I have in providing leadership in physical education for a large school district have caused me to give consideration to the approach needed to achieve acceptable objectives of physical fitness in the secondary boys' physical education program.

Through the cooperation of the San Diego Unified School District, the School of Education and the Physical Education Department of the University of California, Los Angeles, and the Bureau of Health Education, Physical Education and Recreation of the California State Department of Education, I am undertaking a study in the area of physical fitness. Representatives of the President's Council on Physical Fitness have said this information would be most valuable for them also. The purpose of the study is to establish criteria that can be utilized in planning activities in physical education classes for secondary aged boys (12-18) to develop and maintain selected aspects of physical fitness. The selected aspects to be considered are those commonly referred to as cardiorespiratory endurance (stamina), muscular strength and endurance, and flexibility. The information gained will be used in program planning by helping to answer such questions as these: What part of the physical education lesson should be spent on specific fitness activities (such as calisthenics, interval running, weight training, etc.) and what part of the lesson should be spent on sports and games of our culture (such as basketball, volleyball, softball, track, tennis, gymnastics, dance instruction, etc.)? What guidelines should be established in planning the specifics of physical fitness work? What physical fitness activities should be used?

A committee of exercise physiologists nominated you as a person who is qualified to comment on this topic because of your recognized position as a leader and your research in the area of physical fitness. It would be most appreciated if you would answer the enclosed questions and return them in the envelope provided at your earliest convenience. A summary of the answers will be returned to you.

Sincerely yours,

Physical Education Specialist

San Diego Unified School District. Reprinted by permission.

Following Up

This should not come as a big shock, but it is unlikely that you will get 100% return on the initial mailing. A follow-up letter is nearly always needed, and this can be done in many different ways. One approach is to wait about 10 days after the initial mailing and then to send a card to everyone in the sample, stressing the importance of their participation (and apologizing if they have already returned the questionnaires). Approximately 10 days after the card, another letter with another copy of the questionnaire and another self-addressed, stamped envelope should be sent to those who have not responded. This is expensive, of course, but follow-ups are effective. In a number of cases, the person has forgotten to respond, and a mere reminder will prompt a return. Other nonrespondents who had not initially planned to return the questionnaire may be influenced by the researcher's efforts in reiterating the significance of the study and the importance of his or her input.

The follow-up letter should be tactful. The person should not be chastised for failing to respond. The best approach is to write as though the person would have responded had it not been for some oversight or mistake by the investigator (for an example of a follow-up letter, see below).

Sample Follow-Up Letter

Dear _____

I sent a questionnaire regarding criteria for planning physical fitness work in physical education classes to you a few weeks ago and have not heard from you. As you can appreciate, it is important that we obtain response from everyone possible inasmuch as only a few select individuals were contacted. Our school district is planning an immediate study and updating of its physical education program based on the results of this study so it is of vital concern.

The questionnaire was sent during the summer when you may have been away from your office. I have included another copy, however, and it would be most helpful if you could take 30-45 minutes to give your opinions on the information requested.

Thank you so much for your cooperation.

Sincerely yours,

Physical Education Specialist

San Diego Unified School District. Reprinted by permission.

Nonresponse is especially likely when the questionnaire deals with some sensitive area. For example, surveys about program offerings and grading practices are often not returned by schools with inadequate programs and poor grading practices. Thus, the obtained responses are apt to be biased in favor of the better programs and better teachers. Regardless of the nature of the survey, the results from a small return rate (e.g., 10% to 20%) cannot be given much credibility. People who have a particular interest in the topic being surveyed are more likely to respond than people who are less interested. These respondents are self-selected, and the responses are almost invariably biased in ways that are directly related to the purposes of the research (Fowler, 2002). K.E. Green (1991) reported that individuals who responded to the initial mailing had more favorable attitudes toward the topic and also had more positive attitudes about themselves than reluctant respondents.

In fact, in most studies in which more than 20% of the questionnaires are not returned, it is recommended that a sample of the nonrespondents be surveyed. Of course, this is not easy. If people have ignored the initial mailing and one or two follow-ups, the chances are not good that they will respond to another request, but it is worth a try. The preferred technique is to randomly select a small number (e.g., 5% to 10%) of the nonrespondents by using the table of random numbers (this technique is sometimes called "double-dipping"). Then contact is made either by telephone or by a special cover letter (perhaps even via registered mail). After the responses have been obtained, comparisons are made between the answers of the nonrespondents and the answers of the people who responded initially. If the responses are similar, you can assume that the nonrespondents are not different from those who replied. If there are differences, you should either try to get a greater percentage of the nonrespondents or be sure that these differences are noted and discussed in the research report.

To follow up on nonrespondents, you have to know who has not responded. Keeping records of those who have and have not responded may seem to fly in the face of guaranteed anonymity. However, this is not a difficult task. An identifying number written on the questionnaire or return envelope can be used. Of course, it is recommended that you inform the respondents about the use of the number in your cover letter. You might offer an explanation such as the following:

> **Y**our responses will be strictly confidential. The questionnaire has an identification number for mailing purposes only. With this numbering system, I can check your name off the mailing list when your questionnaire is returned. Your name will never be placed on the questionnaire.

Another approach, which works well, is to send a separate postcard (self-addressed and stamped) with the questionnaire. The card contains an identifying number and is to be mailed back separately when the person returns the questionnaire. The card simply states that the person is sending the questionnaire and therefore there is no need for the researcher to send any more reminders. This procedure assures anonymity and yet informs the researcher that the person has responded (Fowler, 2002).

A good return rate is essential. We frequently read journal and newspaper reports of surveys in which something like 2,000 surveys were mailed and 600 were returned. This sounds like a respectable number, but these 600 people represent only a 30% return. They are not a random sample. They are self-selected, and their responses may be quite different from those of the 1,400 who did not respond for whatever reason.

Analyzing the Results and Preparing the Report

These last two steps are discussed in chapter 20, which deals with the results and discussion sections of the research report. The main consideration is that the method of analysis should be chosen in the planning phase of the study. Many questionnaires are analyzed merely by tallying the responses to the various items and reporting the percentage of the respondents who answered one way and the percentage who answered another way. Often, not much in the way of meaningful interpretation can be gained if only a simple tally is given. For example, when the researcher states simply that 18% of the respondents strongly agreed with some statement, 29% agreed, 26% disagreed, 17% strongly disagreed, and 10% had no opinion, the reader's reaction may be, "So what?" Questionnaires, like all surveys, must be designed and analyzed with the same care and scientific insight as experimental studies.

The Delphi Method

Delphi survey method

Survey technique that uses a series of questionnaires in such a way that the respondents (usually experts) reach a consensus about the subject.

round

Stage of the Delphi survey method in which respondents are asked their opinions and evaluations of various issues, goals, and so on.

The **Delphi survey method** uses questionnaires but in a different manner than the typical survey. The Delphi technique uses a series of questionnaires in such a way that the respondents finally reach a consensus about the topic. It is basically a method of using expert opinion to help make decisions about practices, needs, and goals.

The procedures include the selection of experts, or informed people who are to respond to the series of questionnaires. A set of statements or questions is prepared for consideration. Each stage in the Delphi technique is called a **round.** The first round is mostly exploratory. The respondents are asked for their opinions on various issues, goals, and so on. Open-ended questions may be included to allow the participants to express their views and opinions.

The questionnaire is then revised as a result of the first round and sent to the respondents, asking them to reconsider their answers in light of the analysis of all respondents to the first questionnaire. Subsequent rounds are carried out, and the respondents are given summaries of previous results and asked to revise their responses if appropriate. Consensus about the issue is finally achieved through the series of rounds of analysis and subsequent considered judgments. Anonymity is a prominent feature of the Delphi method, and the consensus of recognized experts in the field provides a viable means of confronting important issues. For example, the Delphi method is used sometimes to determine curricular content for programs, to decide on the most important objectives of a program, and to agree on the best solution to a problem.

The Personal Interview

As mentioned earlier, the steps for the interview and the questionnaire are basically the same. The focus in this section is only on the differences.

Preparing for the Interview

The most obvious difference between the questionnaire and the interview is in the gathering of the data. In this respect, the interview is more valid because the responses are apt to be more reliable. Also, there is a much greater percentage of returns.

Participants should be selected using the same sampling techniques as for a questionnaire. Generally, the interview uses smaller samples, especially when a graduate student is doing the survey. Cooperation must be secured by contacting the participants selected for interviewing. If some of those chosen refuse to be interviewed, the researcher must consider possible bias to the results, as was done with nonrespondents in a questionnaire study.

To effectively conduct an interview takes a great deal of preparation. Graduate students sometimes have the impression that anyone can do it. The same procedures used with the questionnaire are followed in preparing the items, with which the interviewer must be very familiar. The researcher must carefully rehearse the interview techniques. One source of invalidity is that the interviewer tends to improve with experience, and thus the results of earlier interviews may differ from interviews conducted later in the study. A pilot study is very important. The interviewer can make sure that the vocabulary level is appropriate and that the questions are equally meaningful given the ages and educational backgrounds of the participants. The interviewer must use careful planning in organizing the questions and visual aids to provide a comfortable flow of presentation and transition of question to question.

Training is required in making initial contact and presenting the verbal "cover letter" by phone. At the meeting, the interviewer must establish rapport and make the person feel at ease. If a tape recorder will be used, permission must be obtained. If a tape recorder will not be used, the interviewer must have an efficient system of coding the responses without consuming too much time and appearing to be taking dictation. The interviewer must not inject his or her own bias into the conversation and certainly should not argue with the respondent. Although the interview holds many advantages over the questionnaire with regard to the flexibility of the questioning, there is also a danger of straying from the questions and getting off the subject. The interviewer must tactfully keep the respondent from rambling, and this requires skill.

The key to getting good information is to ask good questions. Some surveys that employ in-person interviews are quantitative and the questions may be quite similar to a mailed questionnaire. Other surveys deal with qualitative data. In these studies there is less concern with standardization and more emphasis is given to description. Good interviewers in qualitative surveys do not ask yes-or-no questions, they do not ask multiple questions disguised as a single question, and they try to avoid inserting their own points of view. Merriam (1988) defined four major categories of questions: hypothetical, devil's advocate, ideal position, and interpretive. Here are some examples of each:

- Hypothetical: "Let's suppose this is my first day of student teaching. What would it be like?"
- Devil's advocate: "Some people say that professional educational courses are of little value for the student-teaching experience. What would you say to them?"

- Ideal position: "What do you think the ideal student-teaching preparation program should be like?"
- Interpretive: "Would you say that the student-teaching experience is different from what you expected?"

You may recognize these approaches from interviews you have seen on television. The interview has the following advantages over the mailed questionnaire:

- The interview is more adaptable. Questions can be rephrased and clarification can be sought through follow-up questions.
- The interview is more versatile with regard to the personality and receptiveness of the respondent.
- The interviewer can observe how the person responds and can thus achieve greater insight into the sensitivity of the topic and the intensity of feelings of the respondent. This can add considerably to the validity of the results, as respondents' avoiding sensitive topics is one of the greatest threats to validity in questionnaire studies.
- Because each person is contacted before the interview, interviews have a greater rate of return. Moreover, people tend to be more willing to talk than to fill out a questionnaire. A certain amount of ego is involved because a person who is interviewed feels more flattered than one who receives only a list of questions.
- An advantage of in-person interviews is that visual aids such as flash cards can be used to simplify long questions and explain the response lists.

A potential problem in interviews is losing the data. Because of confidentiality restrictions, names may not be placed on the interview data instrument. Consequently, the researcher may later get confused as to which data belong to which respondents. The researcher should use several identifiers for each interview. Oishi (2003) provides an excellent coverage of the dos and don'ts in planning and conducting the personal interview.

I don't think she understands that she's supposed to be doing a *research survey.*

Conducting a Telephone Interview

Interviewing by telephone is becoming more prevalent. Some of the advantages of telephone interviewing over face-to-face interviewing follow:

- Telephone interviewing is less expensive than traveling to visit respondents, especially when researchers use WATS lines. Bourque and Fielder (2003) reported that telephone interviews cost half as much as personal interviews and about twice as much as mail surveys.

- Researchers can conduct a telephone interviews in significantly less time than in doing in-person interviews.

- The interviewer can work from a central location, which facilitates the monitoring and the quality control of the interviews and provides a better opportunity for using computer-assisted interviewing techniques (Borg & Gall, 1989).

- Many people are more easily reached by telephone than by personal visit.

- The telephone interview enables the researcher to reach a wide geographical area; geography is a limitation of the personal interview. This advantage also can increase the validity of the sampling.

- There is some evidence that people respond more candidly to sensitive questions over the telephone than in personal interviews, in which the presence of the interviewer may inhibit some responses.

Using a computer in telephone interviewing eases data collection and analysis. For example, a computer can display the questions for the interviewer to ask, to which the interviewer types the responses obtained. Each response triggers the next question, so the interviewer doesn't have to turn pages and is less likely to ask inappropriate questions. Responses are stored for analysis (thus reducing scoring errors), and the stored responses can then be accessed for statistical analysis.

Among the disadvantages of telephone surveys is that it is increasingly difficult to get a representative sample due to cell phones, answering machines, caller ID, and call-blocking devices. Telemarketing has greatly soured the public on telephone interviews.

A precall letter can be effective in securing cooperation, thereby increasing the response rate. The drawback is that you must have the names, addresses, and telephone numbers of the target population. In some surveys, the researcher uses random-digit dialing (RDD) to select the sample. In this method, a table of random numbers can be used to select the four numbers after the three-number exchange. Both listed and unlisted numbers can be reached in this method.

Much time is consumed in callbacks because of no answer, answering machines, "unwanted" people answering the phone, busy signals, and language barriers. The researcher almost always must hire other people because of the numerous calls that must be made. An important source of bias concerns the time when telephone surveys are done. If conducted during the day, the sample is biased toward the people most likely to be home: homemakers, the unemployed, and the retired. Thus, the interviewers usually must work after 5:00 P.M. Also, it is estimated that one third of the people contacted are not home on the initial call, which means callbacks are necessary (sometimes as many as 10 or 12!).

A commonly cited weakness of telephone surveys has been that the sample is delimited to people who have telephones. However, today a vast majority of people in the United States have telephones. In 2000, 97.25% of all households had telephones (Bourque & Fielder, 2003). The percentage varies somewhat from state to state. Low income people, particularly in inner cities and rural areas, are less likely to have access to telephones. Certainly, if the target population is low-income respondents, the telephone survey is not recommended.

Bourque and Fielder (2003) give a thorough description of the relative advantages, disadvantages, and recommended methodology of the telephone survey.

The Normative Survey

normative survey

Survey method that involves establishing norms for abilities, performances, beliefs, and attitudes.

The **normative survey** is not described in most research methods textbooks. As is implied in the name, this method involves establishing norms for abilities, performances, beliefs, and attitudes. A cross-sectional approach is used: Samples of people of different ages, sexes, and other classifications are selected and measured. The steps in the normative survey are generally the same as in the questionnaire, the difference being the manner in which the data are collected. Rather than asking questions, the researcher selects the most appropriate tests to measure the desired performances or abilities, such as the components of physical fitness.

In any normative survey, it is important that the test be administered in a rigidly standardized manner. Deviations in the way measurements are taken give meaningless results. The researcher collects and analyzes the data from the survey by some norming method, such as percentiles, T scores, or stanines and then constructs norms for the different categories of age, sex, and so on.

The American Alliance for Health, Physical Education, Recreation and Dance (AAHPERD) has sponsored several normative surveys. Probably the most notable was the Youth Fitness Test (see AAHPER, 1958), conducted in response to the furor caused by the results of the Kraus-Weber test (Kraus & Hirschland, 1954), which had revealed that American children were inferior to European children in minimum muscular fitness. The Youth Fitness Test was originally given to 8,500 boys and girls in a nationwide sample. Follow-up testing was done in 1965 and 1975.

In the 1958 AAHPER normative survey, a committee determined a seven-item motor fitness test battery. The University of Michigan Survey Research Center selected a representative national sample of boys and girls in grades 5–12 and then requested the cooperation of each school. They prepared directions for giving the test items and selected and trained physical education teachers in various parts of the country to administer and supervise the testing.

In addition, AAHPERD conducted a sport skills testing project, which established norms for boys and girls of different ages in skills of a number of sports. The National Children and Youth Fitness Study (1985, 1987) was a normative survey that established norms for a health-related physical fitness test.

Sometimes comparisons are made between the norms of different populations. In other studies, the major purpose is simply to establish norms. The primary drawbacks of any normative survey occur in test selection and the standardization of testing procedures. With many test batteries, there is a danger of generalizing on the basis of the test's results. For example, when only one test item is used to measure a particular component (such as strength), it is possible to make unwarranted generalizations.

The standardization of testing procedures is essential for establishing norms. However, when a normative survey involves many testers from different parts of the country, this becomes a potential source of measurement error. Published test descriptions simply cannot address all the aspects of test administration and the ways of handling the many problems of interpreting test protocol that arise. Extensive training of testers is the answer, but this is often logistically impossible.

Summary

The most common descriptive research technique is the survey, which includes questionnaires and interviews. The two are similar except for the method of asking questions. The questionnaire is a valuable tool for obtaining information over a wide geographical area. A good cover letter is important in getting cooperation. One or two follow-ups are often necessary to secure an adequate response. Personal interviews usually yield more valid data because of the personal contact and the opportunity to make sure that respondents understand the questions. Telephone interviews are becoming increasingly popular. They have most of the advantages of personal interviews plus flexibility to involve more participants over a larger geographical area. The Delphi survey technique uses a series of questionnaires in such a way that the respondents eventually reach a consensus about the topic. It is frequently used to survey expert opinion in an effort to make decisions about practices, needs, and goals. The normative survey is designed to obtain norms for abilities, performances, beliefs, and attitudes. The various AAHPERD fitness tests developed over the years are examples of normative surveys. In all surveys, representative sampling is extremely important.

✔ Check Your Understanding

1. Find a thesis or dissertation that used a questionnaire to gather data. Briefly summarize the methodology, such as the procedures used in constructing and administering the questionnaire, selecting participants, following up, and so on.

2. Critique a study that used interviews for gathering data. Discuss the strengths and weaknesses of its description of methods.

Other Descriptive Research Methods

It is better to know some of the questions than all of the answers.—James Thurber

■ ■ ■ ■ ■ ■ ■

In chapter 15, different forms of the survey method of descriptive research were discussed. In this chapter, we address several other descriptive research techniques. One such technique is developmental research. Through cross-sectional and longitudinal studies, researchers investigate the interaction of growth and maturation and of learning and performance variables. The case study, in which the researcher gathers a large amount of information about one or a few participants, is widely used in a number of fields. The job analysis is a technique used to describe the working conditions, requirements, and preparation needed for a particular job. Through observational research, one obtains quantitative and qualitative data about people and situations through observing their behavior. Some research studies employ unobtrusive methods in which the participant is not aware that he or she is being studied. Correlational studies determine and analyze relationships between variables and generate predictions.

Developmental Research

Developmental research is the study of changes in behaviors across years. Although much of the developmental research has focused on infancy, childhood, and adolescence, research on senior citizens and even across the total human life span is increasingly common.

Longitudinal and Cross-Sectional Designs

Developmental research focuses on cross-age comparisons. For example, researchers can compare how far children can jump when they are 6, 8, and then 10 years old, or they can compare adults who are 45, 55, and 65 years old on their knowledge of the effects of obesity on life expectancy. Both of these are developmental studies. The major distinction between the two basic approaches in developmental studies is whether researchers follow the same participants over time (longitudinal design) or whether they select different participants at each age level (cross-sectional design).

   Longitudinal studies are powerful because the changes in behavior across the time span of interest are seen in the same people. However, longitudinal designs are time-consuming. A longitudinal study of children's jumping performance at ages 6, 8, and 10 obviously requires 5 years to complete. That is probably not a wise choice of design for a master's

developmental research

Study of changes in behavior across years.

longitudinal studies

Research in which the same participants are studied over a period of years.

thesis. Longitudinal designs have additional problems besides the time required to complete them. First, over the several years of the study, some of the children are likely to move away when parents change jobs or change schools when school districts are rezoned. In longitudinal studies of senior citizens, some may die over the course of the study. The problem is not knowing whether the sample characteristics remain the same when participants are lost. For example, when children are lost from the sample because parents change jobs, is the sample then composed of children of lower socioeconomic levels because the more affluent parents moved? Furthermore, if obesity is related to longevity, are older people more likely to be less obese and consequently have increased knowledge because the more obese individuals with less knowledge have died? Thus, knowledge about obesity may not change from ages 45 to 65; rather, the sample may change.

Another problem with longitudinal designs is that participants become increasingly familiar with the test items, and the items may cause changes in behavior. The knowledge inventory on obesity may prompt the participants to seek information about obesity, thus changing their knowledge, attitudes, and behaviors. Therefore, the next time they complete the knowledge inventory, they have gained knowledge. However, this gain of knowledge is the result of having been exposed to the test earlier and might not have occurred without that exposure.

cross-sectional study

Research in which samples of participants from different age groups are selected in order to assess the effects of maturation.

cohort problem

Problem in cross-sectional design of whether all the age groups are really from the same population.

Cross-sectional studies are usually less time-consuming to carry out. These studies test several age groups (e.g., 6, 8, and 10) at the same point in time. Although cross-sectional studies are more time efficient than longitudinal studies, a limitation called the **cohort problem** exists: Are all the age groups really from the same population (group of cohorts)? Asked another way, are the environmental circumstances that affect jumping performance for 6-year-olds the same today as when the 10-year-olds were 6, or have physical education programs improved over this 4-year span so that 6-year-olds receive more instruction and practice in jumping than the 10-year-olds did when they were 6? If the latter is true, then we are not looking at the development of jumping performance but rather at some uninterpretable interaction between "normal" development and the effects of instruction. The cohort problem exists in all cross-sectional studies.

Examples of longitudinal developmental studies are Halverson, Roberton, and Langendorfer (1982), who studied the velocity of the overarm throw by children from early elementary school through junior high school, and Nelson, Thomas, Nelson, and Abraham (1986) who investigated sex differences in throwing by children in kindergarten to third grade. Thomas et al. (1983) provided an example of a cross-sectional developmental study, in which they looked at the development of memory for distance information in different age groups. They compared the effects of a practiced strategy for remembering distance at each age level to show that the appropriate use of strategy reduces age differences in remembering distance. Each of these studies suffers from the specific defects associated with the type of developmental design. Both Halverson, Roberton, and Langendorfer (1982) and Nelson et al. (1986) lost participants over the several years of the study. In the latter study, measurements were taken from 100 children in kindergarten at a single school, and 3 years later only 25 children were still at that school. In a recent longitudinal study on children's throwing performance, Robertson and Konczak (2001) reported that 73 children originally were filmed but only 39 children completed the 7-year study. On the other hand, the cross-sectional study by Thomas et al. (1983) could not establish whether their younger children were more familiar with memory strategies than the older children were when they were younger.

Although both longitudinal and cross-sectional designs have some problems, they are the only means available to study development. Thus, both are necessary and important types of research. These two types of design are considered descriptive research. However, either can also be experimental research (chapter 18); that is, an independent variable can be manipulated within an age group. The Thomas et al. (1983) study manipulated the use of memory strategy within each of the three age groups. Thus, age was a categorical variable, whereas strategy was a true independent variable. This point is addressed here because developmental research is not covered in chapter 18 on experimental research.

Methodological Problems of Developmental Research

Whether the developmental research is longitudinal or cross-sectional, several methodological problems exist (for a more detailed discussion, see Thomas, 1984).

Unrepresentative Scores

One of the most common problems is an unrepresentative score. These unrepresentative scores, called outliers, and mentioned in previous chapters, occur in all research but are particularly problematic at the extremes of developmental research (children and senior citizens). Outliers frequently result from shorter attention spans, distraction, and lack of motivation to do the task. The best way to handle these unrepresentative scores is to

1. plan the testing situation within a reasonable time limit that accounts for attention span,
2. set up the testing situation where distractions cannot occur, and
3. be aware of what an unrepresentative score is and retest when one occurs.

The last thing a researcher wants is to use unrepresentative scores. Therefore, outliers not detected at testing should be found when the distribution of the data is studied. There are several ways to test for these extreme and unrepresentative scores (e.g., see Barnett & Lewis, 1978). The developmental researcher should expect and plan to handle outliers in the data set.

Unclear Semantics

A second problem in developmental research involves semantics. Selecting the words to use in explaining a task to various age groups of children is a formidable challenge. If the researcher is not careful, older children will perform better than younger children, primarily because they grasp the idea of what to do more quickly. Although the standard rule in good research is to give identical instructions to all the participants, you must bend this rule for developmental studies with children. The researcher must explain the test in a way that the participants can understand, and the researcher must obtain tangible evidence that children of various ages understand the test before it is conducted. To ensure that the children understand the test, the researcher frequently has them demonstrate the activity to some criterion level of performance before collecting the data.

A good example of the semantics problem involved a group of early elementary school children taking a computer course through a continuing education program at a local college. The children were doing fine until the teacher (a college instructor of computer science) began to write instructions on the blackboard. Then everyone stopped working. Finally, one of the children whispered to the teacher, "Some of us can't read cursive." After the teacher printed the instructions in block letters, all the children happily returned to their computing work (*Chronicle of Higher Education*, 1983).

Lack of Reliability

A third developmental research problem is the lack of reliability in younger children's performances. When obtaining a performance score for a child, it should be a reliable one; that is, if the child is tested again, the performance score should be about the same. Obtaining reliable performance is frequently a problem when testing younger children for many of the same reasons that outliers occur. Of course, making sure that the child understands the task must be the first consideration, and maintaining motivation the second. A task made fun and enjoyable is more likely to elicit a consistent performance. This can be done by the use of cartoon figures, encouragement, and rewards (for some ideas on how cartoon figures can be used to improve motivation on many gross motor tasks, see Herkowitz, 1984). The developmental researcher should maintain frequent reliability checks during testing sessions (for appropriate techniques, see chapter 11).

Statistical Problems

The final developmental research issue to mention is a statistical problem. A frequent means of making cross-age comparisons is ANOVA (see chapter 9), which assumes that the groups being compared have equal variances (spread of scores about the mean). However, researchers often violate this assumption in making cross-age comparisons. Depending on the nature of the task, older children may have considerably larger or smaller variances than younger children. A developmental researcher should be aware of this potential issue and some of the solutions. In particular, pilot work using the tasks of interest in the research should provide insight into this problem.

Protecting Participants

In chapter 5, we discussed the protection of human participants in research. Of course, this protection also pertains to children. Parents or guardians must grant permission for minors to participate in research. Researchers should obtain this permission as they do for adults, except that they give the explanation and consent forms to the parents or guardians. If minors are old enough to understand the methodology, you should also obtain their consent. This means explaining the purpose of the research in terms children can understand. Most public and private schools have their own requirements for approval of research studies.

The normal sequence of events for obtaining permission involves

- planning the research;
- acquiring approval of your university's committee for protection of human participants;
- locating and getting the approval of the school system, the school involved, and the teachers; and
- getting the approval of parents and, when appropriate, students.

You can see that this requires a good deal of paperwork. Thus, beginning the process well before you plan to begin data collection is essential.

The Case Study

case study

Form of descriptive research in which a single case is studied in depth to reach a greater understanding about other similar cases.

In the **case study,** the researcher strives for an in-depth understanding of a single situation or phenomenon. This technique is used in many fields, including anthropology, clinical psychology, sociology, medicine, political science, speech pathology, and various educational areas such as disciplinary problems and reading difficulties. It has been used considerably in the health sciences and to some extent in exercise science, sport science, and physical education.

The case study is a form of descriptive research. Whereas the survey method obtains a rather limited amount of information about many participants, the case study gathers a large amount of information about one or a few participants. Although the study consists of a rigorous, detailed examination of a single case, the underlying assumption is that this case is representative of many other such cases. Consequently, through the in-depth study of a single case, a greater understanding about similar cases is achieved. This is not to say, however, that the purpose of case studies is to make generalizations. On the contrary, drawing inferences about a population from a case study is not justifiable. On the other hand, the findings of a number of case studies may play a part in the inductive reasoning involved in the development of a theory.

The case study is not confined to the study of an individual but can be used in research involving programs, institutions, organizations, political structures, communities, and situations. The case study is used in qualitative research to deal with critical problems of practice and to extend the knowledge base of the various aspects of education, physical education, exercise science, and sport science (qualitative research is discussed in chapter 19).

Information about case study methodology is rather difficult to find. As Merriam (1988) observed, material on case study research strategies can be found everywhere and nowhere. Methodological material on case study research is scattered about in journal articles, conference proceedings, and research reports of the many different fields that use this form of research. Frustration in trying to find substantive material about case study research in an educational setting prompted Merriam (1988) to write her interesting and informative text *Case Study Research in Education*. Yin (2003) presents a comprehensive discussion on design and methods for case study research in the social sciences.

Types of Case Studies

In many ways, case study research is similar to other forms of research. It involves the identification of the problem, the collection of data, and the analysis and reporting of results. As in other research techniques, the approach and the analysis depend on the nature of the research problem. Case studies can be descriptive, interpretive, or evaluative.

Descriptive Studies

A descriptive case study presents a detailed picture of the phenomena but does not attempt to test or build theoretical models. Sometimes, descriptive case studies are historical in nature, and sometimes they are done for the purpose of achieving a better understanding of the present status. Descriptive case studies frequently serve as an initial step or database for subsequent comparative research and theory building (Merriam, 1988).

Interpretive Studies

Interpretive case studies also employ description, but the major focus is to interpret the data in an effort to classify and conceptualize the information and perhaps theorize about the phenomena. For example, a researcher might use the case study approach to better understand the cognitive processes involved in sport.

Evaluative Studies

Evaluative case studies also involve description and interpretation, but the primary purpose is to use the data to evaluate the merit of some practice, program, movement, or event. The efficacy of this type of case study relies on the competence of the researcher to use the available information to make judgments (Guba & Lincoln, 1981). The case study approach enables a more in-depth, holistic approach to the problem than may be possible with survey studies.

Case Study Participants

The selection of participants in a case study depends, of course, on the problem being studied. The individual (or case) may be a person (e.g., student, teacher, coach), a program (e.g., Little League baseball), an institution (e.g., a one-room school), a project, or a concept (e.g., mainstreaming). In most studies, random sampling is not used because the purpose of a case study is not to estimate some population value but to select cases from which one can learn the most. Chien (1981) used the term *purposive sampling*. Goetz and LeCompte (1984) referred to this concept as *criterion-based sampling*. The researcher establishes criteria necessary to include in the study and then finds a sample that meets the criteria. Criteria can include age, years of experience, evidence of level of expertise, and situation and environment. The case may be one classroom that meets certain criteria or a state that is involved in a specific program.

Characteristics of the Case Study

The case study involves the collection and analysis of many sources of information. In some respects, the case study has some of the same features found in historical research.

Although it consists of intensive study of a single unit, the case study's ultimate worth may be that it provides insight and knowledge of a general nature for improved practices. The generalizability of a case study is ultimately related to what the reader is trying to learn from it (Kennedy, 1979). The case study approach is probably most frequently used in trying to understand why something has gone wrong.

Gathering and Analyzing Data

The case study is very flexible with regard to the amount and type of data that are gathered and the procedures used in gathering the data. Thus, the steps in the methodology are not distinct or uniform for all case studies.

Data for case studies can be interviews, observations, or documents. It is not uncommon for a case study to employ all three types of data. A case study involving a child, for example, may include interviews with the child and the child's teachers and parents. The researcher may systematically observe the child in the class or in some other setting. Documents could include medical examination reports, physical performance test scores, achievement test results, grades, interest inventories, scholastic aptitude tests, teachers' anecdotal records, and autobiographies. As noted previously, some case studies are mainly descriptive, others focus on interpretation, and others are evaluative. Some case studies propose and test hypotheses, whereas some attempt to build theories through inductive processes.

According to Yin (2003), analysis of case study data is one of the least developed and most difficult aspects of conducting case studies. Yin maintains that too often the researcher begins a case study without the slightest notion as to how the data are to be analyzed. Analysis of data in a case study is a formidable task because of the nature of the data and the massive amount of information to analyze. According to Merriam (1988), analysis continues and intensifies after the data have been collected. The data must be sorted, categorized, and interpreted. As in any research, the ultimate value of a study rests on the insight, sensitivity, and integrity of the researcher, who is the primary instrument in the collection and analysis of the data in the case study. This is both a strength and a weakness. It is a weakness if the researcher fails to use the appropriate sources of information. The researcher can also be guilty of either oversimplifying the situation or exaggerating the actual state of affairs (Guba & Lincoln, 1981). On the other hand, a competent researcher can use the case study to provide a thorough, holistic account of a complex problem.

Applying Case Study Research in Physical Activities

Several case studies in physical education were directed by H. Harrison Clarke at the University of Oregon (Clarke & Clarke, 1970). The studies dealt with people of low and high fitness levels. Most of the studies sought to discover factors that may contribute to low fitness or lack of strength. A remarkable example of the case study approach and effective follow-up was seen in a demonstration project undertaken by Frederick Rand Roger and Fred E. Palmer (Clarke, 1968). The cause of low fitness in 20 junior high school boys with the lowest physical fitness scores was studied. Information used included somatotype, IQ, academic scholarship, medical history, and status. A follow-up project involved individual attention and special class meetings. Through exercises, improved health habits, medical referrals, and counseling, the students showed vast improvements in fitness, scholarship, and behavior. Two recent examples of case studies in physical education are Werner and Rink (1989), who described the teaching behaviors of four teachers, and Thomas and Thomas (1999), who described the practice patterns of two adult sport experts during their elementary years.

One of the principal advantages of the case study approach is that it can be fruitful in formulating new ideas and hypotheses about problem areas, especially areas for which there is no clear-cut structure or model. The researcher selects the case study method because of the nature of the research questions being asked. The case study, when used effectively, can play an important role in contributing to knowledge in our field.

Job Analysis

A **job analysis** can be considered a type of case study. It is a technique designed to determine the nature of a particular job and the types of training, preparation, skills, working conditions, and attitudes necessary for success in the job.

The procedures in conducting a job analysis vary. The objective is to obtain as much relevant information as possible about the job and the job requirements. One method is to observe someone in the particular occupation. Of course, this is time-consuming and probably bothersome for the person in the job. Nevertheless, the procedure is recommended because the researcher can acquire a kind of vicarious on-the-job experience and gain valuable insight into the whole atmosphere connected with the occupation. A limitation to this method is the lack of sufficient time to observe all facets of the job, particularly seasonal duties and responsibilities.

Questionnaires and interviews are two effective job analysis techniques. They allow people to respond to questions about the kinds of duties they perform, the types and degree of preparation required or recommended for accomplishing their tasks, and the perceived advantages and disadvantages of their jobs.

Systematic job analysis techniques can result in highly specific and quantitative descriptions of the job. For example, a job description for an administrative position reported (1) a percentage breakdown of principal accountabilities; (2) number of hours per day spent sitting, standing, and walking; (3) physical effort requirements for the number of pounds lifted waist high and carried alone and the distance carried; (4) sensory requirements for vision, hearing, and speaking; (5) mental effort requirements; (6) number of hours spent under time pressure and hours spent working rapidly, and (7) percentage of time spent indoors at a desk and in an office. Other requirements and conditions included the minimum academic degree, the necessity of working weekends and evenings, and the ability to use a computer.

The limitations of the job analysis as a research technique include the fallibility of both participants' memory and self-reports. There may be a tendency for participants to accent either the positive or the negative aspects of the job, depending on the time, the circumstances, and the subject. Also, there is a danger that the approach can be too mechanical, thus neglecting some of the more abstract aspects of the job and its requirements.

job analysis

Type of case study that determines the nature of a particular job and the types of training, preparation, skills, and attitudes necessary for success in the job.

Observational Research

Observation is used in a variety of research endeavors. It provides a means of collecting data and is a descriptive method of researching certain problems. In the questionnaire and interview techniques, the researcher relies on self-reports about how the participant behaves or what the participant believes. A weakness of self-reports is that people may not be candid about what they really do or feel and may give what they perceive to be socially desirable responses. An alternative descriptive research technique is for the researcher to observe individuals' behavior and qualitatively or quantitatively analyze the observations. Some researchers claim that this yields more accurate data. There are, of course, several limitations to observational research.

Basic considerations in observational research include the behaviors that will be observed, who will be observed, where the observations will be conducted, and how many observations will be made. Many other considerations are connected to these basic ones. Depending on the problem and the setting, each individual investigation has its own unique procedures. Therefore, only the basic considerations can be discussed in rather general terms.

What Behaviors Will Be Observed?

Deciding what behaviors will be observed relates to the statement of the problem and to the operational definitions. For example, a study on teacher effectiveness must have clearly defined observational measures of teacher effectiveness. Definite behaviors must be observed, for example, the extent to which the teacher asks students questions. Some other aspects of teacher effectiveness include giving individual attention, demonstrating skills, dressing appropriately for activities, and starting class on time. The researcher, in determining what behaviors will be observed, must also limit the scope of the observations to make the study manageable.

Who Will Be Observed?

As with any study, the population from which samples will be drawn must be determined. Will the study focus only on elementary school teachers? Which grades? Will the study include only physical education specialists, or will it also include classroom teachers who teach physical education? There is also the question of the number of participants who will be observed. Will the study include observations of students in addition to teachers? In other words, the researcher must describe precisely who the participants of the study will be.

Where Will the Observations Be Conducted?

The setting for the observations must be considered, in addition to the basic considerations of the size of the sample and the geographical area. Will the setting be unnatural or natural? Using an unnatural setting means bringing the participant to a laboratory, room, or other locale for the observations.

There are some advantages of an unnatural setting in terms of control and of freedom from distractions. For example, a one-way mirror is advantageous for observation, as it removes the influence of the observer on the behavior of the participant. That the participants' behavior is affected by the presence of an observer is also shown in classroom situations. When the observer first arrives, the students (and perhaps the teacher) are curious about the observer's presence. Consequently, they may behave differently than they would if the observer were not there. The teacher may also act differently, possibly by perceiving the observer as a threat or being aware of the purpose of the observations. In any case, the researcher should not make observations on the initial visit. Allowing the students to become gradually accustomed to the observer's presence is best.

Whether a participant will be observed alone or in a group is also related to setting. In a natural setting, such as the playground or classroom, the child may behave more typically, but there will also likely be more extraneous influences on behavior.

How Many Observations Will Be Made?

Many factors determine how many observations the researcher will make. First are the operationally defined behaviors in question and the time constraints of the study itself. For example, if you are studying the amount of activity or participation of students in a physical education class, you must consider several things. In the planning phase of the study, you must decide the type of activity unit and the number of units encompassed in the study. The number of observations for each activity depends on the particular stage of learning in the unit, whether the introductory phase, the practice stage, the playing stage, or so on. This must be specified in the operational definitions, of course, but both the length of the unit and the subsequent length of each stage within each unit play major roles in determining the number of observations that are feasible.

Another factor to consider is the number of observers. If only one person is doing the observations, either the number of persons being observed or the number of observations per person (or both) are restricted. To attempt to generalize "typical" behavior from observing a few individuals on a few occasions is hazardous.

Some types of behavior may not be manifested frequently. Sportsmanship, aggression, leadership, and other traits (as operationally defined) are not readily observable because of the lack of opportunity to display such traits (among other things). The occasion must present itself, and the elements of the situation must materialize so that the participant has the opportunity to react. Consequently, the number of observations is bound to be extremely limited if left to chance occurrence. On the other hand, situations that are contrived to provoke certain behavior are often unsuccessful because of their artificiality.

We cannot say how many observations are necessary but can only warn against too few observations; thus, we recommend a combination of feasibility and measurement considerations. This question is readdressed in the discussion on scoring and evaluating observations.

When Will the Observations Be Made?

You can easily understand that all the basic considerations being discussed are related and overlap with one another. The determination of when to make the observations includes decisions about time of day, day of the week, phase in the learning experience, season, and other time factors.

In our previous example of observing the amount of student activity in a physical education course, different results would be expected if the observations were made at the beginning of the unit than if they were made at the end. Also, allowing the students to get used to the situation so that the observational procedures do not interfere with normal activity is another consideration. In observing student teachers, for example, differences would certainly be expected if some were observed at the beginning of their student-teaching experience and others at the end of the semester.

Graduate students encounter major problems in observational research with regard to time. They find it difficult to spend the time necessary to make a sufficient number of observations to provide reliable results. Furthermore, graduate students usually must gather the data by themselves, making it much more time-consuming than if other observers were available.

How Will the Observations Be Scored and Evaluated?

Researchers employ a number of techniques for recording observational data. The use of computers and other computer-assisted event-recording methods have alleviated many of the technical problems that plagued observational research in the past. Some of the more commonly used procedures for recording observational data include

- narrative, or continual recording;
- tallying, or frequency counting;
- the interval method; and
- the duration method.

In the **narrative,** or **continual-recording, method,** the researcher records in a series of sentences the occurrences that he or she observes as they happen. This is the slowest and least efficient method of recording. The observer must be able to select the most important information to record because everything that occurs in a given situation cannot possibly be recorded. Probably the best use of this technique is in helping to develop more efficient recording instruments. The researcher first uses the continuous method and then develops categories for future recording from the narrative.

The **tallying,** or **frequency-counting, method** involves recording each occurrence of a certain behavior. The researcher must clearly define the behavior and make the frequency counts within a certain time frame, such as the number of occurrences in 10 min or 30 min per session.

The **interval method** is used when the researcher wishes to record whether the behavior in question occurs in a certain interval of time. This method is useful when it is difficult to count individual occurrences. One of the leading standardized systems for interval recording is the Flanders' Interaction Analysis System (Flanders, 1970), in which the observer records the behavior of the student according to 10 specific behavior classifications within each

narrative method

Method of recording in which the researcher describes the observations as they occur.

continual-recording method

See *narrative method.*

tallying method

Method of recording in observational research in which the researcher records each occurrence of a clearly defined behavior within a certain time frame; also called *frequency-counting method.*

frequency-counting method

See *tallying method.*

interval method

Method of recording in observational research, used when it is difficult to count individual occurrences, in which the researcher records whether the behavior in question occurs in a certain interval of time.

time interval. The Academic Learning Time in Physical Education (ALT-PE) is an observational instrument developed by Siedentop and graduate students at Ohio State University for use in physical education (Siedentop, Birdwell, & Metzler, 1979; Siedentop, Trousignant, & Parker, 1982). It entails time sampling in which a child is observed for a specified period of time, during which the child's activities are coded. The Cheffers' Adaptation of the Flanders' Interaction Analysis System (CAFIAS) was developed by Cheffers (1973) to allow systematic observation of physical education classes and classroom situations. The CAFIAS provides a device for coding behavior through a double-category system so that any behavior can be categorized as verbal, nonverbal, or both. The CAFIAS permits the coding of the class as a whole when the entire class is functioning as one unit, of smaller groups when the class is divided, or of individual students when the students are working individually or independently with no teacher influence. Rink and Werner (1989) developed a scoring instrument for teacher observation termed the Qualitative Measures of Teaching Performance Scale (QMTPS).

duration method

Method of recording in observational research in which the researcher uses a stopwatch or other timing device to record how much time a participant spends engaged in a particular behavior.

The **duration method** involves some timed behavior. The researcher uses a stopwatch or other timing device to record how much time an individual spends engaged in a particular behavior. A number of studies have used this method in observing student time on task and off task. In the previous example about the amount of activity in a physical education class, a researcher could simply record the amount of time a student spends in actual participation or the amount of time spent standing in line or waiting to perform. The researcher usually observes a student for a given unit of time (e.g., a class period), starting and stopping a stopwatch as the behavior starts and stops so that cumulative time on task (or off task) is recorded.

Using Videotape for Observation

A potentially valuable instrument for observational research is the videotape. Its greatest advantage is that the researcher need not worry about recording observations at the time the behavior is occurring. Furthermore, it allows the researcher to observe a number of persons at one time. For example, teachers and students can be observed simultaneously, which is difficult in normal observational techniques. In addition, the researcher can replay the videotape as needed to evaluate the behavior and retain a permanent record.

The use of videotape does have some disadvantages. It is expensive, the filming requires a significant degree of technical competence for lighting and positioning, and it may be cumbersome at times to follow the action. The presence of a camera may also alter behavior to the extent that the participants do not behave normally. However, if the disadvantages can be resolved, videotaping can be effective for observational research.

Weaknesses of Observational Research

Problems and limitations of observational research include the following:

- A primary danger in observational research lies in the operational definitions of the study. The behaviors must be carefully defined and observable. Consequently, the actions may be so restricted that they do not depict the critical behavior. For example, teacher effectiveness encompasses many behaviors, and to observe only the number of times the teacher asked questions or gave individual attention may be inadequate samples of effectiveness.

- Effectively using observation forms requires much practice. Inadequate training therefore represents a major pitfall in this form of research. Also, there are difficulties in trying to observe too many things. Often, the observation form is too ambitious for one person to use.

- Certain behaviors cannot be evaluated as finely as some observation forms dictate. A common mistake is to ask the observer to make discriminations that are too precise, thus reducing the reliability of the ratings.

• The presence of the observer almost always affects the behavior of the participants. The researcher must be aware of this and try to reduce the disturbance.

• Generally, observational research is greatly expedited by having more than one observer. Failure to use more than one observer results in decreased efficiency and objectivity.

Unobtrusive Research Techniques

There are multiple methods of gathering information about people other than questionnaires, case studies, and direct observation. Webb and colleagues (1966) discussed different approaches that they termed **unobtrusive measures.** Some examples they mentioned include the replacement rate for floor tiles around museum exhibits as a measure of relative popularity of exhibits. Researchers have assessed the degree of fear caused by telling ghost stories to children by observing the shrinking diameter of the circle of seated children. Researchers have measured boredom by the amount of fidgeting movements in an audience. The rate of library withdrawals of fiction and nonfiction books has been studied to determine the impact of television in communities. Researchers have noted that children demonstrate their interest in Christmas by the size of their drawings of Santa Claus and the amount of distortion in the figures.

In some of the methods just mentioned, the experimenter is not present when the data are being produced. There are conditions, however, when the researcher is present but still acts in a nonreactive manner. *In other words, the individuals are not aware that the researcher is gathering data.* For example, a researcher in psychology measured the degree of acceptance of strangers among delinquent boys by measuring the distance maintained between a delinquent boy and a new boy to whom the researcher introduced the delinquent participant. Sometimes, the researcher intervenes to speed up the action or force the data but in a manner that does not attract attention to the method. In studying the cathartic effect of activity on aggression, Ryan (1970) had an accomplice behave in an obnoxious manner and then measured the amount of electric shock the participants administered to that accomplice and to innocent bystanders. Other researchers have intervened by causing participants to fail or succeed so that responses to winning and losing in competitive situations could be observed.

unobtrusive measures

Measures of behavior taken on individuals who are not aware the researcher is gathering data.

Try as she might, that woman is never going to be unobtrusive!

Levine (1990) described an interesting study on the pace of life using unobtrusive measures involving people in different cities and in different countries. Among the measurements used were (a) the walking speed of randomly chosen pedestrians over a distance of 100 ft (30.5 m), (b) the accuracy of outdoor bank clocks in a downtown area, and (c) the speed with which postal clerks fulfilled a standard request for stamps.

Levine (1990) did a similar study on the pace of U.S. cities. Four unobtrusive measures were used: (a) walking speed for 60 feet (18.3 m), (b) time required for bank clerks to complete a simple request, (c) talking speed of postal clerks in explaining differences between regular, certified, and insured mail, and (d) the proportion of people wearing wristwatches.

Unobtrusive measures also include records such as birth certificates, political and judicial records, actuarial records, magazines, newspapers, archives, and inscriptions on tombstones. An interesting use of city records (Webb et al., 1966) was the analysis of city water pressure as an index of television viewing interest. Immediately after a television show, the water pressure dropped as drinks were obtained and toilets flushed. Mabley (1963) presented data on Chicago's water pressure on the day of an exciting Rose Bowl game that showed a drastic drop in pressure at the time of the game's end.

The ethical issue of invasion of privacy arises in some forms of unobtrusive measures. Informed-consent compliance has placed considerable restraints on certain research practices, such as those that involve entrapment and experiments that aim to induce heightened anxiety.

Correlational Research

correlational research

Research that explores relationships among variables and that sometimes involves prediction of a criterion variable.

Correlational research is descriptive in that it explores relationships that exist among variables. Sometimes predictions are made on the basis of the relationships, but correlation cannot determine cause and effect. The basic difference between experimental research and correlational research is that the latter does not cause something to happen. There is no manipulation of variables or experimental treatments administered. The basic design of correlational research is to collect data on two or more variables on the same people and to determine the relationships among the variables. But, of course, the researcher should have a sound rationale for exploring the relationships.

Different correlational techniques were discussed in chapter 8 through examples of situations that lend themselves to correlational research. The two main purposes for doing a correlational study are to analyze the relationships among variables and to predict.

Steps in Correlational Research

The steps in a correlational study are similar to those used in other research methods. The problem is first defined and delimited. The selection of the variables to be correlated is of critical importance. Many studies have failed in this regard. Regardless of how sophisticated the statistical analysis may be, the statistical technique can deal only with the variables that are entered; as the saying goes, "Garbage in, garbage out." The validity of a study that seeks to identify basic components or factors of fitness hinges on the identification of the variables to be analyzed. A researcher who wishes to discriminate between starters and substitutes in a sport is faced with the crucial task of determining which physiological or psychological variables are the important determinants of success. The researcher must lean heavily on past research in defining and delimiting the problem.

Participants are selected from the pertinent population using recommended sampling procedures. The magnitude and even the direction of a correlation coefficient can vary greatly, depending on the sample used. Remember that correlations show only the degree of relationship between variables, not the cause of the relationship. Consequently, because of other contributing factors, a researcher might obtain a correlation of .90 between two variables in a sample of young children but a correlation of .10 between the same variables in adults, or vice versa. Chapter 8 used examples of how factors such as age can influence certain relationships.

Another aspect of correlation is that the size of the correlation coefficient depends to a considerable extent on the spread of the scores. A sample that is fairly homogeneous in certain traits seldom yields a high correlation between variables associated with those traits. For example, the correlation between a distance run and maximal oxygen consumption with a sample of elite track athletes is almost invariably low because the athletes are so similar; there is not enough variability to permit a high correlation because the scores on the two measures are too uniform. If you included some less-trained runners in the sample, the size of the correlation coefficient would increase dramatically.

In prediction studies, the sample must be representative of the population for whom the study is directed. One of the major drawbacks to prediction studies is that the prediction formulas are often sample specific, which means that a formula's accuracy is greatest (or maybe only acceptable) when it is applied to the particular sample on whom it was developed. Chapter 11 discussed this shrinkage phenomenon as well as cross-validation, which is used to counteract shrinkage.

The collection of data requires the same careful attention to detail and standardization that all research designs do. A variety of methods for collecting data can be used, such as physical performance tests, anthropometric measurements, pencil-and-paper inventories, questionnaires, and observational techniques. The scores must be quantified, however, to be correlated.

Analysis of the data can be performed by a number of statistical techniques. Sometimes the researcher wishes to use simple correlation or multiple correlation to study how variables, either by themselves or in a linear composite of variables, are associated with some criterion performance or behavior. Factor analysis is a data-reduction method that helps determine whether relationships among a number of variables can be reduced to small combinations of factors or common components. Structural analysis is a technique used to test some theoretical model about causal relationships between three or more variables. (Simple correlation, factor analysis, and structural analysis are discussed in chapter 8.)

Prediction studies usually employ multiple regression because the accuracy of predicting some criterion behavior is nearly always improved by using more than one predictor variable. Discriminant analysis is a technique used to predict group membership, and canonical correlation is a method for predicting a combination of several criterion variables from several predictor variables. (Multiple regression and canonical correlation are discussed in chapter 8, and discriminant analysis in chapter 9.)

Limitations of Correlational Research

Limitations of correlational research include those of both planning and analysis. We have already pointed out the importance of the identification of pertinent variables and the selection of proper tests to measure those variables. There should be hypotheses based on previous research and theoretical considerations, rather than simply correlating a set of measurements to see what happens. The selection of an inadequate measure as a criterion in a prediction study is a common weakness in correlational research. For example, a criterion of success in some endeavor is often difficult to define operationally and may be more difficult to measure reliably.

Summary

Descriptive research encompasses many different techniques. In this chapter we described the basic procedures and strengths and weaknesses of six techniques in the category of descriptive research.

Developmental research seeks to study growth measures and changes in behavior over a period of years. In a longitudinal design, the same people are followed over time. When different participants of different ages are sampled, the design is cross-sectional. Of paramount importance in developmental research is whether the participants represent their populations and their performances. A developmental study may involve experimental treatment, with the researcher attempting to determine the interaction of some treatment with age.

In a case study, the researcher attempts to gather a lot of information about one or a few cases. Through in-depth study of a single case, a greater understanding about other similar cases is achieved. Case studies can be descriptive, interpretive, or evaluative. One of the principal advantages of the case study approach is that it can lead to the formulation of ideas and hypotheses about problem areas.

The job analysis is a type of case study in which extensive information is gathered and evaluated about job conditions, duties, and the types of training, preparation, skills, and attitudes that are necessary for success in a certain job.

Observational research is a descriptive technique that involves the qualitative and quantitative analysis of observed behaviors. Unlike the survey method, which relies on self-reports about how a person behaves, observational research attempts to study what a person actually does. Behavior is usually coded by what occurs and when, how often, and how long it occurs. Standardized observation instruments such as the ALT-PE and the CAFIAS are frequently used. Researchers often videotape their participants to record and store the observations for later analysis.

Some studies use unobtrusive measures in which the participants are unaware that the researcher is gathering data. For example, instead of (or in addition to) asking a person how much he or she smokes, the researcher might count the cigarette butts in an ashtray after a certain period. Although the methods can be interesting and innovative, they may be constrained by ethical considerations such as invasion of privacy and entrapment.

Correlational research examines relationships among variables. Sometimes the relationships are used for prediction. Although correlations are often used with experimental research, the study of relationships is descriptive in that it does not involve the manipulation of variables. A major pitfall in correlational research is to assume that because variables are related, one causes another.

✔ Check Your Understanding

1. Locate a cross-sectional study and a longitudinal study in the literature, and answer the following questions about each:

    a. Is the study descriptive or experimental?

    b. What age levels are studied?

    c. What are the independent and the dependent variables?

    d. What statistics are used to make cross-age comparisons?

2. Write an abstract of a case study found in the literature. Indicate the problem, the sources of information used, and the findings.

3. Find an observational study in the literature and write a critique of the article, concentrating on the methodology.

4. Write a brief abstract of a correlational research study.

Physical Activity Epidemiology Research

Barbara E. Ainsworth ■ *Charles E. Matthews*

Health nuts are going to feel stupid someday,
lying in hospitals dying of nothing.
—Redd Foxx

■ ■ ■ ■ ■ ■ ■

The emergence of the mid-twentieth-century heart disease epidemic fueled a number of large-scale observational epidemiologic studies aimed at identifying the determinants of heart disease so that preventive measures could be undertaken to improve the public's health. A series of observational studies was initiated between the late 1940s and early 1960s. A number of them were particularly important to the development of the field of physical activity epidemiology because they were the first to develop methods for measuring physical activity and to systematically study the link between physical activity and a life-threatening disease. These studies, to name a few of them (and important people associated with each), included the Framingham Heart Study, London Busmen/British Civil Servants study (Jeremy Morris), Tecumseh Health Study (Henry Montoye), the College Alumni Study (Ralph Paffenbarger), and early studies from Minnesota (Henry Taylor).

By the early 1980s, researchers had begun to amass solid evidence demonstrating that low levels of physical activity were associated with increased risk of heart disease and overall mortality. Because heart disease is and was the leading cause of death in the United States and the countries of the European Union, physical activity emerged as an important public health concern. Accordingly, the U.S. Public Health Service initiated surveillance programs to quantify the leisure-time physical activity patterns in the U.S. population in the mid-1980s. These programs demonstrated that more than 60% of U.S. adults were not physically active on a regular basis. In the 1990s epidemiologic evidence continued to accumulate, showing that low levels of physical activity were associated with increased risk for a number of important health conditions, including all-cause and cause-specific mortality, cardiovascular disease, osteoporosis, some forms of cancer, and mental health and quality of life problems (U.S. Department of Health and Human Services, 1996). Major initiatives are currently underway to understand the best way to change the behavior of the population at both the individual and community level.

One such initiative is Healthy People 2010, a plan developed by the U.S. Department of Health and Human Services to meet 467 health objectives in 28 focus areas by the year 2010. An index of the leading health indicators (similar to the concept of the leading economic indicators) is being used to monitor progress toward the objectives (U.S. Department of Health and Human Services, 2000).

Objectives for adolescents are to

- increase the proportion who engage in vigorous physical activity that promotes cardiorespiratory fitness 3 or more days per week for 20 or more minutes per occasion, and

- increase the proportion who engage in moderate physical activity for at least 30 minutes on 5 or more of the previous 7 days.

Objectives for adults are to

- increase the proportion who engage regularly, preferably daily, in moderate physical activity for at least 30 minutes per day, and

- increase the proportion who engage in vigorous physical activity that promotes the development and maintenance of cardiorespiratory fitness 3 or more days per week for 20 or more minutes per occasion.

The 2010 objectives for moderate and vigorous activity as compared with the levels of physical activity in 2003 are highlighted in figure 17.1. To reach the 2010 objectives, adolescents need to increase regular vigorous activity by 22% and moderate activity by 10%. Adults have exceeded the moderate activity objective of 30%, whereas they need to increase vigorous activity by 4%.

risk factor

An exposure that has been found to be a determinant of a disease outcome or health behavior.

This mini-history of physical activity epidemiology mirrors the major goals of epidemiologic inquiry that will be described in the remainder of this chapter. Epidemiologic methods are used to provide the scientific backbone for public health endeavors, including (1) quantifying the magnitude of health problems, (2) identifying factors that cause disease (i.e., **risk factors**), (3) providing quantitative guidance for the allocation of public health resources, and (4) monitoring the effectiveness of prevention strategies using population-

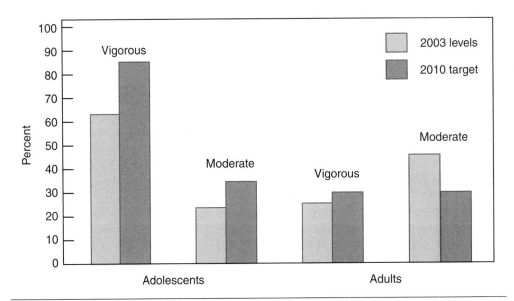

Figure 17.1   Percentages of adolescents and adults who met the Healthy People 2010 physical activity objectives in 2003.

wide surveillance programs (Caspersen, 1989). In this chapter we will define epidemiology, describe methods of measuring physical activity in epidemiologic studies, describe epidemiologic study designs, consider threats to the validity to these designs, and provide an outline for reading and interpreting an epidemiologic study.

Observational Versus Experimental Research

A key to understanding epidemiologic methods comes from considering the difference between observational and experimental research. It has been said that "the ideal circumstances for an epidemiologist are those in which nature contrives to produce the conditions that would have been achievable if an experiment had been conducted" (Rothman, 1986).

Observational research uses existing differences in factors that may cause disease within in a population, such as physical activity, dietary habits, or smoking. A portion of the population chooses to be physically active while others do not. Epidemiologists use these naturally occurring differences in a population to "observe" and therefore understand the effect of these differences on specific disease **outcomes.**

A primary reason that epidemiologic research has become such an important discipline is that it is virtually the only way to obtain a quantitative understanding of the health risks of many behavioral risk factors. This arises from the fact that it would be unethical to conduct true experimental research on health behaviors such as physical inactivity. For example, researchers could not randomize a group of people to either a completely sedentary lifestyle or an active lifestyle, wait 5 to 10 years, and then see how much death and heart disease physical inactivity caused. Although an experimental study such as this would unequivocally demonstrate that physical inactivity is a causal factor in the development of heart disease, the ethical problems of the experiment are obvious.

outcome

The dependent variable in an analysis.

What Is Epidemiology?

Epidemiology has been defined as "the study of the distribution and determinants of health related states or events in specified populations, and the application of this study to the control of health problems" (Last, 1988, 141).

Distribution

The distribution of disease relates to the *frequency* and *patterns* of disease occurrence in a population.

Frequency, or how often the disease occurs, is typically measured as the prevalence, incidence, or mortality rate of a disease.

Disease prevalence refers to the number of people in a given population that have a disease at a particular point in time. For example, in 1996 the prevalence of all cardiovascular diseases in the United States was estimated to be 58,800,000 people, or about 25% of the population (American Heart Association, 1998).

Components of Epidemiologic Research
Distribution
• Frequency—prevalence, incidence, mortality rate
• Patterns—person, place, time
Determinants
• Defined characteristics—associated with change in health
Application
• Translation—knowledge to practice

The frequency of disease occurrence may also be calculated as the rate of new disease or health events, as *incidence* or *morality rates* (i.e., new cases or new deaths from a specific disease within a specified period of time). For example, the mortality rates for cardiovascular disease were 304 per 100,000 men, and 203 per 100,000 women in the United States in 2001 (Centers for Disease Control and Prevention, 2004).

Presentation of the frequency of disease occurrence relative to the number in the population of interest enables comparison of prevalence, incidence, or mortality rates across different populations. For example, the mortality rates clearly indicate that cardiovascular disease mortality is about 50% higher for men than it is for women.

Evaluating the basic *patterns* of disease occurrence within a specified population is often useful for developing hypotheses about risk factors for the disease. Patterns of disease occurrence refer to characteristics related to *person, place,* and *time.* Personal characteristics include demographic factors such as age, gender, and socioeconomic status. Characteristics of place include geographic differences, urban–rural variation, and, particularly important in the history of physical activity epidemiology, differences in types of occupations. Historically, differences in occupational classifications enable researchers to make crude comparisons of occupational physical activity levels and were used in some of the first epidemiologic studies of the relationship between physical activity and heart disease.

Time of disease occurrence refers to annual, seasonal, or daily patterns of occurrence. Quantification of temporal changes in disease rates often lead to hypotheses that generate more detailed examination of the factors that caused this change. A good example of this was the observation that the incidence of upper-respiratory infections (i.e., common colds) in runners increased in the 14-day period after running an ultra-marathon (56K) (Peters, 1983). Of the 141 marathoners studied, 47 (33%) came down with a cold after the race, whereas only 19 (15%) of the 124 people who did not run the marathon had a cold during the same period. Thus, the incidence of colds was more than twice as high among the runners, apparently as a result of the **exposure** of the marathon event. This observation led to intensive research in the area of exercise immunology and ultimately resulted in a greater understanding of the relationship between exercise and the immune system (Nieman, 1994).

exposure

Factors (variables) in epidemiologic studies that are tested for their relationship with the outcome of interest.

Determinants

A determinant is "any factor, whether event, characteristic, or other definable entity, that brings about change in a health condition, or other defined characteristic" (Last, 1988). In physical activity research, the goals are usually to test the hypothesis that activity is or isn't a determinant for a particular disease outcome, or to identify the determinant of physical activity behaviors. Determinants of disease are often called "risk factors" because they increase a person's risk for disease. Epidemiologic studies have been instrumental in identifying risk factors for heart disease, including obesity, high blood pressure, high LDL cholesterol, low HDL cholesterol, and physical inactivity, among others. Identification and surveillance of specific determinants of a particular disease allow for targeted health promotion campaigns that present this new health knowledge to the general public.

Application

Application of the established understanding of the causal factors related to disease is a major goal of public health. Thus, once epidemiologists have identified what causes disease, health educators interact with communities to make them healthier places to live.

Other names for the application of research in a community setting are *translation* and *dissemination.* Successful public health dissemination strategies are those that engage the community using a variety of methods, from motivating people to change their behaviors to affecting public policy. This is referred to as the ecological model for health promotion (McLeroy, Bibeau, Steckler, & Glanz, 1988). Public health disseminators translate knowledge from epidemiologic studies to help increase physical activity among individuals and within social groups and community organizations. Strategies are also used to affect the community environment (e.g., by building walking trails) and encourage policy makers to enact legislation or appropriate funds to enable people to lead physically active lives.

Physical Activity Measurement Definitions

Because there are many ways to describe physical activity, definitions are necessary to increase consistency of measurement and to reduce variability across studies. *Physical activity* is a global term that is defined as all movement that is produced by the contraction of skeletal muscle and that substantially increases energy expenditure. In this context, physical activity may occur on the job, in the home, during leisure and recreational periods, and in getting from place to place (Caspersen, 1989). Physical activity includes all forms of movement done in occupational, exercise, home and family care, transportation, and leisure settings. *Exercise* is planned, structured, and repetitive bodily movement done to improve or maintain one or more components of physical fitness (Caspersen, 1989). *Physical fitness* is a multidimensional concept associated with a set of attributes that people have or achieve that relates to the ability to perform physical activity. Types of physical fitness are cardiorespiratory, muscular, flexibility, metabolic, and motor fitness. *Leisure* is a concept that includes the elements of free choice, freedom from constraints, intrinsic motivation, enjoyment, relaxation, personal involvement, and opportunity for self-expression (Henderson, Bialeschki, Shaw, & Freysinger, 1996). *Volitional* physical activity refers to activities done for a purpose in either structured or unstructured settings. *Spontaneous* physical activity refers to brief periods of movement that results in energy expenditure through fidgeting, gesticulation, or unintentional short accumulated periods of movement.

When physical activity is measured in epidemiologic studies, the frequency, duration, and intensity of volitional physical activities performed at some period in the past are studied in relation to their impact on health.

- *Frequency* relates to the number of times per week or year that one is physically active. For example, in a consensus report from the Centers for Disease Control and Prevention (CDC) and the American College of Sports Medicine (ACSM), physical activity obtained on most days of the week, preferably all days, was recommended to reduce some risks for chronic disease and enhance health-related quality of life (Pate, 1995). According to the ACSM, the minimum frequency of vigorous exercise needed to increase cardiorespiratory fitness is 3 days per week (ACSM, 1994).

> ### CDC-ACSM Moderate Recommendation
> Frequency = 5+ days/week
> Duration = 30+ min/day
> Intensity = 3–6 METs
>
> ### ACSM Vigorous Recommendation
> Frequency = 3+ days/week
> Duration = 20–45 min/event
> Intensity = 60–90% maximal heart rate

- *Duration* refers to the time spent in a specific activity as hours or minutes per session. In the Multiple Risk Factor Intervention Trial (MRFIT) of 12,138 men at high risk for cardiovascular disease, analyses were performed to determine the relationship between physical activity and mortality (Leon, 1987). Physical activity was measured in minutes per day and categorized into three time periods described as short (15 min/day), middle (47 min/day), and long (133 min/day) duration. Results showed lower mortality rates among men who performed leisure-time activity of middle and long duration compared with men in the short-duration group. Based on this study and other epidemiologic studies reviewed by Pate and colleagues (1995), at least 30 minutes per day is a recommended duration of activity to reduce the risks for premature morbidity and mortality.

- *Intensity* refers to the difficulty of an activity and is generally classified as light, moderate, or vigorous in epidemiologic settings. Intensity may be expressed in *absolute* or *relative* terms. The recommended unit for intensity in absolute terms is the metabolic equivalent (MET), which is defined as the ratio of the activity metabolic rate to the resting metabolic rate. One MET is approximately equal to 3.5 ml/kg/min of oxygen consumption or about 1 kcal/kg/hr of energy expenditure for a 60-kg (132-lb) person. To provide consistency in assigning intensity levels to activities, the Compendium of Physical Activities was developed and provides MET intensities for over 500 activities

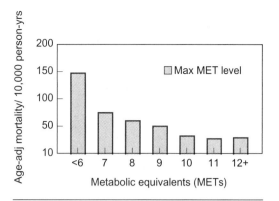

Figure 17.2 Age-adjusted, all-cause mortality rates per 10,000 person-years of follow-up by physical fitness levels.

(Ainsworth, et al., 2000). Expression of physical activity intensity in relative terms provides adjustment in the difficulty of an activity as a result of individual differences. Relative intensity measures include the percent of maximal oxygen uptake (%$\dot{V}O_2$max) or maximal heart rate (% max heart rate), and rating of perceived exertion (RPE). A difficulty in using relative intensities in epidemiologic studies is that comparison of intensity levels for similar activities is unequal between studies. Thus, it is preferable to express physical activity intensity in relative terms when measuring activity for a given person, such as in prescribing exercise programs on an individual basis.

- *Dose* refers to the combination of the frequency, intensity, and duration of physical activity and is expressed as kcal/day, MET-hours/day, minutes on a treadmill graded exercise test, or other units. In the Aerobics Center Longitudinal Study, Blair, et. al. (1989) studied the association between the time to exhaustion on a treadmill graded exercise test and all-cause mortality. They showed a strong inverse dose response relation between maximal cardiorespiratory fitness levels (represented as maximal MET level) and all-cause mortality, as shown in figure 17.2. The greatest reduction in mortality was seen between men in the lowest and next to lowest levels of fitness.

Assessment of Physical Activity

Physical activity can be measured using a variety of methods ranging from direct measurement of the amount of heat a body produces during activity to asking people to rate how active they recall being during the past week or year. From the 1950s to the 1980s, job titles were used to classify physical activity patterns in epidemiologic studies involving occupational physical activity. However, with the changing profile of the labor market, occupational titles no longer reflect the physical requirements of a job, eliminating the use of job titles to classify occupational energy expenditure (Montoye, 1996).

Because of the exceptionally large number of people in many epidemiologic studies, self-administered questionnaires or brief interviews are often used to capture activity spent at work, in exercise, at home, in transportation, and in leisure settings. Questionnaires are classified as global, short recall, and quantitative histories depending on the length and complexity of the items.

> *Physical Activity Surveys*
> Global
> Short recall
> Quantitative history

- *Global* questionnaires are instruments of 1 to 4 items that present a general classification of one's habitual activity patterns. Global surveys are most accurate for classifying people according to their level of vigorous-intensity physical activity. Because these surveys may take little time to administer, less than 2 minutes, they are preferred for use in epidemiologic studies.

- *Short recall* questionnaires generally have 5 to 15 items and reflect recent physical activity patterns (during the past week or month). They are effective for classifying people into categories of activity, such as inactive, insufficiently active, or regularly active based on health guidelines and recommendations for minimal levels of activity. Short recall questionnaires may take from 5 to 15 minutes to complete and are recommended for surveillance activities and descriptive epidemiologic studies designed to assess the prevalence of adults and children who obtain national recommendations for physical activity and health.

- *Quantitative history* questionnaires are detailed instruments that have from 15 to 60 items and reflect the intensity, frequency, and duration of activity patterns in various

From a physical activity viewpoint, the frequency and duration of her games may be okay, but the intensity is zilch.

categories, such as occupation, household, sports and conditioning, transportation, family care, and leisure activities. This type of survey was used by Leon and colleagues (1987) in the MRFIT study. Quantitative history questionnaires allow the researchers to obtain detailed information about physical activity energy expenditure and patterns of activity from the past month to year. However, because of their length and complexity, they may take from 15 to 30 minutes to complete and are usually interviewer administered. Quantitative history questionnaires are appropriate for studies designed to examine issues of dose response in heterogeneous populations with a wide variety of physical activity patterns.

Since their introduction as an objective measure of free-living physical activity in the early 1980s, waist-mounted activity monitors (accelerometers and pedometers) have become a staple of the physical activity assessment repertoire. They have been used extensively in the validation of self-reported physical activity surveys, as outcome measures of physical activity in intervention studies, and in research designed to identify the psychosocial and environmental correlates of physical activity behaviors. The great advantage of objective physical activity measures is that they overcome some of the limitations inherent in self-report methods that necessarily rely on information reported by participants. With activity monitors, there are no reporting errors and no errors introduced by interviewers, and real-time data collection and automated data reduction can provide a rich description of the activity profiles of people and populations. Once the logistics of monitor delivery and retrieval are overcome, activity monitors offer a relatively simple and efficient method of measurement that is suitable for small clinical studies and epidemiologic studies of intermediate size (e.g., fewer than 5,000).

The MTI actigraph (formerly known as the CSA) is a small battery-operated accelerometer-based motion sensor commonly worn on the waist. It provides a computerized record of the intensity and duration of ambulatory movement, presented as movement counts, for up to 28 days. The MTI accelerometer was used in the U.S. National Health and Nutrition Examination Survey IV conducted between 1998 and 2002 to augment physical activity questionnaires in a subsample of participants. Minute-by-minute activity count data for the MTI monitor on two days of monitoring for a middle-aged woman participating in a physical activity intervention study are presented in figure 17.3. Activity count levels

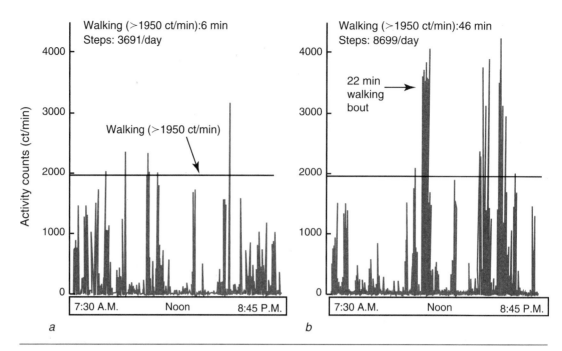

Figure 17.3   MTI actigraph output and interpretation: *(a)* sedentary day, *(b)* active/walking day.

above 1,950 counts per minute reflect purposeful moderate-intensity walking in the range of 3 to 4 mph. Panel A is a relatively sedentary day with few extended bouts of walking. Data summary for this day indicated that only 6 minutes of walking and only 3,691 steps were accumulated. In contrast, Panel B was a more active day in which there was a walking bout (22 minutes) in the middle of the day and several additional shorter episodes of activity later in the day, eventually accumulating 46 walking minutes and 8,699 steps over the course of the day. Objective information such as this can be used to track changes in activity patterns in response to an intervention, or to describe differences in the activity profiles of different populations using the same objective standard. A number of other accelerometer-based activity monitors (e.g., Tritrac, Biotrainer, Mini-mitter) that are comparable to the MTI device are commercially available.

Pedometers also have been used in epidemiologic studies to measure the amount of accumulated steps in free-living settings. Among the less expensive pedometers (e.g., less than $50), the Yamax Digiwalker (Model DW-500) has demonstrated the highest accuracy in controlled field tests for walking (within 2% of counted steps) and jogging (within 3 to 5% of counted steps). Pedometers are used extensively in intervention and health promotion programs by providing feedback to participants about their activity levels, in community

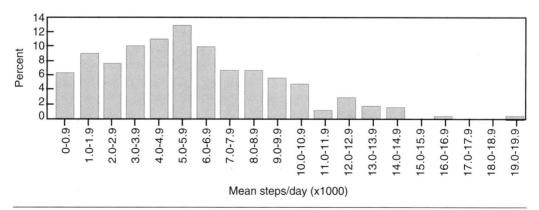

Figure 17.4   Distribution of mean steps per day in a community sample of 209 adults.

Reprinted, by permission, from C. Tudor-Locke, et. al., 2004, "Descriptive epidemiology of pedometer-determined physical activity," *Medicine and Science in Sports and Exercise*, 36 (9): 1567-1573.

trials to measure the impact of studies designed to increase walking behaviors, and to characterize the walking behaviors of community residents. For example, to quantify levels of walking among residents in a small southern community, Tudor-Locke and colleagues (2004) mailed pedometers to 209 adults residing at randomly selected households who agreed to wear the pedometer for 7 consecutive days and write their accumulated steps in a logbook each day. The mean step counts were 5,932 ± 3,664 steps per day. Steps per day were higher in men (7,192 ± 3,596) than in women (5,210 ± 3,518), higher in whites (6,628 ± 3,375) than in nonwhites (4,792 ± 3,874), and inversely related with the body mass index (normal weight = 7,029 ± 2,857; overweight = 5,813 ± 3,441; obese = 4,618 ± 3,359). The distribution of steps in the sample is shown in figure 17.4.

To learn more about the technological aspects and validity of the different pedometers and accelerometers, see Welk (2002) for a review of studies.

Epidemiologic Study Designs

Initial hypotheses are developed and crudely tested using simple descriptive studies that use cross-sectional and ecological study designs. More refined study designs can then be used to carefully test specific hypotheses in analytic studies. Analytic studies use case-control and **cohort** study designs, and mortality or disease incidence are often the dependent variables for the investigations. These two designs, the workhorses of epidemiologic research, have been used for the last 50 years, and a solid body of evidence has been amassed to support their basic methods.

After analytic epidemiologic studies have consistently demonstrated an exposure–disease link, in conjunction with supportive laboratory evidence, experimental studies can then be initiated to test, in a rigorous experimental design, the validity of the observational findings. The design of choice for experimental epidemiologic studies is the randomized controlled trial. Outcomes may be mortality, disease incidence, or an intermediate end point such as blood cholesterol levels or blood pressure. In the case of physical activity intervention research, the dependent outcome variable could be the physical activity level of either a person or a community.

cohort

A specified group of individuals who are followed over a period of time.

Descriptive Epidemiology

Descriptive epidemiology has been defined as "general observations concerning the relationship of diseases to basic characteristics such as age, sex, race, occupation, social class, or geographic location" (Last, 1988). The major objective of descriptive studies is to quantify the magnitude of specific health problems, identify population subgroups that may have higher rates of disease, and develop hypotheses about specific factors that may be determinants of disease. Because descriptive studies use information sources that are readily available or easily obtained, they are inexpensive to conduct, in terms of both the cost of acquiring the data and the time required to complete the investigation.

Cross-Sectional Designs

The cross-sectional design is perhaps the most frequently conducted type of study examining the relationship between physical activity and health outcomes. Frequently, rather than using a disease end point as the main outcome (or dependent variable), known risk factors for a disease are used as intermediate end points. For example, a great deal of the initial work examining the effect of physical activity on cardiovascular disease risk factors (e.g., blood cholesterol or blood pressure) used simple cross-sectional designs. Researchers simply assembled two groups of people, highly active people (usually athletes) and sedentary control participants, and then measured the cardiovascular risk factors of interest. A simple test of the mean differences between groups formed the analysis. Because these studies typically measure the outcome variables at a single point in time, they are relatively simple and easy to conduct.

An example of a cross-sectional study shows the effect of physical activity on estrogen metabolism in postmenopausal women. The dependent variable is the ratio score of two estrogen metabolites, 2-hydroxyestrone:16a-hydroxyestrone (2/16 ratio). Women with low 2/16 ratios were hypothesized to have an increased risk for breast cancer compared to women with high 2/16 ratios. Physical activity was measured using a self-administered questionnaire with leisure-time physical activity dose calculated as MET-hours per day. Results showed an interaction between physical activity and adiposity (measured as body mass index [BMI, kg/m²]), as shown in figure 17.5. Physically active women had higher 2/16 ratios in each of three categories of BMI ranging from low to high. In contrast, inactive women had lower 2/16 levels as their BMI increased. The researchers concluded that a reduction in 2/16 ratios among inactive women with higher BMI levels may be predictive of increased risk for breast cancer.

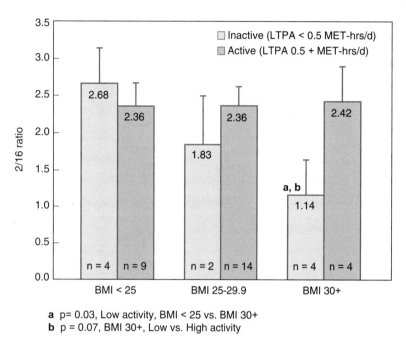

a p= 0.03, Low activity, BMI < 25 vs. BMI 30+
b p = 0.07, BMI 30+, Low vs. High activity

Figure 17.5 Body mass index (BMI) versus 2/16 estrogen ratios in active and inactive postmenopausal women.

An advantage of cross-sectional studies is that the physiological differences of individuals (rather than groups) can be compared, and with planning, investigators can control for factors that could potentially confound the relationship of interest. Confounding factors can be controlled using statistical methods when the size of the study is adequate, or by matching the two comparison groups on an important factor. For example, to control for the confounding effect of body fat on the physical activity blood cholesterol relationship, one could recruit highly active and sedentary people with similar levels of body fat (e.g., ± 2%). Comparisons of the two groups would then be independent of the effects of body fat.

The major limitation of the cross-sectional study design is that the study outcome and exposure are measured at the same point in time. Therefore, it is impossible to know whether physical activity exposure was actually responsible for the effects observed. That is, cross-sectional study designs do not allow for definitive conclusions about cause and effect because the "temporal sequence" (timing) of the relationship between the outcome and the exposure is not known.

A final and often overlooked limitation of cross-sectional studies is that a lack of association in a cross-sectional study may not mean that no longitudinal relationship exists between the two factors being examined. For example, on a cross-sectional basis, there can be little or no relationship between dietary fat intake and blood cholesterol levels. However, in longitudinal studies there is a clear positive relationship between fat intake and cholesterol levels (Jacobs, Anderson, & Blackburn, 1979).

Ecologic Designs

Ecologic studies use existing data sources for both exposure and disease outcomes to compare and contrast rates of disease by specific characteristics of an entire population. Data sources that are typically included in these studies are census data, vital statistics records (for countries, states, counties), employment records, and national figures for health-related information such as food consumption.

For example, Morris, et. al. (1953) reported on a series of studies that used occupational classifications as a surrogate measure of physical activity levels. Perhaps the most famous comparisons were among London busmen working on double-decker buses. Conductors, who had to walk up and down the bus stairs all day every day, were compared with the bus drivers, who sat and drove the bus all day every day. Drivers consistently had heart disease mortality rates that were twice as high as those of conductors. However, also included in this seminal work was an ecologic study of heart disease mortality and occupational physical activity. The investigators tabulated heart disease mortality rates by age and occupational activity level among Englishmen and Welshmen using existing public health statistics for the years 1930 to 1932. The retabulation and classification of these data by occupational types into heavy, intermediate, and light activity categories provided a scientifically crude, but ultimately valid, understanding of the effect of physical activity on heart disease mortality.

Figure 17.6 shows a graded inverse relationship between occupational physical activity and heart disease mortality in each age group. That is, men who were engaged in heavy occupational activity had the lowest death rates, and men in the light occupations had the highest death rates, regardless

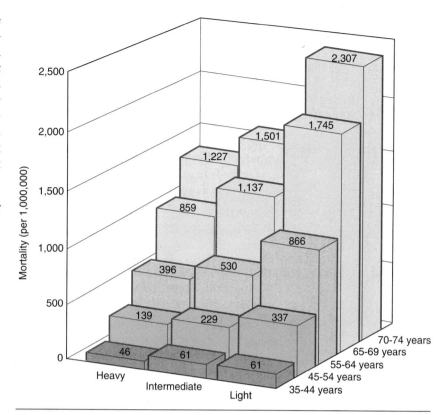

Figure 17.6 Heart disease mortality in heavy, intermediate, and light occupational activity categories by age group among men: England and Wales, 1930–1932. Reprinted with permission from Elsevier (The Lancet, 1953, Vol. 262, No.6796, pp.1111-1120).

of age. Men in intermediate activity categories had death rates that were in between. A clear increase in heart disease mortality with increasing age was also observable.

Like most things that are cheap and easy, ecologic studies have limitations that must be considered. First, the level of analysis is a population group rather than an individual. Therefore, there is no way to link individual physical activity levels to a specific heart disease outcome. Second, ecologic studies are severely limited in their ability to control for the effects of other factors that could obscure the relationship being examined.

Although ecologic analyses can lead to useful and valid estimates of an exposure–disease relationship, as noted in the previous example, they have also been known to produce spurious results. For example, ecologic studies first conducted in the 1970s examining the relationship between dietary fat intake and breast cancer mortality revealed strong linear relationships between this exposure and breast cancer. Women in countries who ate more fat had higher breast cancer mortality rates. However, since that time, a large body of research using methodologically superior study designs has been unable to demonstrate the strong link between fat intake and breast cancer observed in the early ecologic studies (Willett, 1990). Accordingly, ecologic studies should only be considered a first line of investigation aimed at the development of testable hypotheses.

In summary, descriptive studies are useful for developing and crudely testing initial hypotheses about exposure–disease relationships. However, more definitive, scientifically sound results can only come from analytic or experimental study designs.

Analytic Designs

The analytic designs, case-control and cohort studies, are designed to test specific hypotheses regarding causal links between various exposures and mortality and incidence outcomes using purely observational methods.

To understand analytic designs more thoroughly, one must consider a simplified model of the natural history of disease (see figure 17.7). Because most epidemiologic studies

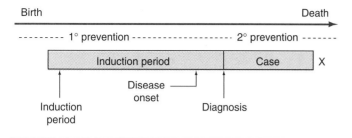

Figure 17.7   Natural history of chronic disease.

that evaluate physical activity as the exposure of main interest are for chronic disease, this model is for chronic disease. Chronic diseases such as heart disease, osteoporosis, or cancer can take 10 to 40 years to develop into a problem sufficient to be diagnosed clinically.

Physical activity epidemiologists are typically interested in how exposure to physical activity early in the disease process alters the course of the natural history of the disease. The disease course could be extended by preventive exposures (an active lifestyle) or hastened by adverse exposures (a sedentary lifestyle).

The "induction period" of a disease is conceptualized as the period between the point at which the exposure of interest alters the physiology of the disease process and the point at which the disease is clinically diagnosed. The clinical diagnosis is always later than the actual disease onset. Once a person is clinically diagnosed with the disease, he or she becomes an incident "case." Primary prevention (1°) refers to prevention of the first occurrence of the disease, and secondary prevention (2°) refers to prevention of disease recurrence. The usual goal of analytic studies is to quantify the effect of exposures that occur during the induction period on eventual disease risk. That is, the goal is to identify predictors of the disease outcome during the early portions of the disease's natural history.

Cohort Studies

The terms *follow-up studies, prospective studies,* and *longitudinal studies* have all been used to describe the cohort study design. In this design, a large disease-free population is defined and assessment of relevant exposures are obtained. Baseline data are used to categorize the cohort into different levels of exposure (i.e., low, medium, high). After this baseline assessment has been made, the follow-up period begins. Because chronic diseases such as colon or breast cancer are relatively rare, the follow-up period can last from as little as 2 years to more than 20 years. At the end of the follow-up period, the number of people within the cohort who died during follow-up and were diagnosed with the disease outcome of interest are tabulated.

Analysis of cohort studies is relatively simple. Because we are interested in how different levels of baseline exposure predict disease occurrence, the basic analysis simply consists of calculating disease rates for the different levels of exposure. For example, mortality rates among people reporting regular exercise would be compared to mortality rates of nonexercisers. In more refined analyses, mortality rates for three to five levels of physical activity may be calculated, and comparisons between the low activity level and each of the higher activity levels could be made.

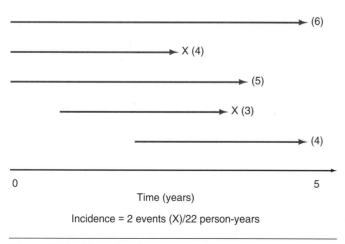

Incidence = 2 events (X)/22 person-years

Figure 17.8   Example of person-years.

Disease rates in cohort studies are often expressed relative to "person-years" of follow-up. A person-year represents 1 year of observation for one person during the follow-up period. For example, in figure 17.8 five people were observed from time 0 to time 5 (6 full years). A subject's person-year contribution to the cohort begins when he or she enters follow-up (at time 0 for the top three people, at time 1 for the fourth person, at time 2 for the fifth person). Follow-up ends for each person when the person has an event (X), when the person is lost for follow-up, or when the follow-up period ends. In the example in figure 17.8, each subject's person-year contribution to the cohort is in parentheses, and the summary measure of the incidence rate is 2 events (X) / 22 person-years of observation.

The summary measures of the exposure–disease relationship in cohort studies are typically expressed in absolute or relative terms. Differences in disease rates between exposure groups expressed on an absolute basis can be obtained by simple subtraction of rates. More frequently, estimates of the exposure–disease relationship are expressed as the **relative risk.** Relative risks are calculated as the ratio of referent category to different exposure levels. The null value, or no effect, of the relative risk is 1.0, and values below 1.0 indicate reduced risk, whereas values above 1.0 indicate increased risk.

relative risk

Measure of association of the exposure–disease relationship, typically employed in cohort studies.

A landmark cohort study in the physical activity epidemiology literature is the College Alumni Health Study. Ralph S. Paffenbarger Jr. and colleagues initiated this study by enrolling 16,936 male Harvard alumni aged 35 to 74 years in 1962 and 1966. At baseline, participants completed mailed questionnaires that included questions about their daily physical activities. Using simple questions about walking, stair climbing, and exercise activities, participants were categorized into six levels of physical activity energy expenditure (i.e., <500, 500-999, 1,000-1,999, 2,000-2,999, 3,000-3,999, 4,000+ kcal/wk). In a follow-up assessment 6 to 10 years later (1972), it was determined that 572 men had experienced their first heart attack (Paffenbarger, Wing, & Hyde, 1978). Rates of fatal, nonfatal, and all first heart attacks were compared in each of the baseline physical activity exposure categories. These data (presented in figure 17.9) clearly indicated that rates for each category of heart attack were inversely related to physical activity index levels, up to about 2,000 kcal/week, when the heart attack rates reached a plateau (or increased slightly).

Stratified comparisons of these data by age group, smoking status, blood pressure level, and adiposity revealed that these potentially confounding factors did not account for this relationship between physical activity and first heart attacks. Thus, the results strongly suggested that the effect of physical activity on heart attack risk was independent of many important factors that could potentially confound this relationship. The summary relative risk estimate of 1.64 was obtained by dividing the heart attack rates for less active men (<2,000 kcal/wk, 57.9 attacks per 10,000 person-years) by the rates for highly active men (2,000+ kcal/wk, 35.3 attacks per 10,000 person-years). That is, less active men were at 1.64 times greater risk of having a heart attack relative to highly active men (Paffenbarger, 1978).

Important contributions from this cohort study include the finding that collegiate sports play did little to provide

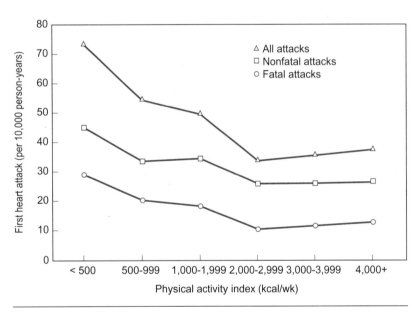

Figure 17.9 Heart attack rates versus physical activity in male Harvard alumni.

From R.S. Paffenbarger, Jr., A.L. Wing, and R.T. Hyde, 1978, "Physical activity as an index of heart attack risk in college alumni," *American Journal of Epidemiology* 180(3): 161-175. Adapted by permission from Oxford University Press.

cardiovascular protection later in life (Paffenbarger, Wing, and Hyde, 1984). That is, to reduce cardiovascular risk the most, participants needed to be active in early adulthood (college) through middle age. Finally, this cohort study demonstrated that beginning moderately vigorous physical activity in middle age can extend the life span (Paffenbarger, et. al., 1993).

Advantages of Cohort Studies

Cohort studies have three primary advantages:

1. The temporal sequence between exposure and outcome is clearly defined. Exposure assessments are obtained before disease onset, so the timing of the exposure in the natural history of the disease is in the correct sequence. This feature is the primary strength of cohort studies.

2. Cohort studies are good for rare exposures. For example, if you wanted to study the effects of a rare exposure such as running, you could specifically recruit a cohort of runners into your study to ensure an adequate number of people in low-, medium-, and high-mileage categories. If you recruited for such a study using a random sample of the adult population of the United States, you would have only 10 to 12% of your study population with your exposure of interest (U.S. Department of Health and Human Services, 1996).

3. Cohort studies are good for understanding the multiple effects of a single exposure. For example, within the Harvard alumni cohort study mentioned earlier, the effects of physical activity on a diverse number of health outcomes were examined, including lung, colon, prostate, and pancreatic cancers, as well as the risk of first heart attack, stroke, hypertension, depression, suicide, Parkinson's disease, and cardiovascular and all-cause mortality.

Disadvantages of Cohort Studies

Cohort studies are difficult and costly to conduct because of the challenges inherent in keeping track of large numbers of people over long periods of time (about 40 years in the case of the College Alumni Health Study). Loss to follow-up can be problematic because large numbers of losses can result in biased estimates of the exposure–disease relationship.

A second limitation of cohort studies is that some disease outcomes are sufficiently rare that even a very large cohort may not produce enough cases for meaningful analyses. For example, it took only 6 to 10 years for the Harvard alumni study ($n = 16,936$) to accumulate 572 cases of first heart attack (or, an incidence of 42.2 per 10,000 men/year). By comparison, in a cohort study of effects of physical activity on the rarer outcome of breast cancer, 14 years of follow-up in 25,624 women were required to accumulate 351 cases of breast cancer (or an incidence of 9.8 per 10,000 women/year).

Case-Control Studies

The case-control study, like the cohort study, aims to identify factors that are causally related to disease outcome. In this design, a population of people with disease (i.e., cases) and without disease (i.e., controls) is recruited into the study over the same period of time. Frequently, the disease-free control subjects are selected to match cases with respect to potentially important confounding factors, such as age or ethnicity. Both cases and controls are then asked about their exposures to potential causal factors using in-person interviews or self-administered questionnaires. Thus, the objective of this retrospective exposure assessment in case-control studies is to identify factors that influenced the natural history of the disease during its induction period (i.e., before disease onset). Analysis of case-control data contrasts the exposure history of the cases to that of controls and typically expresses this relationship as an **odds ratio (OR).** The odds ratio is interpreted as an estimate of the relative risk that would have been calculated in the study group if a cohort study had been conducted. Like the relative risk, the null value of the odds ratio is 1.0. Values above 1.0 typically indicate increased risk, and values less than 1.0 indicate reduced risk.

odds ratio

Measure of association of the exposure–disease relationship, typically employed in case–control studies.

This type of epidemiologic design has been particularly important in studying the relationship between physical activity and relatively rare diseases such as cancer. For example, breast cancer is the leading cancer diagnosis among women in the United States, accounting for 175,000 cases in 1999, or about 30% of all cancers diagnosed in women (Ries, et. al., 1999). Clearly it is a major public health concern, yet the relatively low incidence of the disease makes it a more challenging outcome to investigate using the cohort design.

An interesting example of a case-control study examining the relationship between physical activity and breast cancer was published by Leslie Bernstein and colleagues (Bernstein, et. al., 1994). The purpose of the investigation was to examine the effect of physical activity exposure accumulated from adolescence through adulthood on the risk of breast cancer in women under 40 years of age. Investigators enrolled 545 cases with breast cancer and 545 controls that were free of the disease. All participants were younger (<40 years) residents of Los Angeles and were recruited between 1983 and 1989. Women in the control group were matched to cases according to age, ethnicity, number of pregnancies, and neighborhood of residence. Upon recruitment, each woman was interviewed about lifetime exercise participation and a number of additional breast cancer risk factors. In the interview, women reported each exercise activity in which they participated on a regular basis during their lifetime, the length of time they participated in this activity (in years), and the average duration of their exercise (hours per week). Using these data, women were categorized into five levels of average lifetime exercise exposure. The least active women reported no regular exercise (0 hours per week). Active women were categorized into four categories based on the quartile cutoff points for controls reporting some exercise (>0 hours per week).

Table 17.1 was adapted from the Bernstein article and presents univariate (unadjusted) and multivariate (adjusted) odds ratios for this association. Women who exercised were at lower risk of having breast cancer relative to nonexercisers because the value of the odds ratios is below 1.0. The beneficial effect of exercise was particularly evident for the most active women ($\geq$3.8 hours per week). Women who exercised $\geq$3.8 hours per week were at a 50% reduced risk of breast cancer in comparison to nonexercisers (0 hours per week) in univariate analyses, and a 58% reduced risk after controlling for important confounding factors.

Table 17.1 Odds Ratios for Breast Cancer Associated With Exercise

Lifetime exercise[a]	n cases/ n controls	Univariate OR (95% confidence level)	Multivariate OR (95% confidence level)[b]
None	195/154	1.0	1.0
0.1–0.7	103/90	0.92 (0.65–1.32)	0.95 (0.64–1.41)
0.8–1.6	84/102	0.66 (0.46–0.95)	0.65 (0.45–0.96)
1.7–3.7	103/100	0.82 (0.58–1.17)	0.80 (0.54–1.17)
$\geq$ 3.8	61/99	0.50 (0.34–0.73)	0.42 (0.27–0.64)

[a]Hours per week, from menarche to about 12 months before study entry.

[b]Adjusted for age at menarche, age at first full-term pregnancy, number of full-term pregnancies, months of breast feeding, family history of breast cancer, body mass index at reference date, and oral contraceptive use.

Reprinted from Bernstein et al. 1994, "Physical exercise and reduced risk of breast cancer in young women," *Journal of the National Cancer Institute* 86(18):1371–1372. By permission of Oxford University Press.

The findings in this investigation spurred a number of additional studies of the effect of physical activity on breast cancer risk using both case-control and cohort designs (Friedenreich, et. al., 1998). Two recent reports using the cohort study design (Rockhill et al., 1999; Thune et al., 1997) supported the findings presented by Bernstein and colleagues. That is, high levels of physical activity accumulated during a woman's lifetime appear to reduce the risk of breast cancer.

Advantages of Case-Control Studies

The case-control study design is often the first line of analytic investigation because it can provide valid estimates of exposure–disease relationships in a shorter period of time and with less monetary expense in comparison to the cohort design. Case-control designs

1. are particularly efficient for investigating rare diseases because they obviate the need for very large populations or long follow-up periods to accumulate a useful number of cases for analysis,

2. enable hypothesis testing for multiple exposures for a single disease outcome, and

3. can also be used to acquire exposure information in greater detail because more resources can be used to gather this information as a result of the smaller scale of most case-control studies.

Disadvantages of Case-Control Studies

The major disadvantages of case-control studies are consequences of the fact that exposure information is obtained after the disease has been diagnosed, and that it can be challenging to recruit an appropriate control group. The exposure information and the resulting odds

Systematic errors introduced by differences in the recall accuracy between comparison groups (i.e., between cases and controls).

ratios may be adversely effected by potential problems in each of these areas. First, people who have just been diagnosed with a major disease that is often life threatening may recall their exposures differently than control subjects do, simply because of their recent diagnosis. Bias of this type has been called **recall bias**. For example, in a case-control study examining the physical activity–heart disease relationship, participants with disease (cases) may report that they were less active than they truly were because it is well known that exercise is protective against heart disease. This would increase the apparent difference in physical activity levels between the controls who reported their exposure without a recall bias. In this example, recall bias would bias odds ratios away from the null value of 1.0 and suggest that a stronger relationship existed than was actually present.

Second, for the contrast of exposures histories between cases and controls to provide valid odds ratios, control subjects need to be representative of the population from which the cases were obtained. This means that the control group does not have to be generalizable to the general population, but only to the population that produced the case series. If the exposure histories of the control subjects in the study differ importantly from the unknowable exposure distribution of the cohort from which the cases arose, then **selection biases** may cause odds ratios that are not valid estimates of the true exposure–disease relationship.

Case-control methods have been developed and extensively refined in the last 50 years resulting in a study design that can provide invaluable information about important exposure–disease outcomes in a very efficient manner. However, a full description of the many facets of this study design is well beyond the scope of this chapter. Interested readers are encouraged to consult the epidemiology texts listed in the suggested readings list for a more thorough explanation of these methods.

selection bias

Systematic errors introduced by differences in the characteristics of subjects entering and not entering a study.

Threat to Validity in Analytic Study Designs

The major threats to obtaining a valid understanding of the exposure–disease relationship in analytic studies are methodological biases and confounding factors. Epidemiologic methods, when implemented appropriately, limit the effects of biases and confounding factors. **Bias** in epidemiologic research typically refers to systematic deviations of risk estimates (i.e., odds ratios or relative risk) from their "true" value. Recall and selection biases have previously been discussed in the case-control studies section. Other important biases to consider include observer biases and biases that result from misclassification of exposure. Observer bias can be introduced in studies that use interviewers to collect exposure information but do not blind the interviewers to the disease status of the participants. A result of this lack of blinding can be a systematic bias introduced if the interviewer collects exposure data differently from cases and controls.

bias

Systematic deviation of a calculated (estimated) value from the true value.

Biases that result from the misclassification of physical activity exposure levels often arise from measurement errors introduced when the exposures are measured. Because physical activities are often measured rather crudely, or are assessed at only one point in a person's life, measure of exposure may not adequately reflect the person's true physical activity exposure during the induction period of the disease of interest. Accordingly, the exact relationship between the exposure and the study outcome may not be clearly described by the data collected. If this misclassification of exposure is randomly distributed in the study population, these errors will result in an attenuation in the measures of effect. That is, an actual effect on the disease outcome by the exposure could be less pronounced than it actually is, or the effect may be missed completely. This type of bias is usually referred to as "bias toward the null" (i.e., no effect).

confounding factor

A factor that obscures the true relationship between an exposure–outcome of interest.

A **confounding factor** in epidemiologic studies refers to an additional factor that obscures the observable relationship between an exposure and the outcome. Confounding factors must be associated with both the exposure and the outcome and are identifiable by their effect on the measure of the association. That is, changes in odds ratios or relative risk estimates between unadjusted and adjusted analyses suggest the presence of confounding factors. Confounding factors can be controlled using statistical methods or stratification. For example, if smoking was a confounding factor in the relationship between physical activity and blood cholesterol levels, its confounding effect could be controlled by examining the relationship in smokers and nonsmokers separately.

Experimental Designs With Randomized Trials

Regular physical activity increases cardiorespiratory fitness, promotes weight loss, and if done at regular intervals, reduces the risks for heart attack, colon cancer, and premature mortality (U.S. Department of Health and Human Services, 1996). Experimental designs allow researchers to identify the effects of a specific intervention on a health outcome in a group of people (experimental group) while simultaneously monitoring changes in the same health outcome among people not receiving the intervention (control or comparison group). Randomized trials that are focused on changing health at the individual level are referred to as *clinical trials.* Because denying treatment that is known to cause health benefits is illegal, randomized trials for physical activity often provide a form of physical activity to the comparison group that does not have the same biological effect as the intervention treatment provided to the intervention group, or that compares different types of treatment.

An example of a randomized trial involving a physical activity intervention is Project Active, which is designed to compare the effects of a personal, lifestyle physical activity intervention with a traditional, structured exercise intervention. Subjects were 235 initially sedentary and apparently healthy adults who were randomized into either a lifestyle or a structured intervention group. It was hypothesized that there would be no difference in the physical activity and cardiorespiratory fitness between the two groups at the end of the treatment. Both groups were instructed in programmatically different, but physiologically equivalent, methods to increase their levels of daily physical activity (Kohl, et. al., 1998). Results showed increases in physical activity and cardiorespiratory fitness at the end of the 6-month intensive intervention; however, these changes were similar between the two intervention styles (Dunn, et. al., 1998).

Randomized trials that are focused on changing behaviors in communities are referred to as *community trials.* The rationale for community-level interventions are as follows: (1) targeting everyone may prevent more cases of disease than targeting just high-risk people; (2) environmental modifications may be easier to accomplish than large-scale voluntary behavior change; (3) risk-related behaviors are socially influenced; (4) community interventions reach people in their native habitat; and (5) community interventions can be logistically simpler and less costly on a per-person basis.

In community trials, whole communities are randomized to receive multiple treatments in the form of mass media campaigns, school-based programs, point of purchase interventions (e.g., posting signs in grocery stores indicating brands of food that have less saturated fat), and related activities. The comparison communities do not receive the intervention but serve to show the impact of naturally occurring changes in community behaviors over time. Data collection and analysis include monitoring the extent to which an intervention is actually implemented as intended and the effects of the intervention on community behaviors (process evaluation), individual-level changes in behavior and health outcomes (individual evaluation), and changes in the environment influenced by the intervention (community-level indicator evaluation). Data analysis must account for community-level and individual-level variation in behaviors. An advantage of community trials is that they allow researchers to see the population changes in behaviors as a result of interventions. However, this benefit is tempered by the difficulty of changing individual behaviors and the amount of time it may take (sometimes many years) to see changes in community behaviors.

The simplest design for a community trial involves one intervention and one control community. Examples of these are the North Karelia Project in Finland (Puska, Salonen, & Nissinen, 1983), the Pawtucket Heart Health Project (Carleton, et al., 1995), and the Heart to Heart Project in South Carolina (Goodman, Wheeler, & Lee, 1995), each of which was designed to reduce the risks for cardiovascular disease. Unfortunately, because these study designs had only one intervention and one control community, investigators were unable to compare community-level variation in the observed behaviors. Thus it was unknown whether the changes observed were due to the intervention or naturally occurring changes in the community that would have taken place over time without the intervention.

Subsequent community trials involving physical activity interventions have included multiple intervention and control cities. These include the Stanford Five-City Project (Winkleby, et. al., 1996), which enrolled two interventions and two control communities, and the Minnesota Heart Health Study (Jacobs, et al., 1986), which enrolled three interventions and three control communities located in geographically distant locations in Northern California and Minnesota, respectively.

In the Stanford Five-City Project, intervention activities were designed to reduce cardiovascular disease risk factors and were delivered to two communities from 1979 to 1985. The physical activity interventions were targeted toward increasing moderate and vigorous physical activity in intervention communities and used electronic and print media, individual and community activities, and school-based functions in intervention activities. Results showed modest, but statistically significant changes in physical activity in the intervention communities with men more likely to participate in vigorous activities and women reporting more time spent in moderate-intensity activities. Similar changes were not observed in the control communities.

Reading and Interpreting a Physical Activity Epidemiologic Study

Because epidemiologic studies may involve a variety of study designs and large numbers of people, it is a good idea to follow a guideline when reading and interpreting a research paper. An example will be given using an article written in 2000 by I.M. Lee and R.S. Paffenbarger Jr. titled, "Associations of Light, Moderate, and Vigorous Intensity Physical Activity With Longevity: The Harvard Alumni Health Study (*American Journal of Epidemiology, 151,* 293-299).

Introduction

The introduction should provide a global overview of the rationale for the study, highlighting major research that has shown associations (or lack of associations) between the exposure and outcomes and explaining why this is a health concern worthy of studying. The rationale should also address the biological mechanism for the association between the physical activity and a health outcome. Research questions should reflect findings and questions arising from earlier studies and should state the purpose of the study.

> *Example:* The authors recognized the growing acceptance that a healthy lifestyle includes regular physical activity. However, they noted the inconsistent findings about the optimal intensity of physical activity needed to reduce premature mortality. The purpose of the paper was to present associations between bouts of light-, moderate-, and vigorous-intensity activity and mortality among men enrolled in the Harvard Alumni Health Study.

Materials and Methods

This section should include information about the study design, study population, variables, and analytic methods.

The *study design* should be appropriate to answer the research questions. To describe events in a population (descriptive studies), an ecological or cross-sectional study design is appropriate. However, to identify causal associations that reflect changes in health status across time, prospective cohort studies or experimental study designs are needed. Case-control studies are useful to examine exposure and outcome relationships in small study samples and provide only limited evidence of the existence of a temporal relationship between the exposure and outcome.

The *study population* should be appropriate to answer the research questions asked in the study. Writers should describe their *sample size* to include how many subjects were enrolled in the study, completed the tasks as assigned, and were included in the data analysis. The writers also should provide an explanation of differences in the sample sizes from the start to the end of the study. This information is an important issue for avoiding biased studies. The methods used to recruit the subjects and inclusion or exclusion criteria should be listed.

Characteristics of the study population should include subjects' age, sex, race, income, education, or other factors related to the research questions. Is the study population representative of a group used for generalization?

The *study variables* should be described for the exposure (physical activity), outcome (e.g., mortality), and potential confounders. Writers should identify how the exposure was measured and scored and if there are data that addresses the validity and reliability of the measures. The *outcome* should be identified by a case definition (e.g., ICD-version 10 codes) and a description of how data were obtained, either from hospital records, death certificates, physical exams, or self-reports. Disease outcomes should have a case definition that the authors use to guide the researchers in the classification of the outcome status. *Potential confounders* should be identified with the data collection methods for each variable described. Writers should also describe the quality control measures used to ensure accuracy in the data collection process.

Methods used for *data analyses* should reflect the research questions and hypotheses tested in the study.

> *Example: Study population.* Subjects were described as men who matriculated as undergraduates at Harvard University between 1916 and 1950. Inclusion criteria included having returned a survey sent in 1977 (n = 17,835). Exclusion criteria were the presence of physician-diagnosed cardiovascular disease, cancer, or chronic obstructive pulmonary disease (n = 3,706) or failure to identify information about potential covariate variables on the survey (n = 644). The sample used in analysis was 13,485 men with a mean age of 57.5 + 8.9 years.

> *Example: Exposure.* Physical activity was measured with a mailed survey that had subjects recall the number of stairs climbed, city blocks walked, and sports and recreational activities performed in the past week. A physical activity index (PAI) was expressed as a composite of these activities in kilojoules per week. The survey has a 1-month reliability of r = .72 and a correlation with physical activity records of r = .65. MET levels associated with intensity levels were as follows: light: <4 METs; moderate: 4–5.5 METs; vigorous: >6 METs.

> *Example: Outcome.* Identification of subjects' deaths was obtained from the National Death Index. The causes of death were determined from death certificates.

> *Example: Potential confounders.* Confounder variables were obtained from self-reported information recorded on the mailed health survey. Variables included age, BMI, cigarette smoking, alcohol consumption, and early (<65 years) parental death.

> *Example: Data analyses.* The association between total energy expenditure in kilojoules per week (kJ/wk) and mortality was examined using proportional hazards models.

Results

The text should present a global overview of the results and refer the reader to the tables and figures that show the data. Tables and figures should be self-explanatory and identify the confounding variables controlled in the analyses in a footnote. Statistical comparisons should present p-values or confidence intervals around parameter estimate (i.e., odds ratio, relative risk), and dose response by presenting comparisons of the data among multiple levels.

> *Example: Results.* A total of 2,539 men died between 1977 and 1992. Compared with men who expended <4,200 kJ/wk, those who expended more energy

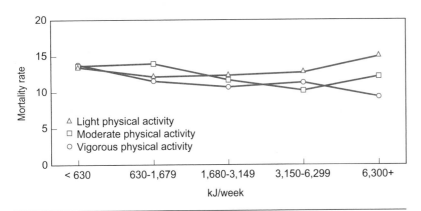

Figure 17.10 Age-adjusted mortality rates (per 1,000) by intensity of physical activity.

From I.M. Lee and R.S. Paffenbarger, Jr., 2000, "Associations of light, moderate, and vigorous intensity physical activity with longevity - The Harvard alumni health study," *American Journal of Epidemiology* 151: 293-299. Used with permission from Oxford University Press.

through regular physical activity reduced their risk for mortality by 20 to 27%. Trends for associations between type and intensity of activity and mortality were significant at $p < .05$ for distance walked, stairs climbed, and moderate and vigorous activity (see figure 17.10). Light activity was not associated with decreases in mortality rate.

Discussion and Conclusions

In the discussion, the writers should highlight the main study findings, compare and contrast the results with similar studies, and relate the findings to data that highlight the magnitude of the problem. Sources of bias that may have influenced the study results should be addressed as potential weaknesses of the study. The criteria for causality should be discussed as a prelude to determining whether the associations and hypotheses tested support a causal association between the physical activity exposure and a health outcome.

> *Example:* The authors concluded that greater energy expenditure is associated with increased longevity. Participation in moderate activities showed a trend toward lower mortality rates, and participation in vigorous activities showed a strong association with lower mortality rates. Participation in light activities was unrelated to mortality rates. Biological mechanisms for the association are supported by other studies showing lower blood pressure, glucose, and insulin levels, higher HDL cholesterol, and enhanced cardiac function and hemostatic factors with regular moderate and vigorous physical activity. Strengths of the study include a detailed physical activity assessment and a long follow-up period. Limitations include the possibility of recall bias in using a self-administered physical activity survey and the inability to account for diseases not measured on the survey. The public health impact of the study shows benefits of moderate- and vigorous-intensity exercise in reducing risks for mortality among men.

Summary

This chapter provides an overview of the research methods used in the study of physical activity epidemiology. Epidemiologic studies are designed to identify the distribution and determinants of physical activity and health-related states in population groups and to apply the research findings to reduce the burden of disease in communities.

✔ Check Your Understanding

1. Describe the differences between observational and experimental research.

2. Briefly describe the strengths and weaknesses of descriptive, analytic, and experimental study designs in examining the relationships between physical activity and health.

3. Compare and contrast the cohort and case-control study designs.

4. Write a review of a physical activity research study published in the *American Journal of Epidemiology,* the *Journal of Clinical Epidemiology,* the *International Journal of Epidemiology,* or *Epidemiology.*

5. Compare and contrast the ways epidemiologists study the distribution, determinants, and applications of health issues.

CHAPTER

18

Experimental and Quasi-Experimental Research

I have seen the future, and it's a lot like the present, but much longer.—Dan Quisenberry (former Major League Baseball player)

■   ■   ■   ■   ■   ■   ■

Experimental research attempts to establish cause-and-effect relationships. That is, an independent variable is manipulated to judge its effect on a dependent variable. However, the process of establishing cause and effect is a difficult one. Three criteria must be present to establish cause and effect:

1. *The cause must precede the effect in time.* For example, the starting gun in a race precedes the runners' beginning the race; the runners' beginning does not cause the starting gun to go off.

2. *The cause and effect must be correlated with each other.* As we have already discussed, just because two variables are correlated does not mean one causes the other; however, cause and effect cannot exist unless two variables are correlated.

3. *The correlation between cause and effect cannot be explained by another variable.* Recall that the relationship between elementary children's academic performance and shoe size was explained by a third variable, age.

We may think of cause and effect in terms of necessary and sufficient conditions (Krathwohl, 1993). For example, if the condition is necessary and sufficient to produce the effect, then it is the cause. However, alternative situations exist (also see the sidebar on p. 322):

- Necessary but not sufficient: It is likely that some related condition produces the effect.
- Sufficient but not necessary: Some alternative condition is likely the cause.
- Neither necessary nor sufficient: Some contributing condition is likely the cause.

Also, remember that cause and effect are not established by statistics. Statistical techniques can only reject the null hypothesis (establish that groups are significantly different) and identify the percentage of variance in the dependent variable accounted for by the independent variable or the effect size. Neither of these procedures establishes cause and effect (they are necessary but not sufficient). Cause and effect can be established only

321

by the application of logical thinking to well-designed experiments. This logical process establishes that no other reasonable explanation exists for the changes in the dependent variable except the manipulation of the independent variable. The application of this logic is made possible by the following:

- The selection of a good theoretical framework
- The use of appropriate participants
- The application of an appropriate experimental design
- The proper selection and control of the independent variable (treatment)
- The appropriate selection and measurement of the dependent variable
- The use of the correct statistical model and analysis
- The correct interpretation of the results

Examples of Cause and Effect in Golf

1. No matter how bad your last shot was, the worst is yet to come. This does not expire until after the 18th hole.

2. Your best round of golf will be followed almost immediately by your worst round ever. (Probability increases based on the number of people you tell about your best round.)

3. Brand new golf balls are water-magnetic. (The magnetic effect increases with the price of the ball.)

4. The higher the golfer's handicap, the more qualified that golfer deems himself/herself to be an instructor.

5. Every par-three hole has a secret desire to humiliate golfers. (The shorter the hole, the greater its desire.)

6. Electric golf carts always fail at the point farthest from the club house.

7. The person you most hate to lose to will always be the one who beats you.

8. The last three holes of a round will automatically adjust your score to what it really should be.

9. Golf balls never bounce off trees back into play. (If one does, the groundskeeper will remove that tree before your next round.)

10. All vows taken on a golf course are valid only until sundown.

In this chapter, we discuss experimental designs by explaining how you can recognize and control sources of invalidity and threats to both internal and external validity. We also explain several types of experimental designs. Before going any further into this chapter, review the following terms (introduced in chapters 1 and 3) that we use throughout the discussion:

- Independent variable
- Dependent variable
- Categorical variable
- Control variable
- Extraneous variable

Sources of Invalidity

All the types of designs we discuss have strengths and weaknesses that pose threats to the validity of the research. The importance of validity was stated well by Campbell and Stanley (1963, 5):

Fundamental . . . is a distinction between internal validity and external validity. Internal validity is the basic minimum without which any experiment is uninterpretable: Did in fact the experimental treatments make a difference in this specific experimental instance? External validity asks the question of generalizability: To what populations, settings, or treatment variables can this effect be generalized?

Both internal validity and external validity are important in experiments. However, they are frequently at odds in research planning and design. Gaining internal validity involves controlling all variables so that the researcher can eliminate all rival hypotheses as explanations for the outcomes observed. Yet in controlling and constraining the research setting to gain internal validity, the researcher places the generalization (external validity) of the findings in jeopardy. In studies with strong internal validity, the answer to the question of to whom, what, or where the findings can be generalized may be very uncertain. This is because in ecologically valid settings (perceived by the participants as intended by the researchers), not everything is controlled and operated in the same way as in the controlled laboratory context. Thus, the researcher must decide: Is it more important to be certain that the manipulation of the independent variable caused the observed changes in the dependent variable, or is it more important to be able to generalize the results to other populations, settings, and so on? We cannot provide an easy answer to that question, which is often debated at scientific meetings and in the literature (e.g., see Martens, 1979, 1987; Siedentop, 1980; Thomas, 1980; Thomas, French, & Humphries, 1986).

To expect any single experiment to meet all research design considerations is unreasonable. A more realistic approach is to identify the specific goals and limitations of the research effort. Is internal validity or external validity the more important issue? Once that is decided, the researcher can plan the research with one type of validity as the major focus while maintaining as much of the other type of validity as possible. Another recourse is to plan a series of experiments in which the first experiment has strong internal validity even at the expense of external validity. If the first experiment confirms that changes in the dependent measure are the result of manipulating the independent variable, subsequent experiments can be designed to increase external validity even at the expense of internal validity. This allows evaluation of the treatment in settings more like the real world. (Review the related discussion of basic and applied research in chapter 1 and Christina, 1989.)

Threats to Internal Validity

Campbell and Stanley (1963) identified eight threats to the internal validity of experiments,[1] and Rosenthal (1966) identified a ninth:

- **History:** events occurring during the experiment that are not part of the treatment
- **Maturation:** processes within the participants that operate as a result of time passing (e.g., aging, fatigue, hunger)
- **Testing:** the effects of one test on subsequent administrations of the same test
- **Instrumentation:** changes in instrument calibration, including lack of agreement within and between observers
- **Statistical regression:** the fact that groups selected on the basis of extreme scores are not as extreme on subsequent testing
- **Selection bias:** choosing comparison groups in a nonrandom manner
- **Experimental mortality:** loss of participants from comparison groups for nonrandom reasons
- **Selection-maturation interaction:** the passage of time affecting one group but not the other in nonequivalent group designs
- **Expectancy:** experimenters' or testers' anticipating that certain participants will perform better

If these threats are uncontrolled, the change in the dependent variable may be difficult to attribute to the manipulation of the independent variable.

History

A history threat to internal validity means that some unintended event occurred during the treatment period. For example, in a study evaluating the effects of a semester of physical education on the physical fitness of fifth graders, the fact that 60% of the children participated in a recreational soccer program would constitute a history threat to internal validity. The soccer program is likely to produce benefits to physical fitness that would be difficult to separate from the benefits of the physical education program.

Maturation

Maturation as a threat to internal validity is most often associated with aging. This threat occurs frequently in designs in which one group is tested on several occasions over a long period. Elementary physical education teachers frequently encounter this source of invalidity when they give a physical fitness test in the early fall and again in the late spring. The children nearly always do better in the spring. The teacher would like to claim that the physical education program was the cause. Unfortunately, maturation is a plausible rival hypothesis for the observed increase; that is, the children have grown larger and stronger and thus probably run faster, jump higher, and throw farther.

Testing

A testing threat is the effect that taking a test once has on taking it again. For example, if a group of athletes are administered a multiple-choice test to evaluate their knowledge about steroids today and again two days later, the athletes will do better the second time even though no treatment intervened. Taking the test once helps in taking it again. The same effect is present in physical performance tests, particularly if the participants are not allowed to practice the test a few times. If a class of beginners in tennis attempts to hit 20 forehand shots delivered to them from a ball machine today and again three days later, they will usually do better the second time. They learned something from performing the test the first time.

Instrumentation

Instrumentation is a problem frequently faced in exercise science research. Suppose the researcher uses a spring-loaded device to measure strength. Unless the spring is calibrated regularly, it decreases in tension with use. Thus, the same amount of applied force produces increased readings of strength compared with earlier readings. Instrumentation problems also occur in research using observers. Unless observers are trained and regularly checked, the same observer's ratings may systematically vary across time or participants (a phenomenon called observer drift), or different observers may not rate the same performance in the same way.

Statistical Regression

Statistical regression can occur when groups are not randomly formed but are selected on the basis of an extreme score on some measure. For example, if someone rates the behavior of a group of children on a playground on an activity scale (very active to very inactive) and two groups are formed—one of very active children and one of very inactive children—statistical regression is likely to occur when the children are next observed on the playground. The children who were very active will be less active (although still active), and the very inactive children will be more active. In other words, both groups will regress (move from the extremes) toward the overall average. This phenomenon reflects only the fact that a participant's score tends to vary about her or his average performance (estimated true score). Extreme scores may simply reflect observation of a performance on the high (or low) side of the participant's typical performance. The next performance is usually not as extreme. Thus, when average scores of extreme groups are compared from one time to the next, the high group appears to get worse, whereas the low group appears to get better. Statistical regression is a particular problem in studies that attempt to compare extreme groups selected on some characteristic, such as highly anxious, fit, or skilled participants versus not very anxious, fit, or skilled participants.

Selection Bias

Selection biases occur when groups are formed on some basis other than random assignment. Thus, when treatments are administered, because the groups were different to begin with, always present is the rival hypothesis that any differences found are due to initial selection biases rather than the treatments. Showing that the groups were not different on the dependent variable at the beginning of the study does not overcome this shortcoming. Any number of other unmeasured variables on which the groups differ might explain the treatment effect. W.R. Borg and Gall (1989) asked important questions that apply to selection (or sampling) bias:

> **D**id the study use volunteers? Use of volunteers is common in research in the study of physical activity. Yet volunteers are often not representative of anyone but other volunteers. They may differ considerably from non-volunteers in motivation for the experimental task and setting.
>
> Are participants extremely nonrepresentative of the population? Often we are unable to select participants at random for our studies, but it is very useful if we at least believe (and can demonstrate) that they represent some larger group from our culture.

Recall the discussion of sampling in chapter 6, particularly the concept of a "good enough" sample.

Experimental Mortality

Experimental mortality refers to the loss of participants. Even when groups are randomly formed, this threat to internal validity can occur. Participants may remain in an experimental group taking part in a fitness program because it is fun, whereas participants in the control group become bored, lose interest, and drop out of the study. Of course, the opposite can occur, too. Participants may drop out of an experimental group because the treatment is too difficult or time-consuming.

Selection-Maturation Interaction

A selection-maturation interaction occurs only in specific types of designs. In these designs, one group is selected because of some specific characteristic, whereas the other group lacks this characteristic. An example is a study of the differences between six-year-olds in two school districts; students in one school district form the experimental group that receives a fitness program, and students in the other are a control group. If the school districts have different admission policies so that the six-year-olds in the experimental group are five months older, it would be difficult to determine whether the fitness program or the fitness program combined with the participants' advanced age produced the observed changes.

Expectancy

One additional threat to internal validity not mentioned by Campbell and Stanley (1963) has been identified. **Expectancy** (Rosenthal, 1966) refers to experimenters' or testers' anticipating that certain participants will perform better. This effect, although usually unconscious on the part of the experimenters, occurs where participants or experimental conditions

expectancy

A threat to internal validity in which the researcher anticipates certain behaviors or results to occur.

are clearly labeled. For example, testers may rate skilled participants better than unskilled participants, regardless of treatment. This effect is also evident in observational studies in which the observers rate posttest performance better than pretest performance because they expect change. If the experimental and control groups are identified, observers may rate the experimental group better than the control even before any treatment occurs. The expectancy effect may influence the participants, too. For example, in a youth-sport study, coaches may actually cause poorer performance by substitutes (compared with starters) because the substitutes realize that the coach treats them differently (e.g., the coach may show less concern about incorrect practice trials).

Any of these nine threats to internal validity may reduce the researcher's ability to claim that the manipulation of the independent variable produced the changes in the dependent variable. We discuss the various experimental designs and how they control (or fail to control) the threats to internal validity later in this chapter.

Threats to External Validity

Campbell and Stanley (1963) identified four threats to external validity, or the ability to generalize results to other participants, settings, measures, and so on:

- **Reactive or interactive effects of testing:** The pretest may make the participant more aware of or sensitive to the upcoming treatment. As a result, the treatment is not as effective without the pretest.
- **Interaction of selection bias and the experimental treatment:** When a group is selected on some characteristic, the treatment may work only on groups possessing that characteristic.
- **Reactive effects of experimental arrangements:** Treatments that are effective in very constrained situations (e.g., laboratories) may not be effective in less constrained settings (more like the real world).
- **Multiple-treatment interference:** When participants receive more than one treatment, the effects of previous treatments may influence subsequent ones.

Reactive or Interactive Effects of Testing

Reactive or interactive effects of testing may be a problem in any design with a pretest. Suppose a fitness program is the experimental treatment. If a physical fitness test is administered to the sample first, the participants in the experimental group might realize that their levels of fitness are low and be particularly motivated to follow the prescribed program closely. However, in an untested population, the program might not be as effective because the participants might be unaware of their low levels of physical fitness.

Interaction of Selection Bias and Experimental Treatment

The interaction of selection bias and the experimental treatment may prohibit the generalization of the results to participants lacking the particular characteristics (bias) of the sample. For example, a drug education program might be quite effective in changing the attitudes of college freshmen toward drug use. This same program would probably lack effectiveness for third-year medical students because they would be much more familiar with drugs and their appropriate uses.

Reactive Effects of Experimental Arrangements

Reactive effects of experimental arrangements mean that the experimental treatment may not be generalizable to real-world situations. These reactive effects are a persistent problem for laboratory-based research (e.g., in exercise physiology, biomechanics, motor behavior). In such research, is the researcher investigating an effect, process, or outcome that is specific to the laboratory and cannot be generalized to other settings? We have referred to this earlier as a problem of ecological validity. For example, in a study employing high-speed cinematography, the skill to be filmed must be performed in a certain place, and joints must be marked for later analysis. Is the skill performed in the same way during participation in a sport?

One specific type of reactive behavior has been labeled the *Hawthorne effect* (Brown, 1954). This effect refers to the fact that participants' performances change when attention is paid to them. This may be a threat to both internal and external validity, as it is likely to produce better treatment effects and reduce the ability to generalize the results.

Multiple-Treatment Interference

Multiple-treatment interference is most frequently a problem when the same participants are exposed to more than one level of the treatment. Suppose participants are going to learn to move to the hitting position in volleyball using a lead step or a crossover step. We want to know which step gets the players in a good hitting position most quickly. If the players are taught both types of steps, learning one might interfere with (or enhance) learning the other. Thus, the researcher's ability to generalize the findings may be confounded by the use of multiple treatments. A better design might be to have two separate groups, each of which learns one of the techniques.

The ability to generalize findings from research to other participants or situations is a question of random sampling (or at least "good enough" sampling) more than any other. Do the participants, treatments, tests, and situations represent any larger populations? Although a few of the experimental designs discussed later control certain threats to external validity, usually the researcher controls these threats through the way he or she selects the sample, treatments, situations, and tests.

Controlling Threats to Internal Validity

Threats to internal and external validity are controlled in different ways by specific techniques. In this section, we describe useful approaches to solving threats to internal validity in the design of experiments. Many threats to internal validity are controlled by making the participants in the experimental and control groups as alike as possible. This is most often done by randomly assigning participants to groups.

Randomization

As mentioned in chapter 6, randomization allows the assumption that the groups do not differ at the beginning of the experiment. The randomization process controls for history up to the point of the experiment; that is, the researcher can assume that past events are equally distributed among groups. It does not control for history effects during the experiment if experimental and control participants are treated at different times or places. The researcher must try to prevent an event (besides the treatment) from occurring in one group but not in the other groups.

Randomization also controls for maturation because the passage of time is equivalent in all groups. Statistical regression, selection biases, and selection-maturation interaction are controlled because they occur only when groups are not randomly formed.

Sometimes ways other than random assignment of participants to groups are used to control threats to internal validity. The matched-pair technique matches pairs of participants who are equal on some characteristic and then randomly assigns each to a different group. The researcher might want very tight control on previous experience in strength training. Thus, participants would be matched on this characteristic and then randomly assigned to the experimental and control groups.

A matched-group technique may also be used. This involves nonrandom assignment of participants to experimental and control groups so that the group means are equivalent on some variable. This is generally regarded as an unacceptable procedure because the groups may not be equivalent on other, unmeasured variables that could affect the outcome of the research.

In within-subjects designs, the participants are used as their own controls. This means that each participant receives both the experimental and the control treatment. In this type of design, the order of treatments should be counterbalanced; that is, half the participants should receive the experimental treatment first and then the control, and the other half should receive the control treatment first and then the experimental treatment. If there are three levels of the independent variable (1 = control, 2 = experimental treatment A, 3 = experimental treatment B), participants should be randomly assigned to one of the six possible treatment orders (1-2-3, 1-3-2, 2-1-3, 2-3-1, 3-1-2, 3-2-1). If the number of treatment levels administered to the same participants is larger than 3 or 4, experimenters can assign a random order of the treatments to each participant rather than counterbalancing treatments.

Placebos, Blind Setups, and Double-Blind Setups

placebo

Method of controlling a threat to internal validity in which a control group receives a false treatment while the experimental group receives the real treatment.

blind setup

Method of controlling a threat to internal validity in which the participant does not know whether he or she is receiving the experimental or control treatment.

double-blind setup

Method of controlling a threat to internal validity in which neither the participant nor the experimenter knows which treatment the participant is receiving.

Avis effect

A threat to internal validity wherein participants in the control group try harder just because they are in the control group.

Other ways of controlling threats to internal validity include placebos and blind and double-blind setups. A **placebo** is used to evaluate whether the observed effect is produced by the treatment or is a psychological effect. Frequently, a control condition is used in which participants receive the same attention from and interaction with the experimenter, but the treatment administered does not relate to performance on the dependent variable.

A study in which participants do not know whether they are receiving the experimental or the control treatment is called a **blind setup** (i.e., the participant is "blind" to the treatment). In a **double-blind setup,** neither the participant nor the tester knows which treatment the participants are receiving. The triple-blind test has also been reported (Day, 1983): The participants do not know what they are getting, the experimenters do not know what they are giving, and the investigators do not know what they are doing. We can only hope the triple-blind technique finds limited use in our field.

All these techniques (except the triple-blind technique) are useful in controlling the Hawthorne effect, expectancy, and halo effects, as well as what we call the **Avis effect** (a recent version of the John Henry effect), or the fact that participants in the control group may try harder simply because they are in the control group.

A good example of the use of these techniques for controlling psychological effects is the use of steroids to build strength in athletes. A number of studies were done to evaluate the effects of steroids. To combat the fact that athletes may get stronger because they think they should when using steroids, a placebo (another pill that looks just like the steroid) is used. In a blind study, the athletes do not know whether they receive the placebo or the steroid. In a double-blind study, the athlete, the person dispensing the steroids or placebos, and the testers do not know which group received the steroids. Unfortunately, these procedures did not work very well in this specific type of study because taking large quantities of steroids made the athletes' urine smell bad. Thus, the athletes knew whether they were receiving the steroid or the placebo.

Uncontrolled Threats to Internal Validity

Three threats to internal validity remain uncontrolled by the randomization process.

Reactive or Interactive Effects

Reactive or interactive effects of testing can be controlled only by eliminating the pretest. However, these effects can be evaluated by two of the designs discussed later in this chapter: pretest-posttest randomized-groups and Solomon four-group.

Instrumentation

Instrumentation problems cannot be controlled or evaluated by any design. Only the experimenter can control this threat to internal validity. In chapter 11, we went into some detail on techniques for controlling the instrumentation threat to develop valid and reliable tests. Of particular significance is test reliability. Whether the measurement is obtained

This is her understanding of the *double blind set-up?*

from a laboratory device (e.g., an oxygen analyzer), a motor performance test (e.g., a standing long jump), an attitude-rating scale (e.g., on feelings about drug use), an observer (e.g., coding percentage of time a child is active), a knowledge test (e.g., on basketball strategy), or a survey (e.g., on available sport facilities), the answer must be consistent. This frequently involves the assessment of test reliability across situations, between and within testers or observers, and within participants. The validity of the instrument (does it measure what it was intended to measure?) must also be established to control for instrumentation problems. The total process of establishing appropriate instrumentation for research is called psychometrics.

One final point about instrumentation is called the halo effect. As we discussed in chapter 11, this occurs in ratings of the same individual on several skills. Raters seeing a skilled performance on one task are likely to rate the participant higher on subsequent tasks, regardless of the level of skill displayed. In effect, the skilled behavior has "rubbed off" (created a "halo") on later performances. There may also be an order effect in observation. Gymnastics and swimming judges often rate earlier performers lower to save room on the rating scale for better performers. Because gymnastics and swimming coaches know this, they always place better performers later in the event order.

Experimental Mortality

Experimental mortality is not controlled by any type of experimental design. The experimenter can control this only by ensuring that participants are not lost (at all, if possible) from groups. Many problems of participant retention can be handled in advance of the research by carefully explaining the research to the participants and emphasizing the need for them to follow through with the project. (During the experiment itself, begging, pleading, and crying sometimes work.)

Controlling Threats to External Validity

External validity is generally controlled by selecting the participants, treatments, experimental situation, and tests to represent some larger population. Of course, random selection (or "good enough" sampling) is the key to controlling most threats to external validity. Remember that more than the participants may be randomly selected. For example, the levels of treatment can be randomly selected from the possible levels, experimental situations can be selected from possible situations, and the dependent variable (test) can be randomly selected from a pool of potential dependent variables.

As previously noted, the ability to generalize the situation is called ecological validity. Although the results of a particular treatment can be generalized to a larger group if the sample is representative, this generalization may apply only to the specific situation in the experiment. If the experiment is conducted under controlled laboratory conditions, then the findings may apply only under controlled laboratory conditions. Frequently, the experimenter hopes that the findings can be generalized to real-world exercise, sport, industrial, or instructional settings. Whether the outcomes can be generalized in this way depends largely on how the participants perceive the study, and this influences the way participants respond to study characteristics. The question of interest here is, Does the study have enough characteristics of real-world settings that participants respond as if they were in the real world? Is there ecological validity? This is not an easy question to answer, and as a result, a number of scholars advocate that more research in kinesiology, physical education, exercise science, and sport science be conducted in field settings (e.g., Costill, 1985; Martens, 1987; Thomas, French, & Humphries, 1986).

Reactive or interactive effects of testing can be evaluated by the Solomon four-group design. Interaction of selection biases and the experimental treatment is controlled by random selection of participants. Reactive effects of experimental arrangements can be controlled only by the researcher (this is again the issue of ecological validity). Multiple-treatment interference can be partially controlled by counterbalancing or randomly ordering the treatments among participants. But the researcher can control whether the treatments still interfere only through knowledge about the treatment rather than the type of experimental design.

Types of Designs

This section (much of which is taken from Campbell & Stanley, 1963) is divided into three categories: preexperimental designs, true experimental designs, and quasi-experimental designs. We use the following notation:

- Each line indicates a group of participants.
- R signifies random assignment of participants to groups.
- O signifies an observation or a test.
- T signifies that a treatment is applied; a blank space in a line where a T appears on another line means that the group is a control.
- $\cdots$ A dotted line between groups means that the groups are used intact rather than being randomly formed.
- Subscripts indicate either the order of observations and treatments (when they appear on the same line) or observations of different groups or different treatments (when they appear on different lines). For example, when the terms T_1 and T_2 appear on different lines, they refer to different treatments; when they appear on the same line, they mean that the treatment is administered more than once to the same group.

preexperimental design

One of three types of research design that control very few of the sources of invalidity and that do not have random assignments of participants to groups: one-shot study, one-group pretest–posttest design, and static group comparison.

Preexperimental Designs

The three designs discussed in this section are called **preexperimental designs** because they control very few of the sources of invalidity. None of the designs has random assignment of participants to groups.

One-Shot Study

In a one-shot study design, a group of participants receives a treatment followed by a test to evaluate the treatment:

T   O

This design fails all the tests of good research. All that can be said is that at a certain point in time this group of participants performed at a certain level. In no way can the level of performance (O) be attributed to the treatment (T).

One-Group Pretest–Posttest Design

The one-group pretest–posttest design, although very weak, is better than the one-shot design. At least we can observe whether any change in performance has occurred:

$$O_1 \qquad T \qquad O_2$$

If O_2 is better than O_1, we can say that the participants improved. For example, Bill Biceps (a qualified exercise instructor) conducted an exercise test at a health club. Participants then trained 3 days per week, 40 min per day, at 70% of their estimated $\dot{V}O_2$max, for 12 weeks. After the training period, participants retook the exercise test and significantly improved their scores. Can Mr. Biceps conclude that the exercise program caused the changes in the exercise test performance he observed? Unfortunately, this design does not allow us to say why the participants improved. Certainly, it could be due to the treatment, but it could also be due to history. Some event other than the treatment (T) may have occurred between the pretest (O_1) and the posttest (O_2); the participants may have exercised at home on the other days. Maturation is a rival hypothesis. The participants may have gotten better (or worse) just as a result of the passage of time. Testing is a rival hypothesis; the increase at O_2 may be the result only of experience with the test at O_1. If the group being tested is selected for some specific reason, then any of the threats involving selection bias can occur. This design is most frequently analyzed by the dependent t test to evaluate whether a significant change occurred between O_1 and O_2.

Static Group Comparison

The static group-comparison design compares two groups, one of which receives the treatment and one of which does not:

$$T \qquad\qquad O_1$$
$$\cdots\cdots\cdots\cdots\cdots\cdots$$
$$O_2$$

However, we do not know whether the groups were not equivalent when the study began, as indicated by the dotted line between the groups. This means that the groups were selected intact rather than being randomly formed. This leaves us unable to determine whether any differences between O_1 and O_2 are because of T or only because the groups differed initially. This design is subject to invalidity because of selection biases and the selection-maturation interaction. A t test for independent groups is used to evaluate whether O_1 and O_2 differ significantly. However, even if they do differ, the difference cannot be attributed to T.

The three preexperimental designs are not valid methods of answering research questions (see table 18.1). They do not represent experiments because the change in the dependent variable cannot be attributed to manipulation of the independent variable. You will not encounter these preexperimental designs in research journals, and we hope that you will not find (or produce) theses and dissertations using these designs. The preexperimental designs represent much wasted effort because little or nothing can be concluded from the findings. If you submit studies using these designs to research journals, you are likely to receive rejection letters similar to one Snoopy (from the *Peanuts* comic strip) received: "Dear Researcher, Thank you for submitting your paper to our research journal. To save time, we are enclosing two rejection letters—one for this paper and one for the next one you send."

True Experimental Designs

The designs discussed in this section are called **true experimental designs** because the groups are randomly formed, allowing the assumption that they were equivalent at the beginning of the research. This controls for past (but not present) history,

true experimental design

Any design used in experimental research in which groups are randomly formed and that controls most sources of invalidity.

Table 18.1 Preexperimental Designs and Their Control of the Threats to Validity

Validity threat	One-shot study	One-group pretest and posttest	Static group
Internal			
History	–	–	+
Maturation	–	–	?
Testing		–	+
Instrumentation		–	
Statistical regression		?	+
Selection	–	+	–
Experimental mortality	–		–
Selection × maturation		–	–
Expectancy	?	?	?
External			
Testing × treatment		–	
Selection biases × treatment	–	–	–
Experimental arrangements		?	
Multiple treatments			

Note. + = strength; – = weakness; blank = not relevent; ? = questionable.

Copyright, 1970, by the American Educational Research Association. Adapted by permission of the publisher.

maturation (which should occur equally in the groups), testing, and all sources of invalidity that are based on nonequivalence of groups (statistical regression, selection biases, and selection-maturation interaction). However, only the experimenter can make sure that nothing happens to one group (besides the treatment) and not the other (present history), that scores on the dependent measure do not vary as a result of instrumentation problems, and that the loss of participants is not different between the groups (experimental mortality).

Randomized-Groups Design

Note that the randomized-groups design resembles static group comparison except that groups are randomly formed:

$$R \quad T \quad O_1$$
$$R \quad \quad O_2$$

If the researcher controls the threats to internal validity that are not controlled by randomization (no easy task), has a sound theoretical basis for the study, and meets the necessary-and-sufficient rule, then this design allows the conclusion that significant differences between O_1 and O_2 are due to T. An independent *t* test is used to analyze the difference between O_1 and O_2. This design as earlier depicted represents two levels of one independent variable. It can be extended to any number of levels of an independent variable:

$$R \quad T_1 \quad O_1$$
$$R \quad T_2 \quad O_2$$
$$R \quad \quad O_3$$

Here, three levels of the independent variable exist, where one is the control and T_1 and T_2 represent two levels of treatment. This design can be analyzed by simple ANOVA, which contrasts the dependent variable as measured in the three groups (O_1, O_2, O_3). For example, T_1 is training at 70% of $\dot{V}O_2$max, T_2 is training at 40% of $\dot{V}O$ max, and the control group is not training. The variables O_1, O_2, and O_3 are the measures of cardiorespiratory fitness (12-min run) in each group taken at the end of the training.

This design can also be extended into a factorial design; that is, more than one independent variable (IV) could be considered. Example 18.1 shows how this works.

Independent variable 1 (IV_1) has three levels (A_1, A_2, A_3), and independent variable 2 (IV_2) has two levels (B_1, B_2). This results in six cells (A_1B_1, A_1B_2, A_2B_1, A_2B_2, A_3B_1, A_3B_2) to which participants are randomly assigned. At the end of the treatments, each cell is tested on the dependent variable (O_1 through O_6). This design is analyzed by a 3×2 factorial ANOVA that tests the effects of IV_1 (F_A), IV_2 (F_B), and their interaction (F_{AB}).

This design can also be extended to address even more independent variables and retain all the controls for internal validity previously discussed. Sometimes this design is used with a categorical independent variable. Look again at example 18.1, and suppose that IV_2 (B_1, B_2) represents two age levels. Clearly, the levels of B could not be randomly formed. The design appears as follows:

$$
\begin{array}{cccc}
 & R & A_1 & O_1 \\
B_1 & R & A_2 & O_2 \\
 & R & A_3 & O_3 \\
\hline
 & R & A_1 & O_4 \\
B_2 & R & A_2 & O_5 \\
 & R & A_3 & O_6 \\
\end{array}
$$

Example 18.1

	IV_2	
	B_1	B_2
A_1	A_1B_1	A_1B_2
A_2	A_2B_1	A_2B_2
A_3	A_3B_1	A_3B_2

IV_1 (labels the rows A_1, A_2, A_3)

R	A_1B_1	O_1
R	A_1B_2	O_2
R	A_2B_1	O_3
R	A_2B_2	O_4
R	A_3B_1	O_5
R	A_3B_2	O_6

Note. Analyzed in a 3×2 factorial ANOVA. F_A = main effect of *A;* F_B = main effect of *B;* F_{AB} = interaction of *A* and *B*.

The levels of A are randomly formed within B, but the levels of B cannot be randomly formed. This no longer qualifies completely as a true experimental design but is frequently used in the study of physical activity. This design is analyzed in a 3×2 ANOVA, but the interpretation of results must be done more conservatively.

Any of the versions of randomized-group design may also have more than one dependent variable. Although the essential design remains the same, the statistical analysis becomes multivariate. Where two or more levels of one independent variable exist and several dependent variables are present, discriminant analysis is the appropriate multivariate statistic. In the factorial versions of this design (two or more independent variables), if multiple dependent variables are used, then MANOVA is typically the most appropriate analysis, although practical concerns (e.g., participant numbers) or theoretical matters may dictate alternative statistical analyses.

Pretest–Posttest Randomized-Groups Design

In the pretest–posttest randomized-groups design, the groups are randomly formed, but both groups are given a pretest as well as a posttest:

$$
\begin{array}{cccc}
R & O_1 & T & O_2 \\
R & O_3 & & O_4 \\
\end{array}
$$

The major purpose of this type of design is to determine the amount of change produced by the treatment; that is, does the experimental group change more than the control group? This design threatens the internal validity through testing, but the threat is controlled because the comparison of O_3 to O_4 in the control group as well as the comparison of O_1 to O_2 in the experimental group includes the testing effect. Thus, although the testing effect cannot be evaluated in this design, it is controlled.

This design is used frequently in the study of physical activity, but its analysis is rather complex. There are at least three common ways to do a statistical analysis of this design. First, a factorial repeated-measures ANOVA can be used. One factor (between subjects) is the treatment versus no treatment, whereas the second factor is pretest versus posttest (within subjects or repeated measures). However, the interest in this design is usually the interaction: Do the groups change at different rates from pretest to posttest? If you choose a repeated-measures ANOVA for this design (in our opinion the best choice), pay close attention to our discussion of univariate and multivariate issues in repeated measures in chapter 9. Another choice is simple ANCOVA, using the pretest for each group (O_1 and O_3) to adjust the posttest (O_2 and O_4). Recall that there are some problems with using the pretest as a covariate (see the discussion of ANCOVA in chapter 9). Finally, the experimenter could subtract each participant's pretest value from the posttest value (called a **difference, or gain, score**) and perform a simple ANOVA (or, with only two groups, an independent t test), using each participant's difference score as the dependent variable. Each of these techniques has strengths and weaknesses, but you will find all three used in the literature.

difference score or gain score

A score that represents the difference (change) from pretest to posttest.

In this design the important question is, Does one group change more than the other group? Although this issue is frequently called the analysis of difference scores, a more appropriate label is the assessment of change. Clearly, in a learning study, the change is expected to be gain. But in an exercise physiology study, the change might be decreased performance caused by fatigue. Regardless, the issues are the same. How can this change be assessed appropriately?

The easiest answer is to obtain a difference score by subtracting the pretest from the posttest. Although this is intuitively attractive, it does have some severe problems. First, these difference scores tend to be unreliable. Second, the level of initial values applies: Participants who begin low in performance can improve more easily than those who begin with high scores. Thus, initial score is negatively correlated with the difference score. How would you like improvement in your tennis performance evaluated if your initial score was high (e.g., 5 successful forehand drives out of 10 trials) in comparison with that of a friend who began with a low score (e.g., 1 out of 10 successful hits)? If you improved to 7 out of 10 on the final test and your friend improved to 5 out of 10 (the level of your initial score), your friend has improved twice as much as you have (a gain of 4 versus 2 successful hits). Yet your performance is still considerably better, and it was more difficult for you to improve. For these reasons, the use of difference scores rarely is a good method for measuring change.

The issues involved in the proper measurement of change are complex, and we cannot treat these issues here. However, much has been written on this topic. We suggest that you read Schmidt and Lee (1999, chapter 10) about this problem in motor learning and performance, or for a classic short book on the topic, see Harris (1963).

This design can also be extended into more complex forms. First, more than two (pretest and posttest) repeated measures can be used. This is common in the areas of exercise physiology, motor behavior, and exercise psychology. Two randomly formed groups of participants in a motor behavior experiment might be measured 30 or more times as they learn a task. The two groups might differ in the information they are given. The design might use as the statistical analysis a 2 (groups) × 30 (trials) ANOVA with repeated measures on the second factor (trials). Remember from chapter 9 that meeting the assumptions for a repeated-measures ANOVA with many repeated measures is very difficult. Thus, in designs like these, trials may be blocked (i.e., several trials averaged, reducing the number of repeated measures) into 3 blocks of 10 trials or 5 blocks of 6 trials.

Sometimes the design is extended in other ways. For example, we could take the design in example 18.1 (a 3 × 2 factorial) and add a third factor of a pretest and a posttest. This would result in a three-way factorial with repeated measures on the third factor. All the versions of this design are subject to the first threat to external validity: reactive or interactive effects of testing. The pretest may make the participant more sensitive to the treatment and thus reduce the ability to generalize the findings to an unpretested population.

Solomon Four-Group Design

The Solomon four-group design is the only true experimental design to specifically evaluate one of the threats to external validity: reactive or interactive effects of testing. The design is depicted as follows:

$$R \quad O_1 \quad T \quad O_2$$
$$R \quad O_3 \quad \quad O_4$$
$$R \quad \quad T \quad O_5$$
$$R \quad \quad \quad O_6$$

This combines the randomized-groups and the pretest–posttest randomized-groups designs. The purpose is explicitly to determine whether the pretest results in increased sensitivity of the participants to the treatment. This design allows a replication of the treatment effect (is $O_2 > O_4$ and is $O_5 > O_6$), an assessment of the amount of change due to the treatment (is $O_2 - O_1 > O_4 - O_3$), an evaluation of the testing effect (is $O_4 > O_6$), and an assessment of whether the pretest interacts with the treatment (is $O_2 > O_5$). Thus, this is a very powerful experimental design. Unfortunately, it is also an inefficient design, as twice as many participants are required. This results in very limited use, especially among graduate students doing theses and dissertations. In addition, no good way exists to analyze this design statistically. The best alternative (one that does not use all the data) is a 2 × 2 ANOVA set up as follows:

	No T	T
Pretested	O_4	O_2
Unpretested	O_6	O_5

Thus, IV_1 has two levels (pretested and not pretested), and IV_2 has two levels (treatment and no treatment). In the ANOVA, the F ratio for IV_1 establishes the effects of pretesting, the F for IV_2 establishes the effects of the treatment, and the F for interaction evaluates the external validity threat of interaction of the pretest with the treatment. Table 18.2 summarizes the control of threats to validity for the true experimental designs.

Quasi-Experimental Designs

Not all research in which an independent variable is manipulated fits clearly into one of the true experimental designs. As researchers attempt to increase external and ecological validity, the careful and complete control of the true designs becomes increasingly difficult, if not impossible. The purpose of **quasi-experimental designs** is to fit the design to settings more like the real world while still controlling as many of the threats to internal validity as possible. The use of these types of designs in kinesiology, physical education, exercise science, sport science, and other areas (e.g., education, psychology, and sociology) has increased considerably in recent years. The most authoritative text on quasi-experimental designs is that by Shadish, Cook, and Campbell (2002).

In quasi-experimental research, the use of randomization to control threats to internal validity is difficult. It makes sense that random assignment cannot be used in many settings. For example, a pedagogy researcher who wants to investigate the impact of a curriculum intervention could not randomly assign children to classes because schools make these

quasi-experimental design

Research designs in which the experimenter tries to fit the design to real-world settings while still controlling as many of the threats to internal validity as possible.

Table 18.2 True Experimental Designs and Their Control of the Threats to Validity

Validity threat	Randomized groups	Pretest–posttest randomized groups	Solomon four-group
Internal			
History	+	+	+
Maturation	+	+	+
Testing	+	+	+
Instrumentation			
Statistical regression	+	+	+
Selection	+	+	+
Experimental mortality	+	+	+
Selection × maturation	+	+	+
Expectancy	?	?	?
External			
Testing × treatment	+	−	+
Selection biases × treatment	?	?	?
Experimental arrangements	?	?	?
Multiple treatments			

Note. + = strength; − = weakness; blank = not relevant; ? = questionable.

Campbell, Donald T. and Julian C. Stanley, *Experimental and Quasi-Experimental Designs for Research.* Copyright © 1963 by Houghton Mifflin Company. Used with permission.

decisions on other criteria that have educational value; no school district would agree to permit the study if it had to change the students' classes. It also would be difficult to randomly assign classes within a school to different treatments, because it is likely that teachers will talk to each other (and even trade ideas from the different curriculum that they felt effective) and reduce the strength of the treatment intervention. Similarly, if a researcher were studying the impact of an exercise program on the aged in a community setting, random assignment would not work because people select where to enroll in classes based on factors in their lives (e.g., convenience, membership, transportation needs) and not on whether it helps a researcher. Asking people to go to another site or to attend class at another time would probably reduce the number of people who agree to participate and increase participant attrition (or, to say it another way, increase experimental mortality).

Reversal Design

The reversal design is used increasingly in school and other naturalistic settings and is depicted as follows:

$$O_1 \quad O_2 \quad T_1 \quad O_3 \quad O_4 \quad T_2 \quad O_5 \quad O_6$$

The purpose here (as with the time series) is to determine a baseline measurement (O_1 and O_2), evaluate the treatment (change between O_2 and O_3), evaluate a no-treatment time period (O_3 to O_4), evaluate the treatment again (O_4 to O_5), and evaluate a return to a no-treatment condition (O_5 to O_6). This design is sometimes called A-B-A-B-A (or sometimes just A-B), where A is the baseline condition and B is the treatment condition.

Lines like A, B, and C in figure 18.1 suggest that the treatment is effective, whereas lines like D, E, and F do not support a treatment effect. Statistical analyses for reversal designs also need to be regression tests of the slopes and intercepts of the lines between various observations.

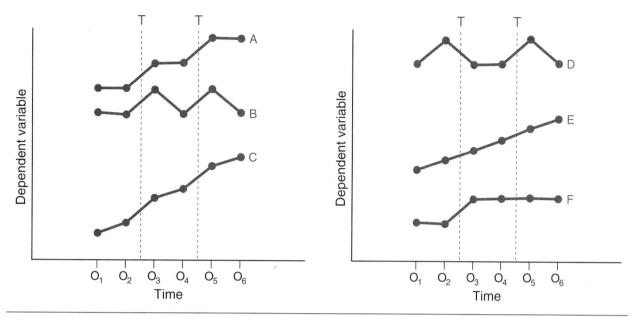

Figure 18.1 Examples of changes across time in reversal designs.

Nonequivalent-Control-Group Design

A design using a nonequivalent control group is frequently used in real-world settings where groups cannot be randomly formed. The design is as follows:

$$O_1 \qquad T \qquad O_2$$

$$\cdots\cdots\cdots\cdots\cdots$$

$$O_3 \qquad\qquad O_4$$

You can recognize this as a pretest–posttest design without randomization. Frequently, researchers compare O_1 and O_3 and declare the groups equivalent if this comparison is not significant. Unfortunately, just because the groups do not differ on the pretest does not mean that they are not different on any number of unmeasured characteristics that could affect the outcome of the research. If the groups differ when O_1 and O_3 are compared, ANCOVA is usually employed to adjust O_2 and O_4 for initial differences. Alternatively, a within (the pretest/posttest comparison—a repeated measure) and between (treatment and control group comparison) two-way ANOVA could be used to analyze whether groups changed from pretest to posttest and whether the change was different for those in the treatment and control group.

Ex Post Facto Design

In its simplest case, ex post facto design is a static group comparison, but with the treatment not under the control of the experimenter. For example, we frequently compare the characteristics of athletes versus nonathletes, highly fit versus unfit individuals, female versus male performers, and expert versus novice performers. In effect, we are searching for variables that discriminate between these groups. Our interest usually resides in the question, Did these variables influence the way these groups became different? Of course, this design cannot answer this question, but it may provide interesting insight and characteristics for manipulation in other experimental designs. This design is also often called a causal comparative design.

The previously mentioned quasi-experimental designs have been frequently reported in research on the study of physical activity. However, several additional designs have considerable potential but have seen less use in our research. We hope the following presentations about two promising designs will increase interest in and use of these designs.

Switched-Replication Design

The switched-replication design (Shadish, Cook, & Campbell, 2002) can be either a true or quasi experiment depending on whether levels are random or intact groups.

Levels (random or intact groups)	Trials				
	1	2	3	4	5
1	O_1 **T**	O_2	O_3	O_4	O_5
2	O_6	O_7 **T**	O_8	O_9	O_{10}
3	O_{11}	O_{12}	O_{13} **T**	O_{14}	O_{15}
4	O_{16}	O_{17}	O_{18}	O_{19} **T**	O_{20}

If participants are randomly assigned to levels 1 to 4, the design is a true experiment. If levels 1 to 4 are different intact groups (e.g., tennis players in college, high school, and two age levels of youth leagues), then the design is quasi-experimental. Any number of levels beyond two can be used, but the number of trials must be one greater than the number of levels.

This design has two strong features: The treatment is replicated several times, and long-term treatment effects can be evaluated. There is no standard statistical analysis for this design, but various ANOVAs with repeated measures could be used. Or the design might be analyzed by fitting regression lines to each level and testing how the slopes and intercepts change.

Many opportunities exist to use this design in our field, yet it is seldom used. This design might be particularly useful in research on sport teams where either different teams or different players within a team could be assigned to the various levels.

Table 18.3 summarizes the threats to validity of quasi-experimental designs discussed so far.

Time-Series Design

The time-series design has only one group but attempts to show that the change that occurs when the treatment is administered differs from the times when it is not. This design may be depicted as follows:

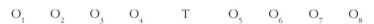

$$O_1 \quad O_2 \quad O_3 \quad O_4 \quad T \quad O_5 \quad O_6 \quad O_7 \quad O_8$$

The basis for claiming that the treatment causes the effect is that a constant rate of change can be established from O_1 to O_4 and from O_5 to O_8 but that this rate of change varies between O_4 and O_5, where T has been administered. For example, in figure 18.2, lines A, B, and C suggest that the treatment (T) results in a visible change between observations, whereas lines D, E, F, and G indicate that the treatment has no reliable effect.

The typical statistical analyses previously discussed do not fit time-series designs very well. For example, a repeated-measures ANOVA with appropriate follow-ups applied to line C in figure 18.2 might indicate that all observations (O_1 to O_8) differ significantly, even though we can visibly see a change in the rate of increase between O_4 and O_5. We do not present the details, but regression techniques can be used to test both the slopes and the intercepts in time-series designs.

Table 18.3   Quasi-Experimental Designs and Their Control of the Threats to Validity

Validity threat	Time series	Nonequivalent control	Reversal	Ex post facto	Switched replication
Internal					
History	−	−	−	?	?
Maturation	+	+	+	?	+
Testing	+	+	+		+
Instrumentation					
Statistical regression	+	−	+		+
Selection	+	+	+	−	?
Experimental mortality	+	+	+		?
Selection × maturation	+	−	+	−	
Expectancy	?	?	?	?	?
External					
Testing × treatment	−	−	?	−	?
Selection biases × treatment	?	?	?	?	?
Experimental arrangements	?	?	?	?	+
Multiple treatments			?		+

Note. + = strength; − = weakness; blank = not relevant; ? = questionable.

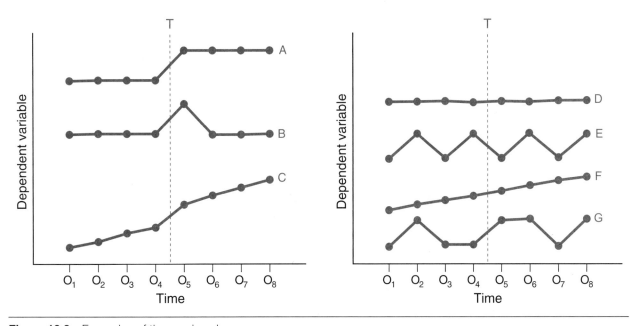

Figure 18.2   Examples of time-series changes.

This type of design appears to control for a number of the threats to internal validity. For example, maturation is constant between observations. Testing effects can also be evaluated, although they could be difficult to separate from maturation. Selection biases also appear to be controlled because the same participants are used at each observation.

Of course, history, instrumentation, and mortality are controlled only to the extent that the researcher controls them. See a humorous example of a time-series design below.

A reminder about using intact groups in quasi-experiments: As we noted in chapter 6 in the section on sampling, the use of intact groups and, particularly, the timing of when a treatment is delivered to a group, influences the analysis and how many groups are needed for a study. This issue has been discussed in length elsewhere (Silverman, 2004; Silverman & Solmon, 1998) and we will not duplicate that whole discussion here. We are compelled to state, however, that when groups receive treatments as a group, the appropriate unit is almost always the *group*. This, then, requires a number of groups to have sufficient power to analyze the data. Those planning quasi-experiments need to consider this early so they are not dealing with the issue after all the data have been collected.

7 ± 2: Miller Must Have Been an Assistant Professor

Summary.—Miller's magical number for memory span (7 ± 2) is humorously brought to task because of its inability to predict the everyday performance of teenagers, graduate students, and full professors. Explanations are provided for the deficits in memory performance of these subgroups during lifespan development.

Miller's (1956) classic paper identifying the memory span as 7 ± 2 items must have omitted using full professors as subjects, at least based on an *n* of 1, namely me. I have frequently heard clinical psychologists accused of going into psychology to study their own problems. Maybe that is a more valid explanation than bringing Miller's work to task. For years my graduate students have said that I study memory for movement because (a) I have little or no memory, and (b) I lose my spatial orientation just walking around the block. However, I prefer to think that Miller's results only apply to a subsample of the population—6- to 10-yr.-old children, college students, and new assistant professors. Six-to 10-yr.-old children never forget anything you promised to do (or even things you said may . . .). However, as soon as they approach the teen years, their memory spans drop to less than 1 unit of information.

Teenagers cannot remember to make their beds as they are getting out of them. Evidently, nothing ever happens to teenagers at school, although I prefer to think that they just cannot recall anything that happened. When these teenagers with little memory capacity go to college, an amazing transition in memory occurs. Between the freshman and senior years, the young adults' memory capacity increases to at least 7 ± 2 units of information. Loosely defined, this means *they know everything;* conversely, parents know nothing.

After working a few years and coming back to graduate school, memory facility is somewhat reduced. Graduate students have a memory capacity of 3 ± 1 units of information: (a) They know that they are graduate students; (b) They remember to pick up their graduate assistantship paycheck; (c) They remember to attend their graduate seminars. The ± 1 refers to the fact that they occasionally remember to do the reading for the seminar (+1) but they sometimes forget to come to class (–1).

As soon as graduate students receive their PhDs, Miller's magical number (7 ± 2) is again a good predictor of memory—new assistant professors know everything. However, movement through the academic ranks gradually reduces capacity to remember until the average capacity of full professors (and parents) is reached, 2 ± 1 units of information. A full professor can remember (a) he or she is a full professor and (b) his or her paycheck comes regularly. The ± 1 refers to the fact that the full professor sometimes remembers that he or she has graduate students (+ 1) but occasionally forgets to pick up the regular paycheck (–1).

Based on the failure of this model to meet the assumptions of stage theory (i.e., one should never regress to an earlier stage), we must assume that this uneven transition in memory states (see figure 1) is environmentally induced.

But how can the environment cause these wide variations in memory performance? First, I believe we have to assume that Miller is correct about the structural maximum of memory

Miller Must Have Been an Assistant Professor   *(continued)*

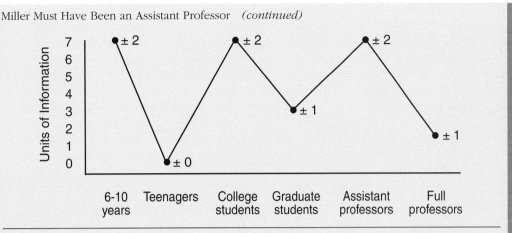

Figure 1. Maximum number of units of information retained in memory across the lifespan.

performance because the deficits are not confined to a single point in the lifespan (i.e., teenagers, graduate students, full professors). The question of interest then becomes, *What causes such a serious depression in memory performance for teenagers, graduate students, and full professors?* Given my extensive study of memory as well as personal experience with all the stages, I can deduce an answer.

For teenagers the drop to about 0 memory with no variance is an interaction of carbonated soft drinks, junk food, and other boys or girls (whichever sex the teenager in question is not). This interaction is not a simple one and has an indirect effect, that is, this interaction causes pimples and the need for braces, both of which are very distracting in and of themselves. Taken in combination with Mom saying, "Wash your face" and "Brush your teeth," all of the teenager's memory capacity is occupied.

The deficit that occurs for graduate students is easy to explain. Three factors are involved. First, the graduate student is expected to work full-time (either as a research or teaching assistant) for one-third or less of normal pay, a very distracting and disconcerting circumstance. Second, the graduate student is expected to study and do research day and night. One's major professor assumes graduate students do not need sleep. Finally, the graduate student's major professor is continually nagging him or her to read this paper, collect these data, write this paper, and work on his or her dissertation. In combination, these items reduce memory capacity.

But why does the full professor who seems to have everything going for him or her show such poor memory performance? Full professors have a living wage (according to some), cars that run, houses with real furniture, tenure, graduate students to do the work, and time to play golf and tennis. What could possibly explain the deficit in memory performance? Answering that question is the easiest of all:

GRADUATE STUDENTS AND TEENAGERS!

Reference

Miller, G. (1956). The magical number seven, plus or minus two: Some limits on our capacity for processing information. *Psychological Review, 63,* 81-97.

Adapted, by permission, 1987, "7+2: Miller must have been an assistant professor," *NASPSPA Newsletter,* 12(1): 10-11.

Single-Subject Design

A single-subject design is exactly what it sounds like—a researcher is looking for the impact of an intervention on a single subject. These designs also are sometimes called $N = 1$ designs because they often have only one subject. There are many designs within this family, and we could include them as a type of quasi-experiment since a researcher using this design is looking for the effect of a treatment without using randomization. We have not done that here because the focus on individual effects instead of group effects makes these designs different. In addition, those who conduct single-subject research look at the changes on graphs and do not analyze the results with statistics.

In our field, single-subject designs are most often used in clinical settings. Examples include observing physical education instruction, pursuing sport psychology work with athletes, studying an outstanding performer (e.g., an Olympic athlete), or looking at the motor function of an individual with a physical impairment (e.g., reaching and grasping by a person with Parkinson's disease). A participant in this type of study is typically measured repeatedly on the task of interest. Many trials are needed to evaluate the influence of the treatment. During some periods a baseline measurement is obtained, and during other periods a treatment is administered. The focus is often on participant variability as well as average values. Quasi-experimental time-series, reversal, and switched-replication designs can work as single-subject designs or group designs. When used with single subjects, these designs are often called A-B or A-B-A-B designs, where A refers to the baseline condition (no treatment) and B refers to when the treatment is administered. Sometimes more than one treatment is administered to the same participant. As in research using a group of participants, counterbalancing the treatment order to separate treatment effects is important. Possible research questions include the following:

- Does the treatment produce the same effect each time?
- Are the effects of the treatment cumulative, or does the participant return to baseline following each treatment period?
- Does the participant's response to the treatment become less variable over multiple treatment periods?
- Is the participant's magnitude of response less sensitive to multiple applications of the treatment?
- Do varying intensities, frequencies, and lengths of treatment produce varying responses?

In figure 18.3, we present a graph with a traditional A-B-A-B design. Note that the A periods are baseline measurements and the B periods are where the intervention occurred. The second A is the reversal, where the treatment was withdrawn.

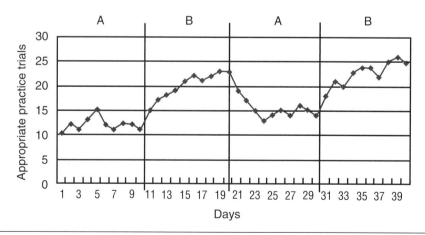

Figure 18.3   Single-subject ABAB design.

The single-subject designs described in this section typically have one subject. Other single-subject designs may have more than one subject undergoing the same intervention. In these studies, the various subjects start the intervention at different points in time by extending the baseline measurement for successive subjects. For example, subject one has baseline measurements conducted over 5 days, subject two over 10 days, and subject three over 15 days. This permits an examination of the intervention at different points in time. There are many permutations of multiple baseline designs, and reversals also could be added to these designs. In figure 18.4 we present one graphic example.

Other quasi-experimental designs exist, but the ones discussed here are the most commonly used. Of course, quasi-experimental designs never quite control internal validity as well as do true experimental designs, but they do allow us to conduct investigations when true experiments cannot be used or when a true experimental design significantly reduces external validity.

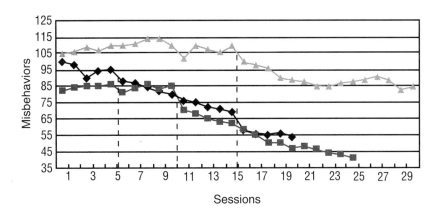

Figure 18.4   Extension of a single-subject design to three subjects over different time spans.

Summary

In experimental research, one or more independent variables (the treatment) are manipulated to assess the effects on one or more dependent variables (the response measured). Research studies are concerned with both internal and external validity. Internal validity requires controlling factors so that the results can be attributed to the treatment. Threats to internal validity include history, maturation, testing, instrumentation, statistical regression, selection biases, experimental mortality, selection-maturation interaction, and expectancy.

External validity is the ability to generalize the results to other participants and to other settings. Four threats exist to external validity: reactive or interactive effects of testing, the interaction of selection biases and the experimental treatment, reactive effects of experimental arrangements, and multiple-treatment interference. Having high degrees of both internal and external validity is nearly impossible. The rigid controls needed for internal validity make it difficult to generalize the results to the real world. Conversely, studies with high external validity are usually weak in internal validity. Random selection of participants and random assignment to treatments are the most powerful means of controlling most threats to internal and external validity.

Preexperimental designs are weak because they can control few sources of invalidity. True experimental designs are characterized by random formation of groups, which allows the assumption that the groups were equivalent at the beginning of the study. The randomized-groups, pretest–posttest randomized-groups, and Solomon four-group designs are examples of true experimental designs.

Quasi-experimental designs are often used when it is difficult or impossible to use true experimental designs or when a true experimental design significantly limits external validity. The time-series design, reversal design, nonequivalent-control-group design, and ex post facto design are the most commonly used. The switched-replication design and single-subject design are potentially useful but less frequently used quasi-experimental designs.

✔ Check Your Understanding

Locate two research papers in refereed journals in your area of interest. One paper should have a true experimental design and the other a quasi-experimental design.

1. Describe the type of design. Draw a picture of it using the notation from this chapter.

2. How many independent variables are there? How many levels of each? What are they?

3. How many dependent variables are there? What are they?

4. What type of statistical analysis was used? Explain how it fits the design.

5. Identify the threats to internal validity that are controlled and uncontrolled. Explain each.

6. Identify the threats to external validity that are controlled and uncontrolled. Explain each.

---

[1] First eight threats from D.T. Campbell and J.C. Stanley. (1963). *Experimental and Quasi-Experimental Designs for Research* (Chicago: Rand McNally), pp. 5–6. Copyright © 1963 by the American Educational Research Associates. Used by permission of Houghton Mifflin Company.

Qualitative Research

Experience enables you to recognize a mistake when you make it again.—Franklin P. Jones

■ ■ ■ ■ ■ ■ ■

Qualitative research in physical education, exercise science, and sport science is still relatively new. Researchers in education began adapting ethnographic research design to educational settings in the United States in the 1970s (Goetz & LeCompte, 1984), and a great deal of qualitative research in physical education and sport science has been conducted steadily since the 1980s. In 1989, Locke was invited to write an essay/tutorial on qualitative research that was published in the *Research Quarterly for Exercise and Sport.*

Qualitative research methods have been employed in anthropology, psychology, and sociology for many years. This general form of research has been called by various names, including ethnographic, naturalistic, interpretive, grounded, phenomenological, subjective, and participant observational research. Although the approaches are all slightly different, each "bears a strong family resemblance to the others" (Erickson, 1986, 119).

For those of you who are unfamiliar with the term *qualitative research,* it might be beneficial to review your courses or readings in anthropology. Nearly everyone has heard of classic ethnographic studies (even if not by that name), such as the famous cultural research Margaret Mead conducted when she lived in Samoa. Mead interviewed the Samoans at length about their society, their traditions, and their beliefs. This is an example of qualitative research. You do not need to live in a remote, exotic land to do it, however. The qualitative research we refer to in this chapter is done mainly in everyday settings, such as schools, gymnasiums, sport facilities, fitness centers, and hospitals.

It could be argued that the term *qualitative research* may be too restrictive in that it seems to denote the absence of anything quantitative; however, this is certainly not the case. Nonetheless, qualitative research seems to be the term used most often in our field. Our discussion of qualitative research focuses on the interpretive method as opposed to the so-called thick, rich description that characterized early research in anthropology, psychology, and sociology. This latter approach involves a very long and detailed account of the entity or incident and was espoused by Franz Boas, who is considered the father of cultural anthropology. In rejecting the armchair speculation that typified the late 19th century work, Boas insisted that the researcher not only collect his or her own data but that it be reported with as little comment or interpretation as possible (cited in Kirk & Miller, 1986).

We believe that the most important feature of qualitative research is the interpretive content rather than an overconcern about procedure. Erickson (1986) argues that the technique of narrative description (sometimes referred to as "writing like crazy") does not necessarily mean that the research being conducted is interpretive or qualitative. It is not our intention to present a comprehensive review of qualitative research, as we do not have the space to do so. Despite our assertion that qualitative research is relatively new in our field, a comprehensive review would still be a major undertaking because we would need to draw from many related fields. Whereas 30 years ago there was almost nothing in print in the area of qualitative methodology, there is now a tremendous amount of literature in the form of books, articles, and monographs. Moreover, there is considerable disagreement among qualitative researchers concerning their methodologies and theoretical presuppositions that would need to be addressed. You can gain a solid appreciation regarding the scope of qualitative research, the different approaches to data collection and analysis, and the subtleties of different approaches to qualitative research in compact form from Locke, Silverman, and Spirduso (2004) or in a more expansive form from Creswell (1998). Readable, single-source texts on conducting qualitative research include those of Bogdan and Biklen (2003), Creswell (2003), Marshall and Rossman (1999), Morse and Richards (2002), and Patton (2002). You will also learn from reading a qualitative study, such as the one in physical education by McCaughtry and Rovegno (2003).

Contrasting Characteristics of Qualitative and Quantitative Research

Much has been written on the debate over the comparative merits of quantitative and qualitative approaches to problem solving, so we do not discuss it here. Locke (1989) observed that the debate in our field comes at a point when the same dispute is winding down in other areas. So instead of relating some arguments about which method is better, let us simply contrast some basic differences between quantitative and qualitative research.

Qualitative research is often depicted as the antithesis of the more traditional quantitative methods, such as experimental and survey research. Quantitative research methods typically involve precise measurements, rigid control of variables (often in a laboratory setting), and statistical analyses. Qualitative research methods generally include field observations, case studies, ethnography, and narrative reports (Locke, Silverman, & Spirduso, 2004). Bogdan and Biklen (2003) provide a detailed contrast of characteristics of qualitative and quantitative research regarding terms, key concepts, theoretical and academic affiliations, goals, design, samples, data, and the nature of research proposals in each paradigm.

Quantitative research tends to focus on analysis (i.e., taking apart and examining components of a phenomenon), whereas qualitative research seeks to understand the meaning of an experience to the participants in a specific setting and how the components mesh to form a whole. Some basic characteristics of qualitative research and experimental quantitative research are contrasted in table 19.1. We would be remiss if we did not note that our contrast is at the extremes, since other types of quantitative research would, likely, not have a large random sample or take place in a laboratory.

Table 19.1 Contrasting Characteristics of Qualitative and Quantitative Research

Research component	Qualitative	Quantitative experiment*
Hypothesis	Inductive	Deductive
Sample	Purposive, small	Random, large
Setting	Natural, real world	Laboratory
Data gathering	Researcher is primary instrument	Objective instrumentation
Design	Flexible, may change	Determined in advance
Data analysis	Descriptive, interpretive	Statistical methods

*These would be different in many other types of quantitative studies.

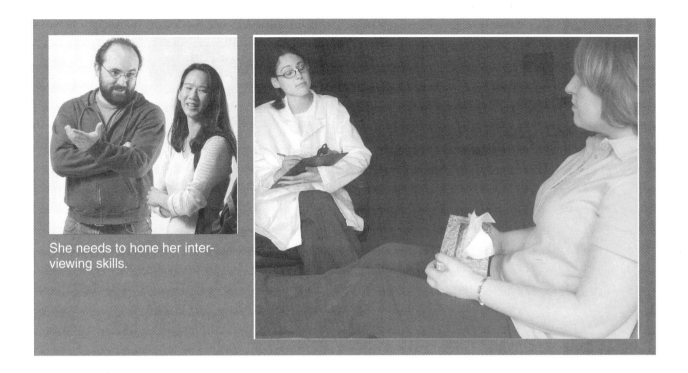

She needs to hone her interviewing skills.

Patton (1987) uses two responses to a questionnaire item to illustrate the depth, detail, and meaning in qualitative methods. In the quantitative approach, a teacher was asked to respond to a statement: "Accountability as practiced in our school system creates an undesirable atmosphere of anxiety among teachers." The choices were strongly agree, agree, disagree, and strongly disagree. She checked "strongly agree." However, her response to an open-ended question about the accountability approach revealed a much more intense, negative feeling about the system. She stressed the atmosphere of fear among the teachers and the strong political motives embedded in the system. She described her feelings of frustration, bitterness, and hatred and condemned accountability as totally counterproductive.

Qualitative research focuses on the "essence" of the phenomenon. One's view of the world varies with one's perception and is highly subjective. The objectives are primarily description, understanding, and meaning. The researcher does not manipulate variables through experimental treatments but takes more interest in process than in product. The researcher observes and gathers data in the field, that is, the natural setting. There are no preconceived hypotheses, which characterize quantitative research. Rather, qualitative research strives to develop hypotheses from the observations. In other words, qualitative research emphasizes induction, whereas quantitative research largely emphasizes deduction.

In quantitative research, the researcher tries very hard to keep out of the data-gathering process by using laboratory measurements, questionnaires, and other so-called objective instruments. The quantitative data are then usually analyzed by statistical formulas, with computations performed by computers. However, in qualitative research, which is very subjective in this sense, the researcher is the primary instrument for data collection and analysis. The researcher interacts with the participants, and the researcher's sensitivity and perception are crucial in procuring and processing the observations and responses. How the researcher manages his or her responses during data collection and analysis influences the quality of the data and conclusions.

Procedures in Qualitative Research

It should go without saying that there are many variations in the way qualitative research is done. Consequently, the procedures outlined in this section should be viewed simply as an attempt to orient readers who are unfamiliar with qualitative research. As with any type of research, the novice will benefit most from reading books that specifically focus on planning research (e.g., Locke, Spirduso, & Silverman, 2000) or qualitative methods, like those listed on page 346, and recently completed qualitative research studies that used particular methods.

Define the Problem

We do not spend much time defining the problem because we have talked about this step before, and it does not differ appreciably from that of other research methods. We emphasize that several possible methods can be used for any research problem. The different methods can yield different information about the problem. Thus, when the researcher decides to use qualitative research as opposed to some other design, the decision is based on what he or she wants to know about the problem.

Formulate Questions and Theoretical Framework

In highlighting some differences between quantitative and qualitative research, we stated that qualitative research usually builds hypotheses and theories in an inductive manner, that is, as a result of the observations. Quantitative research often begins with research hypotheses that are subsequently tested. Thus, most qualitative studies do not have hypotheses as a part of the introduction. There was a time when it was common for those doing qualitative research not to state any hypotheses or questions. It has become more common in recent years for qualitative researchers to state *questions* that will be the focus the study. These questions are not as specific as hypotheses in quantitative research. They do, however, provide the direction that is the research focus. For example, "What is the perspective of children in a competitive youth sport program?" focuses the study and provides a great deal of information about the research direction. Generally, the types of qualitative studies that we address in this chapter start with questions that guide the studies. While it is possible for the researcher to move in other directions and modify questions based on the ongoing research, it is less likely that the original questions will be totally abandoned.

Qualitative researchers, like other researchers, develop a case for why it is important to do a study. This takes place in the introduction of a research paper or as a part of the rationale or significance of the study sections in a thesis or dissertation. One way this occurs is by providing a theoretical framework—a theory that guides the research. A theoretical framework "frames" the study and is used throughout to develop questions, design method, and analyze data. While many qualitative studies use a theoretical framework, it is possible to provide a rationale for a study by using other published research to show how this research fills a gap in our knowledge (Locke, Spirduso, & Silverman, 2000; Shulman, 2003). In either case, in most qualitative research, the questions being asked are situated in previous research and theory and are based on extending what we know about a particular area.

Collect the Data

Several components are involved in collecting data for qualitative research, just as in quantitative research. Training and pilot work are still necessary, and so is selecting participants appropriately. In addition, you need to enter the field setting and become as unobtrusive as possible in your collection of data.

Training and Pilot Work

In qualitative research the investigator is the instrument for collecting and analyzing the data. It is imperative, then, that the researcher be adequately prepared. Certainly course work, fieldwork reports, and interaction with your advisor are helpful, but ultimately the only way to become competent is through hands-on experience. It is important to get this experience with an advisor who has a background and experience with qualitative research. At many universities you can begin this process in a class where students learn and do qualitative research under the direction of an experienced qualitative researcher.

As always, pilot work is essential, and fieldwork experience in a setting similar to that of the proposed study is recommended. Experience in the field setting (e.g., as an instructor, a coach, or a player) is helpful in most respects. However, Locke (1989) made a good point that familiarity tends to spawn "an almost irresistible flood of personal judgments" (p. 7) that, if not recognized and controlled, could become a significant threat to the integrity of the data in a qualitative study.

Selection of Participants

Basically, the selection of participants for a qualitative study is the same as described for the case study in chapter 16. Qualitative research studies do not attempt to make inferences from their participants to some larger population. Rather, the participants are selected because they have certain characteristics. Obviously, there are pragmatic concerns about the location and the availability of participants because, in nearly every case, numerous other sites and people with similar characteristics exist.

Probability sampling is not used, simply because there is no way to estimate the probability of each person's being selected and no assurance that each person has some chance of being included (Chien, 1981). Instead, the selection of participants in qualitative research is purposeful, which in essence means that a sample from which one can learn the most is selected. The researcher may be looking for persons with certain levels of expertise or experience. There are several ways to guide participant selection (e.g., see Marshall & Rossman, 1999, for various alternatives). In essence, the selection of participants in qualitative research involves consideration of where to observe, when to observe, whom to observe, and what to observe (Burgess, 1982).

Entering the Setting

The researcher must have access to the field setting to conduct a qualitative study. Moreover, the researcher must be able to observe and interview the participants at the appropriate time and location. These mundane details may seem somewhat trivial, but nothing is more important than site entry. In any type of research, you must have access to the data, whether it be source material in historical research or participants in an experimental or survey study. However, the problem can be of greater magnitude in qualitative research than in most other methods. The quantitative researcher is simply "borrowing" the participants for a short time for some measurements or taking a little class time to administer a questionnaire. In qualitative research, the investigator is often at the site for weeks or months. This outsider is listening, watching, coding, and videotaping, as well as imposing on the time of the teacher (or coach, etc.) and the students (or players or other participants) for interviews and observations.

We are intentionally belaboring this topic because it is so important. Obviously, it takes diplomacy, personality, and artful persuasion to gain site entry. Frankly, some people just cannot do this. Even when entry is achieved, some studies have failed because the investigator "rubbed people the wrong way" and the participants were not motivated to cooperate fully. Thus, the negotiation of gaining access to the participants in their naturalistic setting is important and complex. It starts with the first contact by telephone or letter, extends through data collection, and continues after the researcher has left the site (Rossman & Rallis, 2003).

Before we discuss some aspects of data collection, we should elaborate a little more on the topic of cooperation with the participants. Rapport is everything. The participants must feel that they can trust you, or else they will not give you the information you seek. Obviously, formal informed consent must be obtained, and the stipulations embodied in the whole concept of informed consent must be observed.

Qualitative research involves a number of ethical considerations simply because of the intensive personal contact with the participants. Thus, the participants need to know that provisions will be followed to safeguard their rights of privacy and to guarantee anonymity.

If, for some reason, it is impossible to keep information confidential, this must be made clear. The researcher must give a great deal of thought to these matters before collecting data, must be able to explain the purpose and significance of the study effectively, and must convey the importance of cooperation in language participants can understand. One of the largest obstacles in the quest for natural behavior and candor on the part of the participants is their suspicion that the researcher will be evaluating them. The researcher must be very convincing in this regard. The most successful studies are those in which the participants feel as though they are a part of the project. In other words, a collaborative relationship should be established.

Methods of Collecting Data

The most common sources of data collection in qualitative research are interviews, observations, and review of documents (Creswell, 2003; Locke, Silverman, & Spirduso, 2004; Marshall & Rossman, 1999). The methodology is planned and pilot-tested before the actual study. Creswell (2003) places the data-collecting procedures into four categories: observations, interviews, documents, and audiovisual materials. He provides a concise table of the four methods, the options within each type, the advantages of each type, and the limitations of each.

We noted previously that the researcher typically has some type of framework (subpurposes perhaps) that determines and guides the nature of the data collection. For example, one phase of the research might pertain to the manner in which expert and nonexpert sport performers perceive various aspects of a game. This phase could involve having the athlete describe his or her perceptions of what is taking place in a specific scenario. A second phase of the study might focus on the interactive thought processes and decisions of the two groups of athletes while they are playing. The data for this phase could be obtained from filming them in action and then interviewing them while they are watching their performances on videotape. Still another aspect of the study could be directed at the knowledge structure of the participants, which could be determined by a researcher-constructed instrument.

You should not expect qualitative data collection to be quick. It is very time intensive. It takes time to collect good data (Locke, Silverman, & Spirduso, 2004) and quick interviews or short observations are unlikely to help you gain more understanding. If you are doing qualitative research, you must plan to be in the environment for enough time to get good data and understand the nuance of what is occurring.

Interviews

The interview is undoubtedly the most common source of data in qualitative studies. The person-to-person format is most prevalent, but occasionally group interviews and focus groups are conducted. Interviews range from the highly structured style, in which questions are determined before the interview, to the open-ended, conversational format. In qualitative research, the highly structured format is used primarily to gather sociodemographic information. For the most part, however, interviews are more open ended and less structured (Merriam, 2001). Frequently, the interviewer asks the same questions of all the participants, but the order of the questions, the exact wording, and the type of follow-up questions may vary considerably.

It requires skill and experience to be a good interviewer. We emphasized earlier that the researcher must first establish rapport with the respondents. If the participants do not trust the researcher, they will not open up and describe their true feelings, thoughts, and intentions. Complete rapport is established over time as people get to know and trust one another. However, one skill in interviewing is being able to ask questions in such a way that the respondent feels that he or she can talk freely.

Kirk and Miller (1986) described their field research in Peru, where they tried to learn how much urban, lower-middle-class people knew about coca, the organic source of cocaine. Coca is legal and widely available there. In their initial attempts to get the people to tell them about coca, they received the same culturally approved answers from all the respondents. It was only after they changed their style to asking less sensitive questions (e.g., "How did you find out you didn't like coca?") that the Peruvians opened up and elaborated on their knowledge of (and sometimes their personal use of) coca. Kirk and Miller made a good point about asking the right questions and the value of using different approaches. Indeed, this is a basic argument for the validity of qualitative research.

Skillful interviewing takes practice. Ways to develop this skill include videotaping your own performance in conducting an interview, observing experienced interviewers, role playing, and critiquing peers. It is very important that the interviewer appear nonjudgmental. This can be difficult in situations where the interviewee's views are quite different from those of the interviewer. The interviewer must be alert to both verbal and nonverbal messages and be flexible in rephrasing and pursuing certain lines of questioning. The interviewer must use words that are clear and meaningful to the respondent and must be able to ask questions so that the participant understands what is being asked. Above all, the interviewer has to be a good listener.

The use of a tape recorder is undoubtedly the most common method of recording interview data because it has the obvious advantage of preserving the entire verbal part of the interview for later analysis. Although some respondents may be nervous to talk while being taped, this uneasiness usually disappears in a short time. The main drawback with tape recording is the equipment's malfunctioning. This is vexing and frustrating when it happens during the interview, but it is devastating when it happens afterward when you are trying to replay and analyze the interview. Certainly, it is wise always to have fresh batteries and to make sure the recorder is working properly early in the interview. It is also recommended that early in the interview you stop and play back some tape to see whether the person is speaking into the microphone loudly and clearly enough and whether you are getting the data. Some participants (especially children) love to hear themselves speak, so playing back the tape for them can also serve as motivation. Remember, however, that machines can always malfunction.

Videotaping seems to be the best method because you preserve not only what the person said but also his or her nonverbal behavior. The drawback to using videotape is that it can be awkward and intrusive and, therefore, is used infrequently. Taking notes during the interview is another common method. Sometimes note taking is used in addition to recording, primarily when the interviewer wishes to note certain points of emphasis or make additional notations. Taking notes without taping prevents the interviewer from being able to record all that is said. It also keeps the interviewer very busy, interfering with his or her thoughts and observations while the respondent is talking. In highly structured interviews and when using some types of formal instrument, the interviewer can more easily take notes by checking off items and writing short responses.

The least preferred technique is trying to remember and write down afterward what was said in the interview. The drawbacks are many, and this method is seldom used.

Focus Groups

Another type of qualitative research technique employs interviews on a specific topic with a small group of people, called a **focus group.** It can be an efficient technique because the researcher can gather information about several people in one session. The group is usually homogeneous, such as a group of students, an athletic team, or a group of teachers.

In his 1988 book *Focus Groups as Qualitative Research,* Morgan discusses the applications of focus groups in social science qualitative research. Patton (2002) argues that focus group interviews may provide quality controls because participants tend to provide checks and balances on one another that can serve to curb false or extreme views. Focus group interviews are usually enjoyable for the participants, and there may be less fear of the interviewer evaluating the individual because of the group setting. The group members get to hear what others in the group have to say, which may stimulate the individuals to rethink their own views.

In the focus group interview, the researcher is not trying to persuade the group to reach consensus. It is an interview. Taking notes can be difficult, but a tape recorder or videotape

focus group

A small group of individuals interviewed concerning a specific topic as a method of qualitative research.

may solve that problem. Certain group dynamics such as power struggles and reluctance to state views publicly are limitations of the focus group interview. The number of questions that can be asked in one session is also limited. Obviously, the focus group should be used in combination with other data-gathering techniques.

Observation

Earlier studies relied on direct observation with note taking and coding of certain categories of behavior. More recently, videotaping has also been used. The videotape can record all of a person's behaviors and preserve them for later analysis. If desired, sounds associated with the observations can simultaneously be recorded, as can comments by the researcher. Newer cameras are lightweight and capable of obtaining remarkably clear pictures in natural lighting.

One major drawback to observation methods is obtrusiveness. A stranger with a camera or pad and pencil is trying to record people's natural behavior. A key word here is *stranger*. The task of a qualitative researcher is to make sure that the participants become accustomed to having the researcher (and, if appropriate, recording device) around. For example, the researcher may want to practice or pretend to film in the setting for at least a couple of days before the initial filming.

In an artificial setting, researchers can use one-way mirrors and observation rooms. In a natural setting, the limitations that stem from the presence of an observer can never be ignored. Locke (1989) observed that most naturalistic field studies are reports of what goes on when a visitor is present. The important question is, How important and limiting is this? Locke suggested ways of suppressing reactivity, such as the visitor's being in the setting long enough so that he or she is no longer considered a novelty and being as unobtrusive as possible in everything from dress to choice of location in a room.

Other Data-Gathering Methods

There are many sources of data in qualitative research, including researcher-constructed behavior-coding inventories and self-reports of knowledge and attitude. The researcher can also develop scenarios, in the form of descriptions of situations or actual pictures, that are acted out for participants to observe. The participant then gives his or her interpretation of what is going on in the scenario. The participant's responses provide his or her perceptions, interpretations, and awareness of the total situation and of the interplay of the actors in the scenario.

Other recording devices include notebooks, narrative field logs, and diaries, in which researchers record their reactions, concerns, and speculations. Printed materials such as course syllabi, team rosters, evaluation reports, participant notes, and photographs of the setting and situations are examples of document data used in qualitative research.

Analysis of the Data

Data analysis in qualitative research is quite different from that in conventional quantitative research. First, analysis is done during and after data collection in qualitative research. During data collection the researcher sorts and organizes data and speculates and develops tentative hypotheses to guide him or her to other sources and types of data. Qualitative research is often done in a manner somewhat similar to multiple-experiment research, in which discoveries made during the study shape each successive phase of the study. Thus, simultaneous data collection and analysis allow the researcher to work more effectively. Analysis then becomes more intensive after the data have been collected (Merriam, 2001). Another difference between quantitative and qualitative data analysis is that qualitative data are generally presented through words, descriptions, and images, whereas quantitative data is typically presented through numbers.

The analysis of data in a qualitative study can take different forms, depending on the nature of the investigation and the defined purposes. Consequently, we cannot go into great depth in discussing analysis without tying it to a specific study. Therefore, we have summarized the general phases of analysis synthesized from descriptions in several qualitative research texts. The general phases include sorting and analysis during data collection, analysis and categorization, and interpretation and theory construction.

Sorting, Analyzing, and Categorizing Data

The simultaneous collection and analysis of data are an important feature of qualitative research. It enables researchers to focus better on certain questions and, in turn, to direct the data collection more effectively. Although the researchers have specific questions in mind when the data collection begins, they will probably shift their focus as the data unfold.

Researchers need to keep in close touch with the data. It is a foolish mistake to wait until after the data are collected to analyze them. Decisions must be made concerning scope and direction, or the researchers may be left with data that are unfocused, repetitious, and overwhelming in the sheer volume of material that needs to be processed (Merriam, 2001). Also, there could be gaps because the researchers may not realize that some needed evidence was not collected.

Researchers typically write many observer comments to stimulate critical thinking about what they are observing. They should not be merely human recording machines. They should try out new ideas and consider how certain data relate to large theoretical, methodological, and substantive issues (Bogdan & Biklen, 1992). However, Goetz and LeCompte (1984) caution that researchers should periodically review the research proposal to make sure the investigation is not straying too far from the original questions that must be addressed in the final report.

Analysis is the process of making sense out of the data. Goetz and LeCompte (1984) recommend that researchers read the data again before analysis to ensure completeness and confirm general analytical categories. This is the beginning of the stages of organizing, abstracting, integrating, and synthesizing, which ultimately permit researchers to report what they have seen and heard. They may develop an outline to search for patterns that can be transformed into categories. Pay attention to how you analyze data, or your conclusions may turn out like the following quotes from church bulletins.

Quotes From Church Bulletins

1. Bertha Belch, a missionary from Africa, will be speaking tonight at Calvary Methodist. Come hear Bertha Belch all the way from Africa.
2. Weight Watchers will meet at 7 p.m. at the First Presbyterian Church. Please use the large double door at the side entrance.
3. Ladies, don't forget the rummage sale. It's a chance to get rid of those things not worth keeping around the house. Don't forget your husbands.
4. Miss Charlene Mason sang "I Will Not Pass This Way Again," giving obvious pleasure to the congregation.
5. The rector will preach his farewell message, after which the choir will sing "Break Forth Into Joy."
6. For those of you who have children and don't know it, we have a nursery downstairs.
7. Next Thursday there will be tryouts for the choir. They need all the help they can get.
8. Barbara remains in the hospital and needs blood donors for more transfusions. She is also having trouble sleeping and requests tapes of Pastor Jack's sermons.
9. Irving Benson and Jessie Carter were married on October 24th in the church. So ends a friendship that began in their school days.

(continued)

Quotes From Church Bulletins   *(continued)*

10. At the evening service tonight, the sermon topic will be "What Is Hell?" Come early and listen to our choir practice.

11. Please place your donation in the envelope along with the deceased person you want remembered.

12. The eighth-graders will be presenting Shakespeare's *Hamlet* in the church basement Friday at 7 p.m. The congregation is invited to attend this tragedy.

13. The ladies of the church have cast off clothing of every kind. They may be seen in the basement on Friday afternoon.

14. The pastor would appreciate it if the ladies of the congregation would lend him their electric girdles for the pancake breakfast next Sunday.

15. Low Self Esteem Support Group will meet Thursday at 7 p.m. Please use the back door.

The qualitative researcher faces a formidable task in sorting the data for content analysis. Obviously, there are many types of categories that can be devised for any given set of data, depending on the problem being studied. For example, a researcher could categorize observations of a physical education class in terms of the teacher's management style, another set of categories could relate to social interaction among the students, another could deal with sex differences in behavior or treatment, and one could be based on verbal and nonverbal instructional behaviors. Categories can range in complexity from relatively simple units of behavior types to conceptual typologies or theories (Merriam, 2001).

Researchers use different techniques for sorting data. Index cards and file folders have been widely used for years. Computer software is now available to store, sort, and retrieve data. Most universities where there are active qualitative researchers have one or two of the most popular qualitative software programs available through their computer center. In addition, many of these universities offer classes (both credit and noncredit) in how to use the software. If you are considering a qualitative study, these classes are excellent opportunities to see the characteristics of the software programs and which programs would be most helpful. Of course, if you have the opportunity to take a class where you collect pilot data and then use the software, it will enhance your eventual efficiency.

Data categorization is a key facet of true qualitative research. Instead of using mere description, the researcher may use descriptive data as examples of the concepts that are being advanced. The data need to be studied and categorized so that the researcher can retrieve and analyze information across categories as part of the inductive process.

Interpreting the Data

When the data have been organized and sorted, the researcher then attempts to merge them into a holistic portrayal of the phenomenon. An acknowledged goal of qualitative research is to vividly reconstruct what happened during the fieldwork. This is accomplished

analytical narrative

A short, interpretive description of an event or situation used in qualitative research.

through the **analytical narrative.** This can consist of a descriptive narrative organized chronologically or topically. However, in our concept of qualitative research, we support the position voiced by Goetz and LeCompte (1984) that researchers who merely describe fail to do justice to their data. As Peshkin (1993) observes, "Pure, straight description is a chimera; accounts that attempt such a standard are sterile and boring" (p. 24). Goetz and LeCompte (1984) maintain that, by leaving readers to their own conclusions, the researcher risks misinterpretation and perhaps trivialization of the data by readers who are unable to make the implied connections. They further suggest that the researcher who can find no implications beyond the data should never have undertaken the study in the first place.

The analytical narrative is the foundation of qualitative research. Researchers (especially novices) are often reluctant to take the bold action needed to assign meaning to the data.

Erickson (1986) suggested that to stimulate analysis early in the process, researchers should force themselves to make an assertion, choose an excerpt from the field notes that substantiates the assertion, and then write a **narrative vignette** that portrays the validity of the assertion. In the process of making decisions concerning the event to report and the descriptive terms to use, the researcher becomes more explicitly aware of the perspectives that are emerging from the data. This awareness thus stimulates and facilitates further critical reflection.

The narrative vignette is one of the fundamental components of qualitative research. As opposed to the typical analysis sections in quantitative research studies (which are about as interesting as watching paint dry), the vignette captures the reader's attention, thereby helping the researcher make his or her point. It gives the reader a sense of "being there." A well-written description of a situation can convey a sense of holistic meaning that is definitely advantageous in providing evidence for the researcher's various assertions. In characterizing qualitative research, Locke (1989) stated that the researcher can describe the physical education scene so vividly that "you can smell the lockers and hear the thud of running feet" (p. 4). Griffin and Templin (1989, 399) provide the following example of a vignette:

> The second period physical education class at Big City Middle School is playing soccer. The teacher has placed two piles of sweatshirts at each end of a large open field to serve as goals. There are no field markings. Four boys run up and down the field following the ball. Several other students stand silently in their assigned positions until the ball comes near, then they move tentatively toward the ball to kick it away. Three girls stand talking in a tight circle near the far end of the field. They are startled when the ball rolls into their group, and two boys yell at them to get out of the way. They do and then regroup after the ball and the boys go to the other side of the field. Two boys, who have not touched the ball during the class, engage in a playful wrestling match near one goal. The teacher stands in the center of the field with a whistle in his mouth. He hasn't said anything since he divided students into teams at the beginning of class. He has blown the whistle twice to call fouls. The students play around him as if he were not there. A bell rings, and all the students drop their pinneys where they are and start toward the school building. Belatedly, the teacher blows his whistle to end the game and begins to move around the field picking up pinneys.
>
> After class, as we walk back to the building, the teacher says, "These kids are wild. If you can just run off some of their energy, they don't get into so much trouble in school. This group especially, not too many smarts (taps his temple), don't get into much game strategy." (He sees a boy and girl from the class standing near the door of the girls' locker room talking.) He yells, "Johnson, get your butt to the shower and stop bothering the ladies." He smiles at me. "You've got to be on them all the time." He looks up and sighs, "Well, two [classes] down, three to go."

We caution you, however, that Locke and other scholars do not hold that richness of detail alone is what makes a narrative vignette valid. Siedentop (1989) warned that whether data are to be trusted should not be based on the narrative skills of the researcher. According to Erickson (1986), a valid account is not simply a description but an analysis: "A story can be an accurate report of a series of events, yet not portray the meaning of the actions from the perspectives taken by the actors in the event. . . . It is the combination of richness and interpretive perspective that makes the account valid" (p. 150). Vignettes are not left to stand by themselves. The researcher should make interpretive connections between narrative vignettes and other forms of description, such as direct quotations and quantitative materials.

Direct quotations from interviews with the participants taken from field notes and audiotapes or videotapes are another form of vignette that enriches the analysis and furnishes documentation for the researcher's point of view. Direct quotations from different individuals can demonstrate agreement (or disagreement) about some phenomenon. Direct quotations from the same people on different occasions can provide evidence that certain events are typical or can demonstrate a pattern or trend in perceptions over time.

narrative vignette

Component of a qualitative research report that gives detailed descriptions of an event, including what people say, do, think, and feel in that setting.

For example, K.R. Nelson (1988), illustrating differences in the thought processes of students taught by expert and novice teachers, used quotations to document the assertion that students of novice teachers tended to think about procedures and organization more than content:

Interviewer: What are you thinking at this point in the lesson?

Student: I didn't know what to do. I thought we were going to run around the gym.

Student: I was thinking are we all going to be in the same group. (p. 58)

In contrast, students' thoughts during classes taught by experts were more related to the lesson content:

Interviewer: What are you thinking at this point in the lesson?

Student: He was showing us how it [heart rate] would change after we did aerobics.

Student: I was thinking about how to . . . uh, make sure I was adding correctly to get the right score and everything like that. (p. 59)

It is usually emphasized in qualitative research that researchers should communicate their perspective clearly to the reader. The narrative's function is to present the researcher's interpretive point in a clear and meaningful manner.

Quantitative Analysis

Although we have tended to emphasize the differences between qualitative and quantitative research, we do not want you to conclude that there are (or should be) no quantitative features in a qualitative study (and vice versa). Qualitative research can employ some quantitative analyses, and mixed-methods research studies, with both qualitative and quantitative aspects, have become more common and more accepted (see, for example, Tashakkori & Teddlie, 1998, 2002).

Frequency tables are not uncommon at all in qualitative studies (e.g., Garcia, 1994). Raw frequencies of occurrences are used to reduce data. This is especially appropriate in studies that use some type of observational instrument that codes designated categories of behavior. The frequencies are often converted to percentages to show the extent of certain behaviors or to make comparative statements.

Gould, Finch, and Jackson (1993) conducted an interesting qualitative study of strategies to handle stress used by champion figure skaters. Their analysis involved the organization of raw data expressed as themes into meaningful categories using inductive and deductive procedures. The 158 raw data themes were organized into 51 first-order themes and 29 second-order themes, which were categorized as subcategories of the 13 highest-level themes, called "general dimensions." The percentages of the national champion skaters using the various general dimensions were reported. An example of the raw data themes and first- and second-order themes included in the general dimension cited by

Table 19.2   Coping Strategies/General Dimensions and Percentage of Skaters Citing Each General Dimension

Coping strategy/general dimension	Percentage
Rational thinking and self-talk	76
Positive focus and orientation	71
Social support	71
Time management and prioritization	65
Precompetitive mental preparation and anxiety management	65
Training hard and smart	65
Isolation and deflection	47
Ignoring	41
Uncategorized strategies	35
Reactive behaviors	29
No coping strategy	24
Striving for a positive working relationship with partner	18
Changing to healthy eating attitudes and behaviors	12

76% (see table 19.2) of the skaters (rational thinking and self-talk) is shown in figure 19.1. The 13 coping strategies/general dimensions and the percentage of skaters citing each are provided in table 19.2.

This is an excellent way of using quantitative analysis to complement the qualitative data. The authors deftly use comments by the skaters to enrich the analyses and to document their assertions regarding the identified dimensions.

The main point we are trying to make here is that qualitative research does not exclude quantitative analysis. One of the negative features or outcomes of arguments that support or defend particular methods is that the reader fails to see points of **convergence** among different methods. The researcher should always be alert and amenable to using any methods that could yield meaningful information. Remember, the major purpose of any analysis is to make the most sense out of the data.

Theory Construction

Our use of the terms *theory* and *theorizing* should not unduly alarm graduate students or discourage them from undertaking qualitative research. We are not talking about developing a model on the scale of the theory of relativity here. **Theorizing,** according to Goetz and LeCompte (1984), is a cognitive process of discovering abstract categories and the relationships among those categories. A **theory** is an explanation of some aspect of practice that permits the researcher to draw inferences about future events. This process is a fundamental tool to develop or confirm explanations. You process information, compare the findings with past experience and sets of values, and then make decisions. The decisions may not be correct, so you may then need to revise the theory or model in such cases.

Data analysis depends on theorizing. The tasks of theorizing are "perceiving; comparing, contrasting, aggregating, and ordering; establishing linkages and relationships; and speculating" (Goetz & LeCompte, 1984, 167). Perceiving involves the consideration of all sources of data and all aspects of the phenomena being studied. Of course, this takes place during as well as after data collection. The perceptual process of determining which specific factors to analyze guides the collection of data.

The tasks of comparing, contrasting, aggregating, and ordering are primary functions in qualitative research. The researcher decides which units are similar or dissimilar and what is important about the differences and similarities. Analytical description cannot occur until the researcher builds the categories of like and unlike properties and carries out a systematic content analysis of the data. Establishing linkages and relationships constitutes a kind of detective work that qualitative researchers do in the theorizing process. The researcher uses both inductive and deductive methods of establishing relationships "while developing a theory or hypothesis that is grounded in the data" (Goetz & LeCompte, 1984, 172).

A theory based on and evolving from data is called a **grounded theory** (Glaser & Strauss, 1976). In applied research, grounded theories are considered the best for explaining observed phenomena, understanding relationships, and drawing inferences about future activities.

Trustworthiness in Qualitative Research

Qualitative researchers do not attempt to provide numerical evidence that their data are reliable and valid. In fact, the terms *reliability* and *validity* are rarely used in qualitative research. That does not mean, however, that the qualitative researcher is not concerned with getting good data and reaching conclusions in which readers can have faith. Both issues are extremely important, and without attention to them, the quality of the research may be suspect. Just as for quantitative data, those reporting qualitative data make a case for the quality of the data *and* the conclusions that were derived from the data analysis.

A variety of terms are used to describe quality in qualitative research. The terms have changed and still are evolving. Lincoln and Guba (1985) used the term **trustworthiness** to describe the overall quality of the results from the study, and this term is most often used. Rossman and Rallis (2003) broke trustworthiness down into two questions: (1) is the study competently conducted? and (2) is it ethically conducted?

convergence

Consistency of results across two or more methodological techniques.

theorizing

Cognitive process of discovering abstract categories and the relationships among those categories.

theory

Explanation of some aspect of practice that permits the researcher to draw inferences about future happenings.

grounded theory

A theory based on and evolving from data.

trustworthiness

A quality achieved in a study when the data collected generally are applicable, consistent, and neutral.

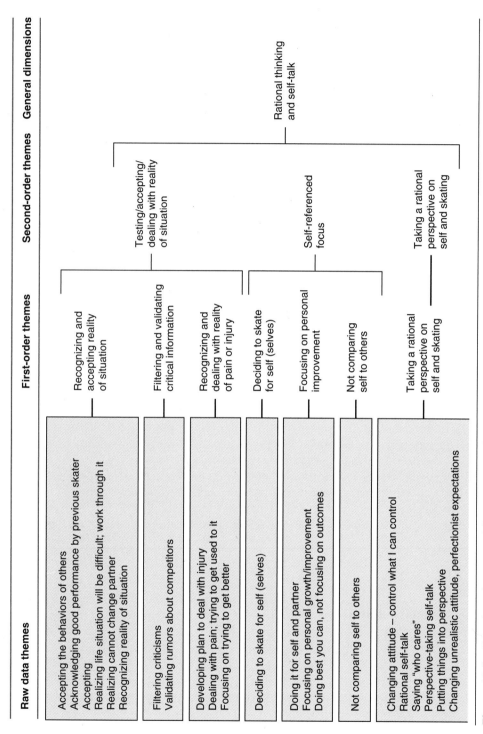

Figure 19.1 Raw data themes and first- and second-order themes making up the general dimension of rational thinking and self-talk.

Reprinted with permission from *Research Quarterly for Exercise and Sport*, Vol. 64, No. 4, p. 464, Copyright 1993 by the American Alliance for Health, Physical Education, Recreation and Dance, 1900 Association Drive, Reston, VA 20191.

Conducting an Ethical Qualitative Study

We will discuss the issue of ethics first. Ethics often plays a prominent role in judging qualitative research. We can think of this in many ways, but it boils down to two major areas. First, the ethical treatment of study participants (Locke et al., 2000) is important in qualitative research. Qualitative researchers often spend a great deal of time with participants, and they should treat them with dignity—from the time they approach them to be a part of the study through the reporting of data. If researchers have promised anonymity, they should maintain it. This does not suggest that those doing quantitative research should not treat participants ethically. The relationships and the type of data are different in qualitative research, however, and often require a more sustained commitment to ethics. Second, as stated by Rossman and Rallis (2003, 63), the research should "contribute in some way to understanding and action that can improve social circumstances." As these authors and others (e.g., Creswell, 1998; Patton 2002) have noted, groups of readers or even individual readers of a research report may have different criteria for determining whether the questions asked, and how they were answered, contribute to our knowledge base and how they can improve what occurs in the setting being studied. As in all fields of study, if researchers ask unimportant or trivial questions, no matter what method they use, the research does not contribute much.

Conducting a Competent Qualitative Study

This brings us to the issue of what constitutes a competently conducted study. Again, we can think of this in a variety of ways. Lincoln and Guba (1985) described four concepts that can be used in qualitative research to think about quality. The first concept is **credibility.** Because the context, participants, and settings are important to interpreting the results of qualitative research, how the researcher understands them and the description presented in a paper are critical to judging other parts of the study. If it is not clear who the participants are and where the study is being completed, readers will have difficulty evaluating the conclusions.

credibility

A quality achieved when the participants and setting of a study are accurately

Transferability is the second concept that is important in evaluating quality and addresses whether the results would be useful to those in other settings or conducting research in similar settings. You may be thinking that this is like generalizability, and in some ways it is, but in other ways it is not. All researchers would like their research to help others, but like much of the quantitative research in our area, most qualitative research does not benefit from random selection from a large population. Transferability is a question of argument and perception. The researcher may present reasons for why a given study might apply in other settings (e.g., many schools operate in a similar fashion, or the participants in an exercise class are similar to those found in many community settings), but ultimately the reader or user of the research must determine whether the study applies to his or her work environment or future research. Again, this often is the case in quantitative research—qualitative researchers just start with that concept up front.

transferability

In qualitative research, whether the results have the potential to be transferred to other settings.

The next concept is **dependability.** In many qualitative studies researchers make changes in the phenomenon being studied or in the methods used based on previous data collection. For a relatively straightforward example, a researcher may use a structured interview to ask questions but change how she asks follow-up questions depending on the previous answer. A second set of interviews might be based on the answers to the first set of interviews. If the researcher asked questions in a lockstep manner without adjustments, the quality of the data would suffer. In this example the researcher could not possibly plan for all possible contingencies. How the researcher deals with the changes determines the dependability of the data.

dependability

Addresses the quality of the data in a qualitative study, including how well the researcher deals with change.

Finally, **confirmability** deals with the issue of researcher bias and, as we discuss later in this section, uses methods that are different from most quantitative research in that reliability statistics are not calculated. Other approaches are used and reported so that the readers have faith in the results of the study.

confirmability

A characteristic of qualitative research that addresses whether another can place faith in the results.

Providing Evidence of Trustworthiness

Researchers can provide evidence of the trustworthiness of a study in a variety of ways based on the preceding categories. We will review the common techniques qualitative researchers use to gather good data and reach objective conclusions. These techniques are used during both data collection and data analysis. As noted earlier, analysis often occurs during data collection, so it is difficult to separate these into temporal categories. Not all researchers will use all techniques in every study, but it is very common for researchers to use multiple methods to increase the trustworthiness of their data and conclusions. The techniques commonly used are as follows:

- *Prolonged engagement.* The researcher must spend enough time to get good data. The collection of qualitative data requires substantial time in a setting so that the researcher develops an in-depth understanding and does not reach superficial conclusions.

- *Audit trail.* The method and focus in qualitative research change during a study. An audit trail describes the changes that occurred and how they influenced the study. These changes are often reported in the method section, where the researcher explains how the changes improved the study.

- *A rich, thick description.* As noted earlier, describing the setting and context is important for credibility. A thorough description in the report is needed so that readers can both understand the study and assess whether the setting and results will transfer to their setting or future research.

- *Clarification of researcher bias.* All researchers come into a study with biases. The management of these biases is particularly important in qualitative research in which the researcher is the instrument of data collection. Evidence that the researcher acknowledged his or her biases and dealt with them is essential. In a move that we believe enhances the quality of research, many qualitative researchers address this directly in the method section by "coming clean" and presenting their biases and explaining how they worked to control them.

- *Triangulation.* This is one of the techniques most often used in qualitative data analysis. It requires the use of three (hence the prefix *tri*) independent sources of data to support a conclusion. For example, a research may use student interview data, teacher interview data, and observation to support a result. Note that this is not simple vote counting (i.e., three people said this in an interview), but the use of independent types of data to support a conclusion.

- *Negative case checking.* This occurs when the researcher looks at instances in which what he or she expected to happen did not. For example, in an exercise class, the researcher notes that as time goes on, many participants seem more upbeat and are integrating the class into their lives. Does this happen for everyone? For those for whom it does not happen, is the phenomenon not as pervasive, did the researcher selectively look at the data to reach a conclusion, or is something different happening for some participants? Negative case checking helps address bias and enables the researcher to investigate the phenomenon further.

- *Member checking.* A member check occurs when the researcher goes back to the participants to share the conclusions and see whether they agree with them. Sometimes an intermediate step occurs where the researcher asks the participants to review interview transcripts, clarify statements, and add anything they think might be missing. That by itself is not true member checking. Presenting the participants with the conclusions goes much further and provides a confirmation of *the analysis*. In some cases in which the participants believe the researcher may have gotten it wrong, the conclusions may be modified to reflect the participants' views. In other cases, the researcher may believe that bias (e.g., when a report does not reflect positively on a person) influenced a participant's evaluation and would then include in the report an analysis of that person's evaluation.

- *Peer debriefing.* As in much of what we do, a new set of eyes can bring new light to a data set and our conclusions. A peer debriefer is another person who examines the data and the conclusions and serves as a devil's advocate, questioning the researcher to see whether the findings still hold up. Often the person performing this has a background in the phenomenon being investigated and expertise in qualitative methods and data analysis. This step can improve the conclusions and how they are presented in the research report.

Writing the Report

There is no standard (or "correct") format for a qualitative research report, just as there are no formats that are rigorously followed for any other type of research. Here we simply mention some main components of a qualitative research report and their placement in the report. Your department or university may have a definite order of components that you must follow.

The components of a qualitative study are similar to those of other conventional research reports. The first part introduces the problem and provides background and related literature. A description of method is an integral part of the report. Although this section is not as extensive in a journal article as in a thesis, it is usually much more extensive than in other forms of research. The reasons are obvious as we explained. The methodology is integrally related to the analysis and is also important in terms of trustworthiness and credibility.

The results and discussion section forms most of the report. If used, charts, tables, and figures are contained in this section and must be integrated into the narrative. A major contributor to the bulk of this section is the description contained in a qualitative study. As we said earlier, narrative vignettes and direct quotations are basic to this type of research. The qualitative study strives to provide enough detail to show the reader that the author's conclusions make sense (Creswell, 2003). The author is faced with the delicate problem of finding a balance between the rich descriptive materials and the analysis and interpretation. Some writers include too much description to illustrate their points, and some use too little. Researchers have suggested that 60% to 70% of the report be descriptive material and that 30% to 40% should be devoted to the conceptual framework (Merriam, 1988). Some

Funny Things That Happened During the Collection of Qualitative Data

1. On two occasions, researcher reported to gymnasium (carrying camera, TV monitor, tape recorder, and notes) to find door locked. Had to walk all the way around the building—once in the rain.

2. Reported to school to collect data to find classes wouldn't be held because of
   a. a doughnut sale
   b. teacher was ill (twice)
   c. holiday music rehearsal

3. Interview with teacher interrupted by principal due to crisis over a parent and a charity clothing sale. Interview postponed.

4. Tape recorder became inoperable during interview.

5. Tape recorder became inoperable while transcribing. Had to repeat interviews.

6. Group tape-recording session had to be rescheduled because of complete loss of control when a third-grade student belched into the microphone.

7. Had to buy own camera because department's camera was inoperable.

8. Broke own camera. Repair took two weeks.

9. Had to use participating school's TV monitor but couldn't find proper cord. Interviews rescheduled.

10. During the making of backup tapes, the VCR with original data tape was stolen. Police nabbed the culprit and data were recovered, but the researcher was comatose.

balance is definitely needed, and the task requires judgment in deciding which evidence to include to illustrate your ideas.

It is doubtful that any graduate student is under the false impression that doing qualitative research is a quick-and-easy technique. To emphasize this point, we call your attention to the box on page 361, which lists some of the trials and tribulations that befell one doctoral student while she was attempting to collect data.

Concluding Remarks

We should point out that qualitative research is by no means confined to the area of pedagogy. Getting that impression is easy because of the ever-increasing volume of literature on qualitative research in educational journals and textbooks. Qualitative research has a great deal of application to the study of the sociological aspect of sport. Bain (1989) observed that two of the earliest qualitative research studies published in the *Research Quarterly for Exercise and Sport* were on this topic. Sage (1989) cited several qualitative studies on sports dealing with Little League baseball, bodybuilding, soccer, surfing, and coaching.

In a provocative essay dealing with the science of human behavior, Martens (1987) questioned the basic assumptions of orthodox science. A longtime critic of the conventional experimental approach to sport psychology as the only way to conduct research, Martens presented a convincing argument and appeal for qualitative paradigms and emphasis on experiential knowledge.

This chapter has focused on interpretive qualitative research. Bain (1989) discussed another approach called **critical theory.** The main difference between the two approaches is found in the research goals. Interpretive research is largely free of value judgments, whereas critical theory research is based on value judgments. In other words, in critical theory the aim is to give the research participants the insight necessary to make choices that improve their lives or empower them. Bain (1989) also stated that critical research is usually grounded in feminism, neo-Marxism, or the empowering pedagogy of Paulo Freire. Each of these theoretical perspectives challenges the status quo and strives for greater equality. Relatively little research in critical theory has been done in exercise, sport science, and physical education. However, a number of obvious issues in sport regarding women, race, and exploitation of athletes lend themselves to this form of research.

It seems advantageous for the physical activity disciplines, professions, and research in general to capitalize on the strengths of both qualitative and quantitative methods rather than to argue about the differences. The quantitative researcher must make many qualitative decisions regarding the question, design, measurements, analytical procedures, and interpretations. Similarly, the qualitative researcher often finds certain quantitative summaries, classifications, and analyses to be useful (Linn, 1986). There is nothing wrong with combining quantitative and qualitative measures. This, in fact, can be thought of as a form of triangulation that enhances the quality of a study. We used the study by Gould et al. (1993) as an example of using quantitative analysis to enrich the meaningfulness of qualitative research. Similarly, McPherson (1999) used qualitative methods in interpreting quantitative analysis results. Sage (1989) pointed out that there is a growing maturity in physical education, exercise science, and sport science with regard to drawing concepts, theories, and methods from all the social sciences and the humanities.

Qualitative research is a legitimate means of addressing certain questions in our field. There has been a remarkable growth of increasingly sophisticated methods to guide qualitative researchers. We should take advantage of the work done in other fields and try to extend the boundaries of knowledge by our own contributions. In 1987, Locke provided a fitting concluding statement regarding the place of qualitative research in our field: It should be done, it will be done, and it is important that it be done well. His prediction certainly has proven to be true.

Page 363 exemplifies a scale that can be used to evaluate the quality of qualitative research studies. It was originally developed by Linda Bain and presented at the 1992 national convention of the American Alliance for Health, Physical Education, Recreation and Dance, and modified by us for this edition of this book.

critical theory

Qualitative research based on value judgements.

Evaluating the Quality of Qualitative Research

Linda L. Bain
California State University, Northridge
AAHPERD, Indianapolis

Definition of the Problem

___ Purpose is clearly stated

___ Focuses on a significant issue

___ Seeks to understand meaning of experiences for the participants

___ Provides holistic view of the setting

Data Collection

___ Researcher has training in methods used

___ Pilot work done in similar setting using similar methods

___ Rationale provided for selection of the sample

___ Researcher has trusting, collaborative relationship with participants

___ Methods for data collection are unobtrusive, where appropriate

___ Data collection procedures provide thorough description of events

___ Prolonged engagement in field

Data Analysis

___ Analysis done during and after data collection

___ Triangulation of data sources and search for convergence

___ Search for negative cases

___ Provides interpretation and theory as well as description of events

___ Provides opportunity for participants to corroborate interpretation (member check)

___ Arranges for peer evaluation of procedures and interpretation

Preparation of Report

___ Complete description of setting

___ Complete description of procedures

___ Includes description of researcher's values, assumptions, and bias and how each was addressed

___ Uses vignettes and quotes to support conclusions and interpretation

General Assessment

___ Internal validity: How much confidence do you have in the quality of the description and interpretation of events in the particular research setting?

___ External validity: What is your assessment of the extent to which the results of this study apply to a different setting with which you are familiar?

Reprinted, by permission, from L.L. Bain, 1992, *Evaluating the quality of qualitative research*, 350.

Summary

Qualitative research methods include field observations, case studies, ethnography, and narrative reports. The researcher gathers data in a natural setting such as a gymnasium, a classroom, a fitness center, or a sport facility.

Qualitative research does not have the preconceived hypotheses that characterize quantitative research. Inductive reasoning is stressed, whereby the researcher seeks to develop hypotheses from observations. The focus is on the "essence" of the phenomena.

The researcher should exhibit sensitivity and perception when collecting and analyzing the data.

We stressed the importance of gaining access to the data in the field setting. Establishing rapport and gaining the participants' trust are essential. The most common methods of collecting data are interviews and observations. Data should be analyzed during and after collection. The researcher must sort and organize the data and develop tentative hypotheses that lead to other sources and types of data.

Data analysis involves organizing, abstracting, integrating, and synthesizing. The analytical narrative is the foundation of qualitative research. The narrative vignette gives the reader a sense of being present for the observation; it conveys a sense of holistic meaning to the situation. It is not unusual for a qualitative study to include quantitative analysis.

The qualitative researcher often attempts to construct a theory through the inductive process to explain relationships among categories of data. A theory that evolves from data is called a grounded theory. In the written report, the qualitative researcher must achieve a balance between the rich description and the analysis and interpretation.

Trustworthiness is used to determine whether the study was competent. Ethical issues are important in determining trustworthiness as are issues related to how the study was conducted (i.e., credibility, transferability, dependability, confirmability). Many techniques are used during data collection and analysis to enhance the quality of the data and the conclusions. Among these are prolonged engagement, keeping an audit trail, providing a thick, rich description, clarification of researcher bias, triangulation, negative case checking, member checking, and peer debriefing. Not all of these can be used in one study, but using most or many is common to enhance the faith readers can place in the conclusions and the transferability to other settings.

Qualitative research is a viable approach to solving problems in our field. It applies to pedagogy in physical education, to exercise science, and to sport science. Answers to the question "What is happening here?" can best be obtained in natural settings through the systematic observation and interactional methodology of qualitative research.

✔ Check Your Understanding

1. Locate a qualitative study and write an abstract of approximately 300 words on the methods used in gathering and analyzing the data (observations, interviews, triangulation, member checking, etc.) and in presenting the results (narrative vignettes, quotations, tables, etc.).

2. Locate a quantitative study and discuss how a qualitative approach (in conjunction with the quantitative approach) could be used to enhance the meaningfulness of the study.

PART
IV

Writing the
Research Report

We have not succeeded in answering all of your

questions. In some ways, we are as confused as you.

However, we have succeeded in raising questions

so that we are confused at a higher level.

—Click and Clack, NPR's Car Talk

■   ■   ■   ■   ■   ■   ■   ■

Part I discussed the research proposal, its purpose, and the structure of the different parts. Parts II and III provided the details needed to understand and conduct research, including statistics, measurement, and types of research. This part completes the research process with instructions on how to prepare the research report. You may also want to refer to chapter 2, which discussed some rules and recommendations for writing the review of literature, for this is an important part of the research report.

Chapter 20 examines all the parts of the research proposal that have already been discussed. In addition, we offer some of our thoughts about the nature of the meeting to review the research proposal. Up until now, we have tried to explain how to understand other research and how to plan your own research. Here we help you organize and write the results and discussion sections (or chapters) about research that you have conducted. We also explain how to prepare tables, figures, and illustrations and where to place them in the research report.

Finally, in chapter 21, we suggest ways of using journal and traditional styles to organize and write theses and dissertations. We also present a brief section on writing for scientific journals and a short discourse on preparing and giving oral and poster presentations.

Completing the Research Process

The biggest liar in the world is "They say."

—Douglas Malloch

■ ■ ■ ■ ■ ■ ■

Completing the process for your research involves writing your proposal and getting it approved as well as carrying out the research and writing it up in the results and discussion section of your thesis. In the following sections, we provide guidelines to help you complete this process. After you have read these guidelines, you can find greater detail in Locke, Spirduso, and Silverman (2000).

Research Proposal

The research proposal contains the definition, scope, and significance of the problem and the methodology that will be used to solve it. If the journal format (advocated in this book and reviewed in detail in chapter 21) is used for the thesis or dissertation, the proposal consists of the introduction and method sections, appropriate tables, figures, and appendixes (e.g., score sheets, cover letters, questionnaires, sample informed-consent forms, and pilot-study data). In a four-section thesis or dissertation (introduction, method, results, and discussion), the proposal consists of the first two sections. In studies using a five-section format, in which the review of literature is the second, the proposal encompasses the first three sections.

One of the goals of this book is to help prepare students to develop a research proposal. We have already discussed the contents of the proposal. Chapters 2, 3, and 4 in this text pertain specifically to the body of the research proposal. Other chapters relate to various facets of planning a study: the hypotheses, measurements, designs, and statistical analyses. Here we attempt to bring the proposal together. We also discuss the proposal meeting and committee actions. Finally, we touch on basic considerations involved in grant proposals, specifically, how they differ from thesis and dissertation proposals.

Developing a Good Introduction

The student's most important task is to convince the committee (whether the proposal committee, a journal reviewer, or a reviewing committee for a granting agency) that the problem is important and worth investigating. The first section of the proposal should do this, and it should also attract the reader's interest to the problem. The review of literature provides background information and a critique of the previous research done on the topic, pointing out weaknesses, conflicts, and areas needing study. A concise statement of the problem informs the reader of the exact purpose, that is, what the researcher intends to do.

Hypotheses or questions are advanced on the basis of previous research and perhaps some theoretical model. Furthermore, operational definitions serve to inform the reader exactly how the researcher is using certain terms (see below for an operational definition of cricket). Operational definitions must describe observable phenomena and must generally relate to the dependent and independent variables. Basic assumptions are also stated and serve to specify certain conditions and premises that must exist for the study to proceed. Limitations and possible shortcomings of the study are acknowledged by the researcher and are generally the result of the delimitations that the investigator imposes. The first section concludes with a statement about the significance of the study, which can be judged from either a basic or an applied research standpoint. The significance section emphasizes contradictory findings and limitations of previous research and the ways in which the proposed study will contribute to further knowledge about the research topic.

Operational Definition of Cricket As Explained to Us by an Australian Friend

You have two sides: one side in the field and one out.

Each man that's in the side that's in goes out, and when he's out he comes in and the next man goes in until he's out.

When they are all out, the side that's out comes in and the side that has been in goes out and tries to get those coming in out.

Sometimes you get men still in and not out.

When both sides have been in and out including the not out, that ends the game!

Chapter 2 of this book concerned the literature review and included a discussion on the inductive and deductive reasoning processes used in developing the problem and formulating hypotheses. Chapter 3 covered the other parts typically required in the introduction section or chapter of a proposal.

Innumerable hours are involved in preparing the first section (introduction) of the proposal, especially in preparing the literature search and the formulation of the problem. The student usually depends heavily on an advisor and on completed studies for examples of format and description. Before we continue, however, we need to mention that you should write your proposal in the future tense. You state that so many participants will be selected and that certain procedures will be carried out. Theoretically, if the proposal is carefully planned and well written, you need only change from future to past tense to have the first two sections or chapters of the thesis or dissertation. Realistically, however, numerous revisions will probably be made between the proposal and the final version. To reiterate, the importance of the study and its contribution to the profession are the main focus of the first section in the proposal, and this constitutes the basis for approval or disapproval.

Describing the Method

The method section of the proposal frequently draws the most questions from committee members in the proposal meeting. In the method section or chapter, the student must clearly describe how the data will be collected to solve the problem set forth in the first section. The student needs to specify who the participants will be and how they will be chosen, how many participants are planned, any special characteristics of importance, how the participants' rights and privacy will be protected, and how informed consent will be obtained. Methods of obtaining measurements are detailed, and the validity and reliability of these measures are documented. Next, the procedures are described. If, for example, the study is a survey, the student discusses the steps in developing the instrument and cover letter, mailing the questionnaires, and following up. If the study is experimental, the treatments (or experimental programs) are described explicitly, along with the control procedures that will be exercised. Finally, the student must explain the experimental design and planned statistical analysis of the data.

We have previously emphasized the importance of conducting pilot studies before gathering data. If pilot work has been done, it should be described, and the results reported. Often the committee members have major concerns about such questions as whether the treatments can produce meaningful changes, whether the measurements are accurate and can reliably discriminate between participants, and whether the investigator can satisfactorily perform the measurements and administer the treatments. The pilot study should provide answers to these questions.

We recommended in chapter 4 that the student use the literature to help determine the methodology. Answers to questions about whether certain treatment conditions are sufficiently long, intense, and frequent to produce anticipated changes can be defended by results of previous studies.

The Proposal Process

We reiterate the contents of the proposal: the introduction (including the review of literature) and the methods to be used. The proposed purpose—in conjunction with pertinent background information, plausible hypotheses, operational definitions, and delimitations—determines whether the study is worthwhile. Consequently, the first section (or chapter) is instrumental in stimulating interest in the problem and establishing the rationale and significance of the study. The committee's decision to approve or disapprove rests primarily with the persuasiveness exhibited in the first section.

Actually, the basic decision about the topic's merit should already have been made before the proposal meeting. The student should consult with the advisor and most (if not all) of the committee members to reach a consensus about the study's worth before the proposal meeting is scheduled. If you cannot convince the majority of the committee that your study is worthwhile, do not convene a formal proposal meeting. You may have a problem if your proposal is returned with a checklist like that below.

Interim Thesis/Dissertation Evaluation

Dear _____:

Greetings! I regret that my busy schedule prohibits me from rendering a detailed written evaluation of your thesis/dissertation. However, I have checked the appropriate actions or comments that apply to your proposal.

___ If at first you don't succeed, try, try again.

___ Don't sell your research methods textbook (Thomas, Nelson, and Silverman, of course); you'll need to take the course again.

___ You were not required to write your paper in a foreign language (Burmese, or whatever it was).

___ I couldn't read beyond the third page; one does not have to eat a whole pie to know it is bad.

___ I hear they are hiring at Sam's Diner.

___ May I have your permission to use your proposal as an example next semester when I teach research methods?

___ Have you paid your tuition and fees yet? If not . . .

___ Please call my secretary and arrange an appointment with me. Consider taking a tranquilizer before you arrive.

___ I have been serving on thesis/dissertation committees for over 15 years and can now honestly say that I've seen it all.

What to Expect of the Proposal Committee

Let us digress a moment to discuss the composition of the proposal committee. The structure of committees and the number of committee members vary from one institution to another. It is probably safe to say that most thesis committees consist of at least three members, and most dissertation committees of at least five. The major and minor professors are included in these numbers at institutions where that occurs, although the master's student is often not required to have a minor. Other members should be chosen on the basis of their knowledge about the subject or their expertise in other aspects of the research, such as design and statistical analysis. Sometimes the institution or department specifies how many members of the committee must be from inside or outside the department. Usually no limit is placed on the maximum number of committee members allowed.

We strongly recommend that the student, with the help and advice of the major professor, get general approval and support for the problem itself from at least two of the three thesis committee members (or three of the five dissertation committee members) before the meeting. This support is tentative, of course, and final approval is contingent on the refinements that might be needed and on the adequacy of the methodology.

Do not wait until the formal committee meeting to plan the study. This planning should be completed beforehand. Some graduate programs have so-called preproposal meetings for brainstorming and informally reaching an agreement on the efficacy of the proposed topic. This kind of meeting functions to garner support from the committee before a great deal of time and effort are wasted on a fruitless endeavor. The student prepares and distributes an outline of the purpose and basic procedures before the meeting. The student should have spent considerable time in consultation with the advisor (and probably at least one other committee member) and should have searched the literature sufficiently to be adequately prepared to present a sound case for the study. The preproposal meeting is not just a "bull session" in which the student is fishing for basic ideas. At the same time, the informality of the occasion does allow a good exchange of ideas and suggestions.

How to Prepare the Formal Proposal

The formal proposal should be carefully prepared. If the proposal contains errors of grammar, spelling, or format, committee members may conclude that the student lacks the interest, motivation, or competence to do the proposed research. With the availability of computers and word processors, no reason exists for a student to present a poorly prepared proposal. Also, spell-checking routines, available with most word-processing software, should always be used. Remember, however, that a spelling checker cannot identify the use of an incorrect word that is spelled correctly, as the following poem by Jerrold H. Zar illustrates:

> I have a spelling checker.
>
> It came with my PC.
>
> It plane lee marks four my review
>
> Miss steaks aye can knot see.

When you are ready to distribute your proposal, print it using a good-quality printer and on good paper. Be sure that copies duplicated for distribution are high quality and that all pages are included (look at each page in each copy—copy machines often do not work perfectly).

Committee members should not ignore errors in proposals with the idea that the student will correct these later. Doing so may lead the student to assume that carelessness is acceptable in data collection or in the final written thesis or dissertation. Copies of the proposal should be given to the committee members well before the meeting. The department or university usually specifies the number of days in advance of the meeting that the proposal should be distributed.

They're going to eat her alive!

What Happens at the Proposal Meeting

In the typical proposal meeting, the student is asked to briefly summarize the rationale for the study, its significance, and the methodology (good visual aids enhance this presentation). The remainder of the session consists of questioning by the committee members (see page 372 for potential questions). If the topic is acceptable, the questions concern mainly the methods and the competence of the student to conduct the study. The student should exhibit tactful confidence in presenting the proposal. A common mistake students make is to be so humble and pliable that they agree to every suggestion made, even those that radically change the study. The advisor should help ward off these "helpful" suggestions, but the student must also be able to respectfully defend the scope of the study and the methodology. If adequate planning has gone into the proposal, the student (with the assistance of the major professor) should be able to recognize useful suggestions and defend against those that seem to offer minimal aid.

A projected schedule of what procedures will be accomplished may or may not be required for a thesis or dissertation proposal. Regardless of whether a time frame is formally required, it is definitely something that a student should address. A common mistake in planning for a thesis or dissertation is underestimating how long each phase of the study will take. Students tend to assume, like Pollyanna, that everyone involved (participants, helpers, committee members, word processors, and university personnel) will drop everything else they are doing to accommodate their study. Moreover, they assume that all phases of the study will proceed without a hitch. Unfortunately, it just doesn't work that way. The simple logistics of accomplishing each aspect of the study is usually more complicated and time-consuming than anticipated. Much grief can be avoided if the student carefully and realistically projects how long the various steps in the research process will take (and then adds some more time).

Once the proposal is approved, most institutions treat it as a contract in that the committee expects the study to be done in the manner specified in the proposal. Moreover, the student can assume that if the study is conducted and analyzed as planned and is well written, it will be approved. If any unforeseen changes are required during the course of the study, they must be approved by the advisor. Substantial changes usually must be reviewed by some or all of the committee members.

Questions by Proposal Committees and Answers by Graduate Students

Committee: This myasthenia gravis, does it affect your memory at all?

Student: Yes.

Committee: In what ways does it affect your memory?

Student: I forget.

Committee: Can you give us an example of something that you've forgotten?

Committee: Were you present when this picture of you using the apparatus was taken?

Student: ?

Committee: How far apart were the participants at the time of the collision?

Student: ?

Committee: You say the stairs the participants climbed went down to a basement?

Student: Yes.

Committee: And these stairs, did they go up also?

Committee: How many participants in your research?

Student: 40

Committee: Men or women?

Student: All women.

Committee: Were there any men?

Committee: Were your participants qualified to give a urine sample?

Student: ?

Committee: After you sacrificed the mouse, was it still alive?

Student: No.

Committee: Did you check for a heart beat?

Student: No.

Committee: So it's possible the mouse was still alive?

Student: No.

Committee: How can you be sure?

Student: Because his brain was sitting in a jar on my desk.

Committee: But the mouse could have been alive nonetheless?

Student: It's possible he could have been alive and serving as a professor somewhere.

Preparing and Presenting Qualitative Research Proposals

The content, procedures, and expectations for the proposal that we have discussed so far largely pertain to quantitative theses and dissertations. Although many similarities exist in the kinds of information presented in quantitative and qualitative research proposals, some salient differences should be mentioned.

Preparing the proposal is much easier if all committee members are familiar with qualitative research. However, if one or more members do not understand the nature of qualitative research and the methodological differences between quantitative and qualitative paradigms, problems may arise. The student who is contemplating doing a qualitative study is strongly urged to read Locke, Spirduso, and Silverman's *Proposals That Work* (2000) or *Designing Qualitative Research* by Marshall and Rossman (1999).

One of the problems that the qualitative researcher may face concerns the possible shifts in focus and methods that may occur during the study. Earlier, we likened the proposal to a contract, whereby the committee members expect the student to carry out the study as specified in the proposal. Unlike quantitative research, however, qualitative research typically undergoes changes in the focus of the research question or the sources of data, the methodology, and the analysis of data. This aspect of qualitative research demands an "open contract" (Locke et al., 2000).

Another difference between the quantitative and qualitative proposal is the literature review. In some qualitative research, closely related studies may be purposefully omitted when planning the study because the researcher does not want to be influenced by the views and perceptions of others. Still another difference is the need for the qualitative researchers to address their own values, biases, and perceptions in order to more clearly understand the context of the research setting.

All committee members, whether familiar or unfamiliar with qualitative research, are interested in methodological concerns: Where will the study take place? Who will be the participants? How will the researcher gain access to the site? What data sources will be used? Does the researcher possess the needed research skills? What pilot work has been done? How will ethical concerns be handled? What strategies will be employed in collecting, sorting, categorizing, and analyzing the data?

It should be apparent by now that there are differences in the quantitative and qualitative proposal format and in the nature of the proposal process. These differences must be recognized and accepted. As with any type of research, the major professor plays a vital role in preparing the student (and, in some cases, the committee members), in showing support, and in helping the student get out of tight spots. Most of all, the student, with the support of the major professor, must be able to assure the committee that the study will be carried out in a competent and scholarly manner.

Writing Proposals for Granting Agencies

All sources of grants, whether governmental agencies or private foundations, require research proposals so that they can decide which projects to fund and how much to award. Granting agencies nearly always publish guidelines for applicants to follow in preparing proposals.

A well-written proposal is everything. The researcher rarely gets a chance to explain or to defend the purpose or procedures. Thus, the decision is based entirely on the written proposal. The basic format for a grant proposal is similar to the thesis or dissertation proposal. However, some additional types of information are required, as are some procedural deviations.

We cannot emphasize too strongly the importance of following the guidelines. Granting agencies tolerate little (if any) deviation from their directions. All sections of the application must be addressed and deadlines strictly followed. Frequently, a statement of intent to submit a proposal is required a month or so before the proposal is filed.

Grant proposals include

- an abstract of the proposed project,
- a statement of the problem and its relevance to one of the granting agency's specified priorities,
- the methodology to be followed,
- a time frame,
- a budget, and
- the curricula vitae of the investigators.

A limited review of literature is often required to demonstrate familiarity with previous research. Occasionally, the funding agency imposes some restrictions on design and methods. For example, it is not uncommon for an agency to prohibit control groups if the treatment is hypothesized to be effective. In other words, the agency may not want anyone to be denied treatment. This can pose some problems for the researcher in scientifically evaluating the outcomes of the project.

Because every granting agency has stipulations regarding the types of programs it can and cannot fund, a detailed budget is required, as is a justification for each budget item. A time frame is usually expected so that the reviewers can see how the project will be conducted and when the various phases will be accomplished. This time frame provides the reviewers with information about the depth and scope of the study, justification for the length of time that support is requested, the contributions of the various personnel, and when the granting agency can expect progress reports.

The competence of the researcher has to be documented. Each researcher must attach a curriculum vitae and often a written statement describing appropriate preparation, experience, and accomplishments. The adequacy of facilities and sources of support must also be addressed. Letters of support are occasionally encouraged. These are included in the appendix. Numerous copies of the proposal are usually required. The proposal is evaluated by a panel of reviewers in accordance with certain criteria regarding the contribution to knowledge, relevance, significance, and soundness of design and methodology.

The preparation of a grant proposal is a time-consuming and exacting process. Several types of information are required, and time is needed to gather the information and state it in the manner prescribed by the guidelines. Applicants are advised to begin preparing the proposal as soon as the guidelines are available.

Finally, it is usually wise to contact the granting agency before preparing and submitting a proposal. Seldom are proposals funded that are submitted without some prior contact with the granting agency. Finding out the agency's interests and needs is a timesaving venture for the researcher. Often, a visit to, or an extended phone conversation with, the research officer is advisable. It helps to look at proposals previously funded by the agency or to seek guidance from researchers who have been funded.

Submitting Internal Proposals

Many colleges and universities (particularly larger research universities) offer internal funding for graduate student research, although these grants usually are not large. These grants typically require a two- to five-page proposal that has the support of the major professor and often the department chair. The contents usually include an abstract, a budget, and a short narrative focusing on the proposed methodology and why the research is important. Notices that internal funding is available are routinely posted or advertised around the college or university. A good place to find internal funding sources is at the office of your graduate school or vice president for research.

Completing Your Thesis or Dissertation

After your proposal meeting, you collect the data to evaluate the hypotheses that you proposed. Of course, you follow the methods that you specified in your proposal very carefully and consult with your major professor if problems arise or changes need to be made. Once the data are collected, you complete the agreed-upon analysis and discuss the outcomes with your major professor (and possibly some committee members, particularly if a member has statistical expertise). Then you are ready to write the results and discussion to complete your research.

Results and Discussion

The final sections of the thesis or dissertation are the results and discussion sections. Results are what you have found, and discussion explains what the results mean. In theses and dissertations, the results and discussion usually are separate sections, although they are sometimes combined (particularly in multiple-experiment papers). We discuss these sections here as separate parts of the research report.

How to Write the Results Section

The results section is the most important part of the research report. The introduction and literature review indicate why you conducted the research, the method explains how you did it, and the results section presents your contribution to knowledge, that is, what you found. The results should be concise and effectively organized and should include appropriate tables and figures.

Because there is no one correct way to present the results section, the results may be organized in various ways. The best way may be to address each of the tested hypotheses; on other occasions, the results may be organized around the independent or dependent variables of interest. Occasionally, you may want to show first that certain standard and expected effects have been replicated before you go on to discuss other findings. For example, in developmental studies of motor performance tasks, older children typically perform better than younger ones. You may want to report the replication of this effect before discussing the other results. When looking at the effects of training on several dependent variables, you may first want to establish that a standard dependent variable known to respond to training did, in fact, respond. For example, before looking at the effects of cardiovascular training as potentially reducing cognitive stress, you need to show that a change occurred in cardiovascular response as a result of the training.

Some items should always be reported in the results. The means and standard deviations for all dependent variables under the important conditions should be included. These are basic descriptive data that allow other researchers to evaluate your findings. The main descriptive data should be presented in one table if possible. Sometimes only the means and standard deviations of important findings are included in the results. However, all the remaining means and standard deviations should be included in the appendix.

The results section should also use tables and figures to display appropriate findings. Figures are particularly useful for percentage data, interactions, or summarizing related findings. Only the important tables and figures should be included in the results; those remaining should be placed in the appendix.

Statistical information should be summarized in the text where possible. Statistics from ANOVA and MANOVA should always be summarized in the text and complete tables relegated to the appendix. Make sure, however, that you include the appropriate statistical information in the text. For example, when giving the F ratio, report the degrees of freedom, the probability, and possibly the effect size: $F(1, 36) = 6.23$, $p < .02$, ES = 0.65. Above all, the statistics reported should be meaningful. Day (1983, 35) reported a classic case that read "33-1/3% of the mice used in this experiment were cured by the test drug; 33-1/3% of the test population were unaffected by the drug and remained in a moribund condition; the third mouse got away."

Sometimes tables are a better way to present this information. If the perfect scientific paper is ever written, the results section will read, "The results are shown in table 1" (Day, 1983, 36). However, this does not mean that the results section should consist mostly of tables and figures. It is disconcerting to have to thumb through eight tables and figures placed between two pages of text. But even worse is to have to turn 50 pages to the appendix to find a necessary table or figure. Read what you have written. Are all the important facts there? Have you provided more information than the reader can absorb?

Do not be redundant and repetitive. A common error is to include a table or figure in the results and then repeat it in the text. It is appropriate to describe tables and figures in a general way or to point out particularly important facts, but do not repeat every finding. Also, be sure that you do not call tables figures, and vice versa. However, as Day (1983) has reported, some writers are so concerned with reducing verbiage that they lose track of antecedents, particularly for the pronoun *it*:

> "**T**he left leg became numb at times and she walked it off. . . . On her second day, the knee was better, and on the third day it had completely disappeared." The antecedent for both *it*s was presumably "the numbness," but I rather think that the wording in both instances was a result of dumbness. (p. 36)

Reporting Statistical Data

A consistent dilemma among researchers, statisticians, and journal editors concerns the appropriate reporting of statistical information for published research papers. In recent years, some progress has been made that involves two issues in particular—always reporting some estimate of the size and meaningfulness of the finding along with the reliability or significance of the finding. Two organizations of importance to our field—American Physiological Society (2004) and American Psychological Association (1999) have now published guidelines regarding these issues.

Following are summaries of general guidelines taken from these two sources:

- Information on how sample size was determined is always important. Indicate the information (e.g., effect sizes) used in the power analysis to estimate sample size. When the study is analyzed, confidence intervals are best used to describe the findings.
- Always report any complications that have occurred in the research including missing data, attrition, and non-response including how these problems were handled in data analysis. "…before you compute *any* statistics, *look at your data*" (*American Psychologist*). Screening your data should always be done (this is not tampering with data) to be sure the measurement make sense.
- Select minimally sufficient analyses—using complicated methods of quantitative analyses may appropriately fit data and lead to useful conclusions, but many designs fit basic and simpler techniques. When they do, these should be the statistics of choice. Your job is not to impress your reader with your statistical knowledge and expertise but to appropriately analyze the research and present it so a reasonably well-informed person can understand it.
- Report actual p values, and confidence intervals are even better. Always report an estimate of the magnitude of the effect. If the measurement units (e.g., maximal oxygen consumption) have real meaning, then reporting them in an unstandardized way such mean difference is useful. Otherwise standardized reporting such as effect size or r^2 is useful. In addition placing these findings in practical and theoretical context adds much to the report.
- Control multiple comparisons through techniques such as Bonferroni.
- Variability should always be reported using the standard deviation. Standard error characterizes the uncertainty associated with a population and is most useful in determining confidence intervals.
- Report your data at the level (e.g., how many decimal places) that is appropriate for scientific relevance.

What to Include in the Discussion Section

Although the results are the most important part of the research report, the discussion is the most difficult to write. There are no cute tricks or clear-cut ways to organize the discussion, but there are some rules that define what to include:

- Discuss your results—not what you wish they were, but what they are.
- Relate your results back to the introduction, previous literature, and hypotheses.
- Explain how your results fit within theory.
- Interpret your findings.
- Recommend or suggest applications of your findings.
- Summarize and state your conclusions with appropriate supporting evidence.

Your discussion should point out both where data support and where they fail to support the hypotheses and important findings. But do not confuse significance with meaningfulness in your discussion. In fact, be especially careful to point out where they may not coincide. In particular, the discussion should point out factual relationships among variables and situations, thus leading to a presentation of the significance of the research. Of course, this is an essential place not to confuse cause and effect with correlation. For example, don't say that a characteristic had an "effect" or "influence" on a variable when you mean that they were related to each other.

The discussion should end on a positive note, possibly a summarizing statement of the most important finding and its meaning. Never end your discussion with a variation of the old standby of graduate students: *More research is needed*. Who would have thought otherwise?

The discussion should also point out any methodological problems that occurred in the research. However, a methodological cop-out to explain the results is unacceptable. *If you did not find predicted outcomes and you resort to methodological failure as an explanation, you did not do sufficient pilot work.*

Graduate students sometimes want their results to sound wonderful and to solve all the problems of the world. Thus, in their discussions they often make claims well beyond what their data indicate. Your major professor and committee are likely to know a lot about your topic and therefore are unlikely to be fooled by these claims. They can see the data and read the results. They know what you have found and the claims that can be made. A much better strategy is to make your points effectively in your discussion and not try to generalize these points into grandiose ideas that solve humanity's major problems. Write so that your limited contribution to knowledge is highlighted. If you make broader claims, knowledgeable readers are likely to discount the importance of your legitimate findings. Your discussion should not sound like the *Calvin and Hobbes* cartoon (by Bill Watterson) in which Calvin said, "I used to hate writing assignments, but now I enjoy them. I realized that the purpose of writing is to inflate weak ideas, obscure poor reasoning, and inhibit clarity."

Another point about writing your discussion is to write so that reasonably informed and intelligent people can understand what you have found. Do not use a thesaurus to replace your normal vocabulary with multisyllabic words and complex sentences. Your writing should not look like the examples in the box on page 378. By translating, you can probably recognize these sentences as some well-known sayings.

Writing the Discussion

Problem Statement: Why did the chicken cross the road?

Method: One chicken observed by several individuals.

Results: Said chicken crossed the road.

Discussion: Following are the explanations given for the chicken crossing the road.

Ralph Nader—That chicken's habitat on the other side of the road had been polluted by unchecked industrial greed. The chicken did not reach the unspoiled habitat on the other side of the road because it was crushed by the wheels of a gas-guzzling SUV.

Dr. Seuss—Did the chicken cross the road?

Did he cross it with a toad?

Yes, the chicken crossed the road,

But why it crossed I've not been told.

Sigmund Freud—The fact that you are at all concerned that the chicken crossed the road reveals your underlying sexual insecurity.

Bill Gates—I have just witnessed *eChicken2005*, which will not only cross roads, but will lay eggs, file your important documents, and balance your checkbook. Internet Explorer is an integral part of *eChicken*.

Bill Clinton—I did not cross the road with *that* chicken. What is your definition of chicken?

Pat Buchanan—To steal the job of a decent, hard-working American.

Captain Kirk—To boldly go where no chicken has ever gone before.

Colonel Sanders—Did I miss one?

Graduate student—Is that regular or extra-crispy?

In Other Words

1. As a case in point, other authorities have proposed that slumbering canines are best left in a recumbent position.
2. It has been posited that a high degree of curiosity proved lethal to a feline.
3. There is a large body of experimental evidence which clearly indicates that smaller members of the genus Mus tend to engage in recreational activity while the feline is remote from the locale.
4. From time immemorial, it has been known that the ingestion of an "apple" (i.e., the pome fruit of any tree of the genus Malus, said fruit being usually round in shape and red, yellow, or greenish in color) on a diurnal basis will with absolute certainty keep a primary member of the health care establishment from one's local environment.
5. Even with the most sophisticated experimental protocol, it is exceedingly unlikely that you can instill in a superannuated canine the capacity to perform novel feats of legerdemain.
6. A sedimentary conglomerate in motion down a declivity gains no addition of mossy material.
7. The resultant experimental data indicate that there is no utility in belaboring a deceased equine.

Your discussion can generally be guided by the following questions taken from the *Publication Manual of the American Psychological Association* (APA, 1994, 19):

- What have I contributed here?
- How has my study helped to resolve the original problem?
- What conclusions and theoretical implications can I draw from my study?

The responses to these questions are the core of your contribution, and readers have a right to clear and direct answers. If after reading your discussion the reader asks, "So what?" then you have failed in your research reporting.

How to Handle Multiple Experiments in a Single Report

Graduate students are conducting more research that involves multiple experiments. These experiments may ask several related questions about a particular problem or may build on one another, with the outcomes of the first leading to questions for the second. This is a positive trend, but it sometimes leads to problems within the traditional (chapter structure) thesis or dissertation format. Chapter 21 discusses the journal and traditional formats for organizing theses and dissertations.

Multiple experiments in journal format typically involve a general introduction and literature review. If the experiments use a common methodology, then a general method section might follow. This is followed by a presentation of each experiment with its own short introduction and citation of a few critical studies, method (specific to this experiment), results, and discussion (sometimes the results and discussion may be combined). It concludes with a general discussion of the series of experiments and their related findings.

Within the traditional framework, multiple experiments are probably best handled by separate chapters. The first chapter includes the introduction, theoretical framework, literature review, a general statement of the research problem, and related definitions and delimitations. Subsequent chapters describe each experiment. Each of these chapters includes a brief introduction, a discussion of the specific problem and hypotheses, and the method, results, and discussion sections. The final chapter is a general discussion in which the experiments are tied together. It contains the features of the discussion previously presented.

How to Use Tables and Figures

Preparing tables and figures is a difficult task. Howard Wainer (1992) wrote one of the best papers on this topic. We begin with a quotation he used.

> **D**rawing graphs, like motor-car driving and love-making, is one of those activities which almost every educator thinks can be done well without instruction. The results are of course usually abominable. (paraphrased, with my [Wainer's] apologies, from Margerison, 1965)

Wainer suggests that tables and figures should allow the reader to answer questions at three levels:

- Basic: Extraction of data
- Intermediate: Trends in parts of the data
- Advanced: Overall questions involving deep structure of the data (seeing trends and comparing groupings)

These levels can be thought of as an ordered effect:

1. Variables by themselves (data)
2. One variable in relation to another
3. The overall comparisons and relationships in the data

Preparing Tables

Getting information from a table is like extracting sunlight from a cucumber, to paraphrase Farquhar and Farquhar (1891).

Remember, tables are for communicating to the reader, not storing data. The first question is, Do you need a table? There is no easy answer, but two characteristics are important: Is the material more easily understood in a table? And does the table interfere with reading the results? Once you decide that you need a table (not all numbers require tables), follow these basic rules:

- Like characteristics should read vertically in the table.
- Headings of tables should be clear.
- The reader should be able to understand the table without referring to the text.

Examples of Poor and Good Tables

Table 20.1 is an example of a useless table; the data could be more easily presented in the text. This table can be handled in one sentence: "The experimental group (M = 17.3, s = 4.7) was significantly better than the control group (M = 12.1, s = 3.9), $t(28)$ = 3.31, p < .05." Table 20.2 is also unnecessary. Of the 10 comparisons among group means, only one was significant. The values in the table are the equivalent of t tests. This table can also be presented in one sentence: "The Scheffé test was used to make comparisons among the age-group means, and the only significant difference was between the youngest

Table 20.1   Useless Table 1

Means, Standard Deviations, and *t* Test for Distance Cartwheeled While Blindfolded

Groups	N	M	s	t
Experimental	15	17.3 m	4.7 m	
				3.31*
Control	15	12.1 m	3.9 m	

*p < .05.

Table 20.2   Useless Table 2

Scheffé's Test for Difference Among Age Levels in Ability to Wiggle Their Ears

Age	7	9	11	13	15
7	–	1.20	1.08	1.79	8.63*
9		–	1.32	1.42	1.57
11			–	1.58	1.01
13				–	0.61
15					–

*p < .05.

Table 20.3   Example of a Useful Table

Characteristics of Users and Nonusers HRFT Pilot Survey

	Users	Nonusers
Gender		
Male	4 (36.4%)	33 (62.3%)
Female	7 (63.6%)	20 (37.7%)
Age		
20–25	1 (09.1%)	2 (03.8%)
25–30	1 (09.1%)	8 (15.1%)
30–35	5 (45.5%)	9 (17.0%)
35–40	1 (09.1%)	7 (13.2%)
40 and over	3 (27.2%)	27 (50.9%)
Type of school		
Elementary	1 (09.1%)	18 (34.0%)
Middle	0 (00.0%)	18 (34.0%)
Junior-senior high	1 (09.1%)	0 (00.0%)
High	9 (81.8%)	17 (32.0%)
Student population		
0–100	0 (00.0%)	0 (00.0%)
100–500	1 (09.1%)	17 (33.3%)
500–1,000	0 (00.0%)	16 (31.4%)
1,000–1,500	1 (09.1%)	1 (02.0%)
Over 1,500	9 (81.8%)	17 (33.3%)

Reprinted with permission from *Research Quarterly for Exercise and Sport,* Vol. 54, pgs. 204-207, Copyright 1983 by the American Alliance for Health, Physical Education, Recreation and Dance, 1900 Association Drive, Reston, VA 20191.

(7-year-olds) and oldest (15-year-olds) groups, $t = 8.63$, $p < .05$."

We have borrowed an example of a useful table (table 20.3) from Safrit and Wood (1983). As you can see, like characteristics appear vertically. Also, an extensive amount of text would be required to present these same results, yet they are easy to understand in this brief table.

Improving Tables

An important question is how to improve tables so that they are more useful, more informative, and easier to interpret. Wainer (1992) offers three good rules for developing tables.

• The columns and rows should be ordered so that they make sense. For example, the row elements are often placed in alphabetical order according to the label for the row (e.g., names, places). This is seldom useful. Order the rows naturally, for instance, by time (e.g., from the past to the future) or by size (e.g., put the biggest or smallest value of mean or frequency first).

• When values go to multiple decimal places, round them off. Two digits are about the most that people can understand, that can be measured with precision, or that anyone cares about. For example, what does a $\dot{V}O_2$max value mean when carried to four decimal places? We don't understand it, we can't measure it that precisely, and no one cares. Sometimes attempts at precision become humorous, bringing to mind the report that the average American family has 2.4 children. (We thought children came only in whole units!) A whole child is the smallest (most discrete) unit of measurement available.

• Use and pay attention to the summaries of rows and columns. The summary data, often provided as the last row or column, are important because these values (sometimes sums, means, or medians) provide a standard of comparison (or usualness). Often setting these values apart in some way (e.g., bold type) is valuable.

Let us try these improvements on an actual table. In our research methods class, we often give students the assignment to find a table in *Research Quarterly for Exercise and Sport* and improve it by applying Wainer's suggestions. Our graduate students have not been hesitant about finding tables from our scholarly work to improve. (Professors: It's not only our work. Offer your students one of your published figures or tables. They will improve it, too.) A good example from a paper by Thomas, Salazar, and Landers (1991) that appeared in *Research Quarterly for Exercise and Sport* was provided by James D. George when he was a doctoral student at Arizona State University (our thanks to Jim for allowing us to use his work). The example on page 381 shows the data as presented by Thomas, Salazar, and Landers (1991). The one on page 382 is George's rearrangement. Observing the improvements in data presentation and understanding is easy. First, the rows have been reordered by the column "ES Info," with all the yes responses followed by all the no responses (it might be just as well to remove the no responses and list the authors' names at the bottom of the table). Then they have been reordered from smallest to largest by the next column, *N* (sample size). Finally, an additional label ("Study's most important effects") was inserted under the "Primary ES" to clarify the meaning of those three columns.

A Fine Table Made Better—The Original Table

Table 2.   Data on articles in Volume 59, 1988

First author	ES info	N^c		Primary ES*		
Doody	no					
Kamen[b]	yes	9	0.64	0.72	0.14	
Alexander	yes	26–48	0.33	0.73*	0.39	
Era	yes	5–6	0.50*	0.10	1.42*	
Kokhonen	yes	9–12	−1.97*	−2.64*	−1.78*	
Farrell	yes	45–368	0.77*	−0.51*	0.37*	
Heinert	no					
Kamen	no[a]	10	1.14	0.81	0.90	
Ober	no					
Simard	yes	7	−1.59*	0.52	−2.71*	
Berger	no					
Stewart	no					
Abernethy	no					
Etnyre	no					
Nelson[b]	yes	13	0.73	1.76	0.85	
Wesson	no					
Housh	yes	20	−0.53*	−2.11*	0.25*	
Hutcheson	yes	34	−0.06	0.63*	−0.30*	

*Comparison of Ms forming ES was significant, $p < .05$.

[a]No significant main effects.

[b]The main effect is significant, but no information is provided regarding the significance of the post hoc comparison.

[c]Per comparison group.

How do these changes relate to Wainer's three levels of questions? The revised table makes clear the sample size and ES information (basic level) for each study (but so did the original table). At the intermediate and advanced levels, however (trends, relationships, and overall structure), the revised table is a considerable improvement. For example, you can more easily observe that sample size and ES are unrelated; that is, neither studies with large samples nor small samples are more likely to produce larger or smaller treatment effects (as estimated by ES).

Another example of the mindless use of numbers is often the reporting of statistical values. Just because computer printouts carry the statistics (e.g., F, r) and probabilities (p) to five or more places beyond the decimal does not mean that the numbers should be reported to that level. Two or (at most) three places are adequate. However, this can result in rather odd probabilities: $t(22) = 14.73$, $p < .000$. Now, $p < .000$ means no chance of error; this cannot occur because if there is no chance of error, how can it be a probability? What happened is that the exact probability was something like $p < .00023$ and the researcher rounded it to $p < .000$. You must not do this. As indicated earlier, we believe that it is more appropriate to report the exact probability (e.g., $p = .025$) and whether this

A Fine Table Made Better—The Better Way

Table 2.　Data on articles in Volume 59, 1988

First author	ES info	N^c	Primary ES* (Study's most important effects)		
Era	yes	5–6	0.10	0.50*	1.42*
Simard	yes	7	–2.71*	–1.59*	0.52
Kamen[a]	yes	9	0.14	0.64	0.72
Kamen	yes[b]	10	0.81	0.90	1.14
Kokhonen	yes	9–12	–2.64*	–1.97*	–1.78*
Nelson[b]	yes	13	0.73	0.85	1.76
Housh	yes	20	–2.11*	–0.53*	0.25*
Hutcheson	yes	34	–0.30*	–0.06	0.63*
Alexander	yes	26–48	0.33	0.39	0.73*
Farrell	yes	45–368	–0.51*	0.37*	0.77*
Abernethy	no				
Berger	no				
Doody	no				
Etnyre	no				
Heinert	no				
Ober	no				
Stewart	no				
Wesson	no				

[a]Per comparison group.

[b]Comparison of Ms forming ES.

[c]The main effect is significant, but no information is provided regarding the significance of the post hoc comparison.

*ES was significant, $p < .05$.

probability exceeded the alpha set for the experiment (e.g., $p < .05$). However, the last digit in the probability must always be 1 or higher. The previous example, $p < .00023$, if reported to three decimals, should read $p < .001$.

Other numbers are frequently used mindlessly as well. In reviewing for a research journal, one of us encountered a study in which children were given a 12-week treatment. The author reported the mean age and standard deviation of the children before and after the 12-week treatment. Not surprisingly, the children had all aged 12 weeks. The author also calculated a t test between the pre- and the posttreatment means for age that was, of course, significant. That is, the fact that the children had aged 12 weeks during the 12-week period was a reliable finding.

Preparing Figures and Illustrations

Many suggestions about table construction also apply to figures and illustrations. A figure is often another way to present a table. Before using a table or a figure, ask, Does the

Table 20.4   Charts and Diagrams

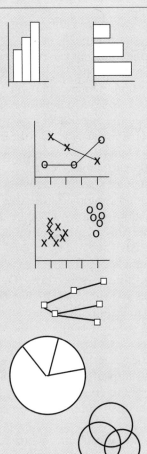

Bar and column charts—Bars (horizontal) are best for comparing amounts; arrange by size, small to large or large to small. Columns (vertical) are good for comparing amounts over time, especially if trends are evident. Shading may be used to distinguish or stack bars or columns.

Curve graph—best for showing change over time; time is horizontal with quantity vertical. Allows more than one curve to be compared. Sometimes area between curves can be shaded to show amount of change; shading, broken lines, symbols, or colored lines can be used to distinguish lines.

Dot graph—Shows patterns of individual scores; each dot represents a score on both the vertical and horizontal axis. Different dots or dot symbols can be used to distinguish groups.

Flow chart—Shows relationships in process; often useful for demonstrating steps in a process where more than one option exists (e.g., if yes, then this; if no, then this).

Pie chart—Circle for pie equals 100%; maximum of five or six segments. Best for showing proportions of segments. Order segments from large to small starting at 12 o'clock; highlight segments with shadings, making smallest segment the darkest.

Schematics—Relations between variables or concepts (e.g., overlap in two correlated variables).

Developed from White, J.V. (1984). *Using charts and graphs: 1000 ideas for visual persuasion.* New Providence, NJ: R.R. Bowker Co.

reader need the actual numbers, or is a picture of the results more useful? A more important question is, Do you need either? Can the data be presented more concisely and easily in the text? Figures and tables do not add scientific validity to your research report. In fact, they may only clutter the results. Day (1983, 56) suggested a reasonable means for deciding whether to use a table or a figure: "If the data show pronounced trends making an interesting picture, use a graph. If the numbers just sit there, with no exciting trend in evidence, a table should be satisfactory."

Several other considerations are important in preparing figures. Selection of the type of figure is somewhat arbitrary, but some distinctions make the choice of one type of figure more appropriate than another (see table 20.4). To evaluate whether you've used a figure appropriately, make sure it

- does not duplicate text,
- contains important information,
- does not have visual distractions,
- is easy to read,
- is easy to understand,
- is consistent with other figures in the text, and
- contains a way to evaluate the variability of the data (e.g., standard deviation bars or confidence intervals).

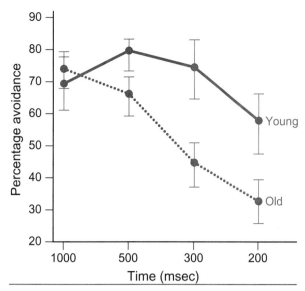

Figure 20.1   Appropriate use of a figure to depict an interaction.

Reprinted with permission from *Research Quarterly for Exercise and Sport,* Vol. 54, pp. 208-218, Copyright 1983 by the American Alliance for Health, Physical Education, Recreation and Dance, 1900 Association Drive, Reston, VA 20191.

Figures are useful for presenting interactions and data points that change over time (or across multiple trials). The dependent variable is placed on the *y* axis, and some independent or categorical variable on the *x* axis. If you have more than one independent variable, how do you decide which to put on the *x* axis? We have already partially answered that question. If time or multiple trials are used, put them on the *x* axis. For example, if a study found an interaction between the dependent variable by age level (7-, 9-, 11-, 13-, and 15-year-olds) and the treatment (experimental versus control), age with five levels is usually the more appropriate choice for the *x* axis. Note that this is a general rule; specific circumstances may dictate otherwise. A good example of the use of a figure to present an interaction is shown in figure 20.1. Note that both age and time are independent variables, so time is placed on the *x* axis.

Figure 20.2 is a good example of a useless figure. The results can be summarized in two sentences: "During acquisition, all three groups reduced their frequency of errors but did not differ significantly from one another. At retention, experimental group 2 further reduced its number of errors, whereas experimental group 1 and the controls remained at the same level." When results of different groups follow a similar pattern, a figure frequently appears cluttered. If figures appear cluttered because standard deviation bars or confidence intervals overlap, try putting bars from one group going upward on the graph and the bars from a group in close proximity downward.

Figures should include error bars representing the variability of each mean data point that is shown. The error bars can be either standard deviations for each mean value (e.g., figure 20.1) or 95% confidence intervals (CI) for each mean value. An advantage of using the 95% CI is that when the 95% CIs for mean values do not overlap on the x-axis at the

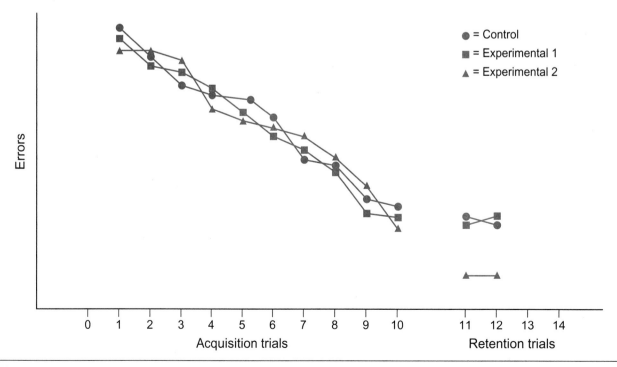

Figure 20.2   Example of a useless figure.

same point, the means are significantly different. This allows an effective interpretation of significant interactions when they appear as a figure. Standard errors should not be used as the representation of variability because they represent the variability of sample means about the true population mean, not variability in an individual sample. The most appropriate use of standard error is to calculate the 95% CI.

One final consideration is the construction of the y axis. In general, between 8 and 12 intervals encompassing the range of values is useful. Do not extend the y axis outside the range of values. This wastes space. Again, consider whether you need a figure at all. Sometimes theses and dissertations include examples such as that in figure 20.3. Looking at that figure, you immediately see a strong and significant interaction between knowledge of results (KR) and goal setting. Now look at the y axis, on which the dependent variable is shown. Note that the scores are given to the nearest hundredth of a second. Actually, there is less than a 500-ms difference among the four groups on a task in which average performance is about 18 s (18,000 ms). In fact, this interaction is not significant and clearly accounts

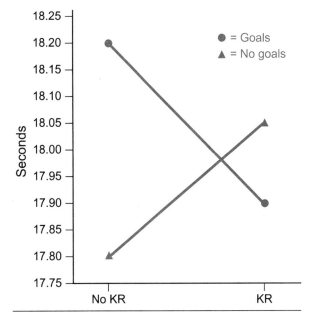

Figure 20.3 A nonsignificant interaction made to appear significant by the scale of the dependent variable.

for little variance. It should merely be reported as nonsignificant with no figure included. The researcher made the interaction appear important by the scale used on the y axis.

Sometimes multiple figures reporting data across several independent variables are used. For example, a comparison of running speed of Chinese, African, and U.S. boys and girls in five age groups might be reported. A single figure would be cluttered with all this data, so several figures (one for each continent of origin, each age, or each sex) might be used. The y axis on each figure should have the same 8 to 12 points for running speed. Otherwise, visual comparisons of the data are difficult.

With current computer technology and software, producing figures from data takes just a matter of seconds. Try different formats for figures to see which looks the best and displays the data most effectively. It is easy to put data in a bar graph and then a line graph and compare how the two display the data. Be creative; you can really improve your presentation of results by using tables and figures effectively.

Illustrations (photographs and line drawings) are also used in research reporting. Most frequently, illustrations are of experimental arrangements and equipment. They should not be used when the equipment is of a standard design or make; a brief description suffices. Any unusual arrangement or novel equipment should be described and either a picture or line drawing included. If specifications and relationships are important to include within the illustration rather than in the text, a line drawing is preferable because it can be more easily labeled.

Remember that tables, figures, and illustrations are generally appropriate for the results chapter but not for the discussion. An exception to this rule is a report of multiple experiments in which the general discussion section could contain a table or figure to display common findings or a summary of several experiments.

To review, when determining whether tables, figures, and illustrations should appear in the text or in the appendix, consider the following recommendations:

- Put important tables, figures, and illustrations in the text and all others in the appendix.
- Try not to clutter the results with too many tables, figures, and illustrations.
- Do not put summary tables for ANOVA and MANOVA in the results. Place the important statistics from these tables in the text, and put the tables in the appendix.

Remember that all journals have prescribed formats and styles for articles submitted to them (e.g., those of the American Psychological Association or American Physiological Society). The instructions for authors usually include directions for preparing tables and figures. Many of these decisions are arbitrary. Read and look at what you have written, then use common sense. Select the tables, figures, and illustrations that are needed to read and understand the results. Everything else goes into the appendix. The appropriate use of tables, figures, and illustrations can add to readers' interest, understanding, and motivation. It is a good idea to get a copy of the *Publication Manual of the American Psychological Association* (APA, 2001) or the style manual used in your department.

Summary

The research proposal describes the definition, scope, and significance of a problem and the methodology to be used to study it. The proposal is essentially the plan for the study. The introduction provides the background and literature on the problem, the problem statement, hypotheses, definitions, assumptions and limitations, and the significance of the study. The method section describes the participants, instrumentation, procedures, and design and analysis.

These two sections are presented by a student to his or her research committee as the plan for inquiry. The student's committee determines the worth of the study, suggests needed alterations, and ultimately must agree that the study should be done. The proposal must be carefully prepared with appropriate pilot work so that the committee is convinced that the student can complete the research plan.

Proposals to granting agencies are similar but are typically required to have specific lengths and formats. Students are advised to talk with outside agencies and more experienced grant writers before preparing and submitting proposals. Many colleges and universities offer internal grants for which graduate students can apply to support thesis and dissertation research.

The results and discussion sections are written after the data have been collected and analyzed. The results tell what you found; the discussion explains what the results mean. The results are the most important part of the research. They represent the unique findings of your study and your contribution to knowledge. The discussion ties the findings back to the literature review, theory, and empirical findings from other studies. Your findings should be interpreted in the discussion. In journal format, the results and discussion are sections in the body of the paper (more on this in chapter 21). In the traditional format, the results are chapter 3 and the discussion is chapter 4. However, in multiple-experiment studies, each experiment may best be reported in a separate section (or chapter) with its own brief introduction, method, results, and discussion. This is often followed by a section (or chapter) of general discussion of the multiple experiments.

Tables should be used in the discussion to summarize and present data when they are more effective than text presentation. Figures and illustrations are also used in the results, most frequently to demonstrate more dramatic findings. Careful construction of tables and figures is important in communicating information to the reader.

☑ Check Your Understanding

1. Find out the steps for writing a thesis or dissertation proposal at your school. List them in chronological order (e.g., select a major professor and committee, prepare a proposal, and get the proposal approved). Explain the process at each step.

2. Select a research report of interest from a refereed journal in your area of interest. Read the paper, but concentrate on the results and discussion. Answer the following questions in a brief report.

   a. Results:

      How is the results section organized?

      Compare the order of reported findings with the introduction, literature review, and statement of the problem. Do you see any relationships?

      How else might the results have been organized? Would this have been better or worse? Why?

b. Tables, figures, and illustrations:

Are there any? How many of each?

Why are they used? Could the data have been reported more easily in the text? When either a table or figure is used, would the other have been as good? Better? Why?

Can you restructure one of the tables or figures to improve it?

c. Discussion:

How is the discussion organized?

Are all results discussed?

Is the discussion accurate in terms of the results?

Are previous findings and theory woven into the discussion?

Are all conclusions and supporting evidence clearly presented?

Did the authors use any methodological cop-outs?

Ways of Reporting Research

He can compress the most words into the smallest idea of any man I ever met.—Abraham Lincoln

■　■　■　■　■　■　■

Our favorite author, Day (1983), provides an appropriate introduction to this chapter on ways of reporting research with the following quotation:

> **S**cientific research is not complete until the results have been published. Therefore, a scientific paper is an essential part of the research process. Therefore, the writing of an accurate, understandable paper is just as important as the research itself. Therefore, the words in the paper should be weighed as carefully as the reagents in the laboratory. Therefore, the scientists must know how to use words. Therefore, the education of a scientist is not complete until the ability to publish has been established. (p. 158)

In chapter 20 we covered the research proposal: how to write the introduction, the literature review, the problem statement, and the methodology for the thesis or dissertation. Then we explained how to organize and write the results and discussion sections. Effectively coordinating all this information into a thesis or a dissertation is the topic of concern in this chapter. We present both the journal style of organization, which we advocate (Thomas, Nelson, & Magill, 1986), and the traditional chapter style. In addition, we include information about writing for publication in research journals, preparing abstracts, and presenting papers orally (including in a poster format). After you've read the chapter, you might read volume 39 (August 1987) of *Quest*. This issue focuses on graduate programs in physical education, exercise science, and sport science. The entire issue is quite informative to potential master's and doctoral students. Articles of particular interest are by Newell (p. 88), Massengale (p. 97), Spirduso (p. 103), Thomas (p. 114), Spirduso and Lovett (p. 129), and King and Bandy (p. 153). Additional articles in this issue focus on specific types of graduate programs (e.g., teacher preparation and sport management).

Basic Writing Guidelines

Day's (1983, 125) rule is this: "Write your thesis to please your major professor, if you can figure out what turns him (or her) on." Once you understand this basic rule, we can offer some guidelines. But in case of doubt or conflict, return to that basic rule.

- Collect all the documents that outline university, graduate school, and departmental policy for theses and dissertations. Then actually read these documents, as someone at some level will eventually check to see whether you have followed them.

- Review the theses and dissertations of past graduate students whose work is well regarded by your institution. Identify common elements in their work, and pattern your work after theirs.

- Allow twice as much time as you think you will need. Remember Murphy's Law, "Whatever can go wrong will go wrong," and its special corollary, "When several things can go wrong, the one that will go wrong is the one that will cause the greatest harm."

The thesis and dissertation have basically the same parts found in any scientific paper: an introduction and a literature review, method, results, and discussion sections. In the traditional format, each of these parts becomes a chapter. Sometimes the introduction and the literature review are separate chapters, leading to five rather than four chapters. This format may vary in historical, philosophic, and qualitative papers (see chapters 12, 13, or 19) or multiple experiments (see chapter 20) or when your major professor says it should.

A Brief Word About Acknowledgments

acknowledgments

Section of a scholarly paper that credits individuals important to the development of the work.

A section seldom mentioned in articles and books on preparing theses and dissertations is the **acknowledgments.** You should acknowledge those people without whom the research would have been impossible. However, acknowledgments as prepared by graduate students take some odd turns. We saw one in which a woman acknowledged her ex-husband. She said that if he had not been so difficult to live with, she would never have gone back to the university and done graduate work. By doing that she found an area in which she had a tremendous interest, and this represented a significant change in her life. That is about the only positive statement we have ever heard about ex-spouses. Some other funny acknowledgments follow:

- My parents, husband, and children provided inspiration and support throughout, but I was able to complete my thesis anyway.

- My major professor, Dr. I.M. Published, coordinated the work and made an occasional contribution.

- Professor B.A. Snobb wants everyone to know he had nothing to do with this thesis.

- Finally, we would like to thank our proofreader, I.D. Best, without whose hell; thes document wld net be pausible.

More seriously, you can acknowledge appropriate individuals, but keep the list short and to the point and do not be mushy. As in other areas, use correct English. In their acknowledgments, graduate students often use "wish" when they mean "want," as in "I wish to acknowledge I.B.A. Fink." Does that mean they might have acknowledged him if his contribution had not been so lousy? The graduate student really wants to acknowledge him. But more appropriately, why not just say it directly: "I acknowledge I.B.A. Fink."

Thesis and Dissertation Format: Traditional Versus Journal

We devote considerable space here to the journal format for thesis and dissertation writing. Our material is partially reproduced from an article two of us wrote with a colleague (Thomas, Nelson, & Magill, 1986). We acknowledge the contribution of Dr. Richard Magill from Louisiana State University, and we appreciate *Quest* (published by Human Kinetics) for allowing us to adapt the article for this book.

Graduate education in the United States, especially at the doctoral level, was modeled after graduate education in Germany. The German university's spirit of the search for knowledge and the concomitant emphasis on productive research were transplanted in large measure in America (Rudy, 1962). Although there have been some changes over the years in the requirements for the doctoral degree, the basic aims and expectations have in essence remained unaltered. The doctoral degree is conferred in recognition of a candidate's scholarship and research accomplishments in a specific field of learning through an original contribution of significant knowledge and ideas (Boyer, 1973).

Stated more simply, research is the foundation of the doctoral program, and the dissertation is the most distinctive aspect of the doctoral degree. It has been reported that the dissertation occupies an average of 39% of the time devoted to obtaining the degree in the fields of biochemistry, electrical engineering, psychology, physics, sociology, and zoology (Porter, et. al., 1982).

A Case for the Journal Format

The major purpose of the thesis and the dissertation is to contribute new knowledge with scientific merit (American Psychological Association, 1959; Berelson, 1960; Porter & Wolfe, 1975). A purpose typically cited in university bulletins is that the thesis or dissertation provides evidence of competency in planning, conducting, and reporting research. In terms of program objectives, the study is a valuable learning experience in that theses and dissertations are functional exercises in executing the steps in the scientific method of problem solving. Even "ABDs" ("all-but-dones," or people who have completed all work for the PhD except the dissertation) acknowledge the contribution that dissertations make to science and the scientific method (Jacks, et. al., 1983).

The importance of the dissemination of research results as an integral part of the research process is well established. It follows that one of the purposes of the thesis or dissertation is to serve as a vehicle to carry the results of independent investigation undertaken by the graduate student. Thus, the dissertation or thesis becomes part of the dissemination process. A.L. Porter et al. (1982) reported that dissertations do make substantial contributions to the knowledge base and that authors' published dissertations are cited more often than other papers that they write.

Despite the potential contribution that the thesis and dissertation can make to a field of study, the fact remains that only about one third to one half of the dissertations (and even fewer theses) become available to the profession through publications (McPhie,

She forgot her PowerPoint disk.

1960; Porter et al., 1982). There are, of course, several reasons why a thesis or dissertation is never published. For one, despite the emphasis placed on research by the institutions, many students do not consider research an important aim. A.L. Porter et al. (1982) reported that 24% of the doctoral recipients surveyed expressed this feeling. Arlin (1977) went so far as to claim that most educators never do another piece of published research after they complete the master's thesis or doctoral dissertation. In addition, job placements fix varying degrees of importance on research and publication. There is also the inescapable fact that not all theses and dissertations are worthy of publication.

Conceivably, another contributing factor in the low rate of publication is the traditional style and format often used in the thesis or dissertation. Granted, the highly motivated new PhD recipient will spend the time and effort to rewrite the dissertation into the proper format for journal review, but the less motivated PhD recipient may not. Why should dissertations and theses be written in a format that requires rewriting before publication if a vital part of the research process is publication of the research study (Day, 1983)?

We contend that the traditional format of dissertations and theses is archaic. Doctoral students (especially those working with productive major professors) may have several scholarly refereed publications by the time the PhD is awarded. Should such people, who have shown that their previous work has scientific merit, be required to go through the ritual of writing a dissertation involving separate chapters that spell out every detail of the research process? It seems more logical to us to have the body of the dissertation prepared in the appropriate format and style for submission to a journal, which is the acceptable model for communicating results of research and scholarly works in many of the arts, sciences, and professions. We suggest restructuring the thesis and dissertation format to one that accomplishes this purpose and we explain the contents of the various parts.

Limitations of Chapter Style

Conventional theses and dissertations typically contain four or five chapters. Traditionally, the chapters are intended to reflect the scientific method for solving problems: developing the problem and formulating the hypotheses, gathering the data, and analyzing and interpreting the results. These steps are usually embodied in chapters such as the introduction (which sometimes contains the literature review; at other times the literature review is a separate chapter), method, results, and discussion chapters.

The thesis or dissertation also has several introductory pages as prescribed by the institution, usually consisting of the title page, acknowledgments, abstract, table of contents, and lists of figures and tables. At the end of the study are the references and one or more appendixes that contain items such as participant consent forms, tabular materials not presented in the text, more detailed descriptions of procedure, instructions to participants, and raw data. A brief biographical sketch (the curriculum vitae) is usually the last entry in the conventional thesis or dissertation.

The chapter format is, of course, steeped in scholastic tradition. In defense of this tradition, the discipline required to accomplish the steps involved in the scientific method is usually viewed as an educationally beneficial experience. Moreover, for the master's student, the thesis is normally the first research effort, and there may be merit in formally addressing such steps as operationally defining terms, delimiting the study, stating the basic assumptions, and justifying the significance of the study.

A more serious limitation of the chapter style relates to the dissemination of the results of the study, that is, publishing the manuscript in a research journal. Considerable rewriting is usually required to publish a thesis or dissertation because journal formats differ from those of theses and dissertations prepared in chapter format. Granted, the information for the journal article is provided in the thesis and the dissertation, but a considerable amount of deleting, reorganizing, and consolidating is necessary to transform the study to journal format.

The student usually would like to publish this product of months (or years) of time and painstaking effort. But in terms of expediency, the chapter format is decidedly counterproductive. The rewriting required may be made more difficult by the fact that the typical new PhD recipient immediately begins a new job that demands considerable time and energy. Unless the new PhD recipient is motivated, the transformation of the conventional dissertation style to journal format may be delayed, sometimes indefinitely. Porter et al. (1982) claimed that new PhD recipients who fail to publish within two years after the awarding of the degree are unlikely to publish later.

The master's thesis is even more unlikely to be published. One reason is that the master's student generally does not consider publishing unless the major professor suggests it. Furthermore, the master's student may not be as well prepared to write for publication as the doctoral student. Thus, the burden for publication is on the major professor, who understandably is often unwilling to spend the additional time necessary to supervise (or do) the rewriting. Thus, most theses are not submitted for publication. Ironically, the time-honored, scholarly, chapter style of the thesis and dissertation actually impedes an integral part of the research process: the dissemination of results.

Structure of the Journal Format

To develop a better model for theses and dissertations, the limitations of the chapter format for reporting must be overcome while maintaining the contents of a complete research report. The format we suggest has three major parts. Preliminary materials include such items as the title page, acknowledgments, and abstract. The body of the thesis and dissertation is a complete manuscript prepared in journal form. Included are the standard parts of a research report, such as the introduction, method, results, discussion, references, figures, and tables. The appendixes often include a more thorough literature review, additional detail about method, and additional results not placed in the body of the thesis or dissertation. However, the journal format, as we describe it, is more appropriate for reporting descriptive (e.g., surveys and correlational studies) and experimental research. The format requires some adjustments for reporting analytical research (e.g., historical and philosophic studies and meta-analyses) and qualitative studies.

Thus, these are the parts of the thesis and dissertation using the journal format:

1.0 Preliminary materials

1.1 Title Page

1.2 Acknowledgments

1.3 Abstract

1.4 Table of contents

1.5 List of tables

1.6 List of figures

2.0 Body of the thesis or dissertation

2.1 Introduction

2.2 Method

2.3 Results

2.4 Discussion

2.5 References

2.6 Tables

2.7 Figures

3.0 Appendixes

3.1 Extended literature review

3.2 Additional methodology

3.3 Additional results

3.4 Other additional materials

4.0 One-page curriculum vitae

How can this format overcome the limitations of the chapter style? For both master's and doctoral students, a manuscript (body of the thesis or dissertation) is developed that is ready for journal submission. All that remains is to add the title page and abstract, and the paper can be sent to a suitable journal.

The advantage of this format for doctoral students should be apparent. Because PhD recipients who fail to publish their dissertations within two years are unlikely to publish afterward, a more functional format encourages publication. Especially when we consider that dissertations appear to make important contributions to knowledge, the evaluation and subsequent publication of that knowledge through refereed journals is an important step to accomplish. Although master's theses are not as likely as dissertations to be published, any format that encourages the publication of quality thesis work is desirable.

One final point before proceeding to the structure of the journal format: Your graduate school probably requires that the thesis or dissertation follow a standard style manual (or at least the style of a journal). The three most common are the *Publication Manual of the American Psychological Association, American Physiological Society,* and *The Chicago Manual of Style* (University of Chicago Press, 1993). This format adapts nicely to any of these styles. University regulations usually do not specify a particular style, but frequently an academic department may adopt one or two specific styles. If the journal format is to be used, a department might want to allow more than one style. For example, many journals reporting exercise physiology and biomechanical studies use *American Physiological Society*. Journals publishing articles in motor behavior, sport psychology and sociology, and professional preparation frequently use the APA manual. Journals that publish articles in history and philosophy of sport frequently use *The Chicago Manual of Style*. It benefits graduate students considerably to have the flexibility to choose the style recommended by the journal to which the paper will be submitted.

Preliminary Materials

Most of the information in the preliminary materials is required by the institution and usually appears at the beginning of the thesis or dissertation. The exact nature of this material is often specified by an institution's graduate school, so make sure you have the necessary information. Generally included are the title page, committee approval form, acknowledgments, dedication, abstract, table of contents, list of tables, and list of figures. Many institutions require that the abstract follow the form for *Dissertation Abstracts International,* which sets a 300-word maximum length for the abstract. We recommend a slightly different length for the abstract. Journals (and the style manuals mentioned previously) typically require abstracts between 100 and 150 words. If the thesis or dissertation writer keeps the abstract between 100 and 150 words, it can serve both requirements.

Body of the Thesis or Dissertation

This section should be a complete research report using the appropriate style for the journal to which it is to be submitted (or the style required by the department in which the work was done). Included are the introduction and literature review, method, results, discussion, references, tables, and figures. The author should keep the length within the bounds set by the journal or the institution. Journals typically set a limit of 20 to 25 pages for single experiments. Most journals allow additional pages for multiple experiments or unusually complex articles. The student and major professor must be rigorous in keeping the paper within a length that a journal will consider. In our experience, maintaining an acceptable length is the most difficult goal to achieve using this format. If the dissertation has multiple parts and will be published as more than one paper, you would include two or more reports in this section.

The thesis or dissertation author should consult several sources in preparing the body of the paper. One source is the journal to which submission is anticipated. Guidelines to authors and instructions regarding the appropriate style manual are typically published in journals (e.g., see each issue of *Research Quarterly for Exercise and Sport*). The author should read a number of similar papers in the selected journal to see how specific issues are handled (tables, figures, unusual citations, multiple experiments). In addition, the thesis or dissertation author should carefully follow the style manual that has been selected.

Of course, the ultimate goal of this part of the thesis or dissertation is to enable the paper to be submitted to a journal as soon as possible. Improved format and better writing will not aid poor-quality research. On the other hand, high-quality research can easily be hidden by a poor format that makes rewriting difficult, poor reporting that omits important information, or weak and boring writing that makes reading tedious. "Thus the scientist must not only 'do' science; he (or she) must 'write' science. Although good writing does not lead to the publication of bad science, bad writing can and often does prevent or delay the publication of good science" (Day, 1983, p. x). Research is not recognized in any formal sense as having been conducted until it has been shared with and evaluated by the academic community. Additionally, there is no law saying that the thesis or the dissertation must be written in such a way that the reader has difficulty staying awake.

Appendixes

In the conventional format, the appendixes serve primarily as a depository for nonessential information. To a degree, this still characterizes what should be put in the appendixes when the journal format is used, but some additional features give them unique worth. Usually, the number and content of the appendixes are determined by the collective agreement of the student, the advisor, and the supervisory committee. In the journal format, we suggest four types of appendixes. Other appendixes could be included, but we see these four as essential.

Each appendix should begin with a description of what is in it and how that information relates to the body of the thesis or dissertation. This enables the reader to get the most use of the information in the appendix.

More Thorough Literature Review

A very important and useful appendix is a review of literature. The introduction in the body of the thesis or dissertation includes a discussion of related research but presents only a minimal amount of information to establish an appropriate background for the one or more studies that follow. One purpose of the thesis or dissertation is to allow the student an opportunity to demonstrate knowledge of the research literature related to the topic. In the body of the journal format, there is limited opportunity for this demonstration because journals typically have concise introductions. A comprehensive literature review should be included as the first appendix to provide an appropriate mechanism for the student to demonstrate knowledge of the relevant literature. We recommend that this review also be written in journal style.

At least two additional purposes are served by including this literature review as an appendix. First, it makes this information available to the student's committee members, some of whom may not be sufficiently familiar with the literature related to the thesis or dissertation. Second, if properly developed, the literature review can go directly to a journal for publication.

The literature review contained in the appendixes can take several forms, the most popular of which is probably the comprehensive narrative that synthesizes and evaluates research. This review links various studies and establishes a strong foundation on which to build the research. It also reveals how the research represented by the thesis or dissertation extends the existing body of knowledge related to the student's research topic.

A second form that the literature review can take is a meta-analysis: a quantitative literature review that synthesizes previous research by analyzing results of many research studies using specified statistical methods (see chapter 14). An example of a meta-analysis related to the study of physical activity can be seen in an article by Thomas and French (1985).

Additional Method Information

A second important appendix presents additional methodological information not included in the body of the thesis or dissertation. Journals encourage authors to provide methodology information that is brief yet sufficient to describe essential details related to the participants, apparatus, and procedures used in the research. The disadvantage of journal articles is that often information sufficient to enable someone to replicate the study is not provided. In the journal format for the thesis or dissertation, this useful additional information becomes an appendix. Information suitable for this section includes more detailed participant characteristics; more comprehensive experimental design information; fuller descriptions (and perhaps photographs) of tests, testing apparatus, or interview protocol; copies of tests, inventories, or questionnaires; and specific instructions given to the participants.

Additional Results Information

The third appendix should include information that is not essential for inclusion in the results section of the body of the thesis or dissertation. Journal editors typically want only summary statements about the analyses and a minimum number of figures and tables. Thus, considerable material concerning the results can be placed in the appendix, such as all means and standard deviations not included in the results section, ANOVA tables, multiple-correlation tables, validity and reliability information, and additional tables and figures.

This information serves several purposes. First, it provides evidence to your committee and advisor that the data were accurately described and analyses properly done. Second, the committee members are given an opportunity to evaluate the statistical analyses and interpretations from the body of the thesis or dissertation. Third, other researchers are provided more detailed data and statistical information should they want it. For example, a researcher may want to include this thesis or dissertation in a meta-analysis. Because the additional results presented in the appendix may have many future uses, their importance to the total thesis or dissertation cannot be overestimated.

Other Additional Materials

A fourth appendix includes information that is not appropriate for the body of the thesis or dissertation or the other appendixes, such as the approval form from the human research committee, individual participants' consent forms, sample data-recording forms, and perhaps the raw data from each participant. Also, detailed descriptions of any pilot work done before the study could be included in this or a separate appendix.

The appendixes in the journal format of a thesis or dissertation provide a mechanism for including information not contained in the journal article that may be significant for the student, the student's committee, and other researchers. They also provide a means for elaborating on some information in the body of the thesis or dissertation. Moreover, the comprehensive literature review provides the student with another possible publication. As a result, the appendixes become meaningful components of the total thesis or dissertation.

One-Page Curriculum Vitae

Many colleges and universities request that the last page of the thesis or dissertation be a one-page curriculum vitae about the student. This should be a professional version of a curriculum vitae and include items such as education and previous work experience.

Examples of Theses and Dissertations Using Journal Format

Numerous theses (e.g., Boorman, 1990; Tinberg, 1993) and dissertations (e.g., French, 1985; Lee, 1982; McPherson, 1987; Yan, 1996) use journal format. These theses and dissertations provide excellent examples of the format discussed here.

The information in an unpublished thesis or dissertation remains in the exclusive domain of a few individuals. We established earlier that a vital part of the research process is the dissemination of knowledge. We maintain that style and format should not impede this process. The proposed journal format accomplishes the traditional objectives of the thesis or dissertation yet facilitates the dissemination of knowledge.

Helpful Hints for Successful Journal Writing

We highly recommend the third edition of a work to which we have frequently referred: *How to Write and Publish a Scientific Paper* (Day, 1998). In our opinion, this book is the researcher's best resource for preparing a paper for submission to a research journal. The book is short, informative, readable, and funny. Although the small section here cannot replace Day's more thorough treatment, we do offer a few suggestions.

First, decide to which journal the research paper is to be submitted. Carefully read its guidelines (select a recent issue, as the guidelines may have changed from older issues) and follow the recommended procedures. Guidelines usually explain the journal's publication style; how to prepare tables, figures, and illustrations; where to submit papers (and how many copies); acceptable lengths; and sometimes estimated review time. Nearly all journals require that manuscripts not be submitted elsewhere simultaneously. To do so is unethical (see chapter 5).

Journals follow a standard procedure for papers submitted. One of us (Thomas) was editor-in-chief of *Research Quarterly for Exercise and Sport* for six years (1983–1989) and another (Silverman) for three years (2002-2005). Here is what happens to a manuscript between submission to the editor and return to the author (the average time of this process for *Research Quarterly for Exercise and Sport* is 75–90 days):

1. The editor checks to see whether the paper is within the scope of the journal, is of appropriate length, and uses correct style. If any of these characteristics are inappropriate, it may be returned to the author.

2. The editor looks through the paper to determine that all appropriate materials are included (e.g., abstract, tables, and figures).

3. The editor reads the abstract, looks at the keywords, and evaluates the reference list to identify potential reviewers.

4. Depending on the journal's size, there may be section editors (*Research Quarterly for Exercise and Sport* has 10 sections: biomechanics, pedagogy, physiology, psychology, etc.) or a board of associate editors. The editor places the paper in the appropriate section or with an appropriate associate editor.

5. The editor (sometimes in consultation with section or associate editors) assigns reviewers (usually two or three).

6. The reviewers and section editor (or associate editor) receive copies of the paper with review forms. They have a specific date by which they should complete the review.

7. The reviewers mail (or e-mail) their reviews to the section editor (or associate editor), who evaluates the paper and the reviews and makes a recommendation to the editor.

8. The editor reads the paper, the reviews, and the evaluations and writes the author concerning the paper's status. Usually the editor outlines the major reasons for the decision.

Smaller, more narrowly focused journals may not have section or associate editors, so reviews go directly to the editor. Publication decisions usually fall into three general categories (some journals have additional subcategories):

- Acceptable (sometimes with varying degrees of revision)
- Unacceptable (sometimes called rejected without bias), which means that the editor will have the paper evaluated again if it is revised
- Rejected, which usually means that the journal will no longer consider the paper

Review criteria are fairly common and are similar to those presented in chapter 2 for a review of published literature. Rates of published to rejected papers vary by the quality of the journal and the area of the paper and are frequently available in an issue of the journal.

Research Quarterly for Exercise and Sport accepts about 20% of the papers submitted (see each June issue for a more detailed description).

If this is your first paper, seek some advice from a more experienced author, such as your major professor or another faculty member. Often papers are rejected because they fail to provide important information. More experienced authors pick up on this immediately. Journal editors and reviewers do not have time to teach you good scientific reporting. Not everyone is fond of journal editors. One wag was reported to have said, "Editors have only one good characteristic—if they can understand a scientific paper, anyone can!" Another said, "Editors are, in my opinion, a low form of life—inferior to the viruses and only slightly above academic deans" (Day, 1983, 80). Teaching good reporting is the responsibility of your major professor, research methods courses, and books such as this one, but it is your responsibility to acquire the skill.

Do not be discouraged if your paper is rejected. Every researcher has had papers rejected. Carefully evaluate the reviews and determine whether the paper can be salvaged. If it can, then rewrite it, taking into account the reviewers' and the editor's criticisms, and submit it to another journal. Do not send the paper out to another journal without evaluating the reviews and making appropriate revisions. Often another journal uses the same reviewers, who will not be happy if you didn't take their original advice. If your research and paper cannot be salvaged, learn from your mistakes.

Reviewers and editors will not always agree in what they say about your paper. This is frustrating because sometimes they tell you conflicting things. Be aware that editors select reviewers for different reasons. For example, in a sport psychology study that develops a questionnaire, the editor may decide that one or more content reviewers and a statistical reviewer are needed to evaluate the work. You would not expect the content and statistical reviewers to discuss the same issues. Studies have been done of the peer review system for scholarly journals. For example, Morrow and colleagues (1992) reported that the *Research Quarterly for Exercise and Sport* had an average reviewer agreement of .37 based on 363 manuscripts between 1987 and 1991. They reported that other behavioral journals had reviewer agreements from a high of .70 in the *American Psychologist* to a low range of .17 to .40 for a group of medical journals. Whether reviewer agreement is to be expected is debatable. For a good overview of the peer review process, see *Behavioral and Brain Sciences, 14*(1), 1991. Our advice to you as a graduate student is to seek help from your advisor and other faculty members in making revisions when you receive reviews of a paper you have submitted.

Occasionally, you may feel that you have received an unfair review. If so, write back to the editor, point out the biases, and ask for another assessment by a different reviewer. Editors are generally open to this type of correspondence if it is handled in a professional way. However, opening with "Look what you and the stupid reviewers said" is unlikely to achieve success (for a humorous example, see page 399). Recognize that appealing to the editor is less likely to be successful if two or more reviewers have agreed on the important criticisms. Editors cannot be experts in every area, and they must rely on reviewers. If you find yourself resorting to this tactic very often, the problem is likely in your work. You do not want to end up like Snoopy in *Peanuts* (by Charles Schulz), who received this note from the journal to which he had submitted a manuscript: "Dear Contributor, We are returning your manuscript. It does not suit our present needs. P.S. We note that you sent your story by first-class mail. Junk mail may be sent third-class."

As a final point in this part, we have noted the infrequency with which most scientific papers are read (of course, none of ours fall into that category). Some writers have speculated that only two to four people read the average paper completely.

Writing Abstracts

Whether you are writing for a journal, using the chapter or journal format for preparing a thesis or dissertation, or are readying a conference paper, you need to include an abstract.

A Letter We've All Wanted to Write

Dear Sir, Madam, or Other:

Enclosed is our latest version of MS #85-02-22-RRRR; that is, the re-re-re-revised revision of our paper. Choke on it. We have again rewritten the entire manuscript from start to finish. We even changed the #@*! running head! Hopefully we have suffered enough by now to satisfy even you and your bloodthirsty reviewers.

I shall skip the usual point-by-point description of every single change we made in response to the critiques. After all, it is fairly clear that your reviewers are less interested in details of scientific procedure than in working out their personality problems and frustrations by seeking some kind of demented glee in the sadistic and arbitrary exercise of tyrannical power over hapless authors like ourselves who happen to fall into their clutches. We do understand that, in view of the misanthropic psychopaths you have on your editorial board, you need to keep sending them papers, for if they weren't reviewing manuscripts they'd probably be out mugging old people or clubbing baby seals to death.

Some of the reviewer's comments we couldn't do anything about. For example if (as reviewer C suggested) several of my recent ancestors were indeed drawn from other species, it is too late to change that. Other suggestions were implemented, however, and the paper has improved and benefited. Thus, you suggested that we shorten the manuscript by 5 pages, and we were able to accomplish this very effectively by altering the margins and printing the paper in a different font with a smaller typeface. We agree with you that the paper is much better this way.

Our perplexing problem was dealing with suggestions 13 through 28 by reviewer B. As you may recall (that is, if you even bother reading the reviews before writing your decision letter), that reviewer listed 16 works that he/she felt we should cite in this paper. These were on a variety of different topics, none of which had any relevance to our work that we could see. Indeed, one was an essay on the Spanish-American War from a high school literary magazine. The only common thread was that all 16 were by the same author, presumably someone whom reviewer B greatly admires and feels should be more widely cited. To handle this, we have modified the Introduction and added, after the review of relevant literature, a subsection entitled "Review of Irrelevant Literature" that discussed these articles and also duly addresses some of the more asinine suggestions in the other reviews. We hope that you will be pleased with this revision and will finally recognize how deserving of publication this work is. If not, then you are an unscrupulous, depraved monster with no shred of human decency. You ought to be in a cage. If you do accept it, however, we wish to thank you for your patience and wisdom throughout this process and to express our appreciation of your scholarly insights. To repay you, we would be happy to review some manuscripts for you; please send us the next manuscript that any of these reviewers submits to your journal.

Assuming you accept this paper, we would also like to add a footnote acknowledging your help with this manuscript and pointing out that we liked the paper much better the way we originally wrote it, but that you held the editorial shotgun to our heads and forced us to chop, reshuffle, restate, hedge, expand, shorten, and in general convert a meaty paper into stir-fried vegetables. We couldn't or wouldn't have done it without your input.

Sincerely,

Trying to Publish but Perishing Anyway

Adapted from the Canadian Society for Psychomotor Learning and Sport Psychology, 1991, "A letter we've all wanted to write," *Bulletin of the Canadian Society for Psychomotor Learning and Sport Psychology.*

Abstracts for each purpose require slightly different orientations, but nearly all have constraints on length and form.

Thesis or Dissertation Abstracts

The abstract for your thesis or dissertation probably has several specific constraints, including constraints on length, form, style, and location. First, consult your university or graduate school regulations, then follow these carefully. For dissertations, the graduate school regulations nearly always include the form for submission to *Dissertation Abstracts International*. The exact headings, length, and margins are provided in a handout available from your graduate school office. In writing the abstract, consider who will read it. It will be located in computer searches by the title and keywords. (The importance of these was discussed in chapters 2 and 3.) Write the abstract so that anyone reading it can decide whether to look at the total thesis or dissertation. Clearly identify the theoretical framework, the problem, the participants, the measurements, and your findings. Do not use all the space writing about your sophisticated statistical analyses or minor methodological problems. Keep jargon to a minimum. People in related areas read the abstract to see whether your work is relevant to theirs. One of us ran across the following sentence in a dissertation abstract recently. Due to jargon (and, some would argue, our poor vocabulary), we could not interpret the sentence:

> Amalgamating the decision maker's inputs is a new and unique decision model that can be classified as a fuzzily parameered, multi-staged, forward and backward chained, Displaced Ideal, two-dimensional attribute model of business-level strategic objectives and the functional-level strategies that realize those objectives.

Abstracts for Published Papers

An abstract for a published paper is much shorter than one for a dissertation, usually between 100 and 150 words. The important consideration is to get to the point: What was the problem? Who were the participants? How did you carry out the research? What did you find? The most useless statement encountered in these abstracts is "Results were discussed." Would anyone expect that results would not be discussed?

Conference Abstracts

Abstracts for conferences are slightly different. Usually you are allowed a little more space because the reviewers must be convinced to accept your paper for presentation. In these abstracts, you should follow these procedures:

1. Write a short introduction to set up the problem statement.
2. State the problem.
3. Describe the methodology briefly, including
   a. participants,
   b. instrumentation,
   c. procedures, and
   d. design and analysis.
4. Summarize the results.
5. Explain why the results are important.

In the sidebar on p. 401 is a humorous example of an abstract developed by one of your authors (J.R. Thomas, 1989).

The results and their importance are a critical part of a conference abstract. If you do this in a nondescript way, the reviewer may conclude that you have not completed the study. This generally leads to rejection. The conference planners cannot turn down other completed research when it is possible that yours will not be finished.

Writing an Abstract

Please use the following style for all abstracts.

I.M. Tenured and U.R. Promoted, JR's South Fork School of Hard Knocks, Dallas, TX 00001

How Motor Skill Research Can Get You a Merit Raise

Current research suggests that merit raises are directly related to the number of papers faculty present and publish and inversely related to the quality of the research . . .

. . . The long-term effect is to increase the number of journals, conferences, and full professors.

Finally, most conferences require that the paper be presented before publication. So if you have a paper in review at a journal, submitting it to a conference that is 8 to 12 months away could be hazardous. Also, many conferences require that the paper not have been previously presented. Be aware of this, and follow regulations. Violating these guidelines will never enhance your professional status, and other scholars will quickly become aware of it.

Making Oral and Poster Presentations

Once your conference paper is accepted, you are faced with presenting it. The presentation is conducted either orally or in a poster session.

How to Give Oral Presentations

Oral presentations usually cause panic among graduate students and new faculty members. There is no way to get over this except to give several papers. But you can help assuage the apprehension. Usually, the time allowed for oral research reports is 10 to 20 minutes, depending on the conference. You will be notified of the time limit when your paper is accepted. Because you must stay within the time limit and there is no way to present a complete report within this limit, what do you do? We suggest that you present the essential features of the report using the following divisions for a 15-min presentation. Visual aids, such as slides, computer presentation software, or overheads, are helpful.

oral presentations

Method of presenting a paper in which the author speaks before a group of colleagues at a conference following this format: introduction, statement of the problem, method, results, discussion, questions.

- Introduction that cites a few important studies: 3 min
- Statement of the problem: 1 min

Don't Use These Quotations From British Sport Commentators in Your Presentation

"Moses Kiptanul—the 19-year-old Kenyan who turned 20 a few weeks ago."

"We now have exactly the same situation we had at the start of the race, only exactly the opposite."

"He's never had major knee surgery on any other part of his body."

"She's not Ben Johnson—but then who is?"

"I owe a lot to my parents, especially my mother and father."

"The Port Elizabeth ground is more of a circle than an oval. It's long and square."

"The racecourse is as level as a billiard ball."

"Watch the time—it gives you an indication of how fast they are running."

"That's inches away from being a millimeter perfect."

"If history repeats itself, I think we can expect the same thing again."

- Method: 3 min
  - Results (present tables and figures; figures are usually better because tables are hard to read): 3 min
  - Discussion (main points): 2 min
  - Questions and discussions: 3 min
  - Total: 15 min

The most frequent errors in oral presentations are spending too much time on method and presenting results poorly (see the box on page 401 for poorly presented statements). Proper use of visual aids is the key to an effective presentation, particularly in the results. Place a brief statement of the problem on a slide, and show it while you talk. A slide of the experimental arrangements reduces much of the verbiage in method. Always use slides to illustrate the results. Pictures of the results (particularly figures and graphs) are much more effective than either tables or a verbal presentation. Keep the figures and graphs simple and concise. Tables and figures prepared for a written paper are seldom effective when converted for visual presentations. They typically contain too much information and are not large enough. Have a pointer available to indicate significant features. Finally, remember our Four Laws of Oral Presentations (below). Practicing your presentation is one solution to these and many other problems. At our universities, we gather graduate students and faculty who are presenting papers at upcoming conferences and conduct practice sessions. Everyone presents her or his paper and has it timed. Then the audience asks questions and offers suggestions to clarify presentations and visual aids. Practice sessions improve the quality of the graduate students' presentations and strengthen their confidence.

Thomas, Nelson, and Silverman's Four Laws of Oral Presentations

1. Something always goes wrong with the LCD projector. More specifically,
   a. unless you check it beforehand, it will not work;
   b. either the electrical cord or the mouse cord (or both) will be too short;
   c. if the projector has worked perfectly for the three previous presenters, the bulb will burn out during your presentation.
2. The screen will be too small for the room.
3. Your paper will be the last one scheduled during the conference. This will result in only the moderator, you, and the previous presenter being there, the latter of whom will leave on completing his or her paper. Or conversely, your paper will be scheduled at 7:30 A.M., at which time you will be hungover and everyone else will still be in bed.
4. At your first presentation, the most prestigious scholar in your field will show up, be misquoted by you, and ask you a question.

Using a Poster Presentation to Its Best Advantage

The poster session is another way to present a conference paper. Poster sessions take place in a large room in which presenters place summaries of their research on the wall or on poster stands. The session is scheduled for a specific period, during which presenters stand by their work while those interested walk around, read the material, and discuss items of interest with the presenters.

We prefer this format to oral presentations. The audience can look at the papers in which they are interested and have more detailed discussions with the authors. Within 75 minutes, 15 to 40 poster presentations can be made available in a large room, whereas only five 15-minute oral presentations can take place in same amount of time. In addition, the audience must sit through several papers, often losing interest or creating a disturbance by arriving or leaving.

Posters allow the audience to pace the way they look at material. We have often heard verbal presentations where we wanted more information or we wished a table or figure presented were left up longer—*that is not an issue in poster sessions*. Participants are more likely to ask questions or comment on points in a poster session where they have more time to consider the material. This often leads to longer and more in-depth conversations that benefit both the presenter and the spectator.

Designing the Poster

Good poster design will maximize all the benefits that poster sessions offer. The poster should be arranged to effectively highlight the important points and eliminate unneeded information. Posters are typically mounted on walls or portable stands. Usually the conference will notify you about how much space you have, but a 4' × 6' space is fairly common. The poster should be arranged in 5 or 6 panels representing the parts of the poster presentation—abstract, introduction and problem statement, methods, results (including figures and tables), discussion, and references. Placing the sections on panels with headings helps readers follow the poster and provides some "white" space that is pleasing to the eye. (If you print the poster on one large sheet, these suggestions still apply.) In figure 21.1, you can see a logical way to arrange the panels using the vertical flow of information (horizontal flow can also be used, but vertical flow is usually preferred). It is often valuable to place tables and figures at the center of the poster and arrange panels around them.

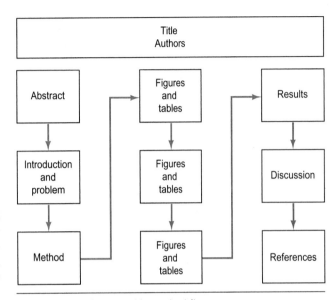

Figure 21.1   Poster with vertical flow.

The tables and figures represent what you have found and should be the central feature of the poster.

Posters will have text but this text should have specific features to increase how easily it can be read and understood:

- Short sentences with simple structure
- To the point text, no filler information
- Use of common words with little jargon
- Little abstraction
- Use of lists with bullets

Design elements that improve posters and make them more understandable include:

- Select a plain and clear typeface (common ones includes Helvetica and Times).
- Use a font size of at least 20 points for text, 28–32 for headings, and 64 for titles.
- Most current evidence suggests posters that use both upper and lower case are more readable.
- Major headings on each panel with section headings inside the panel guide the eye in reading the poster.
- Within sections, use left justification and ragged right rather than fully justified.
- If there is a reasonable choice between a figure and a table, use the figure; if you must use a table, put in minimum information, use large numbers, and use labeling (e.g., shading, colors) to guide the reader to important information; *a table or figure prepared for publication is seldom acceptable for a poster (even when enlarged)*.

- Use color wisely—color can help the poster if it sets off points, unifies sections, and guides the reader's eye; if you can't decide how much color, then under-do color.

If you want your poster to be excellent and attract interested readers, then follow these suggestions, prepare it well in advance, and then have your friends and faculty advisors look at it to make suggestions and comments.

Summary

The journal format for theses and dissertations has the advantage of being the format for journal publication, one of the main ways used to disseminate scholarly work. Yet it retains the essential characteristics of the complete reporting that is so valued in the thesis or dissertation. This format comprises preliminary materials (e.g., title page, abstract), the body (e.g., introduction, method, results, discussion), and a set of appendixes (e.g., extended literature review, additional results). The goal of this reporting style is to promote rapid publication of high-quality research.

Abstracts are frequently used as a form for submitting papers to scholarly meetings so they can be evaluated for possible presentation. Usually, the abstract is limited in length and format as prescribed by the scholarly group to which it is submitted. If the abstract paper is accepted for presentation, the form may be oral or by poster. Oral presentations usually last 10 to 20 min, and poster presentations are typically limited to a specific display space.

✔ Check Your Understanding

1. Writing: Select a study from a journal and write a 150-word abstract in APA style (or whatever style your department uses).

2. Oral presentations: To emphasize the importance of time limits on oral presentations, we suggest that each student give the following presentations in class:

   a. Select a published research study, and present a 2-min summary.

   b. Select another study, but give a 5-min summary.

3. Poster presentation: Prepare a poster presentation of a research study from a journal. Put all posters around the classroom walls. Critique each others' posters (or some percentage of posters for larger classes).

Statistical Tables

Table A.1 Table of Random Numbers

```
22 17 68 65 84    68 95 23 92 35    87 02 22 57 51    61 09 43 95 06    58 24 82 03 47
19 36 27 59 46    13 79 93 37 55    39 77 32 77 09    85 52 05 30 62    47 83 51 62 74
16 77 23 02 77    09 61 87 25 21    28 06 24 25 93    16 71 13 59 78    23 05 47 47 25
78 43 76 71 61    20 44 90 32 64    97 67 63 99 61    46 38 03 93 22    69 81 21 99 21
03 28 28 26 08    73 37 32 04 05    69 30 16 09 05    88 69 58 29 99    35 07 44 75 47

93 22 53 64 39    07 10 63 76 35    87 03 04 79 88    08 13 13 85 51    55 34 57 72 69
78 76 58 54 74    92 38 70 96 92    52 06 79 79 45    82 63 18 27 44    69 66 92 19 09
23 68 35 26 00    99 53 93 61 28    52 70 05 48 34    56 65 05 61 86    90 92 10 70 80
15 39 25 70 99    93 86 52 77 65    15 33 59 05 28    22 87 26 07 47    86 96 98 29 06
58 71 96 30 24    18 46 23 34 27    85 13 99 24 44    49 18 09 79 49    74 16 32 23 02

57 35 27 33 72    24 53 63 94 09    41 10 76 47 91    44 04 95 49 66    39 60 04 59 81
48 50 86 54 48    22 06 34 72 52    82 21 15 65 20    33 29 94 71 11    15 91 29 12 03
61 96 48 95 03    07 16 39 33 66    98 56 10 56 79    77 21 30 27 12    90 49 22 23 62
36 93 89 41 26    29 70 83 63 51    99 74 20 52 36    87 09 41 15 09    98 60 16 03 03
18 87 00 42 31    57 90 12 02 07    23 47 37 17 31    54 08 01 88 63    39 41 88 92 10

88 56 53 27 59    33 35 72 67 47    77 34 55 45 70    08 18 27 38 90    16 95 86 70 75
09 72 95 84 29    49 41 31 06 70    42 38 06 45 18    64 84 73 31 65    52 53 37 97 15
12 96 88 17 31    65 19 69 02 83    60 75 86 90 68    24 64 19 35 51    56 61 87 39 12
85 94 57 24 16    92 09 84 38 76    22 00 27 69 85    29 81 94 78 70    21 94 47 90 12
38 64 43 59 98    98 77 87 68 07    91 51 67 62 44    40 98 05 93 78    23 32 65 41 18

53 44 09 42 72    00 41 86 79 79    68 47 22 00 20    35 55 31 51 51    00 83 63 22 55
40 76 66 26 84    57 99 99 90 37    36 63 32 08 58    37 40 13 68 97    87 64 81 07 83
02 17 79 18 05    12 59 52 57 02    22 07 90 47 03    28 14 11 30 79    20 69 22 40 98
95 17 82 06 53    31 51 10 96 46    92 06 88 07 77    56 11 50 81 69    40 23 72 51 39
35 76 22 42 92    96 11 83 44 80    34 68 35 48 77    33 42 40 90 60    73 96 53 97 86

26 29 13 56 41    85 47 04 66 08    34 72 57 59 13    82 43 80 46 15    38 26 61 70 04
77 80 20 75 82    72 82 32 99 90    63 95 73 76 63    89 73 44 99 05    48 67 26 43 18
46 40 66 44 52    91 36 74 43 53    30 82 13 54 00    78 45 63 98 35    55 03 36 67 68
37 56 08 18 09    77 53 84 46 47    31 91 18 95 58    24 16 74 11 53    44 10 13 85 57
61 65 61 68 66    37 27 47 39 19    84 83 70 07 48    53 21 40 06 71    95 06 79 88 54

93 43 69 64 07    34 18 04 52 35    56 27 09 24 86    61 85 53 83 45    19 90 70 99 00
21 96 60 12 99    11 20 99 45 18    48 13 93 55 34    18 37 79 49 90    65 97 38 20 46
95 20 47 97 97    27 37 83 28 71    00 06 41 41 74    45 89 09 39 84    51 67 11 52 49
97 86 21 78 73    10 65 81 92 59    58 76 17 14 97    04 76 62 16 17    17 95 70 45 80
69 92 06 34 13    59 71 74 17 32    27 55 10 24 19    23 71 82 13 74    63 52 52 01 41

04 31 17 21 56    33 73 99 19 87    26 72 39 27 67    53 77 57 68 93    60 61 97 22 61
61 06 98 03 91    87 14 77 43 96    43 00 65 98 50    45 60 33 01 07    98 99 46 50 47
85 93 85 86 88    72 87 08 62 40    16 06 10 89 20    23 21 34 74 97    76 38 03 29 63
21 74 32 47 45    73 96 07 94 52    09 65 90 77 47    25 76 16 19 33    53 05 70 53 30
15 69 53 82 80    79 96 23 53 10    65 39 07 16 29    45 33 02 43 70    02 87 40 41 45

02 89 08 04 49    20 21 14 68 86    87 63 93 95 17    11 29 01 95 80    35 14 97 35 33
87 18 15 89 79    85 43 01 72 73    08 61 74 51 69    89 74 39 82 15    94 51 33 41 67
98 83 71 94 22    59 97 50 99 52    08 52 85 08 40    87 80 61 65 31    91 51 80 32 44
10 08 58 21 66    72 68 49 29 31    89 85 84 46 06    59 73 19 85 23    65 09 29 75 63
47 90 56 10 08    88 02 84 27 83    42 29 72 23 19    66 56 45 65 79    20 71 53 20 25

22 85 61 68 90    49 64 92 85 44    16 40 12 89 88    50 14 49 81 06    01 82 77 45 12
67 80 43 79 33    12 83 11 41 16    25 58 19 68 70    77 02 54 00 52    53 43 37 15 26
27 62 50 96 72    79 44 61 40 15    14 53 40 65 39    27 31 58 50 28    11 39 03 34 25
33 78 80 87 15    38 30 06 38 21    14 47 47 07 26    54 96 87 53 32    40 36 40 96 76
13 13 92 66 99    47 24 49 57 74    32 25 43 62 17    10 97 11 69 84    99 63 22 32 98
```

```
10 27 53 96 23    71 50 54 36 23    54 31 04 82 98    04 14 12 15 09    26 78 25 47 47
28 41 50 61 88    64 85 27 20 18    83 36 36 05 56    39 71 65 09 62    94 76 62 11 89
34 21 42 57 02    59 19 18 97 48    80 30 03 30 98    05 24 67 70 07    84 97 50 87 46
61 81 77 23 23    82 82 11 54 08    53 28 70 58 96    44 07 39 55 43    42 34 43 39 28
61 15 18 13 54    16 86 20 26 88    90 74 80 55 09    14 53 90 51 17    52 01 63 01 59

91 76 21 64 64    44 91 13 32 97    75 31 62 66 54    84 80 32 75 77    56 08 25 70 29
00 97 79 08 06    37 30 28 59 85    53 56 68 53 40    01 74 39 59 73    30 19 99 85 48
36 46 18 34 94    75 20 80 27 77    78 91 69 16 00    08 43 18 73 68    67 69 61 34 25
88 98 99 60 50    65 95 79 42 94    93 62 40 89 96    43 56 47 71 66    46 76 29 67 02
04 37 59 87 21    05 02 03 24 17    47 97 81 56 51    92 34 86 01 82    55 51 33 12 91

63 62 06 34 41    94 21 78 55 09    72 76 45 16 94    29 95 81 83 83    79 88 01 97 30
78 47 23 53 90    34 41 92 45 71    09 23 70 70 07    12 38 92 79 43    14 85 11 47 23
87 68 62 15 43    53 14 36 59 25    54 47 33 70 15    59 24 48 40 35    50 03 42 99 36
47 60 92 10 77    88 59 53 11 52    66 25 69 07 04    48 68 64 71 06    61 65 70 22 12
56 88 87 59 41    65 28 04 67 53    95 79 88 37 31    50 41 06 94 76    81 83 17 16 33

02 57 45 86 67    73 43 07 34 48    44 26 87 93 29    77 09 61 67 84    06 69 44 77 75
31 54 14 13 17    48 62 11 90 60    68 12 93 64 28    46 24 79 16 76    14 60 25 51 01
28 50 16 43 36    28 97 85 58 99    67 22 52 76 23    24 70 36 54 54    59 28 61 71 96
63 29 62 66 50    02 63 45 52 38    67 63 47 54 75    83 24 78 43 20    92 63 13 47 48
45 65 58 26 51    76 96 59 38 72    86 57 45 71 46    44 67 76 14 55    44 88 01 62 12

39 65 36 63 70    77 45 85 50 51    74 13 39 35 22    30 53 36 02 95    49 34 88 73 61
73 71 98 16 04    29 18 94 51 23    76 51 94 84 86    79 93 96 38 63    08 58 25 58 94
72 20 56 20 11    72 65 71 08 86    79 57 95 13 91    97 48 72 66 48    09 71 17 24 89
75 17 26 99 76    89 37 20 70 01    77 31 61 95 46    26 97 05 73 51    53 33 18 72 87
37 48 60 82 29    81 30 15 39 14    48 38 75 93 29    06 87 37 78 48    45 56 00 84 47

68 08 02 80 72    83 71 46 30 49    89 17 95 88 29    02 39 56 03 46    97 74 06 56 17
14 23 98 61 67    70 52 85 01 50    01 84 02 78 43    10 62 98 19 41    18 83 99 47 99
49 08 96 21 44    25 27 99 41 28    07 41 08 34 66    19 42 74 39 91    41 96 53 78 72
78 37 06 08 43    63 61 62 42 29    39 68 95 10 96    09 24 23 00 62    56 12 80 73 16
37 21 34 17 68    68 96 83 23 56    32 84 60 15 31    44 73 67 34 77    91 15 79 74 58

14 29 09 34 04    87 83 07 55 07    76 58 30 83 64    87 29 25 58 84    86 50 60 00 25
58 43 28 06 36    49 52 83 51 14    47 56 91 29 34    05 87 31 06 95    12 45 57 09 09
10 43 67 29 70    80 62 80 03 42    10 80 21 38 84    90 56 35 03 09    43 12 74 49 14
44 38 88 39 54    86 97 37 44 22    00 95 01 31 76    17 16 29 56 63    38 78 94 49 81
90 69 59 19 51    85 39 52 85 13    07 28 37 07 61    11 16 36 27 03    78 86 72 04 95

41 47 10 25 62    97 05 31 03 61    20 26 36 31 62    68 69 86 95 44    84 95 48 46 45
91 94 14 63 19    75 89 11 47 11    31 56 34 19 09    79 57 92 36 59    14 93 87 81 40
80 06 54 18 66    09 18 94 06 19    98 40 07 17 81    22 45 44 84 11    24 62 20 42 31
67 72 77 63 48    84 08 31 55 58    24 33 45 77 58    80 45 67 93 82    75 70 16 08 24
59 40 24 13 27    79 26 88 86 30    01 31 60 10 39    53 58 47 70 93    85 81 56 39 38

05 90 35 89 95    01 61 16 96 94    50 78 13 69 36    37 68 53 37 31    71 26 35 03 71
44 43 80 69 98    46 68 05 14 82    90 78 50 05 62    77 79 13 57 44    59 60 10 39 66
61 81 31 96 82    00 57 25 60 59    46 72 60 18 77    55 66 12 62 11    08 99 55 64 57
42 88 07 10 05    24 98 65 63 21    47 21 61 88 32    27 80 30 21 60    10 92 35 36 12
77 94 30 05 39    28 10 99 00 27    12 73 73 99 12    49 99 57 94 82    96 88 57 17 91

78 83 19 76 16    94 11 68 84 26    23 54 20 86 85    23 86 66 99 07    36 37 34 92 09
87 76 59 61 81    43 63 64 61 61    65 76 36 95 90    18 48 27 45 68    27 23 65 30 72
91 43 05 96 47    55 78 99 95 24    37 55 85 78 78    01 48 41 19 10    35 19 54 07 73
84 97 77 72 73    09 62 06 65 72    87 12 49 03 60    41 15 20 76 27    50 47 02 29 16
87 41 60 76 83    44 88 96 07 80    85 05 83 38 96    73 70 66 81 90    30 56 10 48 59
```

Adapted, by permission, from R.A. Fisher and F. Yates, 1995, *Statistical tables for biological agricultural and medical research* (Essex, United Kingdom: Pearson Education), 137.

z	One tail π beyond	One tail π remainder	Two tail π beyond	Two tail π remainder	z	One tail π beyond	One tail π remainder	Two tail π beyond	Two tail π remainder
0.00	0.5000	0.5000	1.0000	0.0000	0.45	0.3264	0.6736	0.6527	0.3473
0.01	0.4960	0.5040	0.9920	0.0080	0.46	0.3228	0.6772	0.6455	0.3545
0.02	0.4920	0.5080	0.9840	0.0160	0.47	0.3192	0.6808	0.6384	0.3616
0.03	0.4880	0.5120	0.9761	0.0239	0.48	0.3156	0.6844	0.6312	0.3688
0.04	0.4840	0.5160	0.9681	0.0319	0.49	0.3121	0.6879	0.6241	0.3759
0.05	0.4801	0.5199	0.9601	0.0399	0.50	0.3085	0.6915	0.6171	0.3829
0.06	0.4761	0.5239	0.9522	0.0478	0.51	0.3050	0.6950	0.6101	0.3899
0.07	0.4721	0.5279	0.9442	0.0558	0.52	0.3015	0.6985	0.6031	0.3969
0.08	0.4681	0.5319	0.9362	0.0638	0.53	0.2981	0.7019	0.5961	0.4039
					0.54	0.2946	0.7054	0.5892	0.4108
0.09	0.4641	0.5359	0.9283	0.0717					
0.10	0.4602	0.5398	0.9203	0.0797	0.55	0.2912	0.7088	0.5823	0.4177
0.11	0.4562	0.5438	0.9124	0.0876	0.56	0.2877	0.7123	0.5755	0.4245
0.12	0.4522	0.4378	0.9045	0.0955	0.57	0.2843	0.7157	0.5687	0.4313
0.13	0.4483	0.5517	0.8966	0.1034	0.58	0.2810	0.7190	0.5619	0.4381
					0.59	0.2776	0.7224	0.5552	0.4448
0.14	0.4443	0.5557	0.8887	0.1113					
0.15	0.4404	0.5596	0.8808	0.1192	0.60	0.2743	0.7257	0.5485	0.4515
0.16	0.4364	0.5636	0.8729	0.1271	0.61	0.2709	0.7291	0.5419	0.4581
0.17	0.4325	0.5675	0.8650	0.1350	0.62	0.2676	0.7324	0.5353	0.4647
0.18	0.4286	0.5714	0.8571	0.1429	0.63	0.2643	0.7357	0.5276	0.4713
0.19	0.4247	0.5753	0.8493	0.1507	0.64	0.2611	0.7389	0.5222	0.4778
0.20	0.4207	0.5793	0.8415	0.1585	0.65	0.2578	0.7422	0.5157	0.4843
0.21	0.4168	0.5832	0.8337	0.1663	0.66	0.2546	0.7454	0.5093	0.4907
0.22	0.4129	0.5871	0.8259	0.1741	0.67	0.2514	0.7486	0.5029	0.4971
0.23	0.4090	0.5910	0.8181	0.1819	0.6745	0.25	0.75	0.50	0.50
0.24	0.4052	0.5948	0.8103	0.1897	0.68	0.2483	0.7517	0.4965	0.5035
0.25	0.4013	0.5987	0.8026	0.1974	0.69	0.2451	0.7549	0.4902	0.5098
0.26	0.3974	0.6026	0.7949	0.2051	0.70	0.2420	0.7580	0.4839	0.5161
0.27	0.3936	0.6064	0.7872	0.2128	0.71	0.2389	0.7611	0.4777	0.5223
0.28	0.3897	0.6103	0.7795	0.2205	0.72	0.2358	0.7642	0.4715	0.5285
0.29	0.3859	0.6141	0.7718	0.2282	0.73	0.2327	0.7673	0.4654	0.5346
0.30	0.3821	0.6179	0.7642	0.2358	0.74	0.2296	0.7704	0.4593	0.5407
0.31	0.3783	0.6217	0.7566	0.2434	0.75	0.2266	0.7734	0.4533	0.5467
0.32	0.3745	0.6255	0.7490	0.2510	0.76	0.2236	0.7764	0.4473	0.5527
0.33	0.3707	0.6293	0.7414	0.2586	0.77	0.2206	0.7794	0.4413	0.5587
0.34	0.3669	0.6331	0.7339	0.2661	0.78	0.2177	0.7823	0.4354	0.5646
0.35	0.3632	0.6368	0.7263	0.2737	0.79	0.2148	0.7852	0.4295	0.5705
0.36	0.3594	0.6406	0.7188	0.2812	0.80	0.2119	0.7881	0.4237	0.5763
0.37	0.3557	0.6443	0.7114	0.2886	0.81	0.2090	0.7910	0.4179	0.5821
0.38	0.3520	0.6480	0.7039	0.2961	0.82	0.2061	0.7939	0.4122	0.5878
0.39	0.3483	0.6517	0.6965	0.3035	0.83	0.2033	0.7967	0.4065	0.5935
0.40	0.3446	0.6554	0.6892	0.3108	0.84	0.2005	0.7995	0.4009	0.5991
0.41	0.3409	0.6591	0.6818	0.3182	0.8416	0.20	0.80	0.40	0.60
0.42	0.3372	0.6628	0.6745	0.3255	0.85	0.1997	0.8023	0.3953	0.6047
0.43	0.3336	0.6664	0.6672	0.3328	0.86	0.1949	0.8051	0.3898	0.6102
0.44	0.3300	0.6700	0.6599	0.3401	0.87	0.1922	0.8078	0.3843	0.6157

	One tail		Two tail			One tail		Two tail	
z	π beyond	π remainder	π beyond	π remainder	z	π beyond	π remainder	π beyond	π remainder
0.88	0.1894	0.8106	0.3789	0.6211	1.32	0.0934	0.9066	0.1868	0.8132
0.89	0.1867	0.8133	0.3735	0.6265	1.33	0.0918	0.9082	0.1835	0.8165
0.90	0.1841	0.8159	0.3681	0.6319	1.34	0.0901	0.9099	0.1802	0.8198
0.91	0.1814	0.8186	0.3628	0.6372	1.35	0.0885	0.9115	0.1770	0.8230
0.92	0.1788	0.8212	0.3576	0.6424	1.36	0.0869	0.9131	0.1738	0.8262
0.93	0.1762	0.8238	0.3524	0.6476	1.37	0.0853	0.9147	0.1707	0.8293
0.94	0.1736	0.8264	0.3472	0.6528	1.38	0.0838	0.9162	0.1676	0.8324
0.95	0.1711	0.8289	0.3421	0.6579	1.39	0.0823	0.9177	0.1645	0.8355
0.96	0.1685	0.8315	0.3371	0.6629	1.40	0.0808	0.9192	0.1615	0.8385
0.97	0.1660	0.8340	0.3320	0.6680	1.41	0.0793	0.9207	0.1585	0.8415
0.98	0.1635	0.8365	0.3271	0.6729	1.42	0.0778	0.9222	0.1556	0.8444
0.99	0.1611	0.8389	0.3222	0.6778	1.43	0.0764	0.9236	0.1527	0.8473
1.00	0.1587	0.8413	0.3173	0.6827	1.44	0.0749	0.9251	0.1499	0.8501
1.01	0.1562	0.8438	0.3125	0.6875	1.45	0.0735	0.9265	0.1471	0.8529
1.02	0.1539	0.8461	0.3077	0.6923	1.46	0.0721	0.9279	0.1443	0.8567
1.03	0.1515	0.8485	0.3030	0.6970	1.47	0.0708	0.9292	0.1416	0.8584
1.04	0.1492	0.8508	0.2983	0.7017	1.48	0.0694	0.9306	0.1389	0.8611
1.05	0.1469	0.8531	0.2937	0.7063	1.49	0.0681	0.9319	0.1362	0.8638
1.06	0.1446	0.8554	0.2891	0.7109	1.50	0.0668	0.9332	0.1336	0.8664
1.07	0.1423	0.8577	0.2846	0.7154	1.51	0.0655	0.9345	0.1310	0.8690
1.08	0.1401	0.8599	0.2801	0.7199	1.52	0.0643	0.9357	0.1285	0.8715
1.09	0.1379	0.8621	0.2757	0.7243	1.53	0.0630	0.9370	0.1260	0.8740
1.10	0.1357	0.8643	0.2713	0.7287	1.54	0.0618	0.9382	0.1236	0.8764
1.11	0.1335	0.8665	0.2670	0.7330	1.55	0.0606	0.9394	0.1211	0.8789
1.12	0.1314	0.8686	0.2627	0.7373	1.56	0.0594	0.9406	0.1188	0.8812
1.13	0.1292	0.8708	0.2585	0.7415	1.57	0.0582	0.9418	0.1164	0.8836
1.14	0.1271	0.8729	0.2543	0.7457	1.58	0.0571	0.9429	0.1141	0.8859
1.15	0.1251	0.8749	0.2501	0.7499	1.59	0.0559	0.9441	0.1118	0.8882
1.16	0.1230	0.8770	0.2460	0.7540	1.60	0.0548	0.9452	0.1096	0.8904
1.17	0.1210	0.8790	0.2420	0.7580	1.61	0.0537	0.9463	0.1074	0.8926
1.18	0.1190	0.8810	0.2380	0.7620	1.62	0.0526	0.9474	0.1052	0.8948
1.19	0.1170	0.8830	0.2340	0.7660	1.63	0.0516	0.9484	0.1031	0.8969
1.20	0.1151	0.8849	0.2301	0.7699	1.64	0.0505	0.9495	0.1010	0.8990
1.21	0.1131	0.8869	0.2263	0.7737	1.645	0.05	0.95	0.10	0.90
1.22	0.1112	0.8888	0.2225	0.7775	1.65	0.0495	0.9505	0.0989	0.9011
1.23	0.1093	0.8907	0.2187	0.7813	1.66	0.0485	0.9515	0.0969	0.9031
1.24	0.1075	0.8925	0.2150	0.7890	1.67	0.0475	0.9525	0.0949	0.9051
1.25	0.1056	0.8944	0.2113	0.7887	1.68	0.0465	0.9535	0.0930	0.9070
1.26	0.1038	0.8962	0.2077	0.7923	1.69	0.0455	0.9545	0.0910	0.9090
1.27	0.1020	0.8980	0.2041	0.7959	1.70	0.0446	0.9554	0.0891	0.9109
1.28	0.1003	0.8997	0.2005	0.7995	1.71	0.0436	0.9564	0.0873	0.9127
1.282	0.10	0.90	0.20	0.80	1.72	0.0427	0.9573	0.0854	0.9146
1.29	0.0985	0.9015	0.1971	0.8029	1.73	0.0418	0.9582	0.0836	0.9164
1.30	0.0968	0.9032	0.1936	0.8064	1.74	0.0409	0.9591	0.0819	0.9181
1.31	0.0951	0.9049	0.1902	0.8098	1.75	0.0401	0.9599	0.0801	0.9199

(continued)

| | One tail | | Two tail | | | One tail | | Two tail | |
z	π beyond	π remainder	π beyond	π remainder	z	π beyond	π remainder	π beyond	π remainder
1.76	0.0392	0.9608	0.0784	0.9216	2.20	0.0139	0.9861	0.0278	0.9722
1.77	0.0384	0.9616	0.0767	0.9233	2.21	0.0136	0.9864	0.0271	0.9729
1.78	0.0375	0.9625	0.0751	0.9249	2.22	0.0132	0.9868	0.0264	0.9736
1.79	0.0367	0.9633	0.0734	0.9266	2.23	0.0129	0.9871	0.0257	0.9743
1.80	0.0359	0.9641	0.0719	0.9281	2.24	0.0125	0.9875	0.0251	0.9749
1.81	0.0352	0.9649	0.0703	0.9297	2.25	0.0122	0.9878	0.0244	0.9756
1.82	0.0344	0.9656	0.0688	0.9312	2.26	0.0119	0.9881	0.0238	0.9762
1.83	0.0336	0.9664	0.0672	0.9328	2.27	0.0116	0.9884	0.0232	0.9768
1.84	0.0329	0.9671	0.0658	0.9342	2.28	0.0113	0.9887	0.0226	0.9774
1.85	0.0322	0.9678	0.0643	0.9357	2.29	0.0110	0.9890	0.0220	0.9780
1.86	0.0314	0.9686	0.0629	0.9371	2.30	0.0107	0.9893	0.0214	0.9786
1.87	0.0307	0.9693	0.0615	0.9385	2.31	0.0104	0.9896	0.0209	0.9791
1.88	0.0301	0.9699	0.0601	0.9399	2.32	0.0102	0.9898	0.0203	0.9797
1.89	0.0294	0.9706	0.0588	0.9412	2.326	0.01	0.99	0.02	0.98
1.90	0.0287	0.9713	0.0574	0.9426	2.33	0.0099	0.9901	0.0198	0.9802
1.91	0.0281	0.9719	0.0561	0.9439	2.34	0.0096	0.9904	0.0193	0.9807
1.92	0.0274	0.9726	0.0549	0.9451	2.35	0.0094	0.9906	0.0188	0.9812
1.93	0.0268	0.9732	0.0536	0.9464	2.36	0.0091	0.9909	0.0183	0.9817
1.94	0.0262	0.9738	0.0524	0.9476	2.37	0.0089	0.991	0.0178	0.9822
1.95	0.0256	0.9744	0.0512	0.9488	2.38	0.0087	0.9913	0.0173	0.9827
1.960	0.025	0.975	0.05	0.95	2.39	0.0084	0.9916	0.0168	0.9832
1.97	0.0244	0.9756	0.0488	0.9512	2.40	0.0082	0.9918	0.0164	0.9836
1.98	0.0239	0.9761	0.0477	0.9523	2.41	0.0080	0.9920	0.0160	0.9840
1.99	0.0233	0.9767	0.0466	0.9534	2.42	0.0078	0.9922	0.0155	0.9845
2.00	0.0228	0.9772	0.0455	0.9545	2.43	0.0075	0.9925	0.0151	0.9849
2.01	0.0222	0.9778	0.0444	0.9556	2.44	0.0073	0.9927	0.0147	0.9853
2.02	0.0217	0.9783	0.0434	0.9566	2.45	0.0071	0.9929	0.0143	0.9857
2.03	0.0212	0.9788	0.0424	0.9576	2.46	0.0069	0.9931	0.0139	0.9861
2.04	0.0207	0.9793	0.0414	0.9586	2.47	0.0068	0.9932	0.0135	0.9865
2.05	0.0202	0.9798	0.0404	0.9596	2.48	0.0066	0.9934	0.0131	0.9869
2.054	0.02	0.98	0.04	0.96	2.49	0.0064	0.9936	0.0128	0.9872
2.06	0.0197	0.9803	0.0394	0.9606	2.50	0.0062	0.9938	0.0124	0.9876
2.07	0.0192	0.9808	0.0385	0.9615	2.51	0.0060	0.9940	0.0121	0.9879
2.08	0.0188	0.9812	0.0375	0.9625	2.52	0.0059	0.9941	0.0117	0.9883
2.09	0.0183	0.9817	0.0366	0.9634	2.53	0.0057	0.9943	0.0114	0.9886
2.10	0.0179	0.9821	0.0357	0.9643	2.54	0.0055	0.9945	0.0111	0.9889
2.11	0.0174	0.9826	0.0349	0.9651	2.55	0.0054	0.9946	0.0108	0.9892
2.12	0.0170	0.9830	0.0340	0.9660	2.56	0.0052	0.9948	0.0105	0.9895
2.13	0.0166	0.9834	0.0332	0.9668	2.57	0.0051	0.9949	0.0102	0.9898
2.14	0.0162	0.9838	0.0324	0.9676	2.576	0.005	0.995	0.01	0.99
2.15	0.0158	0.9842	0.0316	0.9684	2.58	0.0049	0.9951	0.0099	0.9901
2.16	0.0154	0.9846	0.0308	0.9692	2.59	0.0048	0.9952	0.0096	0.9904
2.17	0.0150	0.9850	0.0300	0.9700	2.60	0.0047	0.9953	0.0093	0.9907
2.18	0.0146	0.9854	0.0293	0.9707	2.61	0.0045	0.9955	0.0091	0.9909
2.19	0.0143	0.9857	0.0285	0.9715	2.62	0.0044	0.9956	0.0088	0.9912

	One tail		Two tail			One tail		Two tail	
z	π beyond	π remainder	π beyond	π remainder	z	π beyond	π remainder	π beyond	π remainder
2.63	0.0043	0.9957	0.0085	0.9915	3.25	0.0006	0.9994	0.0012	0.9986
2.64	0.0041	0.9959	0.0083	0.9917	3.291	0.0005	0.9995	0.001	0.999
2.65	0.0040	0.9960	0.0080	0.9920	3.30	0.0005	0.9995	0.0010	0.9990
2.70	0.0035	0.9965	0.0069	0.9931	3.35	0.0004	0.9996	0.0008	0.9992
2.75	0.0030	0.9970	0.0060	0.9940	3.40	0.0003	0.9997	0.0007	0.9993
2.80	0.0026	0.9974	0.0051	0.9949	3.45	0.0003	0.9997	0.0006	0.9994
2.85	0.0022	0.9978	0.0044	0.9956	3.50	0.0002	0.9998	0.0005	0.9995
2.90	0.0019	0.9981	0.0037	0.9963	3.55	0.0002	0.9998	0.0004	0.9996
2.95	0.0016	0.9984	0.0032	0.9968	3.60	0.0002	0.9998	0.0003	0.9997
3.00	0.0013	0.9987	0.0027	0.9973	3.65	0.0001	0.9999	0.0003	0.9997
3.05	0.0011	0.9989	0.0023	0.9977	3.719	0.0001	0.9999	0.0002	0.9998
3.090	0.001	0.999	0.002	0.998	3.80	0.0001	0.9999	0.0001	0.9999
3.10	0.0010	0.9990	0.0019	0.9981	3.891	0.00005	0.99995	0.0001	0.9999
3.15	0.0008	0.9992	0.0016	0.9984	4.000	0.00003	0.99997	0.00006	0.99994
3.20	0.0007	0.9993	0.0014	0.9988	4.265	0.00001	0.99999	0.00002	0.99998

From *Biometrika Tables for Statisticians*, Vol. 1, 3rd ed., by E.S. Pearson and H.O. Hartley, 1966, London: Cambridge University Press. Adapted with permission of the Biometrika Trustees.

Table A.3 Critical Values of Correlation Coefficients

	Level of significance for one-tailed test						Level of significance for one-tailed test				
	.05	.025	.01	.005	.0005		.05	.025	.01	.005	.0005
	Level of significance for two-tailed test						Level of significance for two-tailed test				
df = N − 2	.10	.05	.02	.01	.001	df = N − 2	.10	.05	.02	.01	.001
1	.9877	.9969	.9995	.9999	1.000	16	.4000	.4683	.5425	.5897	.7084
2	.9000	.9500	.9800	.9900	.9990	17	.3887	.4555	.5285	.5751	.6932
3	.8054	.8783	.9343	.9587	.9912	18	.3783	.4438	.5155	.5614	.6787
4	.7293	.8114	.8822	.9172	.9741	19	.3687	.4329	.5034	.5487	.6652
5	.6694	.7545	.8329	.8745	.9507	20	.3598	.4227	.4921	.5368	.6524
6	.6215	.7067	.7887	.8343	.9249	25	.3233	.3809	.4451	.4869	.5974
7	.5822	.6664	.7498	.7977	.8982	30	.2960	.3494	.4093	.4487	.5541
8	.5494	.6319	.7155	.7646	.8721	35	.2746	.3246	.3810	.4182	.5189
9	.5214	.6021	.6851	.7348	.8471	40	.2573	.3044	.3578	.3932	.4896
10	.4973	.5760	.6581	.7079	.8233	45	.2428	.2875	.3384	.3721	.4648
11	.4762	.5529	.6339	.6835	.8010	50	.2306	.2732	.3218	.3541	.4433
12	.4575	.5324	.6120	.6614	.7800	60	.2108	.2500	.2948	.3248	.4078
13	.4409	.5139	.5923	.6411	.7603	70	.1954	.2319	.2737	.3017	.3799
14	.4259	.4973	.5742	.6226	.7420	80	.1829	.2172	.2565	.2830	.3568
15	.4124	.4821	.5577	.6055	.7246	90	.1726	.2050	.2422	.2673	.3375
						100	.1638	.1946	.2301	.2540	.3211

Adapted, by permission, from R.A. Fisher and F. Yates, 1995, *Statistical tables for biological, agricultural, and medical research* (Essex, United Kingdom: Pearson Education), 63.

Table A.4   Transformation of r to Z_r

r	z_r	r	z_r	r	z_r	r	z_r	r	z_r
.000	.000	.200	.203	.400	.424	.600	.693	.800	1.099
.005	.005	.205	.208	.405	.430	.605	.701	.805	1.113
.010	.010	.210	.213	.410	.436	.610	.709	.810	1.127
.015	.015	.215	.218	.415	.442	.615	.717	.815	1.142
.020	.020	.220	.224	.420	.448	.620	.725	.820	1.157
.025	.025	.225	.229	.425	.454	.625	.733	.825	1.172
.030	.030	.230	.234	.430	.460	.630	.741	.830	1.188
.035	.035	.235	.239	.435	.466	.635	.750	.835	1.204
.040	.040	.240	.245	.440	.472	.640	.758	.840	1.221
.045	.045	.245	.250	.445	.478	.645	.767	.845	1.238
.050	.050	.250	.255	.450	.485	.650	.775	.850	1.256
.055	.055	.255	.261	.455	.491	.655	.784	.855	1.274
.060	.060	.260	.266	.460	.497	.660	.793	.860	1.293
.065	.065	.265	.271	.465	.504	.665	.802	.865	1.313
.070	.070	.270	.277	.470	.510	.670	.811	.870	1.333
.075	.075	.275	.282	.475	.517	.675	.820	.875	1.354
.080	.080	.280	.288	.480	.523	.680	.829	.880	1.376
.085	.085	.285	.293	.485	.530	.685	.838	.885	1.398
.090	.090	.290	.299	.490	.536	.690	.848	.890	1.422
.095	.095	.295	.304	.495	.543	.695	.858	.895	1.447
.100	.100	.300	.310	.500	.549	.700	.867	.900	1.472
.105	.105	.305	.315	.505	.556	.705	.877	.905	1.499
.110	.110	.310	.321	.510	.563	.710	.887	.910	1.528
.115	.116	.315	.326	.515	.570	.715	.897	.915	1.557
.120	.121	.320	.332	.520	.576	.720	.908	.920	1.589
.125	.126	.425	.337	.525	.583	.725	.918	.925	1.623
.130	.131	.330	.343	.530	.590	.730	.929	.930	1.658
.135	.136	.335	.348	.535	.597	.735	.940	.935	1.697
.140	.141	.340	.354	.540	.604	.740	.950	.940	1.738
.145	.146	.345	.360	.545	.611	.745	.962	.945	1.783
.150	.151	.350	.365	.550	.618	.750	.973	.950	1.832
.155	.156	.355	.371	.555	.626	.755	.984	.955	1.886
.160	.161	.360	.377	.560	.633	.760	.996	.960	1.946
.165	.167	.365	.383	.565	.640	.765	1.008	.965	2.014
.170	.172	.370	.388	.570	.648	.770	1.020	.970	2.092
.175	.177	.375	.394	.575	.655	.775	1.033	.975	2.185
.180	.182	.380	.400	.580	.662	.780	1.045	.980	2.298
.185	.187	.385	.406	.585	.670	.785	1.058	.985	2.443
.190	.192	.390	.412	.590	.678	.790	1.071	.990	2.647
.195	.198	.395	.418	.595	.685	.795	1.085	.995	2.994

Reprinted, by permission, from A.L. Edwards, 1967, *Statistical methods,* 2d ed. (New York: Holt, Reinhart, and Winston, Inc.), 427.

Table A.5 Critical Values of *t*

df	Level of significance for one-tailed test					
	.10	.05	.025	.01	.005	.0005
	Level of significance for two-tailed test					
	.20	.10	.05	.02	.01	.001
1	3.078	6.314	12.706	31.821	63.657	636.619
2	1.886	2.920	4.303	6.965	9.925	31.598
3	1.638	2.353	3.182	4.541	5.841	12.941
4	1.533	2.132	2.776	3.747	4.604	8.610
5	1.476	2.015	2.571	3.365	4.032	6.859
6	1.440	1.943	2.447	3.143	3.707	5.959
7	1.415	1.895	2.365	2.998	3.499	5.405
8	1.397	1.860	2.306	2.896	3.355	5.041
9	1.383	1.833	2.262	2.821	3.250	4.781
10	1.372	1.812	2.228	2.764	3.169	4.587
11	1.363	1.796	2.201	2.718	3.106	4.437
12	1.356	1.782	2.179	2.681	3.055	4.318
13	1.350	1.771	2.160	2.650	3.012	4.221
14	1.345	1.761	2.145	2.624	2.977	4.140
15	1.341	1.753	2.131	2.602	2.947	4.073
16	1.337	1.746	2.120	2.583	2.921	4.015
17	1.333	1.740	2.110	2.567	2.898	3.965
18	1.330	1.734	2.101	2.552	2.878	3.922
19	1.328	1.729	2.093	2.539	2.861	3.883
20	1.325	1.725	2.086	2.528	2.845	3.850
21	1.323	1.721	2.080	2.518	2.831	3.819
22	1.321	1.717	2.074	2.508	2.819	3.792
23	1.319	1.714	2.069	2.500	2.807	3.767
24	1.318	1.711	2.064	2.492	2.797	3.745
25	1.316	1.708	2.060	2.485	2.787	3.725
26	1.315	1.706	2.056	2.479	2.779	3.707
27	1.314	1.703	2.052	2.473	2.771	3.690
28	1.313	1.701	2.048	2.467	2.763	3.674
29	1.311	1.699	2.045	2.462	2.756	3.659
30	1.310	1.697	2.042	2.457	2.750	3.646
40	1.303	1.684	2.021	2.423	2.704	3.551
60	1.296	1.671	2.000	2.390	2.660	3.460
120	1.289	1.658	1.980	2.358	2.617	3.373
∞	1.282	1.645	1.960	2.326	2.576	3.291

Adapted, by permission, from R.A. Fisher and F. Yates, 1995, *Statistical tables for biological agricultural and medical research* (Essex, United Kingdom: Pearson Education), 46.

Table A.6 Critical Values of F

Degrees of freedom for denominator of F (n_2)

n_1 degrees of freedom (for numerator of F)

n_2	1	2	3	4	5	6	7	8	9	10	11	12	14	16	20	24	30	40	50	75	100	200	500	∞
1	161 / 4,052	200 / 4,999	216 / 5,403	225 / 5,625	230 / 5,764	234 / 5,859	237 / 5,928	239 / 5,981	241 / 6,022	242 / 6,056	243 / 6,082	244 / 6,106	245 / 6,142	246 / 6,169	248 / 6,208	249 / 6,234	250 / 6,258	251 / 6,286	252 / 6,302	253 / 6,323	253 / 6,334	254 / 6,352	254 / 6,361	254 / 6,366
2	18.51 / 98.49	19.00 / 99.00	19.16 / 99.17	19.25 / 99.25	19.30 / 99.30	19.33 / 99.33	19.36 / 99.34	19.37 / 99.36	19.38 / 99.38	19.39 / 99.40	19.40 / 99.41	19.41 / 99.42	19.42 / 99.43	19.43 / 99.44	19.44 / 99.45	19.45 / 99.46	19.46 / 99.47	19.47 / 99.48	19.47 / 99.48	19.48 / 99.49	19.49 / 99.49	19.49 / 99.49	19.50 / 99.50	19.50 / 99.50
3	10.13 / 34.12	9.55 / 30.82	9.28 / 29.46	9.12 / 28.71	9.01 / 28.24	8.94 / 27.91	8.88 / 27.67	8.84 / 27.49	8.81 / 27.34	8.78 / 27.23	8.76 / 27.13	8.74 / 27.05	8.71 / 26.92	8.69 / 26.83	8.66 / 26.69	8.64 / 26.60	8.62 / 26.50	8.60 / 26.41	8.58 / 26.35	8.57 / 26.27	8.56 / 26.23	8.54 / 26.18	8.54 / 26.14	8.53 / 26.12
4	7.71 / 21.20	6.94 / 18.00	6.49 / 16.69	6.39 / 15.98	6.26 / 15.52	6.16 / 15.21	6.09 / 14.98	6.04 / 14.80	6.00 / 14.66	5.96 / 14.54	5.93 / 14.45	5.91 / 14.37	5.87 / 14.24	5.84 / 14.15	5.80 / 14.02	5.77 / 13.93	5.74 / 13.83	5.71 / 13.74	5.70 / 13.69	5.68 / 13.61	5.66 / 13.57	5.65 / 13.52	5.64 / 13.48	5.63 / 13.46
5	6.61 / 16.26	5.79 / 13.27	5.41 / 12.06	5.19 / 11.39	5.05 / 10.97	4.95 / 10.67	4.88 / 10.45	4.82 / 10.27	4.78 / 10.15	4.74 / 10.05	4.70 / 9.96	4.68 / 9.89	4.64 / 9.77	4.60 / 9.68	4.56 / 9.55	4.53 / 9.47	4.50 / 9.38	4.46 / 9.29	4.44 / 9.24	4.42 / 9.17	4.40 / 9.13	4.38 / 9.07	4.37 / 9.04	4.36 / 9.02
6	5.99 / 13.74	5.14 / 10.92	4.76 / 9.78	4.53 / 9.15	4.39 / 8.75	4.28 / 8.47	4.21 / 8.26	4.15 / 8.10	4.10 / 7.98	4.06 / 7.87	4.03 / 7.79	4.00 / 7.72	3.96 / 7.60	3.92 / 7.52	3.87 / 7.39	3.84 / 7.31	3.81 / 7.23	3.77 / 7.14	3.75 / 7.09	3.72 / 7.02	3.71 / 6.99	3.69 / 6.94	3.68 / 6.90	3.67 / 6.88
7	5.59 / 12.25	4.74 / 9.55	4.35 / 8.45	4.12 / 7.85	3.97 / 7.46	3.87 / 7.19	3.79 / 7.00	3.73 / 6.84	3.68 / 6.71	3.63 / 6.62	3.60 / 6.54	3.57 / 6.47	3.52 / 6.35	3.49 / 6.27	3.44 / 6.15	3.41 / 6.07	3.38 / 5.98	3.34 / 5.90	3.32 / 5.85	3.29 / 5.78	3.28 / 5.75	3.25 / 5.70	3.24 / 5.67	3.23 / 5.65
8	5.32 / 11.26	4.46 / 8.65	4.07 / 7.59	3.84 / 7.01	3.69 / 6.63	3.58 / 6.37	3.50 / 6.19	3.44 / 6.03	3.39 / 5.91	3.34 / 5.82	3.31 / 5.74	3.28 / 5.67	3.23 / 5.56	3.20 / 5.48	3.15 / 5.36	3.12 / 5.28	3.08 / 5.20	3.05 / 5.11	3.03 / 5.06	3.00 / 5.00	2.98 / 4.96	2.96 / 4.91	2.94 / 4.88	2.93 / 4.86
9	5.12 / 10.56	4.26 / 8.02	3.86 / 6.99	3.63 / 6.42	3.48 / 6.06	3.37 / 5.80	3.29 / 5.62	3.23 / 5.47	3.18 / 5.35	3.13 / 5.26	3.10 / 5.18	3.07 / 5.11	3.02 / 5.00	2.98 / 4.92	2.93 / 4.80	2.90 / 4.73	2.86 / 4.64	2.82 / 4.56	2.80 / 4.51	2.77 / 4.45	2.76 / 4.41	2.73 / 4.36	2.72 / 4.33	2.71 / 4.31
10	4.96 / 10.04	4.10 / 7.56	3.71 / 6.55	3.48 / 5.99	3.33 / 5.64	3.22 / 5.39	3.14 / 5.21	3.07 / 5.06	3.02 / 4.95	2.97 / 4.85	2.94 / 4.78	2.91 / 4.71	2.86 / 4.60	2.82 / 4.52	2.77 / 4.41	2.74 / 4.33	2.70 / 4.25	2.67 / 4.17	2.64 / 4.12	2.61 / 4.05	2.59 / 4.01	2.56 / 3.96	2.55 / 3.93	2.54 / 3.91
11	4.84 / 9.65	3.98 / 7.20	3.59 / 6.22	3.36 / 5.67	3.20 / 5.32	3.09 / 5.07	3.01 / 4.88	2.95 / 4.74	2.90 / 4.63	2.86 / 4.54	2.82 / 4.46	2.79 / 4.40	2.74 / 4.29	2.70 / 4.21	2.65 / 4.10	2.61 / 4.02	2.57 / 3.94	2.53 / 3.86	2.50 / 3.80	2.47 / 3.74	2.45 / 3.70	2.42 / 3.66	2.41 / 3.62	2.40 / 3.60
12	4.75 / 9.33	3.88 / 6.93	3.49 / 5.95	3.26 / 5.41	3.11 / 5.06	3.00 / 4.82	2.92 / 4.65	2.85 / 4.50	2.80 / 4.39	2.76 / 4.30	2.72 / 4.22	2.69 / 4.16	2.64 / 4.05	2.60 / 3.98	2.54 / 3.86	2.50 / 3.78	2.46 / 3.70	2.42 / 3.61	2.40 / 3.56	2.36 / 3.49	2.35 / 3.46	2.32 / 3.41	2.31 / 3.38	2.30 / 3.36
13	4.67 / 9.07	3.80 / 6.70	3.41 / 5.74	3.18 / 5.20	3.02 / 4.86	2.92 / 4.62	2.84 / 4.44	2.77 / 4.30	2.72 / 4.19	2.67 / 4.10	2.63 / 4.02	2.60 / 3.96	2.55 / 3.85	2.51 / 3.78	2.46 / 3.67	2.42 / 3.59	2.38 / 3.51	2.34 / 3.42	2.32 / 3.37	2.28 / 3.30	2.26 / 3.27	2.24 / 3.21	2.22 / 3.18	2.21 / 3.16
14	4.60 / 8.86	3.74 / 6.51	3.34 / 5.56	3.11 / 5.03	2.96 / 4.69	2.85 / 4.46	2.77 / 4.28	2.70 / 4.14	2.65 / 4.03	2.60 / 3.94	2.56 / 3.86	2.53 / 3.80	2.48 / 3.70	2.44 / 3.62	2.39 / 3.51	2.35 / 3.43	2.31 / 3.34	2.27 / 3.26	2.24 / 3.21	2.21 / 3.14	2.19 / 3.11	2.16 / 3.06	2.14 / 3.02	2.13 / 3.00
15	4.54 / 8.68	3.68 / 6.36	3.29 / 5.42	3.06 / 4.89	2.90 / 4.56	2.79 / 4.32	2.70 / 4.14	2.64 / 4.00	2.59 / 3.89	2.55 / 3.80	2.51 / 3.73	2.48 / 3.67	2.43 / 3.56	2.39 / 3.48	2.33 / 3.36	2.29 / 3.29	2.25 / 3.20	2.21 / 3.12	2.18 / 3.07	2.15 / 3.00	2.12 / 2.97	2.10 / 2.92	2.08 / 2.89	2.07 / 2.87
16	4.49 / 8.53	3.63 / 6.23	3.24 / 5.29	3.01 / 4.77	2.85 / 4.44	2.74 / 4.20	2.66 / 4.03	2.59 / 3.89	2.54 / 3.78	2.49 / 3.69	2.45 / 3.61	2.42 / 3.55	2.37 / 3.45	2.33 / 3.37	2.28 / 3.25	2.24 / 3.18	2.20 / 3.10	2.16 / 3.01	2.13 / 2.96	2.09 / 2.89	2.07 / 2.86	2.04 / 2.80	2.02 / 2.77	2.01 / 2.75
17	4.45 / 8.40	3.59 / 6.11	3.20 / 5.18	2.96 / 4.67	2.81 / 4.34	2.70 / 4.10	2.62 / 3.93	2.55 / 3.79	2.50 / 3.68	2.45 / 3.59	2.41 / 3.52	2.38 / 3.45	2.33 / 3.35	2.29 / 3.27	2.23 / 3.16	2.19 / 3.08	2.15 / 3.00	2.11 / 2.92	2.08 / 2.86	2.04 / 2.79	2.02 / 2.76	1.99 / 2.70	1.97 / 2.67	1.96 / 2.65
18	4.41 / 8.28	3.55 / 6.01	3.16 / 5.09	2.93 / 4.58	2.77 / 4.25	2.66 / 4.01	2.58 / 3.85	2.51 / 3.71	2.46 / 3.60	2.41 / 3.51	2.37 / 3.44	2.34 / 3.37	2.29 / 3.27	2.25 / 3.19	2.19 / 3.07	2.15 / 3.00	2.11 / 2.91	2.07 / 2.83	2.04 / 2.78	2.00 / 2.71	1.98 / 2.68	1.95 / 2.62	1.93 / 2.59	1.92 / 2.57

F-distribution table (continued). Two values per cell (upper row value / lower row value).

df (denom.)																								
19	4.38 / 8.18	3.52 / 5.93	3.13 / 5.01	2.90 / 4.50	2.74 / 4.17	2.63 / 3.94	2.55 / 3.77	2.48 / 3.63	2.43 / 3.52	2.38 / 3.43	2.34 / 3.36	2.31 / 3.30	2.26 / 3.19	2.21 / 3.12	2.15 / 3.00	2.11 / 2.92	2.07 / 2.84	2.02 / 2.76	2.00 / 2.70	1.96 / 2.63	1.94 / 2.60	1.91 / 2.54	1.90 / 2.51	1.88 / 2.49
20	4.35 / 8.10	3.49 / 5.85	3.10 / 4.94	2.87 / 4.43	2.71 / 4.10	2.60 / 3.87	2.52 / 3.71	2.45 / 3.56	2.40 / 3.45	2.35 / 3.37	2.31 / 3.30	2.28 / 3.23	2.23 / 3.13	2.18 / 3.05	2.12 / 2.94	2.08 / 2.86	2.04 / 2.77	1.99 / 2.69	1.96 / 2.63	1.92 / 2.56	1.90 / 2.53	1.87 / 2.47	1.85 / 2.44	1.84 / 2.42
21	4.32 / 8.02	3.47 / 5.78	3.07 / 4.87	2.84 / 4.37	2.68 / 4.04	2.57 / 3.81	2.49 / 3.65	2.42 / 3.51	2.37 / 3.40	2.32 / 3.31	2.28 / 3.24	2.25 / 3.17	2.20 / 3.07	2.15 / 2.99	2.09 / 2.88	2.05 / 2.80	2.00 / 2.72	1.96 / 2.63	1.93 / 2.58	1.89 / 2.51	1.87 / 2.47	1.84 / 2.42	1.82 / 2.38	1.81 / 2.36
22	4.30 / 7.94	3.44 / 5.72	3.05 / 4.82	2.82 / 4.31	2.66 / 3.99	2.55 / 3.76	2.47 / 3.59	2.40 / 3.45	2.35 / 3.35	2.30 / 3.26	2.26 / 3.18	2.23 / 3.12	2.18 / 3.02	2.13 / 2.94	2.07 / 2.83	2.03 / 2.75	1.98 / 2.67	1.93 / 2.58	1.91 / 2.53	1.87 / 2.46	1.84 / 2.42	1.81 / 2.37	1.80 / 2.33	1.78 / 2.31
23	4.28 / 7.88	3.42 / 5.66	3.03 / 4.76	2.80 / 4.26	2.64 / 3.94	2.53 / 3.71	2.45 / 3.54	2.38 / 3.41	2.32 / 3.30	2.28 / 3.21	2.24 / 3.14	2.20 / 3.07	2.14 / 2.97	2.10 / 2.89	2.04 / 2.78	2.00 / 2.70	1.96 / 2.62	1.91 / 2.53	1.88 / 2.48	1.84 / 2.41	1.82 / 2.37	1.79 / 2.32	1.77 / 2.28	1.76 / 2.26
24	4.26 / 7.82	3.40 / 5.61	3.01 / 4.72	2.78 / 4.22	2.62 / 3.90	2.51 / 3.67	2.43 / 3.50	2.36 / 3.36	2.30 / 3.25	2.26 / 3.17	2.22 / 3.09	2.18 / 3.03	2.13 / 2.93	2.09 / 2.85	2.02 / 2.74	1.98 / 2.66	1.94 / 2.58	1.89 / 2.49	1.86 / 2.44	1.82 / 2.37	1.80 / 2.33	1.77 / 2.29	1.74 / 2.23	1.73 / 2.21
25	4.24 / 7.77	3.38 / 5.57	2.99 / 4.68	2.76 / 4.18	2.60 / 3.86	2.49 / 3.63	2.41 / 3.46	2.34 / 3.32	2.28 / 3.21	2.24 / 3.13	2.20 / 3.05	2.16 / 2.99	2.11 / 2.89	2.06 / 2.81	2.00 / 2.70	1.96 / 2.62	1.92 / 2.54	1.87 / 2.45	1.84 / 2.40	1.80 / 2.32	1.77 / 2.29	1.74 / 2.23	1.72 / 2.19	1.71 / 2.17
26	4.22 / 7.72	3.37 / 5.53	2.98 / 4.64	2.74 / 4.14	2.59 / 3.82	2.47 / 3.59	2.39 / 3.42	2.32 / 3.29	2.27 / 3.17	2.22 / 3.09	2.18 / 3.02	2.15 / 2.96	2.10 / 2.86	2.05 / 2.77	1.99 / 2.66	1.95 / 2.58	1.90 / 2.50	1.85 / 2.41	1.82 / 2.36	1.78 / 2.28	1.76 / 2.25	1.72 / 2.19	1.70 / 2.15	1.69 / 2.13
27	4.21 / 7.68	3.35 / 5.49	2.96 / 4.60	2.73 / 4.11	2.57 / 3.79	2.46 / 3.56	2.37 / 3.39	2.30 / 3.26	2.25 / 3.14	2.20 / 3.06	2.16 / 2.98	2.13 / 2.93	2.08 / 2.83	2.03 / 2.74	1.97 / 2.63	1.93 / 2.55	1.88 / 2.47	1.84 / 2.38	1.80 / 2.33	1.76 / 2.25	1.74 / 2.21	1.71 / 2.16	1.68 / 2.12	1.67 / 2.10
28	4.20 / 7.64	3.34 / 5.45	2.95 / 4.57	2.71 / 4.07	2.56 / 3.76	2.44 / 3.53	2.36 / 3.36	2.29 / 3.23	2.24 / 3.11	2.19 / 3.03	2.15 / 2.95	2.12 / 2.90	2.06 / 2.80	2.02 / 2.71	1.96 / 2.60	1.91 / 2.52	1.87 / 2.44	1.81 / 2.35	1.78 / 2.30	1.75 / 2.22	1.72 / 2.18	1.69 / 2.13	1.67 / 2.09	1.65 / 2.06
29	4.18 / 7.60	3.33 / 5.42	2.93 / 4.54	2.70 / 4.04	2.54 / 3.73	2.43 / 3.50	2.35 / 3.33	2.28 / 3.20	2.22 / 3.08	2.18 / 3.00	2.14 / 2.92	2.10 / 2.87	2.05 / 2.77	2.00 / 2.68	1.94 / 2.57	1.90 / 2.49	1.85 / 2.41	1.80 / 2.32	1.77 / 2.27	1.73 / 2.19	1.71 / 2.15	1.68 / 2.10	1.65 / 2.06	1.64 / 2.03
30	4.17 / 7.56	3.32 / 5.39	2.92 / 4.51	2.69 / 4.02	2.53 / 3.70	2.42 / 3.47	2.34 / 3.30	2.27 / 3.17	2.21 / 3.06	2.16 / 2.98	2.12 / 2.90	2.09 / 2.84	2.04 / 2.74	1.99 / 2.66	1.93 / 2.55	1.89 / 2.47	1.84 / 2.38	1.79 / 2.29	1.76 / 2.24	1.72 / 2.16	1.69 / 2.13	1.66 / 2.07	1.64 / 2.03	1.62 / 2.01
32	4.15 / 7.50	3.30 / 5.34	2.90 / 4.46	2.67 / 3.97	2.51 / 3.66	2.40 / 3.42	2.32 / 3.25	2.25 / 3.12	2.19 / 3.01	2.14 / 2.94	2.10 / 2.86	2.07 / 2.80	2.02 / 2.70	1.97 / 2.62	1.91 / 2.51	1.86 / 2.42	1.82 / 2.34	1.76 / 2.25	1.74 / 2.20	1.69 / 2.12	1.67 / 2.08	1.64 / 2.02	1.61 / 1.98	1.59 / 1.96
34	4.13 / 7.44	3.28 / 5.29	2.88 / 4.42	2.65 / 3.93	2.49 / 3.61	2.38 / 3.38	2.30 / 3.21	2.23 / 3.08	2.17 / 2.97	2.12 / 2.89	2.08 / 2.82	2.05 / 2.76	2.00 / 2.66	1.95 / 2.58	1.89 / 2.47	1.84 / 2.38	1.80 / 2.30	1.74 / 2.21	1.71 / 2.15	1.67 / 2.08	1.64 / 2.04	1.61 / 1.98	1.59 / 1.94	1.57 / 1.91
36	4.11 / 7.39	3.26 / 5.25	2.86 / 4.38	2.63 / 3.89	2.48 / 3.58	2.36 / 3.35	2.28 / 3.18	2.21 / 3.04	2.15 / 2.94	2.10 / 2.86	2.06 / 2.78	2.03 / 2.72	1.98 / 2.62	1.93 / 2.54	1.87 / 2.43	1.82 / 2.35	1.78 / 2.26	1.72 / 2.17	1.69 / 2.12	1.65 / 2.04	1.62 / 2.00	1.59 / 1.94	1.56 / 1.90	1.55 / 1.87
38	4.10 / 7.35	3.25 / 5.21	2.85 / 4.34	2.62 / 3.86	2.46 / 3.54	2.35 / 3.32	2.26 / 3.15	2.19 / 3.02	2.14 / 2.91	2.09 / 2.82	2.05 / 2.75	2.02 / 2.69	1.96 / 2.59	1.92 / 2.51	1.85 / 2.40	1.80 / 2.32	1.76 / 2.22	1.71 / 2.14	1.67 / 2.08	1.63 / 2.00	1.60 / 1.97	1.57 / 1.90	1.54 / 1.86	1.53 / 1.84
40	4.08 / 7.31	3.23 / 5.18	2.84 / 4.31	2.61 / 3.83	2.45 / 3.51	2.34 / 3.29	2.25 / 3.12	2.18 / 2.99	2.12 / 2.88	2.07 / 2.80	2.04 / 2.73	2.00 / 2.66	1.95 / 2.56	1.90 / 2.49	1.84 / 2.37	1.79 / 2.29	1.74 / 2.20	1.69 / 2.11	1.66 / 2.05	1.61 / 1.97	1.59 / 1.94	1.55 / 1.88	1.53 / 1.84	1.51 / 1.81
42	4.07 / 7.27	3.22 / 5.15	2.83 / 4.29	2.59 / 3.80	2.44 / 3.49	2.32 / 3.26	2.24 / 3.10	2.17 / 2.96	2.11 / 2.86	2.06 / 2.77	2.02 / 2.70	1.99 / 2.64	1.94 / 2.54	1.89 / 2.46	1.82 / 2.35	1.78 / 2.26	1.73 / 2.17	1.68 / 2.08	1.64 / 2.02	1.60 / 1.94	1.57 / 1.91	1.54 / 1.85	1.51 / 1.80	1.49 / 1.78
44	4.06 / 7.24	3.21 / 5.12	2.82 / 4.26	2.58 / 3.78	2.43 / 3.46	2.31 / 3.24	2.23 / 3.07	2.16 / 2.94	2.10 / 2.84	2.05 / 2.75	2.01 / 2.68	1.98 / 2.62	1.92 / 2.52	1.88 / 2.44	1.81 / 2.32	1.76 / 2.24	1.72 / 2.15	1.66 / 2.06	1.63 / 2.00	1.58 / 1.92	1.56 / 1.88	1.52 / 1.82	1.50 / 1.78	1.48 / 1.75

Degrees of freedom for denominator of F

(continued)

Table A.6 (continued)

Degrees of freedom for denominator of F

n_1 degrees of freedom (for numerator of F) — upper value = .05 level; lower value = .01 level

n_2	1	2	3	4	5	6	7	8	9	10	11	12	14	16	20	24	30	40	50	75	100	200	500	∞
46	4.05 / 7.21	3.20 / 5.10	2.81 / 4.24	2.57 / 3.76	2.42 / 3.44	2.30 / 3.22	2.22 / 3.05	2.14 / 2.92	2.09 / 2.82	2.04 / 2.73	2.00 / 2.66	1.97 / 2.60	1.91 / 2.50	1.87 / 2.42	1.80 / 2.30	1.75 / 2.22	1.71 / 2.13	1.65 / 2.04	1.62 / 1.98	1.57 / 1.90	1.54 / 1.86	1.51 / 1.80	1.48 / 1.76	1.46 / 1.72
48	4.04 / 7.19	3.19 / 5.08	2.80 / 4.22	2.56 / 3.74	2.41 / 3.42	2.30 / 3.20	2.21 / 3.04	2.14 / 2.90	2.08 / 2.80	2.03 / 2.71	1.99 / 2.64	1.96 / 2.58	1.90 / 2.48	1.86 / 2.40	1.79 / 2.28	1.74 / 2.20	1.70 / 2.11	1.64 / 2.02	1.61 / 1.96	1.56 / 1.88	1.53 / 1.84	1.50 / 1.78	1.47 / 1.73	1.45 / 1.70
50	4.03 / 7.17	3.18 / 5.06	2.79 / 4.20	2.56 / 3.72	2.40 / 3.41	2.29 / 3.18	2.20 / 3.02	2.13 / 2.88	2.07 / 2.78	2.02 / 2.70	1.98 / 2.62	1.95 / 2.56	1.90 / 2.46	1.85 / 2.39	1.78 / 2.26	1.74 / 2.18	1.69 / 2.10	1.63 / 2.00	1.60 / 1.94	1.55 / 1.86	1.52 / 1.82	1.48 / 1.76	1.46 / 1.71	1.44 / 1.68
55	4.02 / 7.12	3.17 / 5.01	2.78 / 4.16	2.54 / 3.68	2.38 / 3.37	2.27 / 3.15	2.18 / 2.98	2.11 / 2.85	2.05 / 2.75	2.00 / 2.66	1.97 / 2.59	1.93 / 2.53	1.88 / 2.43	1.83 / 2.35	1.76 / 2.23	1.72 / 2.15	1.67 / 2.06	1.61 / 1.96	1.58 / 1.90	1.52 / 1.82	1.50 / 1.78	1.46 / 1.71	1.43 / 1.66	1.41 / 1.64
60	4.00 / 7.08	3.15 / 4.98	2.76 / 4.13	2.52 / 3.65	2.37 / 3.34	2.25 / 3.12	2.17 / 2.95	2.10 / 2.82	2.04 / 2.72	1.99 / 2.63	1.95 / 2.56	1.92 / 2.50	1.86 / 2.40	1.81 / 2.32	1.75 / 2.20	1.70 / 2.12	1.65 / 2.03	1.59 / 1.93	1.56 / 1.87	1.50 / 1.79	1.48 / 1.74	1.44 / 1.68	1.41 / 1.63	1.39 / 1.60
65	3.99 / 7.04	3.14 / 4.95	2.75 / 4.10	2.51 / 3.62	2.36 / 3.31	2.24 / 3.09	2.15 / 2.93	2.08 / 2.79	2.02 / 2.70	1.98 / 2.61	1.94 / 2.54	1.90 / 2.47	1.85 / 2.37	1.80 / 2.30	1.73 / 2.18	1.68 / 2.09	1.63 / 2.00	1.57 / 1.90	1.54 / 1.84	1.49 / 1.76	1.46 / 1.71	1.42 / 1.64	1.39 / 1.60	1.37 / 1.56
70	3.98 / 7.01	3.13 / 4.92	2.74 / 4.08	2.50 / 3.60	2.35 / 3.29	2.23 / 3.07	2.14 / 2.91	2.07 / 2.77	2.01 / 2.67	1.97 / 2.59	1.93 / 2.51	1.89 / 2.45	1.84 / 2.35	1.79 / 2.28	1.72 / 2.15	1.67 / 2.07	1.62 / 1.98	1.56 / 1.88	1.53 / 1.82	1.47 / 1.74	1.45 / 1.69	1.40 / 1.62	1.37 / 1.56	1.35 / 1.53
80	3.96 / 6.96	3.11 / 4.88	2.72 / 4.04	2.48 / 3.56	2.33 / 3.25	2.21 / 3.04	2.12 / 2.87	2.05 / 2.74	1.99 / 2.64	1.95 / 2.55	1.91 / 2.48	1.88 / 2.41	1.82 / 2.32	1.77 / 2.24	1.70 / 2.11	1.65 / 2.03	1.60 / 1.94	1.54 / 1.84	1.51 / 1.78	1.45 / 1.70	1.42 / 1.65	1.38 / 1.57	1.35 / 1.52	1.32 / 1.49
100	3.94 / 6.90	3.09 / 4.82	2.70 / 3.98	2.46 / 3.51	2.30 / 3.20	2.19 / 2.99	2.10 / 2.82	2.03 / 2.69	1.97 / 2.59	1.92 / 2.51	1.88 / 2.43	1.85 / 2.36	1.79 / 2.26	1.75 / 2.19	1.68 / 2.06	1.63 / 1.98	1.57 / 1.89	1.51 / 1.79	1.48 / 1.73	1.42 / 1.64	1.39 / 1.59	1.34 / 1.51	1.30 / 1.46	1.28 / 1.43
125	3.92 / 6.84	3.07 / 4.78	2.68 / 3.94	2.44 / 3.47	2.29 / 3.17	2.17 / 2.95	2.08 / 2.79	2.01 / 2.65	1.95 / 2.56	1.90 / 2.47	1.86 / 2.40	1.83 / 2.33	1.77 / 2.23	1.72 / 2.15	1.65 / 2.03	1.60 / 1.94	1.55 / 1.85	1.49 / 1.75	1.45 / 1.68	1.39 / 1.59	1.36 / 1.54	1.31 / 1.46	1.27 / 1.40	1.25 / 1.37
150	3.91 / 6.81	3.06 / 4.75	2.67 / 3.91	2.43 / 3.44	2.27 / 3.14	2.16 / 2.92	2.07 / 2.76	2.00 / 2.62	1.94 / 2.53	1.89 / 2.44	1.85 / 2.37	1.82 / 2.30	1.75 / 2.20	1.71 / 2.12	1.64 / 2.00	1.59 / 1.91	1.54 / 1.83	1.47 / 1.72	1.44 / 1.66	1.37 / 1.56	1.34 / 1.51	1.29 / 1.43	1.25 / 1.37	1.22 / 1.33
200	3.89 / 6.76	3.04 / 4.71	2.65 / 3.88	2.41 / 3.41	2.26 / 3.11	2.14 / 2.90	2.05 / 2.73	1.98 / 2.60	1.92 / 2.50	1.87 / 2.41	1.83 / 2.34	1.80 / 2.28	1.74 / 2.17	1.69 / 2.09	1.62 / 1.97	1.57 / 1.88	1.52 / 1.79	1.45 / 1.69	1.42 / 1.62	1.35 / 1.53	1.32 / 1.48	1.26 / 1.39	1.22 / 1.33	1.19 / 1.28
400	3.86 / 6.70	3.02 / 4.66	2.62 / 3.83	2.39 / 3.36	2.23 / 3.06	2.12 / 2.85	2.03 / 2.69	1.96 / 2.55	1.90 / 2.46	1.85 / 2.37	1.81 / 2.29	1.78 / 2.23	1.72 / 2.12	1.67 / 2.04	1.60 / 1.92	1.54 / 1.84	1.49 / 1.74	1.42 / 1.64	1.38 / 1.57	1.32 / 1.47	1.28 / 1.42	1.22 / 1.32	1.16 / 1.24	1.13 / 1.19
1000	3.85 / 6.66	3.00 / 4.62	2.61 / 3.80	2.38 / 3.34	2.22 / 3.04	2.10 / 2.82	2.02 / 2.66	1.95 / 2.53	1.89 / 2.43	1.84 / 2.34	1.80 / 2.26	1.76 / 2.20	1.70 / 2.09	1.65 / 2.01	1.58 / 1.89	1.53 / 1.81	1.47 / 1.71	1.41 / 1.61	1.36 / 1.54	1.30 / 1.44	1.26 / 1.38	1.19 / 1.28	1.13 / 1.19	1.08 / 1.11
∞	3.84 / 6.64	2.99 / 4.60	2.60 / 3.78	2.37 / 3.32	2.21 / 3.02	2.09 / 2.80	2.01 / 2.64	1.94 / 2.51	1.88 / 2.41	1.83 / 2.32	1.79 / 2.24	1.75 / 2.18	1.69 / 2.07	1.64 / 1.99	1.57 / 1.87	1.52 / 1.79	1.46 / 1.69	1.40 / 1.59	1.35 / 1.52	1.28 / 1.41	1.24 / 1.36	1.17 / 1.25	1.11 / 1.15	1.00 / 1.00

■ = .05 level; ■ = .01 level.

Reprinted, by permission, from G.W. Snedecor and W.G. Cochran, 1980, *Statistical methods*, 7th ed. (Ames, IA: Iowa State University Press).

Table A.7 Critical Values of Chi Square

Probability under H_0 that $\chi^2 \geq$ chi square

df	.99	.98	.95	.90	.80	.70	.50	.30	.20	.10	.05	.02	.01	.001
1	.00016	.00063	.0039	.016	.064	.15	.46	1.07	1.64	2.71	3.84	5.41	6.64	10.83
2	.02	.04	.10	.21	.45	.71	1.39	2.41	3.22	4.60	5.99	7.82	9.21	13.82
3	.12	.18	.35	.58	1.00	1.42	2.37	3.66	4.64	6.25	7.82	9.84	11.34	16.27
4	.30	.43	.71	1.06	1.65	2.20	3.36	4.88	5.99	7.78	9.49	11.67	13.28	18.46
5	.55	.75	1.14	1.61	2.34	3.00	4.35	6.06	7.29	9.24	11.07	13.39	15.09	20.52
6	.87	1.13	1.64	2.20	3.07	3.83	5.35	7.23	8.56	10.64	12.59	15.03	16.81	22.46
7	1.24	1.56	2.17	2.83	3.82	4.67	6.35	8.38	9.80	12.02	14.07	16.62	18.48	24.32
8	1.65	2.03	2.73	3.49	4.59	45.53	7.34	9.52	11.03	13.36	15.51	18.17	20.09	26.12
9	2.09	2.53	3.32	4.17	5.38	6.39	8.34	10.66	12.24	14.68	16.92	19.68	21.67	27.88
10	2.56	3.06	3.94	4.86	6.18	7.27	9.34	11.78	13.44	15.99	18.31	21.16	23.21	29.59
11	3.05	3.61	4.58	5.58	6.99	8.15	10.34	12.90	14.63	17.28	19.68	22.62	24.72	31.26
12	3.57	4.18	5.23	6.30	7.81	9.03	11.34	14.01	15.81	18.55	21.03	24.05	26.22	32.91
13	4.11	4.76	5.89	7.04	8.63	9.93	12.34	15.12	16.98	19.81	22.36	25.47	27.69	34.53
14	4.66	5.37	6.57	7.79	9.47	10.82	13.34	16.22	18.15	21.06	23.68	26.87	29.14	26.12
15	5.23	5.98	7.26	8.55	10.31	11.72	14.34	17.32	19.31	22.31	25.00	28.26	30.58	37.70
16	5.81	6.61	7.96	9.31	11.15	12.62	15.34	18.42	20.46	23.54	26.30	29.63	32.00	39.29
17	6.41	7.26	8.67	10.08	12.00	13.53	16.34	19.51	21.62	24.77	27.59	31.00	33.41	40.75
18	7.02	7.91	9.39	10.86	12.86	14.44	17.34	20.60	22.76	25.99	28.87	32.35	34.80	42.31
19	7.63	8.57	10.12	11.65	13.72	15.35	18.34	21.69	23.90	27.20	30.14	33.69	36.19	43.82
20	8.26	9.24	10.85	12.44	14.58	16.27	19.34	22.78	25.04	28.41	31.41	35.02	37.57	45.32
21	8.90	9.92	11.59	13.24	15.44	17.18	20.34	23.86	26.17	29.62	32.67	36.34	38.93	46.80
22	9.54	10.60	12.34	14.04	16.31	18.10	21.34	24.94	27.30	30.81	33.92	37.66	40.29	48.27
23	10.20	11.29	13.09	14.85	17.19	19.02	22.34	26.02	28.43	32.01	35.17	38.97	41.64	49.73
24	10.86	11.99	13.85	15.66	18.06	19.94	23.34	27.10	29.55	33.20	36.42	40.27	42.98	51.18
25	11.52	12.70	14.61	16.47	18.94	20.87	24.34	28.17	30.68	34.38	37.65	41.57	44.31	52.62
26	12.20	13.41	15.38	17.29	19.82	21.79	25.34	29.25	31.80	35.56	38.88	42.86	45.64	54.05
27	12.88	14.12	16.15	18.11	20.70	22.72	26.34	30.32	32.91	36.74	40.11	44.14	46.96	55.48
28	13.56	14.85	16.93	18.94	21.59	23.65	27.34	31.39	34.03	37.92	41.34	45.42	48.28	56.89
29	14.26	15.57	17.71	19.77	22.48	24.58	28.34	32.46	35.14	39.09	42.56	46.69	49.59	58.30
30	14.95	16.31	18.49	20.60	23.36	25.51	29.34	33.53	36.25	40.26	43.77	47.96	50.89	59.70

Adapted, by permission, from R.A. Fisher and F. Yates, 1995, *Statistical tables for biological agricultural and medical research* (Essex, United Kingdom: Pearson Education), 47.

A Brief Historical Overview of Research in Physical Activity in the United States

■ ■ ■ ■ ■ ■ ■

Early Physical Education Research

Postwar Research

All the types of research we talk about in this book didn't just happen by chance. Research in the United States has systematically evolved since the mid-1800s. We present here a chronicle of that evolutionary process. Our information came from several secondary sources (Hackensmith, 1966; Lee, 1983; Leonard & Affleck, 1947; Van Dalen & Bennett, 1971). Particularly helpful were M. Lee's summaries of research (Lee, 1983, chaps. 6, 9, 13, and 16).

The subdisciplines in the study of physical activity (e.g., biomechanics, exercise physiology, motor behavior, sport psychology, sport sociology, sport pedagogy, adapted physical activity, sport history, sport philosophy) all began at various times and in various fields of study. Not until the 1960s did research in these subdisciplines coalesce under a common area, the study of physical activity. The following account represents the general field of physical education, which developed into the study of physical activity. In addition to knowing the history of the field of physical activity, graduate students should know more about the intellectual history of their specific subdiscipline. Massengale and Swanson (1996) provide a history of the subdisciplines of adapted physical education, exercise physiology, biomechanics, motor behavior (including motor learning, control, and development), sport psychology, sport pedagogy, sport sociology, sport history, and sport philosophy. We urge you to read the chapter from their book that most closely aligns with your interest. In addition, the 75th anniversary issue of *Research Quarterly for Exercise and Sport* provides interesting analyses of research and publishing in our field.

Early Physical Education Research

The beginnings of systematic research in physical education in the United States closely followed the establishment in 1854 of the first department of hygiene and physical education at Amherst College. Although John Hooker was the first director of this department, the appointment of Edward Hitchcock (MD, Harvard Medical School, 1853) as director in 1861 represents the beginnings of research efforts in physical education. Dr. Hitchcock frequently made anthropometric measurements and used chin-ups to assess arm strength. Over the next 20 years, the list of measurements taken was extended considerably. The tabulation of these measurements was first published in the *Anthropometric Manual* (1887, revised editions in 1889, 1893, and 1900), which represents one of the first data-based research publications of physical education in the United States.

Then, Dudley Sargent (MD, Yale Medical School, 1878) established a private gymnasium in New York, where he began taking a series of bodily measurements of participants. Two years later, he was appointed assistant professor at Harvard and director of the new Hemenway Gymnasium. All entering freshmen were given an examination, including both strength tests and anthropometric measures. Dr. Sargent collected more than 50,000 anthropometric measurements of individuals during his career. In fact, lifelike statues of typical American youths were constructed on the basis of his measurements and displayed at the 1893 Chicago World's Fair. This might be considered the first formal research presentation in physical education in the United States. Sargent also led in the development of strength testing, and he used these tests to determine membership of Harvard's athletic teams. Following Sargent's lead, several women's colleges (Bryn Mawr, Mount Holyoke, Radcliffe, Rockford, Vassar, and Wellesley) established programs of anthropometric measurements and strength tests of students in the 1880s.

When a new gymnasium was constructed at Johns Hopkins University in 1883, Edward Hartwell (PhD, Johns Hopkins; MD, Medical College of Ohio) was appointed its director. There, in 1897, he began the use of survey research in physical education through his evaluations of gymnastics in the United States.

To measure throwing, running speed, and distance jumped, Dr. Luther Gulick devised the very first achievement test in 1890. Called the Pentathlon Test, it was initially created for the Athletic League of the YMCA of America but was further developed in the early 1900s for the Public School Athletic League in New York.

In the late 1800s and early 1900s, several books reported, summarized, or influenced early research in physical education. Among those were Blaikie's *How to Get Strong and How to Stay So* (1879), which reported Sargent's early work and influenced Harvard (Blaikie was a Harvard alumnus) to employ Sargent. In Berlin in 1885, DuBois-Reymond published *Physiology of Exercise*, which was translated into English in *Popular Science Monthly*. Following Sargent's lead in anthropometric measurement, Seaver published *Anthropometric and Physical Examinations* in 1896.

The beginning of the 20th century saw a continuing interest in physical education research. In particular, an interest arose in tests of physical achievement and cardiovascular efficiency. Physicians (both inside and outside the physical education profession) showed increased interest in classifying exercise intensity based on cardiovascular function. At Springfield College (Massachusetts), James McCurdy (1895–1935) began studying changes occurring during adolescence in heart function and blood pressure. The first of the widely used cardiovascular efficiency tests was developed in 1917 by Schneider at Connecticut Wesleyan College and was used in World War I for evaluating the fitness of aviators.

Physical achievement testing advanced considerably in the early 1900s, and in 1922 the American Physical Education Association set up a committee under McCurdy's direction to develop motor ability tests. The two most recognized tests of the 1920s were Sargent's Physical Test of Man and Roger's Physical Fitness Index. By the 1930s many schools and colleges were administering physical achievement tests. In fact, this increase in motor performance and fitness testing probably led to the establishment in 1930 of *Research Quarterly* by the American Physical Education Association (for a review of the history of *Research Quarterly*, see Park, 1980). This was the first journal established specifically to report research in physical education.

The American Academy of Physical Education was established in 1926 with the purpose of electing outstanding scholars and leaders in the field to a limited and select membership. The academy attempts to use the research knowledge developed by scholars in the field to influence the direction that the study of physical activity takes. The organization continues today under the name American Academy of Kinesiology and Physical Education, with a maximum at one time of 125 elected active fellows. Since 1926 more than 400 outstanding U.S. scholars have been elected to membership. In addition, the academy has over 75 international fellows representing more than 20 other countries.

In the 1930s, tests and measurements became the most active research area in physical education. Many physical fitness and motor ability tests were developed and widely used in both public schools and colleges and universities. Names of particular importance in test development were David Brace and C.H. McCloy.

The progressive education movement of the 1930s, frequently characterized as "teach what the child wants to learn," had a significant impact on both academic education and physical education. Physical educators concentrated on making the curriculum fun. This resulted in questionnaire research designed to discover what children liked to do. However, a statement by McCloy (published in 1961, a year after his death) seemed to put everything into perspective: "I hope the next fifty years will cause physical education to . . . seek for facts, proved objectively; to question principles based on average opinions of people who don't know but are all anxious to contribute their average ignorance to form a consensus of uninformed dogma."

As testing became more popular in the schools during the 1930s, the use of true–false knowledge tests for physical education activities increased considerably. With the advent of new statistical techniques, both physical and knowledge tests gained increased scientific rigor and provided standards for physical education. Sometimes, however, the emphasis on numbers got out of hand: "We lived in one long orgy of tabulation. . . . Mountains of fact were piled up, condensed, summarized, and interpreted by the new quantitative technique. The air was full of normal curves, standard deviations, coefficients of correlation, regressive equations" (Rugg, 1941, 182).

Postwar Research

The outbreak of World War II brought renewed interest in physical fitness because large numbers of the men drafted did not meet the minimum standards for physical fitness. In fact, one third of the men examined were found unfit for service, and even those accepted generally lacked adequate levels of physical fitness. Many physical education leaders were assigned to test and condition servicemen. Thus, many advances were made in both testing and training in exercise physiology.

World War II also led to the development of the area known today as motor learning and control. Many psychologists were involved in developing pilot training programs. Considerable motor coordination is involved in learning to control an airplane, and this led to many studies of motor skill performance and learning. Unfortunately, when the war ended, most of the interest in motor skill acquisition was lost as experimental psychologists returned to their interest in cognitive performance.

The 1950s brought a renewed interest in physical fitness. The 1953 publication of the Kraus-Weber test results revealed the poor record of American children when compared with European children. Although the test does not assess many of the factors considered important in fitness today (e.g., cardiovascular endurance and strength), the fact that nearly 60% of the American children failed compared with less than 9% of the European children attracted national attention. As a result, President Eisenhower called a special White House conference in 1956. This conference and several subsequent ones led to the establishment of the President's Council on Youth Fitness and the development of the AAHPERD Fitness Test. This boosted research in exercise physiology and physical activity tests and measurements.

In the 1960s, research in physical education began to expand rapidly. Franklin Henry's historic memory drum theory (Henry & Rogers, 1960) launched a renewed interest in motor learning and control. Henry (1964) contributed further with his paper about physical education as a discipline. This led to the identification of a knowledge base for physical education that was frequently called kinesiology, human movement, or physical activity. Research expanded dramatically as exercise physiology, biomechanics, motor learning and control, motor development, sport psychology, and sport sociology began to develop as subdisciplines and to produce knowledge about movement. The *Research Quarterly* was expanded, and such new journals as *Medicine and Science in Sport* and the *Journal of Motor Behavior* emerged.

The 1970s saw continued interest in research in the study of physical activity and renewed interest in research within the profession of physical education. Observational techniques were developed and refined that allowed accurate assessment of teaching behavior and student–teacher interaction. The area of classroom observation called *research on teaching* was developed for use in the gymnasium and playing fields in physical education. Research on curriculum theory in physical education, measurement and evaluation, and the history of sport and physical education received fresh attention.

Many researchers from the discipline of physical activity began to establish specific professional groups outside the traditional affiliations these researchers had held in AAHPERD. Such groups as the American College of Sports Medicine, the North American Society for Sport History, the North American Society for the Psychology of Sport and Physical Activity, and several others contributed to the research base. Responding to these inroads into membership and activities, AAHPERD created various academies, elevated the Research Consortium to its research arm, and changed the name of *Research Quarterly* on its 50th anniversary to *Research Quarterly for Exercise and Sport*. For its first issue under the new name (Safrit, 1980), the editor and the advisory committee solicited papers from the various subdisciplines in the study of physical activity: biomechanics, exercise physiology, psychology of sport, measurement and research design, sociology of sport, motor development, and motor behavior.

This increased emphasis on research in both the study of physical activity (see Massengale & Swanson, 1966) and the professional base of physical education can lead only to increased knowledge and higher scholarly standards. While the discipline based on the study of physical activity continues to undergo growing pains and an identity crisis (e.g., What shall we call ourselves—kinesiology, exercise and sport science, movement science?), high-quality graduate programs, research, and scholarship are evident. For example, a special issue of *Quest* was published in the fall of 1987 devoted completely to graduate education. New scholarly journals have evolved, particularly associated with specialized subareas of the discipline (e.g., *Journal of Exercise and Sport Psychology, Journal of Sport Sociology, Pediatric Exercise Science, Journal of Sport Management*).

School- and community-based exercise and sport skill programs are now refining their goals (developing physical fitness and motor skills) based on a solid knowledge of physical activity and on increased sophistication in planning and teaching children and adults. New scholarly journals have evolved in these areas, such as the *Journal of Teaching in Physical Education,* to accommodate the growth in knowledge produced. Each of these factors should result in a more knowledgeable, physically fit, and skillful population during the next 20 to 30 years.

APPENDIX C

Sample Consent Forms

Application for the Conduct of Research Involving Human Subjects
Arizona State University
University Human Subjects Research Review Committee

The Arizona State University Human Subjects Research Review Committee reviews all requests to conduct research involving human subjects. In completing the following application, be advised that the persons reviewing it may be entirely unfamiliar with the field of study involved. Present the request in typewritten form and in nontechnical terms understandable to the committee. It is the investigator's responsibility to give information about research procedure that is most likely to entail risk *but not to express judgment about the risk.* Please submit a copy of your complete proposal, an informed-consent/assent form as subjects will view it, and a curriculum vitae or biographical sketch.

Principal investigator/director:	Department/center:	Date of request:

Type of review: New ☐   Renewal ☐   Continuation ☐

Exempt ☐   Identify by numbers that apply (see page 2) _____

Expedite ☐   Identify by numbers that apply (see page 4) _____

If Renewal or Continuation, are there any substantive changes?   Yes ☐   No ☐

Project title:

Agency submitted to:	Submission date:	Location of project:

1. General purpose of the research:

2. Data obtained by: Mail ☐   Telephone ☐   Interview ☐

   Observation ☐   Experiment ☐   Secondary Source ☐

   Other (explain) _____

3. Project description: The committee must have sufficient information, nontechnical and detailed, about what will happen with/to subjects to evaluate/estimate the risks. Assurance from the investigator, no matter how strong, will not substitute for a description of the transaction between investigator and subject. *If a questionnaire is used, attach a copy.*

When visual or auditory stimuli, chemical substances, or other measures might affect the health of subjects, a statement from a qualified person or other appropriate documentation will aid in evaluating the nature of any risk created. In questionable cases, the committee will require such documentation.

---

4. Subject selection: Will subjects be less than 18 years of age?   Yes ☐   No ☐

How many subjects will participate? _____   Male ☐   Female ☐   Age ____ to _____

Will subjects be students at Arizona State University?   Yes ☐   No ☐

Source: _____

---

5. How will subjects be selected, enlisted, or recruited?

---

6. How will subjects be informed of procedures, intent of the study, and potential risks to them?

---

7. What steps will be taken to allow subjects to withdraw at any time without prejudice?

---

8. How will subjects' privacy be maintained and confidentiality guaranteed?

Attachments: Please indicate those items we can expect to find as attachments.

Complete proposal ☐   Informed-consent form (as subjects will view it) ☐

Questionnaire ☐   Assent form (as child will view it) ☐

Other instrumentation ☐   Curriculum vitae or biographical sketch ☐

Other documentation _____

In making this application, I certify that I have read and understand the *Policies and Procedures for Projects that Involve Human Subjects*, and that I intend to comply with the letter and spirit of the university policy. Significant changes in the protocol will be submitted to the committee for written approval prior to these changes being put into practice. Informed-consent/assent records of the subjects will be kept for at least (3) years after the completion of the research.

Signatures: *Principal investigator (faculty)*	*Department chair*	Date

---

This application has been reviewed by the Arizona State University IRB:

Full Board Review ☐   Exempt ☐   Expedite ☐   Categories: _____

Approved ☐   Deferred ☐   Disapproved ☐

Project requires review more often than annual ☐   every _____ months.

Renewal or continuation ☐   Approved with no substantive changes ☐

Disapproved ☐   Approved with changes attached ☐

Third-party verification sought ☐

Comments, modification/conditions for approval, or reason for disapproval:

Signature:

Chair of the University Committee   Date

Sample Form C.2

Informed-Consent Form for Adults

Read and address each numbered element of this model form in developing an informed-consent form for the proposed human research study. The items numbered and in quotations are to be included in the consent form. PLEASE NUMBER THE CONSENT FORM FOR SUBMISSION TO THE COMMITTEE. You may request the numbering be waived during data collection. The consent form must be written in lay language and must be typewritten. The language may be further simplified to meet the needs of a specific population. Add additional comments when appropriate.

1. *"Investigator's name,* who is *title/position,* has requested my participation in a research study at this institution. The title of the research is *title of research.*" [Place title of project at top of all pages of consent form.]

2. "I have been informed that the purpose of the research is to . . ." [Describe the justification for the research. If appropriate, include the number of subjects involved and why the subject is included.]

3. "My participation will involve . . ." [Describe the subject's participation and identify those aspects of participation that are experimental. Indicate the expected duration of the subject's participation.]

4. "I understand there are foreseeable risks or discomforts to me if I agree to participate in the study. The possible risks are . . ." "Possible discomforts include . . ." [Any foreseeable risks or discomforts are to be explained/described.]

    OR

    "There are no foreseeable risks or discomforts."

5. "I understand that there are alternative procedures available. Alternative procedures include . . ." [Describe any alternative procedures to be included in language the subject can understand.]

    OR

    "There are no feasible alternative procedures available for this study."

6. "I understand that the possible benefits of my participation in the research are . . ." [Describe the benefits of participation, or lack of benefits, to the individual as well as to society.]

7. "I understand that the results of the research study may be published but that my name or identity will not be revealed. In order to maintain confidentiality of my records, *name of investigator* will . . ." [Indicate specifically how the investigator will keep the names of the subjects confidential, the use of subject codes, how this information will be secured, and who will have access to the confidential information. "Confidentiality will be maintained" is not acceptable.]

8. "I understand that in case of injury I can expect to receive the following treatment or care which will be provided at my expense:" [If *more* than minimal risk of foreseeable injury is anticipated, describe the facilities, medical treatment, or services that will be made available in the event of injury or illness to a subject. Description may include on- and off-campus services.]

    OR

    "I have been advised that the research in which I will be participating does not involve more than minimal risk." [If the research will *not* involve more than minimal risk, briefly explain how or why the investigator determined that the subject would be exposed to no more than minimal risk.]

9. "I have been informed that I will be compensated for my participation as follows:" [If compensation is to be provided to the subject, include amount of compensation, method of payment, and schedule for payment including whether payment will be made in increments or in one lump sum.]

    OR

    "I have been informed that I will not be compensated for my participation."

10. "I have been informed that any questions I have concerning the research study or my participation in it, before or after my consent, will be answered by *name of individual, address, and telephone number.*" [This refers to the principal investigator. In the event the investigator is a student, the name of the doctoral or thesis advisor (responsible faculty member) must be included.]

11. "I understand that in case of injury, if I have questions about my rights as a subject/participant in this research, or if I feel I have been placed at risk, I can contact the Chair of the Human Subjects Research Review Committee." [This information must be included in all consent forms.]

12. "I have read the above information. The nature, demands, risks, and benefits of the project have been explained to me. I knowingly assume the risks involved and understand that I may withdraw my consent and discontinue participation at any time without penalty or loss of benefit to myself. In signing this consent form, I am not waiving any legal claims, rights, or remedies. A copy of this consent form will be given to me."

Subject's signature _____ Date _____

Other signature (if appropriate) _____ Date _____

13. "I certify that I have explained to the above individual the nature and purpose, the potential benefits, and possible risks associated with participation in this research study, have answered any questions that have been raised, and have witnessed the above signature."

14. "These elements of informed consent conform to the Assurance given by Arizona State University to the Department of Health and Human Services to protect the rights of human subjects."

15. "I have provided the subject/participant a copy of this signed consent document."

Signature of investigator _____ Date _____

Sample Form C.3

Informed-Consent Form for Minors

The elements of the informed-consent form for adults are used with the following variations:

1. *"Investigator's name,* who is *title/position* at Arizona State University, has requested my minor child's (ward's) participation in a research study at this institution. The title of the research is *title of research."*

2. [same as adult]

3. "My child's (ward's) participation will involve . . ."

4.–5. [same as adult]

6. "I understand that the possible benefits of my child's (ward's) participation in the research are . . ."

7. "I understand that the results of the research study may be published but that my child's (ward's) name or identity will not be revealed. In order to maintain confidentiality of my child's (ward's) records, *name of investigator* will . . ."

8. "I understand that in case of injury I can expect the following treatment or care to be provided at my expense . . ."

OR

"I have been advised that the research in which my child (ward) will be participating does not involve more than minimal risk."

9. "I have been informed that compensation for participation is as follows . . ."

OR

"I have been informed that I will not be compensated for my child's (ward's) participation."

10.–15. [same as adult]

Subject's signature _____ Date _____
(father, mother, legal guardian, or legally authorized official)

Other signature _____ Date _____

Child Assent Form

Language must be simplified as appropriate for the age-group used as subjects, such as:

I, _____, understand that my parents (mom and dad) have given permission (said it's okay) for me to take part in a project about _____

_____

done by _____

I am taking part because I want to, and I have been told that I can stop at any time I want to and I won't get in trouble (nothing bad will happen to me if I want to stop).

_____
Signature

OR

I, _____, understand that my parents have given permission for me to participate in a study concerning _____

_____

under the direction of _____

My involvement in this project is voluntary, and I have been told that I may withdraw from participation in this study at any time without penalty and loss of benefit to myself.

_____
Signature

Date filed: _____   Project no.: _____

Animal Protocol Review
Arizona State University Animal Care and Use Committee

Please read "Instructions for Completing Animal Protocol Review."

I. A. A single member of the university faculty and/or principal investigator is considered the responsible individual.

   Name: _____   Campus phone: _____

   B. University position and department: _____

   C. Project/program title: _____

   _____

   D. ___ Nonfunded research   ___ Teaching   ___ Grant/contract

   E. Protocol type: New ___   Renewal ___   Revision ___   Previous no. ___

   F. Research (see 1, 2, 3, and 4 below): ___

   Teaching (see 5 and 6 below): ___

   1. Granting agency: _____   Deadline: _____

   2. Coinvestigator(s): _____

   3. Graduate student: _____   Phone: _____

   4. Thesis/research degree program: _____

   _____

   5. Course title, schedule: _____

   6. Teaching assistant/laboratory instructor(s): _____

   _____

II. A. List species:

   B. Where will you prefer animals be housed? _____

   C. Does project require a waiver of provision(s) of PHS policy?

   No ___ Yes ___   (If yes, attach a Request of Waiver statement justifying the need for the waiver.)

   Review date(s): _____   Approval date: _____

   Signature, Chair: _____

III. A. *Abstract of planned use of animals.* Write a brief yet complete description of the planned use of animals. (Use additional pages if necessary.) Use language understandable to a layperson. One complete copy of a grant proposal or graduate student research proposal must be attached to the original.

   B. *Rationale for involving animals and the appropriateness of the species and number used* (include potential contribution the species may generate and state briefly why living vertebrates are required rather than some alternate model).

(continued)

Sample Form C.5 (continued)

IV. Animals to be used:

    A. Species (give both scientific and common names for unusual species):

       Is this a threatened or endangered species? Yes ___ No ___

    B. Estimated number: Per year _____ Entire study _____

       Are these estimated number of animals maximum for the entire study?

                                    Yes ___   No ___

    C. Sex and age (or weight range): _____

    D. Source (e.g., purchased, institutional breed, transferred from another study, donated, captured from wild):

       _____

    E. Will animals stay in investigator's lab at any time?   Yes ___   No ___

       If so, how long? _____

       If greater than 24 hours, state justification: _____

       _____

    All facilities or laboratories that house animals for longer than 24 hours must be inspected and approved according to DHEW policy.

V. Major categories of use:

    Please check those applicable and attach appropriate sections concerning methodology from your grant proposal or a brief description of the methodology to be used in nonfunded project or course.

    Yes   No

    ___   ___   a. Harvest tissue, blood, etc.

    ___   ___   b. Immunization for antibody production; describe antigen adjuvant used, route of immunization, method of obtaining blood.

    ___   ___   c. Physiologic measurements; if surgery is necessary, see "n."

    ___   ___   d. Dietary manipulations (e.g., caloric restrictions, specific constituent restriction).

    ___   ___   e. Pharmacologic/toxicologic material used, route of administration, etc.

    ___   ___   f. Immunologic studies.

    ___   ___   g. Behavioral studies.

    ___   ___   h. Irradiation; include Radioisotope Approval Form.

    ___   ___   i. Biohazardous materials (carcinogens, chemicals, etc.); include Biohazard Approval Form.

    ___   ___   j. Infectious agents; include Biohazard Approval Form.

    ___   ___   k. *Trauma, injury, burning, freezing, electric shock.

    ___   ___   l. *Environmental stress (e.g., temperature, long-term restraint, forced exercise, nutritional distress).

    ___   ___   m. Other:_____

    ___   ___   n. Surgery: *If "Yes," complete Item VI below.*

                       *Submit necessary justifications(s).

VI. Surgical procedures:

   A. Survival* _____ Nonsurvival _____
       *All survival surgery will be performed under aseptic conditions.

   B. Where is surgery to be performed?

   C. Person performing surgery; person's qualifications: _____

       _____

   D. Anesthetic regimen:

       Drug: _____ Dose: _____

       Route: _____ Duration: _____

       Monitoring procedures:

       Who will administer/monitor anesthesia, and what are their qualifications?

       _____

       If anesthetics are not used, justify:_____

       _____

   E. Postoperative pain or distress:

       Is postoperative pain or distress anticipated? Yes ___ No ___

       Will analgesics/tranquilizers be used? Yes ___ No ___

       Drug: _____ Dose: _____

       Route: _____ Duration: _____

       If pain/distress are anticipated but analgesics/tranquilizers will not be used, attach Justification of Pain or Distress statement.

       _____

   F. Postoperative care:

       Intensive care required? Yes ___ No ___

       What time period? _____

       Who will provide care? _____

       Postoperative routine:

       Who will provide? _____

       What monitoring will be performed? _____

       What drugs administered? _____

       Antibiotics (type/dosage/frequency)? _____

       Special care to be provided: _____

       _____

       Person(s) to contact in case of emergency:

       _____

       Phone: Office _____ Home _____

       What postoperative complications may be expected? _____

       _____

       Method of treatment: _____

       _____

(continued)

    G. Multiple surgical procedures: Will individual animals be subjected to more than one surgical procedure?

       ___ Yes   ___ No

       If yes, provide justification.

    H. Have all personnel on this protocol been certified in the federal mandated training requirement?

       ___ Yes   ___ No

VII. Euthanasia

    A. What is (are) method(s) of euthanasia?

       Chemical/gas: Agent _____   Dose _____

                          Agent _____   Dose _____

       Physical: ___   Cervical dislocation (mice, immature rats)

                     ___   Decapitation (rodents)

                     ___   Captive bolt

                     ___   Exsanguination under anesthesia

                     ___   Other* _____

                     ___   *Scientific justification (references if possible):

               _____

    B. Name any qualifications of person(s) performing euthanasia:

       _____

       _____

VIII. Assurance:

The information contained herein is accurate to the best of my knowledge. Procedures involving animals will be carried out humanely, and all procedures will be performed by or under the direction of trained or experienced persons. Any revisions to animal care and use in this project will be promptly forwarded to the Animal Care and Use Committee for review. *Revised protocols will not be used until committee clearance is received. The use of alternatives to animal models has been considered and found to be unacceptable at this time.*

_____   _____

Signature (individual listed on I.A. and graduate student,   Date
if applicable)

IX. Additional approval:

The department chair and the college dean must sign the protocol after approval by the Animal Care and Use Committee.

_____   _____

Chair   Date

_____   _____

Dean   Date

References

Adams, J.A. (1971). A closed-loop theory of motor learning. *Journal of Motor Behavior, 3,* 111–150.

Adelman, M.L. (1986). *A sporting time: New York City and the rise of modern athletics, 1820–1870.* Urbana: University of Illinois Press.

Ainsworth, B.E., Haskell, W.L., Whitt, M.C., Irwin, M.L., Swartz, A.M., Strath, S.J., O'Brien, W.L., Bassett, D.R., Jr., Schmitz, K.H., Emplaincourt, P.O., Jacobs, D.R., Jr., & Leon, A.S. (2000). Compendium of physical activities: An update of activity codes and MET intensities. *Medicine and Science in Sports and Exercise, 32* (9 Suppl), S498–504.

American Alliance for Health, Physical Education, Recreation and Dance. (1980). *AAHPERD health related physical fitness test manual.* Reston, VA: Author.

American Association for Health, Physical Education and Recreation. (1958). *AAHPER youth fitness test manual.* Washington, DC: Author.

American College of Sports Medicine. (1994). ACSM's guidelines for exercise testing and prescription. Baltimore: Williams and Wilkins.

American Heart Association. (1998). *1999 heart and stroke statistical update.* Dallas: American Heart Association.

American Psychological Association. (1959). *Graduate education in psychology.* Washington, DC: Author.

American Psychological Association. (1994). *Publication manual of the American Psychological Association* (4th ed.). Washington, DC: Author.

American Psychological Association (1999). Statistical methods in psychology journals: Guidelines and explanations. *American Psychologist, 54,* 594–604.

American Psychological Association (2001). Publication manual of the American Psychological Association (5th ed.). Washington, DC: Author.

American Physiological Association. (2004). Guidelines for reporting statistics in journals published by the American Physiological Association. *Journal of Applied Physiology, 97,* 457–459.

Anshel, M.H., & Marisi, D.Q. (1978). Effects of music and rhythm on physical performance. *Research Quarterly, 49,* 109–115.

APA Statement on Authorship of Research Papers. (1983, Sept. 14). *Chronicle of Higher Education 27,* 7.

Appleby, J. (1978). Modernization theory and the formation of modern social theories in England and America. *Comparative Studies in Society and History, 20,* 261.

Appleby, J., Hunt, L., & Jacob, M. (1994). *Telling the truth about history.* New York: Norton.

Arlin, M. (1977). One-study publishing typifies educational inquiry. *Educational Researcher, 6* (9), 11–15.

Atkinson, R.F. (1978). *Knowledge and explanation in history.* Ithaca, NY: Cornell University Press.

Bain, L.L. (1989). Interpretive and critical research in sport and physical education. *Research Quarterly for Exercise and Sport, 60,* 21–24.

Barnes, B. (1982). *T.S. Kuhn and social science.* New York: Columbia University Press.

Barnett, V., & Lewis, T. (1978). *Outliers in statistical data.* New York: Wiley.

Baumgartner, T.A. (1989). Norm-referenced measurement: Reliability. In M.J. Safrit & T.M. Wood (Eds.), *Measurement concepts in physical education and exercise science* (pp. 45–72). Champaign, IL: Human Kinetics.

Baumgartner, T.A., & Jackson, A.S. (1991). *Measurement for evaluation in physical education* (4th ed.). Dubuque, IA: Wm. C. Brown.

Baxter, N. (1993–94, winter). Is there another degree in your future? Choosing among professional and graduate schools. *Occupational Outlook Quarterly*, 19–49.

Behnke, A.R., & Wilmore, J.H. (1974). *Evaluation and regulation of body build and composition.* Englewood Cliffs, NJ: Prentice Hall.

Bender, T. (1986, June). Wholes and parts: The need for synthesis in American history. *Journal of American History, 73,* 120–136.

Bentham, J. (1970). *Introduction to the principles of morals and legislation.* London: Athalone Press.

Berelson, B. (1960). *Graduate education in the United States.* New York: McGraw-Hill.

Bernstein L., Henderson B.E., Hanisch R., Sullivan-Halley J., & Ross R.K. (1994). Physical exercise and reduced risk of breast cancer in young women. *Journal of the National Cancer Institute, 86,* 1403–1408.

Betz, N.E. (1987). Use of discriminant analysis in counseling psychology research. *Journal of Counseling Psychology, 34,* 393–403.

Biddle, S.J.H., Markland, D., Gilbourne, D., Chatzisarants, N.L.D., & Sparkes, A.C. (2001). Research methods in sport and exercise psychology: Quantitative and qualitative issues. *Journal of Sport Sciences, 19,* 777–809.

Blair, S.N. (1993). Physical activity, physical fitness, and health (1993 C.H. McCloy Research Lecture). *Research Quarterly for Exercise and Sport, 64,* 365–376.

Blair, S.N., Kohl, H.W., III, Paffenbarger, R.S., Clark, D.G., Cooper, K.H., & Gibbons, L.W. (1989). Physical fitness and all-cause mortality: A prospective study of healthy men and women. *JAMA, 262,* 2395–2401.

Bogdan, R.C., & Biklen, S.K. (1992). *Qualitative research for education: An introduction to theory and methods.* Boston: Allyn & Bacon.

Bogdan, R.C., & Biklen, S.K. (2003). *Qualitative research for education: An introduction to theories and method* (4th ed.). Boston: Allyn & Bacon.

Boorman, M.A. (1990). Effect of age and menopausal status on cardiorespiratory fitness in masters women endurance athletes. Master's thesis, Arizona State University, Tempe.

Borg, G.A. (1962). *Physical performance and perceived exertion.* Lund, Sweden: Gleerup.

Borg, W.R., & Gall, M.D. (1989). *Educational research* (5th ed.). New York: Longman.

Bouchard, C., Shepard, R.J., & Stephens, T. (Eds.). (1994). *Physical activity, fitness, and health.* Champaign, IL: Human Kinetics.

Bourque, L.B., & Fielder, E.P. (2003). *How to conduct in-person interviews for surveys* (2nd ed.). Thousand Oaks, CA: Sage.

Boyer, C.J. (1973). *The doctoral dissertation as an informational source: A study of scientific information flow.* Metuchen, NJ: Scarecrow Press.

Brown, J.A.C. (1954). *The social psychology of industry.* Middlesex, England: Penguin.

Buchowski, M.S., Darud, J.L., Chen, K.Y., & Sun, M. (1998). Work efficiency during step aerobic exercise in female instructors and noninstructors. *Research Quarterly for Exercise and Sport, 69,* 82–88.

Burgess, R.G. (Ed.). (1982). *Field research: A source book and field manual.* London: Allen & Unwin.

Campbell, D.T., & Stanley, J.C. (1963). *Experimental and quasi-experimental designs for research* (pp. 5–6). Chicago: Rand McNally.

Cardinal, B.J., & Thomas, J.R. (in press). The 75th anniversary of *Research Quarterly for Exercise and Sport:* An analysis of status and contributions. *Research Quarterly for Exercise and Sport.*

Carlberg, C.C., Johnson, D.W., Johnson, R., Maruyama, G., Kavale, K., Kulik, C., Kulik, J.A., Lysakowski, R.S., Pflaum, S.W., & Walberg, H. (1984). Meta-analysis in education: A reply to Slavin. *Educational Researcher, 13*(4), 16–23.

Carleton, R.A., Lasater, T.M., Assaf, A.R., Feldman, H.A., McKinaly, S., & the Pawtucket Heart Health Program Writing Group. (1995). The Pawtucket Heart Health Program: Community changes in cardiovascular risk factors and projected disease risk. *American Journal of Public Health, 85,* 777–785.

Carron, A.V., Hausenblas, H.A., & Mack, D. (1996). Social influence and exercise: A meta-analysis. *Journal of Sport and Exercise Psychology, 18,* 1–16.

Carron, A.V., Widmeyer, W.N., & Brawley, L.R. (1985). The development of an instrument to assess cohesion in sport teams: The group environment questionnaire. *Journal of Sport Psychology, 7,* 244–266.

Cartmell, M. (1993). *A view to a death in the morning: Hunting and nature through history.* Cambridge, MA: Harvard University Press.

Caspersen, C.J. (1989). Physical activity epidemiology: Concepts, methods, and applications to exercise science. *Exercise and Sport Sciences Reviews, 17,* 423–474.

Cheffers, J.T.F. (1973). The validation of an instrument design to expand the Flanders' system of interaction analysis to describe nonverbal interaction, different varieties of teacher behavior and pupil responses (Doctoral dissertation, Temple University, Philadelphia, 1972). *Dissertation Abstracts International, 34,* 1674A.

Chien, I. (1981). Appendix: An introduction to sampling. In L.H. Kidder (Ed.), *Selltiz, Wrightsman, and Cook's research methods in social relations* (4th ed.). New York: Holt, Rinehart & Winston.

Christina, R.W. (1989). Whatever happened to applied research in motor learning? In J.S. Skinner et al. (Eds.), *Future directions in exercise and sport science research* (pp. 411–422). Champaign, IL: Human Kinetics.

Clarke, H.H. (Ed.). (1968, December). *Physical Fitness Newsletter, 14* (4).

Clarke, H.H., & Clarke, D.H. (1970). *Research processes in physical education, recreation, and health.* Englewood Cliffs, NJ: Prentice Hall.

Clifford, C., & Feezell, R. (1997). *Coaching for character: Reclaiming the principles of sportsmanship.* Champaign: Human Kinetics.

Cohen, J. (1988). *Statistical power analysis for the behavioral sciences* (2nd ed.). New York: Academic Press.

Cohen, J. (1990). Things I have learned (so far). *American Psychologist, 45,* 1304–1312.

Cohen, J. (1994). The earth is round ($p < .05$). *American Psychologist, 49,* 997–1003.

Cohen, J., & Cohen, P. (1983). *Applied multiple regression in behavioral research.* New York: Holt, Rinehart & Winston.

Cohen, J., Cohen, P., West, S.G., & Aiken, L.S. (2003). *Applied multiple regression/correlation analysis for the behavioral sciences* (3rd ed.). Mahwah, NJ: Erlbaum.

Conover, W.J. (1971). *Practical nonparametric statistics.* New York: Wiley.

Cooper, H., & Hedges, L.V. (Eds.). (1994). *The handbook of research synthesis.* New York: Sage Foundation.

Coorough, C., & Nelson, J.K. (1997). The dissertation in education from 1950 to 1990. *Educational Research Quarterly, 20*(4), 3–14.

Costill, D.L. (1985). Practical problems in exercise physiology research. *Research Quarterly for Exercise and Sport, 56,* 378–384.

Crase, D., & Rosato, F.D. (1992). Single versus multiple authorship in professional journals. *Journal of Physical Education, Recreation and Dance, 63*(7), 28–31.

Creswell, J.W. (1998). *Qualitative inquiry and research design: Choosing among the five traditions.* Thousand Oaks, CA: Sage.

Creswell, J.W. (2003). *Research design: Qualitative, quantitative and mixed methods approaches* (2nd ed.). Thousand Oaks, CA: Sage.

Cronbach, L. (1951). Coefficient alpha and the internal structure of tests. *Psychometrika, 16,* 297–334.

Davidson, M.L. (1972). Univariate versus multivariate tests in repeated-measures experiments. *Psychological Bulletin, 77,* 446–452.

Day, R.D. (1983). *How to write and publish a scientific paper* (2nd ed.). Philadelphia: ISI Press.

Day, R.D. (1988). *How to write and publish a scientific paper* (3rd ed.). Phoenix: Oryx Press.

Dennett, D. (1991). *Consciousness explained.* Boston: Little, Brown.

Denton, J.J., & Tsai, C. (1991). Two investigations into the influence of incentives and subject characteristics on mail survey responses in teacher education. *Journal of Experimental Education, 59,* 352–366.

Dillman, D.A. (1978). *Mail and telephone survey: The total design method.* New York: Wiley.

Dolgener, F.A., Hensley, L.D., Marsh, J.J., & Fjelstul, J.K. (1994). Validation of the Rockport Fitness Walking Test in college males and females. *Research Quarterly for Exercise and Sport, 65,* 152–158.

Drowatzky, J.N. (1993). Ethics, codes, and behavior. *Quest 45,* 22–31.

Drowatzky, J.N. (1996). *Ethical decision making in physical activity research.* Champaign, IL: Human Kinetics.

Dunn, A.L., Garcia, M.E., Marcus, B.H., Kampert, J.B., Kohl, H.W., III, & Blair, S.N. (1998). Six-month physical activity and fitness changes in Project Active, a randomized trial. *Medicine and Science in Sports and Exercise, 30,* 1076–1083.

Edge, D.M., & Claxton, D.B. (2000). 21st century literature searching in physical education. *Journal of Physical Education, Recreation and Dance, 71*(6), 49–52.

Edwards, A.L. (1957). *Techniques of attitude and scale construction.* New York: Appleton-Century-Crofts.

Erickson, F. (1986). Qualitative methods in research on teaching. In M.C. Wittrock (Ed.), *Handbook of research on teaching* (3rd ed., pp. 119-161). New York: Macmillan.

Fahlberg, L.L., & Fahlberg, L.A. (1994). A human science for the study of movement: An integration of multiple ways of knowing. *Research Quarterly for Exercise and Sport, 65,* 100–109.

Farquhar, A.B., & Farquhar, H. (1891). *Economic and industrial delusions: A discourse of the case for protection.* New York: Putnam.

Feltz, D.L., & Landers, D.M. (1983). The effects of mental practice on motor skill learning and performance: A meta-analysis. *Journal of Sport Psychology, 5,* 25–57.

Fernandez-Balboa, J-M. (Ed.). (1997). *Critial postmodernism in human movement, physical education, and sport.* Albany: State University of New York.

Fine, M.A., & Kurdek, L.A. (1993). Reflections on determining authorship credit and authorship order on faculty–student collaborations. *American Psychologist, 48,* 1141–1147.

Fink, A. (2003). *How to sample in surveys* (2nd ed.). Thousand Oaks, CA: Sage.

Flanders, N.A. (1970). *Analyzing teaching behavior.* Reading, MA: Addison-Wesley.

Fowler, F.J., Jr. (1988). *Survey research methods.* Newbury Park, CA: Sage.

Fowler, F.J. Jr. (2002). *Survey research methods* (3rd ed.). Thousand Oaks, CA: Sage.

Fowler, H.W., & Fowler, F.G. (1954). *The king's English.* Oxford: Oxford University Press.

Fraleigh, W. (1984). *Right actions in sport: Ethics for contestants.* Champaign, IL: Human Kinetics.

Franks, B.D., & Huck, S.W. (1986). Why does everyone use the .05 significance level? *Research Quarterly for Exercise and Sport, 57,* 245–249.

French, K.E. (1985). The relation of knowledge development to children's basketball performance. Unpublished doctoral dissertation, Louisiana State University, Baton Rouge.

French, K.E., & Thomas, J.R. (1987). The relation of knowledge development to children's basketball performance. *Journal of Sport Psychology, 9,* 15–32.

Friedenreich, C.M., Thune, I., Brinton, L.A., & Albanes, D. (1998). Epidemiologic issues related to the association between physical activity and breast cancer. *Cancer, 83,* 600–610.

Garcia, C. (1994). Gender differences in young children's interactions when learning fundamental motor skills. *Research Quarterly for Exercise and Sport, 65,* 213–225.

Gardner, R. (1989). On performance-enhancing substances and the unfair advantage argument. *Journal of the Philosophy of Sport, 16,* 59–73. Reprinted in *Philosophic inquiry in sport,* 2nd ed., edited by W.J. Morgan and K.V. Meier (Champaign, IL: Human Kinetics, 1995), 222–231.

Gerlach, L.R. (1994, summer). Not quite ready for prime time: Baseball history, 1983–1993. *Journal of Sport History, 21,* 103–137.

Giddens, A. (1984). *The constitution of society.* Berkeley: University of California.

Gill, D.L., & Deeter, T.E. (1988). Development of the Sport Orientation Questionnaire. *Research Quarterly for Exercise and Sport, 59,* 191-202.

Glaser, B.G., & Strauss, A.L. (1976). *The discovery of grounded theory.* Chicago: Aldine.

Glass, G.V. (1976). Primary, secondary, and meta-analysis. *Educational Researcher, 5,* 3–8.

Glass, G.V. (1977). Integrating findings: The meta-analysis of research. *Review of Research in Education, 5,* 351–379.

Glass, G.V., McGaw, B., & Smith, M. (1981). *Meta-analysis in social research.* Beverly Hills, CA: Sage.

Glass, G.V., & Smith, M.L. (1979). Meta-analysis of research on the relationship of class-size and achievement. *Evaluation and Policy Analysis, 1,* 2–16.

Glassford, R.G. (1987). Methodological reconsideration: The shifting paradigms. *Quest, 39*(3), 295–312.

Goetz, J.P., & LeCompte, M.D. (1984). *Ethnography and qualitative design in educational research.* Orlando, FL: Academic Press.

Goodman, R.M., Wheeler, F.C., & Lee, P.R. (1995). Evaluation of the Heart to Heart Project: Lessons from a community-based chronic disease prevention project. *American Journal of Health Promotion, 9,* 443–445.

Goodrich, J.E., & Roland, C.G. (1977). Accuracy of published medical reference citations. *Journal of Technical Writing and Communication, 7,* 15–19.

Gorn, E. (1986). *The manly art: Bare-knuckle prize fighting in America.* Ithaca, NY: Cornell University Press.

Gould, D., Finch, L.M., & Jackson, S.A. (1993). Coping strategies used by national champion figure skaters. *Research Quarterly for Exercise and Sport, 64,* 453–468.

Grabe, S.A., & Widule, C.J. (1988). Comparative bio-mechanics of the jerk in Olympic weight lifting. *Research Quarterly for Exercise and Sport, 59,* 1–8.

The great foot race. (1835, June). *American Turf Register and Sporting Magazine, 6,* 518–520.

Green, K.E. (1991). Reluctant respondents: Differences between early, late, and nonresponders to a mail survey. *Journal of Experimental Education, 59,* 268–276.

Green, S.B. (1991). How many subjects does it take to do a regression analysis? *Multivariate Behavioral Research, 26,* 499–510.

Greenockle, K.M., Lee, A.M., & Lomax, R. (1990). The relation between selected student characteristics and activity patterns in required high school physical education class. *Research Quarterly for Exercise and Sport, 61,* 59–69.

Griffin, P., & Templin, T.J. (1989). An overview of qualitative research. In P.W. Darst, D.B. Zakrajsek, & V.H. Mancini (Eds.), *Analyzing physical education and sport instruction* (2nd ed., pp. 399–410). Champaign, IL: Human Kinetics.

Gruneau, R. (1983). *Class, sports, and social development.* Amherst: University of Massachusetts Press.

Guba, E.G., & Lincoln, Y.S. (1981). *Effective evaluation.* San Francisco: Jossey-Bass.

Guttmann, A. (1978). *From ritual to record: The nature of modern sport.* New York: Columbia University Press.

Haase, R.F., & Ellis, M.V. (1987). Multivariate analysis of variance. *Journal of Counseling Psychology, 34,* 404–413.

Hackensmith, C.W. (1966). *History of physical education.* New York: Harper & Row.

Hagen, R.L. (1997). In praise of the null hypothesis statistical test. *American Psychologist, 52,* 15–24.

Hall, J.R. (1984, fall). Temporality, social action, and the problem of quantification in historical analysis. *Historical Methods, 17,* 206–218.

Halverson, L.E., Roberton, M.A., & Langendorfer, S. (1982). Development of the overarm throw: Movement and ball velocity changes by seventh grade. *Research Quarterly for Exercise and Sport, 53,* 198–205.

Hamlyn, D.W. (1990). *In and out of the black box.* Oxford: Basil Blackwell.

Handlin, O. (1979). *Truth in history* (p. 120). Cambridge, MA: Belknap Press.

Hardy, S. (1997, September). Sport in urbanizing America: A historical review. *Journal of Urban History, 23,* 675–708.

Harris, C. (1963). *Problems in measuring change.* Madison: University of Wisconsin Press.

Harris, R.J. (1985). *A primer of multivariate statistics* (2nd ed.). Orlando, FL: Academic Press.

Harwell, M.R. (1990). A general approach to hypothesis testing for nonparametric tests. *Journal of Experimental Education, 58,* 143–156.

Hedges, L.V. (1981). Distribution theory for Glass's estimator of effect size and related estimators. *Journal of Educational Statistics, 6,* 107–128.

Hedges, L.V. (1982a). Estimation of effect size from a series of independent experiments. *Psychological Bulletin, 92,* 490–499.

Hedges, L.V. (1982b). Fitting categorical models to effect sizes from a series of experiments. *Journal of Educational Statistics, 7,* 119–137.

Hedges, L.V. (1984). Estimation of effect size under nonrandom sampling: The effects of censoring studies yielding statistically insignificant mean differences. *Journal of Educational Statistics, 9,* 61–85.

Hedges, L.V., & Olkin, I. (1980). Vote counting methods in research synthesis. *Psychological Bulletin, 88,* 359–369.

Hedges, L., & Olkin, I. (1983). Regression models in research synthesis. *American Statistician, 37,* 137–140.

Hedges, L.V., & Olkin, I. (1985). *Statistical methods for meta-analysis.* New York: Academic Press.

Helmstadter, G.C. (1970). *Research concepts in human behavior.* New York: Appleton-Century-Crofts.

Henderson, J. (1990, March 1). When scientists fake it. *American Way,* 56–62, 100–101.

Henderson, K.A., Bialeschki, M.D., Shaw, S.M., & Freysinger, V.J. (1996). *Both gains and gaps: Feminist perspectives on women's leisure.* State College, PA: Venture.

Herkowitz, J. (1984). Developmentally engineered equipment and playgrounds. In J.R. Thomas (Ed.), *Motor development during childhood and adolescence.* Minneapolis: Burgess.

Hollinger, D. (1973, April). T.S. Kuhn's theory of science and its implications for history. *American Historical Review, 78,* 370–393.

Humphrey, S.R. (1980). The historian, his documents, and elementary modes of historical thought. *History and Theory, 19,* 1–20.

Hunter, J.E., & Schmidt, F.L. (1990). *Methods of meta-analysis: Correcting error and bias in research findings.* Newbury Park, CA: Sage.

Husserl, E. (1962). *Ideas: General introduction to pure phenomenology* (W.R. Boyce, Trans.). New York: Collier Books.

Hyde, J.S. (1981). How large are cognitive gender differences? A meta-analysis using v^2 and d. *American Psychologist, 36*(8), 892–901.

Hyland, D. (1990). *Philosophy of sport.* New York: Paragon House.

Jacks, P., Chubin, D.E., Porter, A.L., & Connally, T. (1983). The ABCs of ABDs: An interview study of incomplete doctorates. *Improving College and University Teaching, 31,* 74–81.

Jackson, A.W. (1978). The twelve-minute swim as a test for aerobic endurance in swimming. Unpublished doctoral dissertation, University of Houston.

Jacobs, D., Anderson, J., & Blackburn, H. (1979). Diet and serum cholesterol: Do zero correlations negate the relationship? *American Journal of Epidemiology, 110,* 77–87.

Jacobs, D.R., Jr., Luepker, R.V., Mittelmark, M.B., Folsom, A.R., Pirie, P.L., Mascioli, S.R., Hannan, P.J., Pechacek, T.F., Bracht, N.F., Carlaw, R.W., et al. (1986). Community-wide prevention strategies: Evaluation design of the Minnesota Heart Health Program. *Journal of Chronic Disease, 39,* 775–778.

Jarausch, K.H., & Hardy, K.A. (1991). *Quantitative methods for historians: A guide to research, data, and statistics.* Chapel Hill: University of North Carolina Press.

Johnson, B.L., & Nelson, J.K. (1986). *Practical measurements for evaluation in physical education* (4th ed.). Minneapolis: Burgess.

Johnson, R.L. (1979). The effects of various levels of fatigue on the speed and accuracy of visual recognition. Unpublished doctoral dissertation, Louisiana State University, Baton Rouge.

Jones, E.R. (1988, winter). Philosophical tension in a scientific discipline: So what else is new? *NASP-SPA Newsletter, 14*(1), 10–16.

Joynt, C.B., & Rescher, N. (1961). The problem of uniqueness in history. *History and Theory, 2,* 150–162.

Kavale, K., & Mattson, P.D. (1983). One jumped off the balance beam: Meta-analysis of perceptual-motor training. *Journal of Learning Disabilities, 16,* 165–173.

Keating, X.D., & Silverman, S. (2004). Physical education teacher attitudes toward fitness test scale: Development and validation. *Journal of Teaching in Physical Education, 23,* 143–161.

Kendall, M.G. (1959). Hiawatha designs an experiment. *American Statistician, 13,* 23–24.

Kennedy, J.J. (1983). *Analyzing qualitative data: Introductory loglinear analysis for behavioral research.* New York: Praeger.

Kennedy, M.M. (1979). Generalizing from single case studies. *Evaluation Quarterly, 3,* 661–679.

King, H.A., & Bandy, S.J. (1987). Doctoral programs in physical education: A census with particular reference to the status of specializations. *Quest, 39*(2), 153–162.

Kirk, D. (1992). *Defining physical education: The social construction of a school subject in postwar Britain.* London: Falmer Press.

Kirk, J., & Miller, M.L. (1986). *Reliability and validity in qualitative research.* Newbury Park, CA: Sage.

Kirk, R.E. (1982). *Experimental design: Procedures for the behavioral sciences* (2nd ed.). Belmont, CA: Brooks/Cole.

Kohl, H.W., III, Dunn, A.L, Marcus, B.H., & Blair, S.N. (1998). A randomized trial of physical activity interventions: Design and baseline data from Project Active. *Medicine and Science in Sports and Exercise, 30,* 275–283.

Kokkonen, J., Nelson, A. G., & Cornwell, A. (1998). Acute muscle stretching inhibits maximal strength performance. *Research Quarterly for Exercise and Sport, 69,* 411–415.

Krathwohl, D.R. (1993). *Methods of educational and social science research.* New York: Longman.

Kraus, H., & Hirschland, R.P. (1954). Minimum muscular fitness tests in school children. *Research Quarterly, 25,* 177–188.

Kretchmar, R.S. (1997). Philosophy of sport. In J.D. Massengale & R.A. Swanson (Eds.), *The History of Exercise and Sport Science* (pp. 181–202). Champaign, IL: Human Kinetics.

Kretchmar, R.S. (2005). *Practical philosophy of sport and physical activity* (2nd ed.). Champaign, IL: Human Kinetics.

Kroll, W.P. (1971). *Perspectives in physical education.* New York: Academic Press.

Kruskal, W., & Mosteller, F. (1979). Representative sampling, III: The current statistical literature. *International Statistical Review, 47,* 245–265.

Kuhn, T.S. (1970). *The structure of scientific revolutions.* Chicago: University of Chicago Press.

Kulinna, P.H., & Silverman, S. (1999). The development and validation of scores on a measure of teachers' attitudes toward teaching physical activity and fitness. *Educational and Psychological Measurement, 59,* 507–517.

Lane, K.R. (1983). Comparison of skinfold profiles of black and white boys and girls ages 11–13. Unpublished master's thesis, Louisiana State University, Baton Rouge.

Last, J.M. (1988). *A dictionary of epidemiology.* New York: Oxford University Press.

Lawson, H.A. (1993). After the regulated life. *Quest, 45*(4), 523–545.

Lee, I.M., & Paffenbarger, R.S., Jr. (2000). Associations of light, moderate, and vigorous intensity physical activity with longevity: The Harvard Alumni Health Study. *American Journal of Epidemiology, 151,* 293–299.

Lee, M. (1983). *A history of physical education and sports in the U.S.A.* New York: Wiley.

Lee, T.D. (1982). On the locus of contextual interference in motor skill acquisition. Unpublished doctoral dissertation, Louisiana State University, Baton Rouge.

Leon, A.S., Connett, J., Jacobs, D.R., Jr., & Rauramaa, R. (1987). Leisure-time physical activity levels and risk of coronary heart disease and death: The Multiple Risk Factor Intervention Trial. *JAMA, 258,* 2388–2395.

Leonard, F.G., & Affleck, G.B. (1947). *A guide to the history of physical education* (3rd ed.). Philadelphia: Lea & Febiger.

Levine, R.V. (1990, September/October). The pace of life. *American Scientist, 78,* 450–459.

Light, R.J., Singer, J.D., & Willett, J.B. (1994). The visual presentation and interpretation of meta-analysis. In H. Cooper and L.V. Hedges (Eds.), *The handbook of research synthesis* (pp. 439–453). New York: Russell Sage Foundation.

Lincoln, Y.S., & Guba, E.G. (1985). *Naturalistic inquiry.* Newbury Park, CA: Sage.

Linn, R.L. (1986). Quantitative methods in research on teaching. In M.C. Wittrock (Ed.), *Handbook of research on teaching* (3rd ed., pp. 92-118). New York: Macmillan.

Lipsey, M.W. (1990). *Design sensitivity: Statistical power for experimental research.* Thousand Oaks, CA: Sage.

Locke, L.F. (1987). The question of quality in qualitative research. In J.K. Nelson Ed., *Proceedings of the fifth measurement and evaluation symposium* (pp. 31–36). Baton Rouge: Louisiana State University Press.

Locke, L.F. (1989). Qualitative research as a form of scientific inquiry in sport and physical education. *Research Quarterly for Exercise and Sport, 60,* 1-20.

Locke, L.F., Spirduso, W.W., & Silverman, S.J. (1993). *Proposals that work: A guide for planning dissertations and grant proposals* (3rd ed.). Newbury Park, CA: Sage.

Locke, L.F., Spirduso, W.W., & Silverman, S.J. (2000). *Proposals that work* (4th ed.). Thousand Oaks, CA: Sage.

Locke, L.F., Spirduso, W.W., & Silverman, S.J. (2003). *Proposals that work* (5th ed.). Thousand Oaks, CA: Sage.

Locke, L.F., Silverman, S.J., & Spirduso, W.W. (2004). *Reading and understanding research* (2nd ed.). Thousand Oaks, CA: Sage.

Loland, S. (2002). *Fair play in sport: A moral norm system.* London/New York: Routledge.

Looney, M.A., Feltz, C.S., & VanVleet, C.N. (1994). The reporting and analysis of research findings for within subjects designs: Methodological issues for meta-analysis. *Research Quarterly for Exercise and Sport, 65,* 363–366.

Lord, F.M. (1969). Statistical adjustments when comparing preexisting groups. *Psychological Bulletin, 72,* 336–337.

Mabley, J. (1963, January 22). Mabley's report. *Chicago American,* p. 62.

Mandelbaum, M. (1977). *The anatomy of historical knowledge.* Baltimore: Johns Hopkins University Press.

Margerison, T. (1965, January 3). Review of *Writing Technical Reports* (by B.M. Copper). *Sunday Times.* In R.L. Weber (Compiler) & E. Mendoza (Ed.), *A random walk in science* (p. 49). New York: Crane, Russak.

Marrou, H.-I. (1967). *The meaning of history* (Robert J. Olsen, Trans.) (pp. 76–77). Baltimore: Johns Hopkins University Press.

Marsh, H.W., Marco, I.T., & Aþçý, F.H. (2002). Cross-cultural validity of the physical self-description questionnaire: Comparison of factor structures in Australia, Spain, and Turkey. *Research Quarterly for Exercise and Sport, 73,* 257–270.

Marshall, C., & Rossman, G.B. (1999). *Designing qualitative research* (3rd ed.). Thousand Oaks, CA: Sage.

Martens, R. (1973, June). People errors in people experiments. *Quest, 20,* 16–20.

Martens, R. (1977). *Sport competition anxiety test.* Champaign, IL: Human Kinetics.

Martens, R. (1979). About smocks and jocks. *Journal of Sport Psychology, 1,* 94–99.

Martens, R. (1987). Science, knowledge, and sport psychology. *Sport Psychologist, 1,* 29–55.

Massengale, J.D. (1987). Current status of graduate physical education: Program demography and the issue of program rating. *Quest, 39*(2), 97–102.

Massengale, J.D., & Swanson, R.A. (Eds.). (1996). *History of exercise and sport science.* Champaign, IL: Human Kinetics.

Matt, K.S. (1993). Ethical issues in animal research. *Quest, 45*(1), 45–51.

Matthews, P.R. (1979). The frequency with which the mentally retarded participate in recreation activities. *Research Quarterly, 50,* 71–79.

Mattson, D.E. (1981). *Statistics: Difficult concepts, understandable explanations.* St. Louis: Mosby.

Maxwell, J.A. (2004). *Qualitative research design: An interactive approach* (2nd ed.). Thousand Oaks, CA: Sage.

McCaughtry, N., & Rovegno, I. (2003). Development of pedagogical content knowledge: Moving from blaming students to predicting skillfulness, recognizing motor development, and understanding emotion. *Journal of Teaching in Physical Education, 22,* 355–368.

McCloy, C.H. (1930). Professional progress through research. *Research Quarterly, 1,* 63-73.

McCullagh, P., & Meyer, K.N. (1997). Learning versus correct models: Influence of model type on the learning of a free-weight squat lift. *Research Quarterly for Exercise and Sport, 68,* 56–61.

McDonald, R.P. (1999). Test theory: A unified treatment. Mahwah, NJ: Lawrence Erlbaum.

McLeroy, K.R., Bibeau, D., Steckler, A., & Glanz, K. (1988). An ecological perspective on health promotion programs. *Health Education Quarterly, 15,* 351–377.

McNamara, J.F. (1994). *Surveys and experiments in education research.* Lancaster, PA: Nechnomic.

McNamee, M. (1998). Celebrating trust: Virtues and rules in the ethical conduct of sports coaches. In M. McNamee & S.J. Parry (Eds.), *Ethics and Sport* (pp. 148–168). London/New York: Spon.

McPherson, S.L. (1987). The development of children's expertise in tennis: Knowledge structure and sport performance. Unpublished doctoral dissertation, Louisiana State University, Baton Rouge, LA.

McPherson, S.L. (1999). Expert–novice differences in performance skills and problem representations of youth and adults during tennis competition. *Research Quarterly for Exercise and Sport, 70,* 233–251.

McPherson, S.L., & Thomas, J.R. (1989). Relation of knowledge and performance in boys' tennis: Age and expertise. *Journal of Experimental Child Psychology, 48,* 190–211.

McPhie, W.E. (1960). Factors affecting the value of dissertations. *Social Education, 24,* 375–377, 385.

Megill, A. (1989, June). Recounting the past: "Description," explanation, and narrative in historiography. *American Historical Review, 94,* 627–653.

Merleau-Ponty, M. (1964). *The primacy of perception.* Evanston, IL: Northwestern University Press.

Merriam, S.B. (1988). *Case study research in education.* San Francisco: Jossey-Bass.

Merriam, S.B. (2001). *Qualitative research and case study approaches in education.* San Francisco: Jossey-Bass.

Merriam-Webster. (1989). *Webster's ninth new collegiate dictionary.* Springfield, MA: Merriam-Webster.

Micceri, T. (1989). The unicorn, the normal curve, and other improbable creatures. *Psychological Bulletin, 105,* 156–166.

Midgley, M. (1994). *The ethical primate: Humans, freedom and morality.* London/New York: Routledge.

Milligan, J.D. (1979). The treatment of an historical source. *History and Theory, 18,* 177–196.

Montoye, H., Kemper, H., Saris, W., & Washburn, R. (1996). *Measuring physical activity and energy expenditure.* Champaign, IL: Human Kinetics.

Mood, D.P. (1989). Measurement methodology for knowledge tests. In M.J. Safrit & T.M. Wood (Eds.), *Measurement concepts in physical education and exercise science* (pp. 251–270). Champaign, IL: Human Kinetics.

Morgan, D.L. (1988). *Focus groups as qualitative research.* Thousand Oaks, CA: Sage

Morgenstern, N.L. (1983). Cogito ergo sum: Murphy's refutation of Descartes. In G.H. Scherr (Ed.), *The best of* The Journal of Irreproducible Results (p. 112). New York: Workman Press.

Morland, R.B. (1958). A philosophical interpretation of the educational views held by leaders in American physical education. *Health, Physical Education and Recreation Microform Publications, 1,* (October 1949–March 1965): PE394.

Morris, J.N., Heady, J.A., Rattle, P.A.B., Roberts, C.G., & Parks, J.W. (1953). Coronary heart disease and physical activity of work. *Lancet, 2,* 1053–1057.

Morrow, J.R., Jr., Bray, M.S., Fulton, J.E., & Thomas, J.R. (1992). Interrater reliability of 1987–1991. *Research Quarterly for Exercise and Sport* reviews. *Research Quarterly for Exercise and Sport, 63,* 200–204.

Morrow, J.R., & Frankiewicz, R.G. (1979). Strategies for the analysis of repeated and multiple measure designs. *Research Quarterly, 50,* 297–304.

Morse, J.M., & Richards, L. (2002). *Read me first for a user's guide to qualitative methods.* Thousand Oaks, CA: Sage.

National Children and Youth Fitness Study. (1985). *Journal of Physical Education, Recreation and Dance, 56*(1), 44–90.

National Children and Youth Fitness Study II. (1987). *Journal of Physical Education, Recreation and Dance, 56*(9), 147–167.

Nelson, J.K. (1988, March). Some thoughts on research, measurement, and other obscure topics. LAHPERD Scholar Lecture presented at the LAHPERD Convention, New Orleans.

Nelson, J.K. (1989). Measurement methodology for affective tests. In M.J. Safrit & T.M. Wood (Eds.), *Measurement concepts in physical education and exercise science* (pp. 229–248). Champaign, IL: Human Kinetics.

Nelson, J.K., Thomas, J.R., Nelson, K.R., & Abraham, P.C. (1986). Gender differences in children's throwing performance: Biology and environment. *Research Quarterly for Exercise and Sport, 57,* 280–287.

Nelson, J.K., Yoon, S.H., & Nelson, K.R. (1991). A field test for upper body strength and endurance. *Research Quarterly for Exercise and Sport, 62,* 436–447.

Nelson, K.R. (1988). *Thinking processes, management routines, and student perceptions of expert and novice physical education teachers.* Unpublished doctoral dissertation, Louisiana State University, Baton Rouge.

Newell, K.M. (1987). On masters and apprentices in physical education. *Quest, 39*(2), 88–96.

Newell, K.M., & Hancock, P.A. (1984). Forgotten moments: A note on skewness and kurtosis as influential factors in inferences extrapolated from response distributions. *Journal of Motor Behavior, 16,* 320–335.

Nieman, D.C. (1994). Exercise, upper respiratory tract infection, and the immune system. *Medicine and Science in Sports and Exercise, 26B,* 128–139.

Nietzsche, F. (1967). *The will to power* (W. Kaufmann & R. Hollingdale, Trans.). New York: Vintage Books.

Nunnaly, J.C. (1978). *Psychometric theory* (2nd ed.). New York: McGraw-Hill.

Oakman, R.L. (1984). *Computer methods for literary research.* Athens: University of Georgia Press.

Oishi, S.M. (2003). *How to conduct in-person interviews for surveys* (2nd ed.). Thousand Oaks, CA: Sage.

Paffenbarger, R., Hyde, R., Wing, A., Lee, I., Jung, D., & Kampert, J. (1993). The association of changes in physical activity level and other lifestyle characteristics with mortality among men. *New England Journal of Medicine, 328,* 538–545.

Paffenbarger, R.S., Hyde, R.T., Wing, A.L., & Steinmetz, C.H. (1984). A natural history of athleticism and cardiovascular health. *JAMA, 252,* 491–495.

Paffenbarger, R.S., Jr., Wing, A.L., & Hyde, R.T. (1978). Physical activity as an index of heart attack risk in college alumni. *American Journal of Epidemiology, 108,* 161–175.

Park, R.J. (1980). The *Research Quarterly* and its antecedents. *Research Quarterly for Exercise and Sport, 51,* 1–22.

Park, R.J. (1986, spring). Hermeneutics, semiotics, and the nineteenth-century quest for a corporeal self. *Quest, 38,* 33–49.

Park, R.J. (1994, spring). A decade of the body: Researching and writing about the history of health, fitness, exercise and sport, 1983–1993. *Journal of Sport History, 21,* 59–82.

Pate, R.R., Pratt, M., Blair, S.N., Haskell, W.L., Macera, C.A., Bouchard, C., Buchner, D., Ettinger, W., Heath, G.W., King, A.C., et al. (1995). Physical activity and public health: A recommendation from the Centers for Disease Control and Prevention and the American College of Sports Medicine, [review]. *JAMA 273,* 402–407.

Patton, M.Q. (1987). *How to use qualitative methods in evaluation*. Newbury Park, CA: Sage.

Patton, M.Q. (2002). *Qualitative research and evaluation methods* (3rd ed.). Thousand Oaks, CA: Sage.

Payne, V.G., & Morrow, J.R., Jr. (1993). Exercise and $\dot{V}O_2$max in children: A meta-analysis. *Research Quarterly for Exercise and Sport, 64,* 305–313.

Pedhazur, E.J. (1982). *Multiple regression in behavioral research: Explanation and prediction* (2nd ed.). New York: Holt, Rinehart & Winston.

Pedhazur, E.J. (1997). *Multiple regression in behavioral research* (3rd ed.). New York: Harcourt Brace College Publishers.

Peshkin, A. (1993). The goodness of qualitative research. *Educational Researcher, 22(2),* 23–29.

Peters, E.M., & Bateman, E.D. (1983). Ultramarathon running and upper respiratory tract infections: An epidemiological survey. *South African Medical Journal, 64,* 582–584.

Pinker, S. (1997). *How the mind works*. New York/London: Norton.

Pinker, S. (1999). *Words and rules: The ingredients of language*. New York: HarperCollins.

Pinker, S. (2002). *The blank slate: The modern denial of human nature*. New York: Viking/Penguin Group.

Polanyi, M. (1958). *Person knowledge: Toward a post-critical philosophy*. Chicago: University of Chicago Press.

Pope, S.W. (1997). *Patriotic games: Sporting traditions in the American imagination 1876–1926*. New York: Oxford University Press.

Porter, A.C., & Raudenbush, S.W. (1987). Analysis of covariance: Its model and use in psychological research. *Journal of Counseling Psychology, 34,* 383–392.

Porter, A.L., Chubin, D.E., Rossini, F.A., Boeckmann, M.E., & Connally, T. (1982, September/October). The role of the dissertation in scientific careers. *American Scientist,* 475–481.

Porter, A.L., & Wolfe, D. (1975). Utility of the doctoral dissertation. *American Psychologist, 30,* 1054–1061.

Porter, D.H. (1981). *The emergence of the past: A theory of historical explanation*. Chicago: University of Chicago Press.

Punch, M. (1986). *The politics and ethics of fieldwork*. Beverly Hills, CA: Sage.

Punch, K. F. (2003). *Survey research: The basics*. Thousand Oaks, CA: Sage.

Puri, M.L., & Sen, P.K. (1969). A class of rank order tests for a general linear hypothesis. *Annals of Mathematical Statistics, 40,* 1325–1343.

Puri, M.L., & Sen, P.K. (1985). *Nonparametric methods in general linear models*. New York: Wiley.

Puska, P., Salonen, J.T., & Nissinen, A. (1983). Change in risk factors for coronary heart disease during 10 years of community intervention programme (North Karelia Project). *British Medical Journal, 287,* 1840–1844.

Realist, B.A. [G. Benford]. (1982, March). How to write a scientific paper. *Omni,* 130.

Ridley, M. (2003). *Nature via nurture: Genes, experience, & what makes us human*. New York: Harper Collins.

Ries, L.A.G., Kosary, C.L., Hankey, B.F., Miller, B.A., Clegg, L., & Edwards, B.K. (Eds). (1999). SEER Cancer Statistics Review, 1973–1996. Bethesda, MD: National Cancer Institute.

Riess, S.A. (1994). From pitch to putt: Sport and class in Anglo-American sport. *Journal of Sport History, 21,* 138–184.

Rink, J., & Werner, P. (1989). Qualitative measures of teaching performance scale. In P. Darst, D. Zakrajsak, & P. Mancini (Eds.), *Analyzing physical education and sports instruction* (2nd ed., pp. 269–276). Champaign, IL: Human Kinetics.

Roberts, G.C. (1993). Ethics in professional advising and academic counseling of graduate students. *Quest, 45,* 78–87.

Roberts, T. (1998). Sporting practice protection and vulgar ethnocentricity: Why won't Morgan go all the way? *Journal of the Philosophy of Sports, XXV,* 71–81.

Robertson, M.A., & Konczak, J. (2001). Predicting children's overarm throw ball velocities from their developmental levels in throwing. *Research Quarterly for Exercise and Sport, 72,* 91–103.

Rockhill, B., Willett, W.C., Hunter, D.J., Manson, J.E., Hankinson, S.E., & Colditz, G.A. (1999). A prospective study of recreational physical activity and breast cancer risk. *Archives of Internal Medicine, 159,* 2290–2296.

Rosen, S. (1989). *The ancients and the moderns: Rethinking modernity.* New Haven, CT: Yale University Press.

Rosenau, P.M. (1992). *Post-modernism and the social sciences: Insights, inroads, and intrusions.* Princeton, NJ: Princeton University Press.

Rosenthal, R. (1966). Sport, art, and particularity: The best equivocation. *Journal of the Philosophy of Sport, 13,* 49–63.

Rosenthal, R. (1979). The file-drawer problem and tolerance for null results. *Psychological Bulletin, 86,* 638–641.

Rosenthal, R. (1991). Cumulating psychology: An appreciation of Donald T. Campbell. *Psychological Science, 2,* 213, 217–221.

Rosenthal, R. (1994). Parametric measures of effect size. In H. Cooper & L.V. Hedges (Eds.), *The handbook of research synthesis* (pp. 231-244). New York: Russel Sage Foundation.

Rosnow, R.L., & Rosenthal, R. (1989). Statistical procedures and the justification of knowledge in psychological science. *American Psychologist, 44,* 1276–1284.

Rossman, G.B., & Rallis, S.F. (2003). *Learning in the field: An introduction to qualitative research.* Thousand Oaks, CA: Sage.

Rothman, K.J. (1986). *Modern epidemiology.* Boston: Little Brown.

Rudy, W. (1962). Higher education in the United States, 1862–1962. In W.W. Brickman & S. Lehrer, (Eds.), *A century of higher education: Classical citadel to collegiate colossus* (pp. 20–21). New York: Society for the Advancement of Education.

Rugg, H. (1941). *That men may understand.* New York: Doubleday, Doran.

Ryan, E.D. (1970). The cathartic effect of vigorous motor activity on aggressive behavior. *Research Quarterly, 41,* 542–551.

Sachtleben, T.R., Berg, K.E., Cheatham, J.P., Felix, G.L., & Hofschire, P.J. (1997). Serum lipoprotein patterns in long-term anabolic steroid users. *Research Quarterly for Exercise and Sport, 68,* 110–115.

Safrit, M.J. (Ed.). (1976). *Reliability theory.* Washington, DC: American Alliance for Health, Physical Education and Recreation.

Safrit, M.J. (Ed.). (1980). *Research Quarterly for Exercise and Sport, 51*(1).

Safrit, M.J., Cohen, A.S., & Costa, M.G. (1989). Item response theory and the measurement of motor behavior. *Research Quarterly for Exercise and Sport, 60*(4), 325–335.

Safrit, M.J., & Stamm, C.L. (1980). Reliability estimates for criterion-referenced measures in the psychomotor domain. *Research Quarterly for Exercise and Sport, 51*(2), 359–368.

Safrit, M.J., & Wood, T.M. (1983). The health-related fitness test opinionnaire: A pilot survey. *Research Quarterly for Exercise and Sport, 54,* 204–207.

Sage, G.H. (1989). A commentary on qualitative research as a form of inquiry in sport and physical education. *Research Quarterly for Exercise and Sport, 60,* 25–29.

Sammons, J.T. (1994, winter). Race and sport: A critical, historical examination. *Journal of Sport History, 21,* 203–278.

Scheffé, H. (1953). A method for judging all contrasts in analysis of variance. *Biometrika, 40,* 87–104.

Schein, E.H. (1987). *The clinical perspective in fieldwork.* Newbury Park, CA: Sage.

Scherr, G.H. (Ed.). (1983). *The best of* The Journal of Irreproducible Results (p. 152). New York: Workman Press.

Schmidt, R.A. (1975). A schema theory of discrete motor skill learning. *Psychological Review, 82,* 225–260.

Schmidt, R.A. (1988). *Motor control and learning.* Champaign, IL: Human Kinetics.

Schmidt, R.A., & Lee, T.D. (1999). *Motor control and learning* (3rd ed.). Champaign, IL: Human Kinetics.

Schumacker, R.E., & Lomax, R.G. (2004). *A beginner's guide to structural equation modeling* (2nd ed.). Mahwah, NJ: Erlbaum.

Schutz, R.W. (1989). Qualitative research: Comments and controversies. *Research Quarterly for Exercise and Sport, 60,* 30–35.

Schutz, R.W., & Gessaroli, M.E. (1987). The analysis of repeated measures designs involving multiple dependent variables. *Research Quarterly for Exercise and Sport, 58,* 132–149.

Searle, J.R. (1980). Mind, brains and programs. *The Behavioral and Brain Sciences, 3,* 417–457.

Serlin, R.C. (1987). Hypothesis testing, theory building, and the philosophy of science. *Journal of Counseling Psychology, 34,* 365–371.

Shadish, W.R., Cook, T.D., & Campbell, D.T. (2002). *Experimental and quasi-experimental designs for generalized causal inference.* Boston: Houghton Mifflin.

Shafer, R.J. (1980). *A guide to historical method* (3rd ed.). Homewood, IL: Dorsey Press.

Sheets-Johnstone, M. (1999). *The primacy of movement.* Amsterdam/Philadelphia: John Benjamins.

Shore, E.G. (1991, February). Analysis of a multi-institutional series of completed cases. Paper presented at Scientific Integrity Symposium. Harvard Medical School, Boston.

Shulman, L.S. (2003, April). Educational research and a scholarship of education. Charles DeGarmo Lecture presented at the annual meeting of the American Educational Research Association, Chicago.

Siedentop, D. (1980). Two cheers for Rainer. *Journal of Sport Psychology, 2,* 2–4.

Siedentop, D. (1989). Do the lockers really smell? *Research Quarterly for Exercise and Sport, 60,* 36–41.

Siedentop, D., Birdwell, D., & Metzler, M. (1979, March). A process approach to measuring teaching effectiveness in physical education. Paper presented at the American Alliance for Health, Physical Education, Recreation and Dance national convention, New Orleans.

Siedentop, D., Trousignant, M., & Parker, M. (1982). *Academic learning time—Physical education: 1982 coding manual.* Columbus: Ohio State University, School of Health, Physical Education, and Recreation.

Siegel, S. (1956). *Nonparametric statistics for the behavioral sciences.* New York: McGraw-Hill.

Silverman, S. (2004). Analyzing data from field research: The unit of analysis issue. *Research Quarterly for Exercise and Sport, 74,* iii–iv.

Silverman, S. (Ed.). (2005). *Research Quarterly for Exercise and Sport: 75th Anniversary Issue, 76*(supplement).

Silverman, S., & Keating, X.D. (2002). A descriptive analysis of research methods classes in department of kinesiology and physical education in the United States. *Research Quarterly for Exercise and Sport, 73,* 1–9.

Silverman, S., & Solmon, M. (1998). The unit of analysis in field research: Issues and approaches to design and data analysis. *Journal of Teaching in Physical Education, 17,* 270–284.

Silverman, S., & Subramaniam, P.R. (1999). Student attitude toward physical education and physical activity: A review of measurement issues and outcomes. *Journal of Teaching in Physical Education, 19,* 97–125.

Simon, R. (2004). *Fair play: The ethics of sport.* Boulder, CO: Westview Prentice Hall.

Singer, P. (1995). *How are we to live? Ethics in an age of self-interest.* Amherst, NY: Prometheus.

Singer, R.N., Hausenblas, H.A., & Janelle, C.M. (Eds.). (2001). *Handbook of sport psychology* (2nd ed.). New York: Wiley.

Slavin, R.E. (1984a). Meta-analysis in education: How it has been used. *Educational Researcher, 13*(4), 6–15.

Slavin, R.E. (1984b). A rejoinder to Carlberg et al. *Educational Researcher, 13*(4), 24–27.

Smith, M.L. (1980). Sex bias in counseling and psychotherapy. *Psychological Bulletin, 87,* 392–407.

Snyder, C.W., Jr., & Abernethy, B. (Eds.). (1992). *The creative side of experimentation.* Champaign, IL: Human Kinetics.

Sparling, P.B. (1980). A meta-analysis of studies comparing maximal oxygen uptake in men and women. *Research Quarterly for Exercise and Sport, 51,* 542–552.

Spirduso, W.W. (1987). Graduate program ranking in physical education. *Quest, 39*(2), 103–112.

Spirduso, W.W., & Lovett, D.J. (1987). Current status of graduate education in physical education: Program demography. *Quest, 39*(2), 129–141.

Spray, J.A. (1987). Recent developments in measurement and possible applications to the measurement of psychomotor behavior. *Research Quarterly for Exercise and Sport, 58,* 203–209.

Spray, J.A. (1989). New approaches to solving measurement problems. In M.J. Safrit & T.M. Wood (Eds.), *Measurement concepts in physical education and exercise science* (pp. 229–248). Champaign, IL: Human Kinetics.

Stamm, C.L., & Safrit, M.J. (1975). Comparison of significance tests for repeated measures ANOVA design. *Research Quarterly, 46,* 403–409.

Starkes, J.L., & Allard, F. (Eds.). (1993). *Cognitive issues in motor expertise.* Amsterdam: North Holland.

Stock, W.A. (1994). Systematic coding for research synthesis. In H. Cooper & L.V. Hedges (Eds.), *The handbook of research synthesis.* New York: Sage Foundation.

Struna, N.L. (1986, spring). E.P. Thompson's notion of "context" and the writing of physical education and sport history. *Quest, 38,* 24–27.

Struna, N.L. (1988, December). Sport and society in early America. *International Journal of the History of Sport, 5,* 292–311.

Struna, N.L. (1996a). *People of prowess: Sport, leisure, and labor in early Anglo-America.* Urbana: University of Illinois Press.

Struna, N.L. (1996b). Sport history. In J. Massengale & R. Swanson (Eds.), *History of exercise and sport science,* Chapter 9. Urbana: University of Illinois Press.

Stull, G.A., Christina, R.W., & Quinn, S.A. (1991). Accuracy of references in the *Research Quarterly for Exercise and Sport. Research Quarterly for Exercise and Sport, 62,* 245–248.

Subramaniam, P.R., & Silverman, S. (2000). The development and validation of an instrument to assess student attitude toward physical education. *Measurement in Physical Education and Exercise Science, 4,* 29–43.

Sundgot-Borgen, J. (1994). Risk and trigger factors for the development of eating disorders in female athletes. *Medicine and Science in Sports and Exercise, 26,* 414–419.

Tashakkori, A., & Teddlie, C. (1998). *Mixed methodology.* Thousand Oaks, CA: Sage.

Tashakkori, A., & Teddlie, C. (2002). *Handbook of mixed methods in social and behavioral research.* Thousand Oaks, CA: Sage.

Tew, J. (1988). Construction of a sport specific mental imagery assessment instrument using item response and classical test theory methodology. Unpublished doctoral dissertation, Louisiana State University, Baton Rouge.

Tew, J., & Wood, M. (1980). *Proposed model for predicting probable success in football players.* Houston, TX: Rice University Press.

Thomas, J.R. (1977). A note concerning analysis of error scores from motor-memory research. *Journal of Motor Behavior, 9,* 251–253.

Thomas, J.R. (1980). Half a cheer for Rainer and Daryl. *Journal of Sport Psychology, 2,* 266–267.

Thomas, J.R. (Ed.). (1983). Publication guidelines. *Research Quarterly for Exercise and Sport, 54,* 219–221.

Thomas, J.R. (Ed.). (1984). *Motor development during childhood and adolescence.* Minneapolis: Burgess.

Thomas, J.R. (Ed.). (1986). Editor's viewpoint: Research notes. *Research Quarterly for Exercise and Sport, 57,* iv–v.

Thomas, J.R. (1987). Are we already in pieces, or just falling apart? *Quest, 39*(2), 114–121.

Thomas, J.R. (1989). An abstract for all seasons. *NASPSPA Newsletter, 14*(2), 4–5.

Thomas, J.R., & French, K.E. (1985). Gender differences across age in motor performance: A meta-analysis. *Psychological Bulletin, 98,* 260–282.

Thomas, J.R., & French, K.E. (1986). The use of meta-analysis in exercise and sport: A tutorial. *Research Quarterly for Exercise and Sport, 57,* 196–204.

Thomas, J.R., French, K.E., & Humphries, C.A. (1986). Knowledge development and sport skill performance: Directions for motor behavior research. *Journal of Sport Psychology, 8,* 259–272.

Thomas, J.R., & Gill, D.L. (Eds.). (1993). The academy papers: Ethics in the study of physical activity [special issue]. *Quest, 45*(1).

Thomas, J.R., Lochbaum, M.R., Landers, D.M, & He, C. (1997). Planning significant and meaningful research in exercise science: Estimating sample size. *Research Quarterly for Exercise and Sport, 68,* 33–43.

Thomas, J.R., & Nelson, J.K. (2001). *Research methods in physical activity* (4th ed.). Champaign, IL: Human Kinetics.

Thomas, J.R., Nelson, J.K., & Magill, R.A. (1986, spring). A case for an alternative format for the thesis/dissertation. *Quest, 38,* 116–124.

Thomas, J.R., Nelson, J.K., & Thomas, K.T. (1999). A generalized rank-order method for non-parametric analysis of data from exercise science: A tutorial. *Research Quarterly for Exercise and Sport, 70,* 11–23.

Thomas, J.R., Salazar, W., & Landers, D.M. (1991). What is missing in $p < .05$? Effect size. *Research Quarterly for Exercise and Sport, 62,* 344–348.

Thomas, J.R., Thomas, K.T., Lee, A.M., Testerman, E., & Ashy, M. (1983). Age differences in use of strategy for recall of movement in a large scale environment. *Research Quarterly for Exercise and Sport, 54,* 264–272.

Thomas, K.T., Keller, C.S., & Holbert, K. (1997). Ethnic and age trends for body composition in women residing in the U.S. southwest: II. Total fat. *Medicine and Science in Sport and Exercise, 29,* 90–98.

Thomas, K.T., & Thomas, J.R. (1999). What squirrels in the trees predict about expert athletes. *International Journal of Sport Psychology, 30,* 221–234.

Thompson, E.P. (1972, spring). Anthropology and the discipline of historical context. *Midland History, 3,* 41–55.

Thompson, E.P. (1978). *The poverty of theory and other essays* (pp. 28–29). New York: Monthly Review Press.

Thune, I., Brenn, T., Lund, E., & Gaard, M. (1997). Physical activity and the risk of breast cancer. *New England Journal of Medicine, 336,* 1269–1275.

Tinberg, C.M. (1993). The relation of practice time to coaches' objectives, players' improvement, and level of expertise. Master's thesis, Arizona State University, Tempe.

Tinsley, H.E., & Tinsley, D.J. (1987). Uses of factor analysis in counseling psychology research. *Journal of Counseling Psychology, 34,* 414–424.

Tolson, H. (1980). An adjustment to statistical significance: v^2. *Research Quarterly for Exercise and Sport, 51,* 580–584.

Toothaker, L.E. (1991). *Multiple comparisons for researchers.* Newbury Park, CA: Sage.

Tran, Z.V., Weltman, A., Glass, G.V., & Mood, D.P. (1983). The effects of exercise on blood lipids and lipoproteins: A meta-analysis. *Medicine and Science in Sports and Exercise, 15,* 393–402.

Tuckman, B.W. (1978). *Conducting educational research* (2nd ed.). New York: Harcourt Brace Jovanovich.

Tudor-Locke, C., Ham, S.A., Macera, C.A., Ainsworth, B.E., Kirtland, K.A., Reis, J.P., & Kimsey, C.D. (2004) Descriptive epidemiology of pedometer-determined physical activity. *Medicine and Science in Sports & Exercise, 36,* 1567–1573.

Turner, A.P., & Martinek, T.J. (1999). An investigation into teaching games for understanding effects on skill, knowledge, and game play. *Research Quarterly for Exercise and Sport, 70,* 286–296.

University of Chicago Press. (1993). *The Chicago manual of style* (14th ed.). Chicago: Author.

U.S. Department of Health and Human Services. (1996). *Physical activity and health: A report of the Surgeon General.* Atlanta, GA: U.S. Dept of Health and Human Services, Centers for Disease Control and Prevention, National Center for Chronic Disease Prevention and Health Promotion.

U.S. Department of Health and Human Services. (2000). Healthy People 2010 (Conference Edition in Two Volumes). Washington, DC: January 2000.

Van Dalen, D.B., & Bennett, B.L. (1971). *A world history of physical education: Cultural, philosophical, comparative* (2nd ed.). Englewood Cliffs, NJ: Prentice Hall.

Vealey, R.S. (1986). The conceptualization of sport-confidence and competitive orientation: Preliminary investigation and instrument development. *Journal of Sport Psychology, 8,* 221–246.

Verducci, F.M. (1980). *Measurement concepts in physical education.* St. Louis: Mosby.

Vertinsky, P. (1990). *The eternally wounded woman: Women, exercise, and doctors in the late nineteenth century.* Manchester: Manchester University Press.

Vertinsky, P. (1994, spring). Gender relations, women's history and sport history: A decade of changing enquiry, 1983–1993. *Journal of Sport History, 21,* 1–58.

Vockell, E.L. (1983). *Educational research.* New York: Macmillan.

Wainer, H. (1992). Understanding graphs and tables. *Educational Researcher, 21*(1), 14–23.

Wallace, B.A. (2000). *The taboo of subjectivity: Toward a new science of consciousness.* New York: Oxford University Press.

Webb, E.J., Campbell, D.T., Schwartz, R.D., & Sechrest, L. (1966). *Unobtrusive measures: Nonreactive research in the social sciences.* Chicago: Rand McNally.

Weiss, M.R., Bredemeier, B.J., & Shewchuk, R.M. (1985). An intrinsic/extrinsic motivation scale for the youth sport setting: A confirmatory factor analysis. *Journal of Sport Psychology, 7,* 75–91.

Weiss, M.R., McCullagh, P., Smith, A.L., & Berlant, A.R. (1998). Observational learning and the fearful child: Influence of peer models on swimming skill performance and psychological responses. *Research Quarterly for Exercise and Sport, 69,* 380–394.

Weiss, P. (1969). *Sport: A philosophic inquiry.* Carbondale: Southern Illinois University Press.

Werner, P., & Rink, J. (1989). Case studies of teacher effectiveness in physical education. *Journal of Teaching in Physical Education, 8,* 280–297.

White, P.A. (1990). Ideas about causation in philosophy and psychology. *Psychological Bulletin, 108,* 3–18.

Willett, W. (1990). *Nutritional epidemiology.* New York: Oxford University Press.

Wilson, A. (1981, summer). Inferring attitudes from behavior. *Historical Methods, 14,* 143–144.

Wilson, J.Q. (1993). *The moral sense.* New York: The Free Press/Macmillan.

Winkleby, M.A., Taylor, C.B., Jatulis, D., & Fortmann, S.P. (1996). The long-term effects of a cardiovascular disease prevention trial: The Stanford Five-City Project. *American Journal of Public Health, 86,* 1773–1779.

Wood, T.M., & Safrit, M.J. (1987). A comparison of three multivariate models for estimating test battery reliability. *Research Quarterly for Exercise and Sport, 58,* 150–159.

Yan, J.H. (1996). Development of motor programs across the lifespan: Arm movement control. Doctoral dissertation, Arizona State University, Tempe.

Yin, R.K. (2003). *Case study research: Design and methods* (3rd ed.). Thousand Oaks, CA: Sage.

Young, D.R, Haskell, W.L, Taylor, C.B., & Fortmann, S.P. (1996). Effect of community health education on physical activity knowledge, attitudes, and behavior: The Stanford Five-City Project. *American Journal of Epidemiology, 144,* 264–274.

Zelaznik, H.N. (1993). Ethical issues in conducting and reporting research: A reaction to Kroll, Matt, and Safrit. *Quest, 45,* 62–68.

Ziv, A. (1988). Teaching and learning with humor: Experiment and replication. *Journal of Experimental Education, 57,* 5–15.

Author Index

Subject Index

About the Authors

Jerry R. Thomas, EdD, is professor and chair of the department of health and human performance at Iowa State University. Besides writing the previous editions of this book, Thomas has authored more than 200 publications, 87 of which are data-based refereed publications, with numerous contributions in research methods. Awarded the C.H. McCloy Lecturer in 1999, based on his career research production, Thomas has served as editor in chief for *Research Quarterly for Exercise and Sport* and as a reviewer for most major research journals in kinesiology and numerous journals in psychology. He has also served as president of the American Academy of Kinesiology and Physical Education, of the American Alliance for Health, Physical Education, Recreation and Dance (AAHPERD) Research Consortium, and of the North American Society for Psychology of Sport and Physical Activity. He was named an AAHPERD Alliance Scholar in 1990 and NASPSPA Distinguished Scholar in 2003, based on lifetime achievement in research.

Jack K. Nelson, EdD, is professor emeritus in the department of kinesiology at Louisiana State University. Nelson conducted and published research and taught research methods for 35 years. He has advised more than 50 doctoral dissertations and more than 50 masters' theses focused on the research process. In addition, he has more than 80 publications and has served as editor of research publications. A fellow in the Research Consortium, he has been a member of AAHPERD, the American Educational Research Association, and the American College of Sports Medicine. He has also served as president of the Association for Research, Administration, Professional Councils and Societies (now AAALF) and as vice president of AAHPERD.

Stephen Silverman, EdD, has taught and written about research methods for over 20 years. He is a professor of education at Teachers College, Columbia University, and he has conducted research on teaching in physical education focusing on how children learn motor skills and develop attitudes. He has published more than 50 research articles in addition to many books and book chapters. Silverman is a fellow of both the American Academy of Kinesiology and Physical Education and the AAHPERD Research Consortium. A former coeditor of the *Journal of Teaching in Physical Education* and current editor in chief of the *Research Quarterly for Exercise and Sport,* Silverman was the American Educational Research Association Physical Education Scholar Lecturer and a Research Consortium Scholar Lecturer and Weiss Lecturer for AAHPERD.

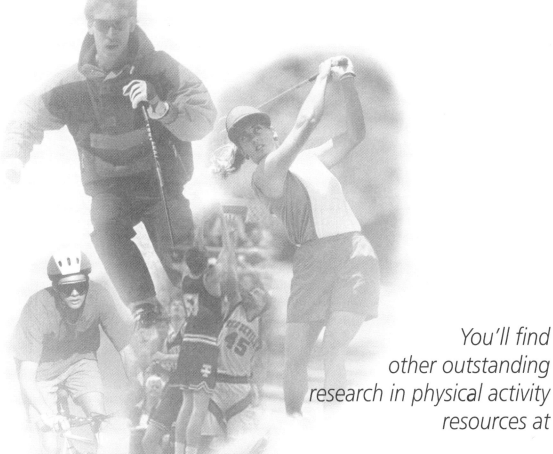